The Scramble for the Amazon and the "Lost Paradise" of Euclides da Cunha

The Scramble for the Amazon and the "Lost Paradise" of Euclides da Cunha

SUSANNA B. HECHT

The University of Chicago Press
Chicago and London

SUSANNA B. HECHT is professor in the Luskin School of Public Affairs, Institute of the Environment, and Department of Geography at the University of California, Los Angeles. She is coauthor of *Fate of the Forest*, also published by the University of Chicago Press.

The University of Chicago Press, Chicago 60637
The University of Chicago Press, Ltd., London

Printed in the United States of America

22 21 20 19 18 17 16 15 14 13   1 2 3 4 5

ISBN-13: 978-0-226-32281-0 (cloth)
ISBN-10: 0-226-32281-5 (cloth)

Library of Congress Cataloging-in-Publication Data

Hecht, Susanna B.
The scramble for the Amazon and the "Lost paradise" of Euclides da Cunha / Susanna B. Hecht.
pages ; cm
Includes bibliographical references and index.
ISBN-13: 978-0-226-32281-0 (hardcover : alkaline paper)
ISBN-10: 0-226-32281-5 (hardcover : alkaline paper) 1. Cunha, Euclides da, 1866–1909. 2. Cunha, Euclides da, 1866–1909—Travel—Amazon River Region. 3. Brazil—History—1889- 4. Amazon River Region—History—19th century. 5. Rio Branco, José Maria da Silva Paranhos Júnior, Barão do, 1845–1912. I. Title.
F2537.C97554H33 2013
981'.1—dc23

2012027399

♾ This paper meets the requirements of ANSI/NISO Z39.48-1992 (Permanence of Paper).

Constantly braiding, diverging, intersecting—
just like the Purús in its great meanders of life:
it is to my sister Holly Ebel and my brother Thomas Hecht
that I dedicate this book.

Let them fantasize, of winds leaping
From wine-skins and of amorous Calypso;
Harpies who foul their own banquets;
Pilgrimages to the underworld;
However they polish and decorate
With metaphor and such empty fables
My own tale in its naked purity
Outdoes all boasting and hyperbole.

LUIS VAZ DE CAMÕES, *The Lusiads*

In that indefinable compound—the Brazilian—I came upon something that was stable, a point of resistance reminiscent of the integrating molecule in the initial stages of crystallizations. And it was natural enough that, once having admitted the bold and inspiring conjecture that we are destined to national unity, I should have seen in those sturdy *caboclos* the bedrock of our race.

EUCLIDES DA CUNHA,
Introduction to the 1905 edition of *Os Sertões*

Contents

Acknowledgments

This book begins in the Amazon basin, where I first encountered da Cunha's prose. I felt that no one had seen or written on the Amazon as luminously, and I was annoyed and upset that so little was known about him and his Amazon life. In some ways my entanglement with him was a natural outgrowth of the study of nineteenth-century naturalists that I had begun as a graduate student under the great historian of environmental philosophy and geographic thought Clarence Glacken. I suppose this volume can be seen in some way as revisions of a paper I began with him. Another long-standing debt is to the late Hilgard O'Reilly Sternberg, my dissertation chair, whose interest in the Belgian history in the Amazon was my first intimation of the scramble. I also thank Barbara Weinstein, doyenne of the Amazon rubber economy. Our conversations as graduate students at the Museu Goeldi still echo throughout this work, as well as those with activist scholar Marie Allegretti. Other analysts of the Upper Amazon, including Francisco Santos-Granero, Peter Gow, and Miguel Pinedo-

Vásquez, recast my thinking about the landscape and history of the Peruvian Purús.

This book's formal life actually began in the garage of neighbor Casey Kelley after my house was destroyed in the Northridge earthquake. I resided in her Airstream trailer and used her capacious garage as my office, surrounded by boxes of my worldly life while I watched my rubbled house go from dumpster to reconstruction. In many ways Casey is the patroness of this work, all the more so since she did this not from neighborly obligation but from the very goodness of her heart. There were others who continued along, encouraging, providing the moral support, meals, and solace that were necessary for the reconstruction of a life and a house and the writing of a book. My neighbors Charlie Lockwood, Dan and Martha Seeger, and Uschi Obermeyer provided dog walks and the pleasures of daily life. Robert Wayne and Blair Van Valkenberg, Judith Carney, Richard Rosimov, Stephanie Pincetl, and Jonathan Katz fed and cheered me on. Joe and Karen Paff and Tally Wren of Petrolia, California, were happily gossipy, and Joe especially always demanded that one keep up with Schopenhauer. Craig Storper provided a lot of telephonage and was always happy to talk long hours with me about everything from craft to lawsuits. Longtime confrere Michael Storper provided shelter from the storm when the winters made Casey's garage too cold, and his always effective jaundiced eye cauterized any lapse into sentimentality. Sassan Saatchi and Leila Farzani kept cooking and throwing parties that made all our lives better. Melissa Savage was an encouraging and loyal friend.

Alberto Lourenço Pereira was an assiduous critic of the early rough translations and did much to correct some really terrible bloopers. Whatever faults remain are entirely due to my own shortcomings. John Comaroff encouragingly read the earlier translations and was an unflagging supporter. The later materials were kindly reviewed (and edited) by Charles Mann, who was also a great champion of the work and its most generous critic and analyst. He has been an active intellectual companion in the long slog (or Great March) involved in a project of this type and was a continuous sounding board when this work seemed to only echo back into earlier centuries. The continuing conversations about our our parallel writing efforts were part of the pleasures of an endeavor such as this. Christine Padoch was enthusiastic when I flagged, and Michael Storper yelled at me to get the damn thing finished. Mike Heckenberger read many of the earlier chapters and always encouraged a deeper framing of the history. The eminent da Cunha scholar Leo Bernucci also was a companion in the disquieting affliction of passion for Euclides.

There were other supporters who continued to believe that I would actually finish this in spite of all evidence to the contrary; these include Candace Slater.

The posse from the University of Texas at Austin also deserves my warmest thanks, no one more than Chandler Stolp. He had to endure unremitting conversations about historical lacunae and did so with tremendous cheer and kindness for most of a decade. Robert Wilson, with whom I ran a student and research exchange to Brazil in the midst of all this, was also always a delight and always up for a good dinner and lively company. Seth Garfield's historical ethnographies of the World War II rubber period were a helpful counterpoint, and he has remained a generous critic. My trips to Texas were always a pleasure due to the Teresa Lozano Long Institute of Latin American Studies, the extraordinary group of scholars that it embraces, and the superb Benson Library.

The late Alexander Cockburn must be thanked for his appreciation of da Cunha and also for the important apprenticeship in the craft of writing and structuring a book. Our joint book *Fate of the Forest* was peppered with da Cunha quotes, and Alexander's gift of an early edition of the Herndon-Gibbon book (with maps!) alerted me to how much Lardner Gibbon had been excised from modern editions.

There was also solidarity throughout this project from Euclides's most important Amazon heir: Lucio Flavio Pinto, Amazonian journalist in the most radical and combative tradition, son of the land and lover of it. The place has no more potent defender. Like Euclides, he was born to be the "Jeremiah of his times." He was an indefatigable supporter of this work and a tremendous friend over the years.

There were cheers from the blood kin. Holly and David Ebel and their sons, Greg, Jon, and Kip, and the grandnieces and nephews provided ample amusement, while my brother Tom, his wife Jane Burwell, and the feisty Sara were also pleased to house, wine, and dine me when the situation demanded.

There are people who no longer walk the planet who had profound influence on this work through their innovative reading of Brazilian nineteenth-century social history and Euclides da Cunha. These are critic Robert Ventura, UCLA historian Bradford Burns, the master historians of Amazonia Leandro Tocantins and Arthur César Ferreira Reis, and da Cunha biographer Frederic Amory. Anyone who works on da Cunha owes an enormous debt to the Brazilian da Cunha scholar Walnice Nogueira Galvão, who has tirelessly made many of the archival materials more widely available.

Research of this scope that takes this long has been supported by many institutions and grants, including the UCLA Committee on Research, the UCLA Latin American Center, and an American Council of Learned Societies Senior Research Fellowship. I thank that most generous institution, the Center for Advanced Studies in the Behavioral Sciences at Stanford University, where much of the basic research was developed and benefited from the presence of historians Richard White, Walter Johnson, and Kathy Morrison, with their attention to both empire and hidden social history. Another fellowship at the Shelby Cullom Davis Center at Princeton University, whose theme "Utopias" deeply influenced this book, provided a wonderful environment for taking intellectual risks. There Lauren Benton, Jennifer Wurzel, Gyan Prakash, and Jeremy Adelman provided a supportive environment, and the Firestone Library, a legacy of the global rubber economy, offered up remarkable resources. To these all I owe an immense debt of gratitude. But my experience at the remarkable Institute of Advanced Study in Princeton was crucial. Wonderful colleagues there provided ongoing critiques and commentaries that helped refine many of the ideas. Especially I thank Joan Wallach Scott, Kristen Ghodsee, Rosalind Morris, Farsana Shaik, C. K. Lee, Jennifer Pitts, David Scott, Steven Feierman, and the rest of the "Third World Now" seminar, the last one convened by Clifford Geertz. They helped me frame the Scramble in ways I could never have imagined and provided the rich, embracing intellectual life that nourished this effort in its final phases. A John Simon Guggenheim Foundation fellowship helped out immensely in the final phases of this book.

A work such as this depends on the quality of libraries and librarians. I have been exceedingly fortunate to have had access to the great Latin American research library collections in the United States at Stanford, Texas, and Princeton, and that premier collection, largely assembled by Dr. Lawrence Lauerhaus, at UCLA. I often felt that Larry had assembled the UCLA collection just for me, a feeling I know many other scholars share. I suspect there is no better gathering of nineteenth-century arcana on the rubber economy of the Upper Amazon than that which resides at my own institution. The tireless Meleiza Figueroa fetched endless volumes from the UCLA libraries with tremendous good cheer and wrangled the Endnote programs to manage the realms of accents and foreign languages. Lee Mackey often felt as if he needed a burro for the library forays. I was always thrilled when exactly the book I needed appeared as if by magic out of the tattered Trader Joe's grocery bags that hauled hundreds of volumes from the expansive stacks to my cramped office. Donald Sloane, the pope of Young Research Library Circulation, provided necessary indulgences from time to time, and was extremely understanding about late returns.

This effort was not produced on the shoulders of giants but by following the spoor of geniuses, da Cunha and Rio Branco, and I merely hobbled behind in their footsteps. The daily life of producing this book occurred through a network that seemed to be a kind of trampoline, so whenever I began to descend and to despair, it would always loft me upward. My gratitude and debt to those "on my side" is as immense as the Amazon itself.

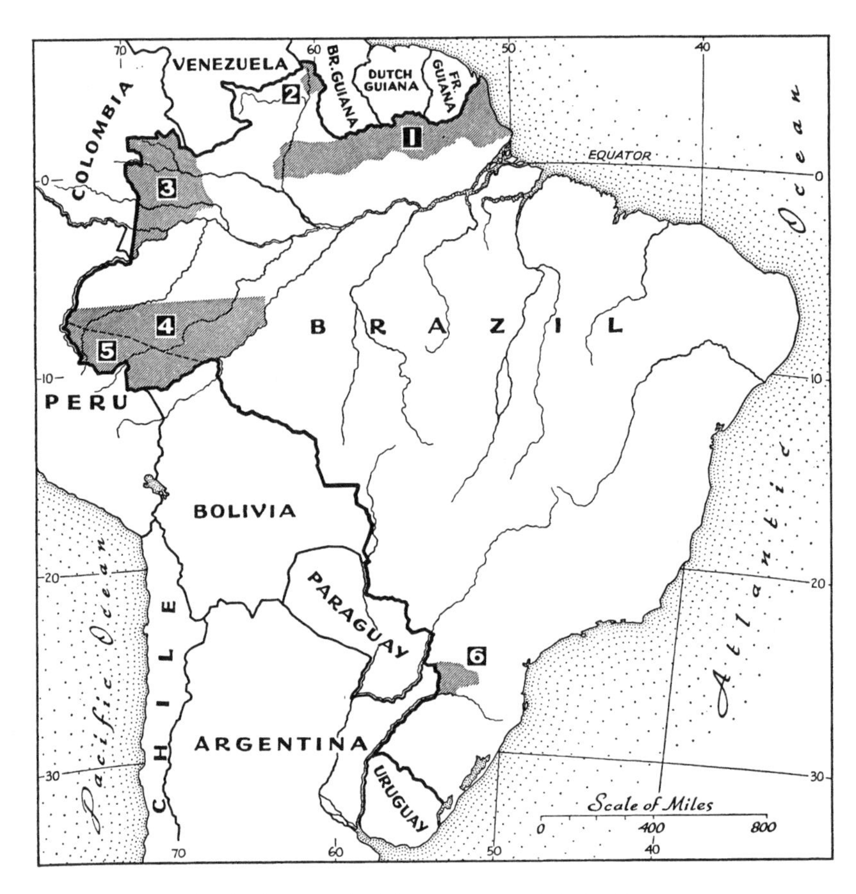

Map 1. Map of territorial settlements made by Rio Branco.

PART 1

Os Sertões

The Pre-Amazonian Life of Euclides da Cunha

1

A Short Prelude

From *Os Sertões* to *As Selvas*

Euclides Who? Narrating the Brazilian Nation

Euclides da Cunha is not known as a player in the pivotal moments of the creation of the first Brazilian Republic. His exploits as a rebel firebrand, as an Amazon explorer, as an intimate of Brazil's greatest diplomat, the Baron of Rio Branco, and as a central ideologue and field surveyor in Brazil's "Scramble" for western Amazonia, are largely forgotten. His Amazon time is especially obscure, given only footnotes and short chapters in his biographies, in spite of the fact that most of the last years of his life were taken up with urgent Amazon concerns. Da Cunha resides in, indeed dominates, a quite different realm: Latin American literature. He is regarded as one of the greatest writers Brazil has produced, and among the most luminous stylists to ever have written in Portuguese, due to his masterpiece *Os Sertões* (*Rebellion in the Backlands*), published in 1902.

Da Cunha's ideas shaped the rhetoric of national politics from the first moments of revolutionary insurrection to his last days in Rio Branco's court. Da Cunha was deeply engaged in the political, ideological, and geographical construction of the young Brazilian republic, whether in uprisings in Rio, in rebellions in the Northeastern backlands, or in imperial contest on Amazon frontiers. His biography mirrors the signal events of the Brazilian republic in three key periods. He is situated at the heart of the revolution that overthrew Emperor Pedro II—through his mentors such as Benjamin Constant Magalhães,[1] his own dramatic actions, and his exile to São Paulo where, as a journalist, he became an antimonarchist propagandist for the young republic. Later, after the relatively bloodless revolution, da Cunha married Ana de Ribeiro, daughter of General Solon Ribeiro, one of the instigators of the republican coup. This further helped da Cunha's revolutionary standing, but set into play a deep domestic unhappiness whose denouement in a modest Rio suburb was a defining scandal of Brazil's Belle Époque.

Os Sertões

The Canudos rebellion in 1897 was a pivotal moment for da Cunha and the republic. In the outback of Bahia (the "Sertões" of the his title), a motley group of peasants, natives, ex-slaves, and devout followers of the backland prophet Antonio Conselheiro had successfully defeated several Brazilian military campaigns. The movement was portrayed as millenarian, backward, royalist, and implacably allied against the godless, secular nature of the first republic. Devoted to the pious Princess Isabel (who had ended slavery in 1888) and protected by *jagunços* (the fighters of the Northeastern outback), Canudos was emblematic of all the atavism that Brazil, in its yearning for stature among modern states, wanted to forget.

The republic, faced with multiple insurgencies, deemed it essential to quash any resistance to the state in the most decisive military and symbolic ways and to quell any rumors about restoration of Dom Pedro or his family. The importance of a victory at Canudos for the military regime that had barely come to power cannot be overemphasized. The frail legitimacy on which the republic rested, the ideological importance of the military as the great unifier of the nation, and the rationalist triumph over "superstitious hordes" were necessary if Brazil was to position itself as an enlightened state rather than a backward nation led by a constitutional monarch. Da Cunha was sent to the outback as an "embedded reporter" and aide to General Bittancourt to report on events for the newspaper *Estado do São Paulo*.

Civilization's triumph over barbarism—the usual colonial story, glossed with the fashionable ideologies of conquest and racism that infused most imperial tracts of the day—seemed to frame *Os Sertões*, but by the end of the book these received ideas were in rubble, just like the rebel city. *Os Sertões* chronicles the suppression of Canudos, providing an extraordinary account of military campaigns (and guerrilla warfare). The book is celebrated not just for the brilliance of the writing but also because da Cunha's authorial voice, initially sneeringly superior, insisting on the inevitable victory of Brazil's coastal civilization and the white race over the pathetic barbarism of the mestizo backlanders, increasingly empathizes with the doomed rebels, with their vibrant life and culture unfolding in the dry forests and in their "mudwalled Jerusalem." Resisting the assaults of the republic (and handily defeating three of the four expeditions against them), the backwoodsmen held off a brutal siege by five thousand troops. Brazil's well-armed militia hardly advanced in three months. The blockade (and the latest German weaponry) eventually took their toll and the republican army triumphed, but only after house-to-house combat extraordinary in its lethality on both sides. Those who had given themselves up as prisoners of war were slaughtered, while the huddling women and children of Canudos were sent off to brothels or dismal lives of indentured servitude. The conflict was one of the major wars of its age anywhere on the planet and was followed avidly in the national and European press.[2] In Brazil, it was meant as an object lesson for other insurgencies. Even so, other uprisings, like the Contestado Rebellion in Santa Caterina (1914–17), took their inspiration from the Canudos resistance.[3]

Da Cunha began his book as a narrative on the triumph of white culture and the inevitable extinction of degenerate races by the more advanced ones. He proposed his work as a kind of nostalgic ethnography of a vanishing world, crumbling before inevitable civilization. At the end, with Canudos's Vasa Barris River literally running red, the prisoners of war rotting in a mass grave, the head of Antonio Conselheiro, the "King of *jagunços*," impaled on a pike, led the returning battalions back to a deliriously cheering Salvador. But the larger meaning of barbarism and the taste for annihilation remained in the air. *Os Sertões* is an unhappy epic of a backwoods polity, an autochthonous, desperate, and brave population who were, as he would describe it, "the bedrock of our race."[4] This was the revelation of da Cunha's book for Brazilians, who at the time could barely take their eyes off the Atlantic with its arriving boatfuls of new books and stylish fashions.

Os Sertões was a literary sensation, an icon of "Brazilianness" in sea of derivative novels. Da Cunha was soon considered among the most illustrious writers of his time—indeed of any time—eclipsing virtually all of his cohort, with the possible exception of his friend the great mulatto novelist and satirist Machado de Assis. Euclides, Machado, and a handful of literary critics such as his friends Silvio Romero and José Verissimo (who will appear later in these pages) were concerned to create a Brazilian literature, a *national* literature that did not obtain its style or inspiration from mostly Francophone pretensions.[5]

The battles of Canudos were da Cunha's *Iliad*. He transformed what had been framed as straightforward military repression against a backward, racially degenerate monarchist uprising into a more complex, essentially anticolonial rebellion. The revolt was, as he wrote it, the ambiguous expression of an autonomous national culture, a culture that he and his literary circle felt could be described by Brazilian writers only in Brazilian idioms, since Europe had no words for their experience. This hybrid culture bred in the backlands landscape was also largely invisible to the coastal elites (inveterate and avid Europhiles) and the masters of the new republic.[6] As da Cunha put it: "What we know of the *sertões* is little more than its rebarbative etiology, *desertus*. . . . We could easily inscribe on large swathes of our own maps our searing ignorance and dread: Here be Dragons. . . . Our own geography remains an unwritten book."[7] His next task was the inscription of another enormous backland, a "new geography" of Amazonia, into Brazil's national destiny.

As Selvas: Explorer, Scholar, and Paladin of the Amazon

It was in the clashing imperialisms of Peru and Brazil, the last great contest in the Amazon Scramble, that da Cunha's next role in the formation of the young republic unfolded. What was at stake in these arbitrations was, in da Cunha's view, 720,000 square kilometers of Amazonia, the source of one of the most valuable global commodities at the time, rubber. As an aide to the Baron Rio Branco, José Maria de Panhanos, da Cunha mapped one of the longest tributaries of the Amazon, the Purús, and developed the nationalist/imperial narrative that would shape the boundary mediations between Brazil, Bolivia, and Peru by unveiling the hidden histories of Amazonian conquest and settlement. Had Peru won the arbitrations it would have become an Amazonian superpower—a kind of Brazil. Instead, Brazil with its documents, maps, essays. and arguments, largely prepared by da Cunha, prevailed, giving us the map of the Amazon we know today, as he brought into focus unseen worlds—places, he would argue, that were the most Brazilian of Brazil.

Da Cunha would write a great deal of the social geography of Amazonia's least-known western territories through the formal surveys of the Purús River, the recovery of the region's hidden history in documents, treaties, maps, and oral accounts, in his own rural sociology and political commentary on the disputed territory that had been called for centuries "the Land of the Amazons." This was his *Odyssey* to the *Iliad* of Canudos. The ravaged northeasterners would reemerge in the Amazon through the great *sertanejo** diaspora, as they fled El Niño droughts and the residues of slavery and migrated to the watery forests and the labor-starved rubber economy.[8]

As Selvas: The Jungles

Considering the continental size of Amazonia and the fact that most European polities had colonies in South America, the lack of attention that has been paid to competing Amazonian imperialisms is surprising. The famously uncertain or creative boundary lines, the endless contestation over lands and labor, the contending ambitions over mythical or actual bounty kept adventurers, ecclesiastics, crowns, and spies riveted on greater Amazonia. The rise of steam travel and the opening of the Amazon to international trade made exploration vastly easier, made possible the explosive commerce in latexes, and intensified the politics of the Scramble.

Most scientific travelers of the nineteenth century had agendas beyond the sale of their collections, the advancement of science, or simple exploration. Even a casual rereading of these works places them at the heart of imperial trajectories.[9] Naturalists, surveyors, and adventurers were part of what Joseph Conrad called "Geography Militant," the colonial enterprise concerned with both science and conquest.[10] Strikingly, most nineteenth-century naturalists portrayed Amazonia's "reality" to European readers as an untrammeled wildness, a "Land without History," as da Cunha would title (with great irony) a famous set of essays. Recent scholarship, as well as da Cunha's own reports, reveals far more complex regional economies, sociologies, and histories. This moment of Euro-American imperialism and Atlantic globalization had no better or more eloquent scientific or social observer than da Cunha.

Da Cunha's Amazon writing addresses topics a century ahead of his time: "everyday" forms of state formation, environmentalism, political ecology, comparative imperialisms, social history "from below," political cartography, and comparative social history. He wrote on tropical geomorphology, and he remains a premier historical ethnographer. These contributions would be difficult to grasp, though, if there were no unifying thread of Brazil, the Amazon, his time, his style, and history. Placing da Cunha's work in the global context of "Scramble for Amazonia" helps clarify the purposes of his prose.

* A *sertanejo* is a person from the Sertão, the backlands.

The Scramble for the Amazon

Another protagonist of this book is Brazil's greatest diplomat, the Baron of Rio Branco. Rio Branco incorporated an area the size of France into the young republic and secured the Brazilian boundary of more than twelve thousand miles, most of it in Amazonia. His diplomatic talents had been honed in legations to England, France, Germany, and the United States. Sent initially overseas as a kind of pleasant sinecure during the last decades of Emperor Pedro II, he lived in Liverpool, the most important European port for tropical commerce. As a member of the Brazilian delegation there, he was well placed to note the prominence of Amazonian latex in the world economy, an insight that eluded many southern Brazilians, entranced as they were by the wealth of their coffee production. Other postings sent him to Paris and, especially relevant, to Berlin during the "Scramble for Africa": the European powers, inflamed by highly public colonial exploits such as those of Stanley in the Congo, inter-European rivalries, and more than a century of covert and detailed resource surveys, parceled out a continent among themselves, heedless of the local inhabitants.[11] These lessons about the imperial ambitions and practices of European powers were not lost on Rio Branco. Later stints in Bern and in New York also proved highly strategic.

Rio Branco was also alert to the territorial ambitions of the other hemispheric republics, the United States to the north and Peru and Bolivia to the west. Rio Branco had spent time in Washington preparing for a territorial adjudication between Brazil and Argentina that was mediated by US president Grover Cleveland, and there had seen firsthand American methods of continental consolidation. By 1902, when Rio Branco took over as the minister of foreign affairs, North Americans had had more than half a century of enterprise and programs aimed at the Amazon. The Baron observed with a combination of admiration and apprehension the US adventures in the Spanish American West (where a third of Mexico became US territory), its filibustering hemispheric forays into Central America and the expropriation of Panama from Colombia.

Brazil confirmed its national boundaries in Amazonia by thwarting the imperial ambitions of France, Britain, the United States, Belgium, Bolivia, and Peru. In this Scramble for the Amazon—a territorial grab that occurred at the same time and involved many of the same actors as the Scramble for Africa—da Cunha, as Rio Branco's agent, carried out remarkable feats of physical exploration, political maneuverings, and discursive construction. Da Cunha deployed every imperial idea on offer—historical conquest, manifest destiny, historical settlement, national mythologies, social and historical cartography in an imaginary of a new tropical, Brazilian civilization to defy other colonial ambitions. The Scramble for the Amazon has received little attention, even from Amazonian scholars, yet the final outcome is reflected in every modern map of South America. The Brazilian national territory emerged from its mostly uncharted physical and cultural tropical backlands in the span of less than two decades as Baron Rio Branco fixed Brazil's formal sovereignty over most of the basin, thus ending four hundred years of contest.

Tropical Hot Spots

When Rio Branco came to power in 1902, two guerrilla wars unfolded in the Purús watershed: one between Brazil and Bolivia over the Acre Territories and the other on the upper Purús and parts of the Juruá Rivers between Peruvian *caucho* gatherers and Brazilian rubber tappers. The Treaty of Petrópolis between Bolivia and Brazil (1903) gave Brazil the Bolivian Territories of Acre and its unimaginably valuable rubber forests. But Peru claimed these same terrains on the basis of earlier Spanish colonial claims, arguing that Bolivia had no legal rights over them and so surrender of those lands to Brazil was illegal. While mediations between Peru and Bolivia over these territorial claims would last for most of a decade, the Purús-Juruá remained engulfed in guerrilla warfare. A modus vivendi was signed to neutralize the conflict zones between Peru and Brazil while various documents and boundary surveys could be worked out. This was why da Cunha headed up the great rubber River Purús on a binational survey commission of Brazilians and Peruvians to the demilitarized zones. The issue was extremely sensitive: if Bolivia had had no right to relinquish Acre, Rio Branco would have made a devastating, costly and unconscionable diplomatic blunder.[12]

Acting as a kind of Lewis and Clark to Rio Branco's Jefferson, the reconnaissance team was supposed to travel up to the remote headwaters of the Purús River to map the "unknown heart" of the Amazon. Da Cunha's job would be to ascertain where and what were the lands that should be handed over to Brazil, based on patterns of discovery, settlement, treaties, historical cartography, the nationalities of current settlers, and a huge diplomatic dossier. He was also charged with describing anything interesting he might see along the way. The latter included economic and social vignettes, natural history, and the intrigues unfolding in the expedition itself.

Da Cunha's travels were fraught with disasters, and he lived in perpetual anguish and near starvation throughout the voyage. Beset with shipwrecks, humiliating penury, disease, and setbacks, he and his men ascended the Purús under the worst possible conditions. The binational survey expedition that da Cunha led with Captain Alexandre Buenaño became a bitter rivalry as they advanced to the headwaters. Their personal animosity mimicked the politics of the larger international intrigue and grew more acute as they traveled through the conflict zones. This voyage (and its rationale) contrasted remarkably with the placid flotillas of explorers who were marking out an "uninhabited" Amazon of "Nature," just as the region was flooded by hundreds of thousands of migrants for the expanding latex economies. The Amazon was the object of treaties, armed incursions, and roving battalions for much of the nineteenth century and deeply integrated into global commodity circuits. European and hemispheric players were prowling in the environs, looking for an excuse to intervene. While travelers marveled at the grandeur of the "untrammeled" Amazonian forests, at the time it was a geopolitical hot spot.

Having been a journalist of the Paraguyan War, Rio Branco knew how deadly campaigns could be in the tropical swamps of Brazil's western borders and wanted to avoid the inevitable disasters of far-flung tropical battlegrounds. Rio Branco much preferred the "Chancellery wars" of negotiations, where well-crafted historical arguments and nationalist rhetoric stood a good chance of success.[13]

Imperial Ideologies: Da Cunha and the Anti-imperium

As historical commentary, da Cunha's writings were to shape the arbitration over the disputed frontiers. The Peruvian argument was essentially a bureaucratic one based on administrative domains in the earlier Viceroyalty of Peru. Da Cunha used his travel, research, and cartography to recast the debate in idioms that reworked frontier Amazonia away from treaty histories or conquest by the transient mixed-blood slavers from São Paulo known as the *bandeirantes,* to settlement rooted in the less glamorous but courageous every day actions of the *sertanejos.* In Brazilian national mythology, the *bandeirantes* had claimed Amazonia through their monumental travels of discovery and plunder. They were, however, ephemeral presences in Amazonia no matter how important they were symbolically. Rather than fashioning a "boys' own" story of bravura and far-flung ruffians, da Cunha cast the story of Brazilian claims quite differently through the efforts of his backlanders, while still keeping São Paulo's (and Brazil's) national heroes on the stage. It was important to do this because the *bandeirantes* were a powerful entrepreneurial element in São Paulo's emerging modernist identity, in contrast to the indolent oligarchs who still dominated Brazil's external image.[14]

Da Cunha saw Amazonia's true conquerors in the modest, impoverished, and beaten-down *sertanejos* he first saw in Canudos, who incarnated the "everyday forms" of nation creation, transforming an "unknown swamp" into an economic engine of the new Brazilian republic. He saw South American explorers and settlers as Amazonia's true discoverers, placing the region within a South American practices of *national* exploration, regional imperialism, and settlement rather than the typical narrative of "discovery," scientific or otherwise, by external colonial powers.[15] The *sertanejos*' little huts and modest farms occupied these lands, their unorthodox warfare defeated formal armies, and in the end, for da Cunha, it was they, more than any former treaty, slaving gang, or glorious military enterprise, who were responsible for transferring enormous, contested swaths of Amazon terrain from other claimants to Brazil. These were the unsung colonials that shaped, were shaped by, and sprang from the "womb of the waters"—the Amazon itself. This ethnographic and historical argument was the crux of da Cunha's national and imperial narrative.

Before da Cunha set sail to the Purús, he viewed Amazonia as a place of nature, "unready for Man,"[16] repeating the received ideas of the time. As he traveled into the rubber country of the Purús, his maps and writings unveiled an inhabited landscape, one that included desperate debt peons, tormented Indians, *caucho* collectors, toiling yeoman families, British engineers, German boatmen, Parisian fortune hunters, and pitiless rubber barons. This vibrant humanized countryside supported Brazilian imperial ideologies, where colonial claims grew from settlement and use. As in *Os Sertões*, da Cunha begins in cliché and ends up in a world of earnest work and a tropical homeland in Amazonia's deep interior. This discovery was completely at odds with the northern European tropicalisms that favored an "empty" world, a largely natural, wild place inhabited by indolent primitives, a *tabula rasa* waiting for the industrious enterprises, colonists, and civilizing missions of the imperial North and whiter races.

An Empire of Nature

Patricia Seed, a historical anthropologist, points out in her illuminating study of ideologies of conquest that in northern European laws, classification of places as "wasteland," wilderness, an unused emptiness, was central for legally justifying external claims to tropical territory.[17] The numerous, mostly Anglophone nineteenth-century naturalists, cartographers, and adventurers whose writings on Amazonia have so much defined it as "portentous nature" contributed to this colonial view, although today they are read more for their environmental aesthetics than their imperial pronouncements.

What is remarkable, though usually unnoted (except by da Cunha), is that these observers largely traveled by commercial steamship on the main channel, stopped at the regular wooding stations, collected specimens in roughly the same places (or had the locals do it while they prepared them), and, ironically, were crafting the ideas of the forest primeval crammed with exotica on regular, British-owned steamboat routes at the moment when the region was becoming the migratory goal for more than million people and the export platform for close to half of the world's latex, one of the most valuable global commodities at the time.[18] Da Cunha himself derided these explorer writings and had little patience for most of the explorers—"they stood merely at her threshold . . . and never knew Amazonia";[19] he devoted his own texts to the exploits of the ecclesiastic, administrative, and diasporic travels that unfolded everywhere in Amazonia's deep interior rather than adulating what were in many respects nineteenth-century spy trips, reconnaissance, and colonial ecotourism.

There were other scrambles at play besides the Brazilian story that I tell here: Peru was actively engaged in contests with Ecuador, Colombia, as well as Bolivia, the Venezuela-Colombia boundaries were complicated, and some adjudications, such as that between Brazil and Great Britain, have been masterfully described elsewhere.[20] Yet none had such an eloquent observer, and none were as significant: the Purús was the largest and last episode of the Brazilian side of the Scramble.

The Amazon and Artistic Ambitions

Da Cunha saw his survey position as a significant professional coup, but he also realized that it would provide him with the foundations for a new literary work. He intended to title his new work "O Paraíso Perdido" (Paradise Lost) and for it to be a companion volume to *Os Sertões*. The fragments he wrote on the Amazon contain many structural elements that parallel his novel—the landscape description, the social relations, the regional history, and detailed discussions of ecological adaptation.

His Amazonian writing extends the saga of the ravaged northeasterners to their next economic and geographic phase. While *Os Sertões* ends with the backlanders who challenged traditional oligarchs and the new republic being crushed by them, his Amazon epic would chronicle the making of Brazil's tropical empire by those same *sertanejos*. What he remarks in another context could easily be applied to those migrants: "What I had first seen as a desperate crawl and grasping was in fact a leap of triumph."[21] They would, he felt, scrawl Brazil's true destiny over the continent and, more importantly, evolve a new tropical civilization of mixed-blood pioneers, a New World counterweight to white European imperialism. The "bronzed titans" who had been driven down in defeat in the Northeast materialized in the Amazon as the masters of its (and Brazil's) destiny. In this sense, the ideological transformation in *Os Sertões* from the fashionable racial clichés about the *sertanejos* to their reincarnation as bedrock of Brazil is the point of departure for his Amazonian oeuvre.

Fragments

Da Cunha's Amazon material was, by his own assessment, to be his masterwork[22] and his second "volume of revenge." Yet he never finished it, because he was shot to death by his wife's lover. Most of the attention that has fallen on his Amazon days focuses on a few polished elements that were jumbled together and published posthumously as *À margem da história* (At the Margin of History).[23] There are newspaper columns, the formal survey reports with their rich historical additions, his essays on the Peruvian and Bolivian claims (*Peru vs Bolivia*, 1908), and an extensive correspondence that addresses his travel to the headwaters of the Purús and the politics (and personal disaster) of his return. But these are unintelligible and simple fragments without a unifying theme. And there certainly was a unifying theme. Virtually all of da Cunha's Amazonian writing can be seen as, and indeed, much of it was explicitly produced for, a nationalist/imperial narrative to justify Brazil's claims in the upper Amazon. That he completed his task by linking it to his own dreams of an emergent tropical civilization through the moving revelations of social history and nature writing speaks to his artistic ambitions and his brilliance.

His Amazon work integrated his practical concerns (an income), artistic impulses, political yearnings, and geopolitics with his longing for a new type of Brazilian society and nation that valued its "bronzed backwoods titans" as much as its gilded coastal elites. It revealed his desire to show the force and beauty of the forest and to uncover a lost social history obscured by the meanders of the rivers, languishing in dusty archives, and couched in the twangs of his *sertanejo* informants. The moving elements of *Os Sertões*—his empathetic descriptions of land and life, his attention to history, folklore, and landscape—are all present in his Amazonian work.

As a man of letters and part of the rebellious literary coterie of Brazil's Belle Époque, he was interested in national themes rather than the artistic possibilities of the transposed French genre of the novel of manners or the imitations of Alexander von Humboldt's extravagant style. Euclides felt the "lexicon of science" was most evocative of the region because most exact. Of the prose of botanist Jacques Huber he writes, "The eloquence and brilliance were imparted by the extraordinary display surrounding him. . . . I . . . could not find in myself the choice vocabulary which to describe it, but he could, using only the language drawn from the austere lexicon of technical diction."[24] On the writing of his friend and land surveyor Alfredo Rangel (and he might as well have been writing about himself): "The dreamer aligned his way on compass coordinates. . . . And his most touching insights were written on the last pages of field notebooks."[25] And indeed, da Cunha's own notebook of studies for the unwritten "Paraiso perdido" engage mostly scientific annotations from the British Royal Geographical Society, on the work of the first formal explorer of the Purús River, William Chandless.

Who is Euclides, and why is he there? What is really going on behind the powerful texts, petty bitching about bureaucracies, complaints about the climate, and yearning for friends and family? His aim, of course, is to come up with a truly Brazilian narrative, a Brazilian epic and a New World Lusiad. He is not given to the romantic sublime or the picturesque although he is possibly the best writer on Amazon nature, nor is he a chronicler of oddities suitable for retelling in fashionable salons. It wasn't exotic to him: it was *Brazilian*. He is part of the "writing of empire," as the cliché has it, but his position is so different from most of the genre that, like *Os Sertões*, it becomes something else entirely, genius in many ways. We can perhaps understand his Amazon prose on its own, but it seems to me that there is no possible means of assessing Amazonia's impact on his own thought and life, and his broader impact on "nation and narration," without having more context, without understanding what it was about his times, Amazonia, and his active role in it that made him much more than a literary figure. And indeed, his life was a desperate tragedy that also relates to his Amazon times.

The book is organized to provide some of the early biographical context for his work, but most writers cover this period and discuss da Cunha, *Os Sertões*, and his circle in great detail for the period 1897–1903, which is not my purpose.[26] The biographical material—parts I and V—brackets the materials on the scramble and most of da Cunha's writing (parts II, III, and IV), a world that opened up for him in 1904 and ended, along with his life, in 1909.

That time in the Amazon is remote to us, but it was an extraordinary moment in tropical history when the machinery of industrial revolution, its transport systems, and thousands of innovations for daily life were dependent on tropical latex, scratched out from distant trees and processed by debt peons and slaves over a smoky fire. What transpired in those Amazonian *selvas* was momentous—an expression of planetary economics and imperial ambition, as well as everyday practices of nation formation, of political imagination and possibility, and it reveals how global processes transformed and were transformed by the aspirations that stretched from royal palaces to the humblest tapper hovels.

2

The Unlikely Protagonist

Euclides da Cunha, the mestizo grandchild of a Portuguese slaver and a Kararí Indian, was born in Bahia to Manuel and Eudóxia da Cunha in 1866. He was a feeble child, "stunned by the unpleasant surprise of being born."[1] His mother died when he was three, and thereafter he was raised mostly by fussy and distracted female relatives who were alternately doting and indifferent. He was shunted from household to household according to the fortunes of Manuel da Cunha, a minor coffee grower. Although not entirely unconcerned about his children, da Cunha had domestic arrangements that were as shaky as his commercial interests, and that meant the sickly child remained quasi-orphaned among his handsome cousins in the comfortable homes of his extended family. Young Euclides moved among villas and boarding schools, introverted and shy.

He was neurasthenic and frail in his youth, in part because he was tubercular, and indeed, that disease had carried off his mother. His later adventures however, revealed extraordinary physical capacities, since he was able to survive military

Figure 2.1. Euclides da Cunha at age ten.

campaigns in the desert bush of the Northeast (an enterprise that killed thousands) and a grueling Amazon foray. He was frequently ill as both child and adult with lung hemorrhages and the periodic fevers of malaria, and was exhausted by insomnia. What few photographs we have of him show a pale, intense child, and later a slender man, slickly mustachioed in the dandyish style of the day, with the lambent eyes of a nocturnal animal.

Da Cunha's father, inhabiting the treacherous and tenuous realms of the

Brazilian middle class, was highly sensitive to the precarious forms of ascension and patronage in the last days of empire. Absent the funds for European study or to send his studious offspring to one of the prominent national universities (playgrounds for sons of well-off planters), there were two possibilities for advancement: the military and the seminary. By temperament the austere Euclides was probably more suited to the seminary, and all his life he yearned to teach in one of the new academies of higher learning that the Brazilian state was just beginning to develop.[2] But the practical predilections and economic concerns of his father suggested civil engineering rather than theology as the boy's pragmatic career choice. Manuel da Cunha opted for what was the only sure way to secure a degree and the positions, patronage, and prestige that usually went with it: he sent him to the main technical institutes of the time. The youngster had brilliance in language, natural history, and math, and soon he would be the protégé of the Brazilian revolutionary, and author of the country's first constitution, Benjamin Constant Magalhães.

Euclides initially encountered Constant at the Colégio Aquino, a polytechnic high school run, ironically, by Jesuit clerics committed to the secular ideas of Auguste Comte, whose political philosophy and evolutionary social theory were sweeping South America and whose ideological influence was decisive in Republican Brazil.[3] When Constant, a devotee of Comte and one of the people most instrumental in advancing these ideas within the military and new republic, moved to the War College at Praia Vermelha on Rio's back bay, so did the intellectually enthralled Euclides. Since fees and board were provided there, this scholarly path was gratifying to the elder da Cunha. And so the youngster was trundled off to learn an applied technical trade at this strange institution that was half barracks and half boarding school. Euclides found himself in the upper echelons of military privilege, within an institution that would successfully challenge the monarchy and shape the ideological frameworks, political institutions, and national identity of Brazil for much of the twentieth century. These Praia Vermelha connections structured the contours of da Cunha's entire life and positioned him within the military cohort of Brazil's elite in a moment of transitional national politics—something that his social standing on its own never could have done for him.

Da Cunha came of age in a period of cataclysmic transformation of the three central institutions that underpinned Brazilian life at the time: the monarchy, slavery, and the military. By the end of the last decade of the nineteenth century, Emperor Pedro II was gone, slavery was abolished, and the military had shifted from a commanded institution—one that had spent most of the nineteenth century chasing down runaway slaves, quelling local rebellions, and building infrastructure—to a key political actor in the creation and rule of the modern Brazilian state, where it took charge, on and off, for much of the twentieth century.

Precarious Times

Da Cunha's world began in the twilight of the Bragança imperial monarchy, an empire buckling under internal stresses and the distracted indifference of Emperor Pedro II, who had come to power at the age of fifteen.[4] His grandfather, Dom João VI, a royal fugitive, had abandoned Portugal after Napoleon invaded the Iberian Peninsula in 1807. Dom João packed up his entire court and loaded them on thirty-three galleons to ultimately disembark in their sweaty silks and drooping satins on the muddy quays of Rio, blinking in the glare of the tropical light, gawped at by the locals. Though the place looked like a tropical paradise and was vast, rich, and sensual, it was inhabited, to the monarch's mind, by barbarian races, intractable slaves, an indolent elite, and ambitious upstarts. Thirteen years later, he impulsively abandoned the throne to his child king, Dom Pedro I. It seemed a small matter to pack up the court favorites, the jewels, and a few mementos and sail back home, leaving as his legacies the beautiful Jardim Botânico in Rio, an ambiguous constitution, and a Brazilian native son to incarnate his regime.[5]

Pedro's reign was characterized by rebellion and political dissensions at almost every turn, torn between monarchic ideologies and revolutionary thought emanating from Europe, and within the government, between federalist impulses that seemed only to inflame successional movements and centralizing pressures that provoked powerful antimonarchic sentiments. These, along with an inability to control the apparatus of government the realm's oligarchs, and the difficulty of consolidating national territory, all undermined Pedro's rule. He was sent to exile in Portugal in 1831. Like his father, Pedro I left ruling offspring, but until Pedro II took power in 1840, the state was managed by a set of highly unpopular regents.[6]

The regents enacted harsh centralizing laws and established the institutions and powers of an absolutist state to combat a nation rife with insurgency, as regional elites resisted imperial demands and as the constant slave insurrections roiled through its shantytowns and hinterlands. This legislation remained intact for almost fifty years, until the military coup of 1889 finally overthrew the monarchy.[7] The laws had four main goals: to restore and maintain the order and power of the emperor's favorites, the Northeastern oligarchs and those with special court licenses; to uphold an export-led system of production, based on slavery, with scant attention to the masses of its urban or rural population;[8] to enfeeble liberal and federalist movements, inspired by the French and American Revolutions, that tended to favor regional oligarchs at the expense of the Crown's favorites; and finally, to protect local economies by opposing free trade and market ideologies flowing out of Britain and the United States. The monarchist policies were inflexible, mediated by an old guard that controlled a vast and corrupt system of patronage even as these policies became flashpoints of political dissent and economic stranglehold.[9]

Patronage and Its Problems

The Bragança court in Brazil, enmeshed in its own pressing intrigues and domestic disputes, focused mostly on patronage and petty bureaucracy.[10] This circle was at odds with, and largely oblivious to, the enormous socioeconomic and political upheavals that occurred as the structure and location of Brazil's economic dynamos shifted from the slave-based sugarcane estates and depleted mines of Minas Gerais to the booming southern coffee lands surrounding Rio and São Paulo, the latter's incipient industrial systems, and wealth beginning to flow from Amazonian forests. Economic power fell away from the old Northeastern sugar oligarchs and increasingly swung to an emerging coterie of urban professionals, bureaucrats, entrepreneurs, and minor industrialists. As the nineteenth century wore on, the divergence between the royal *corte* and those who commanded economic engines of the country became more acute. The new economic groups from thriving provinces like São Paulo increasingly differed from the old oligarchs in their views on immigration law, tariffs, land policy, and government subsidies. The lack of representation felt by this rising economic coalition became more galling: their appetites for power and influence went unattended even as the hand of the Crown in their affairs seemed "mostly to grasp and meddle."[11] Amazonian elites, well beyond the imperial system, were commanded more by global demands and local administrators than an indifferent emperor and, as we will see, resisted royal incursions. New elites found themselves shut out from the delightful sinecures of the Brazilian "Versailles" of the Braganças and from meaningful political expression. They faced obstructive economic policies even as the taxes from their successful ventures paid the imperial bills.[12]

The structural change in the economy coincided with processes of urbanization stimulated by international migration and expanded trade and industry, and had the effect of reinforcing the coastal towns and cities, such as Rio and São Paulo, as centers of political and cultural life rather than simple entrepôts.[13] These rising cities provided fertile ground for modern European ideas and nourished the ambition of local elites to shake Brazil out of its torpor so as to vie with Europe on equal economic and cultural terms as a modern state, rather than as a threadbare outpost of the declining Lusophone empire. The modernizing coteries, attentive to European intellectual fashions, vulnerable to the European (especially British) fiscal pressures, and alert to the liberal if not revolutionary trends in France and the United States, increasingly chafed under the conservative directives emanating from imperial Rio and craved more power and local autonomy in the face of parasitic dynasties maintained by royal favor. The regime's rigidities left few options. By the 1870s, republican manifestos and liberal parties were promoting policies that would ultimately undermine the traditional oligarchs, curtail government interference in the private sector, and enhance local powers. Their political agenda, influenced by liberal politics and ideologies of Europe and the United States, included direct elections of state governors, autonomy of the judiciary, religious freedom, secularization, freedom of the press, universal suffrage, and gradual elimination of slavery. Times were turbulent indeed.Dom Pedro's attention, buffered as it was by the comforts of his tropical life and his sycophantic courtiers, was apt to wander in later years when the subject at hand strayed much beyond his library or his lovers.[14] Gouty, diabetic, and, like the family nemesis Napoleon, epileptic, he increasingly preferred to stay with his domestic circle and loyal retainers enjoying the comforts of his lovely palace in Petrópolis, situated comfortably in the mountains away from Rio's heat and pestilence. For him, power was vested in divine right, imperial tradition, and a gracious life in the environs of Rio, although the experience of his progenitors certainly ought to have provided a cautionary tale.[15] As well-bred carriage horses clopped through Rio's leafy suburbs, visitors described Rio as "exotic," "lush," and "exuberant" and were simultaneously attracted and repelled by the oddly decadent and old-fashioned air of this very pretty metropolis.[16]

Rather than struggle with the inconveniences of domestic travel, Pedro II took long trips to the United States, the Mideast, and Europe, visiting friends, relatives, and his old amours. Although he did make one successful trip to the Northeast and visited Porto Alegre for a time during the Paraguayan War, he preferred, on the whole, not to visit his country's immense interior. Pedro II, who adored natural history, was acquainted with his great hinterland mainly through accounts of its biotic marvels. He enjoyed his realm mostly by proxy, content to view it through the eyes of visitors, expeditions sponsored by European royals or American tycoons, or through scholarly books.[17] He was thrilled to meet with Amazon explorers, such as the delightful Agassiz family—Louis Agassiz, who helped establish Harvard's natural history museum, which now bears his name, his wife Elizabeth (later one of the presidents of

Radcliffe College), and the team of young student geologists and naturalists from Harvard and Cornell that traveled with them in the Thayer and Morgan commissions. These included young scientists who would later become giants of tropical geology such as Frederick Hartt and Orville Derby but also well-connected college students such as William James (the eminent religious philosopher/psychologist and brother of the novelist Henry James). Hartt and Derby were the pioneers of Brazilian geology and biogeography and with their local counterparts, especially the geographer Teodoro Sampaio, provided the intellectual foundations for the extensive geomorphologic and biogeographic studies of the Sertão (backlands) and the Amazon *selva* on which Euclides da Cunha would later rely.

Dom Pedro II remained largely indifferent to the turmoil brewing just beyond the velvet curtains of his carriages—his country's dubious and increasingly conflictive boundaries and ever more intractable populations. The period after the end of the Paraguayan War was marked by rising internal tension as slavery began to unravel and federalist movements gained strength. In these years Pedro became increasingly diabetic and demented. He left the country for years at a time, delegating his rule to his daughter, Isabel, the highly religious regent princess, and her husband, Gaston d'Orléans, Count d'Eu (one of the nephews of the Belgian master imperialist King Leopold). Neither enjoyed the political allegiance of the military nor the power to control the Brazilian political apparatus very effectively during Pedro II's long absences.[18]

Slavery and Insurrection

The genteel life of the imperial court stood in contrast to the discord that rumbled continuously and ominously in Brazil's slums, backlands, and distant tropical forests. There were real questions whether the immense territory could ever cohere into one nation, given the frailty of its political institutions and its divergent histories and dubious boundaries. Controversies in territorial governance were matched by the resistance and rebellions that attended the slave economies, based on both native and imported African populations. Slavery was the key economic institution until 1888, with profound impacts that stretched well past that date.[19] If the monarchy was Brazil's salient political institution, slavery was its defining economic one.

Portuguese slavery was well established in the fifteenth century along the African coast, and Portugal itself was home to more than 100,000 African and Mediterranean slaves prior to its forays into the New World.[20] Indeed, the first African slaves to Brazil came from Portugal, where their Christian credentials could be assured and the proper taxes paid on them.[21] The techniques developed for African slave trade were adapted to the New World and were simple in their essentials: stimulate indigenous warfare through manipulation of warring factions and trade in commodities and weapons. Prisoners of war and other captives were then shunted into the commerce of human chattel.[22] Slavery was hardly a novelty in the Mediterranean, since it predated Roman days, but the integration of humans into international trade routes as bulk commodities was largely a Portuguese innovation in this phase of globalization.

Slavery in Brazil, whether of natives or of Africans, was the first real national institution: it crucially shaped early forms of governance, since the legal frameworks were largely structured to control territories with implied rights to peoples within them. While slavery of native groups was theoretically forbidden by the mid-sixteenth century, exceptions were framed within the idea of "just war" to allow bondage of captives: paganism, practices of cannibalism, and aggressions against Portuguese populations were handy justifications for raiding tribes and shackling them to colonial enterprise.[23]

For New World natives, the allure of metals (iron being as interesting to the indigenes as gold was to the Europeans) coupled with the long history of internecine warfare among the various coastal and interior groups produced systems of enslavement based on war captives and built upon indigenous forms of enslavement.[24] Natives would soon, unfortunately, find themselves transformed from mercantilists and traders into merchandise themselves.[25] As indigenous groups fled or died due to disease or became more intractable, groups of gun-slinging *bandeirantes* were formed. These were Portuguese and native "half-breeds"—*mamelucos* in the racial classification of the day—who carried out expeditions into the interior for gold and slaves. Slaving forays began early, in the mid-sixteenth century, and continued through the rubber period even into the twentieth century, so though natives were formally "outside" of slave commerce, they remained covert commodities within it for more than four hundred years. The *bandeirantes* came to occupy a formidable place in the mythology of the Brazilian nation, since it was their flags (*bandeiras*) that "claimed" the great interior as Brazilian terrain. They have remained iconic to this day. The subsidized corporate groups that opened up the Amazon with cattle ranches in the 1970s and 1980s referred to themselves as *os novos bandeirantes*.—the new *bandeirantes*.

While there is no question about the catastrophic nature of the contact between Old World diseases and New World populations, it is also important to recognize that the "Indian wars" were not mere conquest by microbes.[26] Native groups fought endlessly and resisted the encroachments, and there were many defeats of the Europeans in what became deeply protracted warfare lasting centuries.[27] Indigenous peoples did not just evaporate: they fought, they fled, they interbred with the invaders and their fellow slaves. And they also died by the millions. The African trade moved into place to fill the labor void.

While disease and slavery made natives seem only ephemeral actors in the drama of contact, they were not the only ones whose lives ended quickly. Slavery in Brazilian plantations and mining was especially lethal: Katia Mattoso, one of the foremost scholars of Brazilian slavery, states that the average life expectancy for black slaves was on the order of six years after arrival in Brazil. As historical demographers of slavery have noted, about two people died for every slave who made it through the Middle Passage—the journey across the Atlantic Ocean.[28] African slavery shunted more than six million captives (of whom 44 percent were from the Bantu groups of Congo and Angola) to Brazil by the time of the abolition of slavery in 1888. The toll on both sides of the Atlantic in terms of anguish and social disruption was unimaginable.[29]

Contrary to idea that African captives were an undifferentiated black population—a notion often coupled with the assumption that these captives had been "cultureless" forest primitives—slaves arrived as ethnicities and "nationalities" and often regrouped along ethnic, religious, political, and linguistic lines, reflecting the full complexity of West African polities. African slaves brought and adapted to the New World their knowledge systems of agricultural production, animal management, literatures, craft skills, metallurgy, warfare, healing, spiritual ideas, religious institutions, writing and institutional practices. Slaves were not simply passive elements in the bondage system but reinvented numerous institutional forms for collaboration and resistance. Further, they actively restructured the content of many Brazilian institutions and landscapes to make their lot as bearable as possible.[30] They re-created their lives within the modalities of bondage and liberation that unfolded in Brazil.

African slaves changed the New World ecologies. They created dense palm environments, moved ritual plants for their religious observances, and introduced the food plants of rice, *guandule* (black-eyed peas), watermelons, okra, and oil palm. From their mattress pallets and hay bales for feeding the animals on their trans-Atlantic travels came molasses grass (*Melinis* sp) and guinea grass (*Panicum maximum*), invasive species that would support the creole cattle that became the vanguard of European conquest and caused Latin America's profoundest and most extensive ecological change.[31]

Slaves reinvented African institutions where feasible and recast existing ones within an African register whenever they could: "Christian" brotherhoods, burial groups, cooperative work gangs, religious societies, and apprentice groups worked to support identity, help ensure daily survival, nourish spiritual life—and all this social infrastructure could become platforms for resistance.[32] These could range from "everyday forms of resistance"—pilfering, foot dragging, petty sabotage, and "intractability"—to full-blown rebellion and murder. Slaves also fled to the margins to create independent lives and polities. These, the most dramatic form of resistance, were known as *quilombos* or *mocambos*—fugitive communities. These were widespread in Brazil's interior, with hidden communities stretching from the Atlantic to the Andes.[33]

While the plantation model with its bullwhip and overseers dominates the imagination of slavery today (perhaps because of the overriding influence of *Gone with the Wind* in American iconography), protoindustrial sugar slavery was but one form of bondage, and hardly the most common. Slavery in Brazil was overwhelmingly the realm of a master with a handful of slaves, and there were a host of economies where the activities of slaves could not be monitored to the same degree as on the ordered cane plantations. There were small-scale slave producers of provisions and tobacco, slave cowboys from the cattle and horse traditions of what is today the interior of Nigeria and Sudan, and specialized slave "folk geologists" from Ghana's Mina ports—the "Gold Coast" for the mines of Minas Gerais, Mato Grosso, and Goiás. There were black slaves in the cacao plantations of the Amazon; there were Africans and indentured natives who rowed upriver to collect all the extraction products for the international markets, including quinine, turtle fat, brazil nuts, and the parrots and monkeys that everyone loved. Nothing could be done without slaves.

In urban areas, many owners had only one or two slaves: a domestic helper and a laboring servant who could be hired out. These *ganhadores* or "earners" were sources of household income among less well-off middle classes. They were stevedores, litter bearers, day laborers, as well as artisans such as blacksmiths (taking advantage of the extensive metallurgical traditions of West Africa), carpenters, masons, and so on. In order to earn, they had to have freedom of movement, and thus, like mining, extraction, and livestock, this was a slavery without overseers.[34] Slaves with special skills or those who worked hard could, in principle, buy their freedom. Both urban and rural slaves sought to use agricultural plots, curing skills, artisanal skills, sexual attachments, and family loyalty for self-purchase or manumission. But many fled to *quilombos*, where they developed autonomous communities with their own systems of government and livelihood, and where they were relentlessly hunted. *Quilombos* can be usefully understood as the main form of colonization of Brazil's interior. Slaves were reconstructing the terms of their bondage and were active participants in abolition politics through complex forms of domestic influence, politics, and overt resistance.[35]

As Euclides was growing up, abolition was in the air, and like the monarchy, slavery was considered an institution that blocked the country's advancement to a modern state and economy. European and especially British pressure weighed heavily in the realm of international geopolitics.[36] Da Cunha, like much of Brazil's urban intelligentsia, abhorred the "dreaded institution." Its atavistic taint seemed out of place in a modernizing and industrial economic world. In school journals and adolescent publications da Cunha regularly denounced the "hideous practice," writing poems and screeds against it. Yet he was fluent in the racialist ideologies that were part of the scientific training at Praia Vermelha; they saturate his masterpiece, *Os Sertões*. Brazil was the last major nation to free its slaves—an act carried out by Princess Isabel in 1888 while Dom Pedro was traveling. The protracted survival of slavery and its highly mestizoed population with "unpromising" racial characteristics raised questions about Brazil's capacity to "develop."[37] Although slavery had begun to crumble with prohibitions on international trade (1850), and a series of laws meant to gradually unravel its force (particularly the Free Womb Law of 1871, designed by the first Baron of Rio Branco, and the law for the manumission of aged slaves of 1886), it still structured the content of economic relations.

In most agricultural areas, abolition produced a vast and sudden proletariat whose access to plots for subsistence farming was not necessarily assured by their former masters. In the Northeastern backlands, the retraction of the sugar and cotton economies and the limited number of cowboys needed for the extensive cattle estates left a deracinated population with few rights to land or in law, a "nonslave" yet not exactly free class of workers, a population vulnerable to the vagaries of climate and the caprices of the rich. These were the origins of some of the insurgents da Cunha would meet in the mountains of Bahia and the forests of the Amazon, around whom he would structure his *Iliad* and *Odyssey*: *Os Sertões* and his fragments of *Paraíso Perdido*.

Lineages of the Modernizing Military

Slavery and monarchy—the structuring institutions of the Brazilian state in the nineteenth century—were held in place by the glue of the military, an institution that became more complex as the century wore on. By the time da

Cunha studied at the War College, it was a breeding ground for insurrection, actively engaged in abolition and conspiring for the expulsion of the emperor. But the military in Brazil was an exceedingly contradictory organization in the last half of the nineteenth century: it had transformed itself from a kind of penal institution inhabited by criminals and reprobates to the embodiment of disciplined masculinity and modern nationalism.

Da Cunha's credentials as a member of the revolutionary Republican Army served him all his life: he married into an influential military family and moved in and out of military assignments throughout his career. Sent to Praia Vermelha to take advantage of its technical training, he could not be indifferent to the highly divergent social strands that coalesced in the military and gave it legitimacy in a period of such monumental transitions in Brazilian society. New ideologies of the military elevated it to the key institution for nation building because of its ability to integrate regional and racial differences in a functioning bureaucracy and fighting machine, and as an incubator of modernist thought.[38]

The military was racially complex. In the early and mid 1800s the army had essentially functioned as a form of indentured incarceration, but the experience of the Paraguay War (which will be discussed below) was decisive in changing the view of blacks within the military, and the self-image of the military itself. The army had always been a venue for some social ascension for Brazil's poor, and its officer class provided an honorable profession for Brazil's petty bourgeoisie. The military command clearly ascribed to racial hierarchies, but until abolition, it was an institution that could take a slave and make him free. Runaway slaves might billet themselves into the army in a different city. The dragoons that swept through Brazil's hamlets snatching everyone of color often seized slaves and, at the end of their tour of duty, manumitted them. Military impressments, however, could also take free men of color and make them slaves. To escape the eight-year obligation of involuntary service, a regular stream of deserters fled into the backlands and frontiers, and often into *quilombos*.[39]

The impressments for army service, which grabbed fathers, husbands, and sons to fight against secession movements and insurgencies or to chase after fugitive slaves, did not endear the monarchy to the populace at large. Historian Peter Beattie describes how women would set upon the conscriptors, drive them away, and then burn their records. Cartoons showed villages full of animals while men hid in and peered out from forests. This resistance became more intense as other bureaucratic actions, usually associated with taxes and other emerging forms of state control like censuses and land demarcation, seemed to extract more and deliver less in return. A strong antimonarchist streak existed among the *praças*, the "grunts" drafted against their will into the Brazilian army, which only intensified over time.[40]

The military came to be seen as the "integrating molecule" of Brazil. Its leaders were deeply aware of the idea of a "nation in arms" unfolding in the United States and Europe and took inspiration from models of land-based militarism for galvanizing cultural, economic, and political unity.[41] The military sought to take regional identities such as "Bahian" and make them "Brazilian." These shifts brought powerful new forms of allegiance that transcended the regionalism and racism (though certainly not eliminating them) that had historically defined Brazilian society. The armed forces saw itself as a populist, integrative model for the nation, an institution that could "teach" the rest of the country about modern nationhood.[42] This approach to identity, rather than the usual ones embedded in place, position, or patronage, was especially evident in the officer class; its "philosopher cadets," an emergent group of technocrats allied to Republican elites. They were especially impatient with the archaic political and economic structures of monarchist Brazil.

This military, which Euclides would encounter, and which would exile Dom Pedro II, was shaped in the crucible of a largely forgotten war, one that actually was a prelude for the Scramble for the Amazon, since large of areas of Paraguayan (142,000 km^2) and Bolivian lands (97,000 km^2) shifted to Brazil in this conflict.

The Paraguayan War (1864–70)

The Paraguayan War had complex outcomes for Brazil. A distant war now known mainly through novels about the Paraguayan leader's tempestuous Irish mistress, Eliza Lynch, it was the largest and bloodiest war in South American history, claiming the lives of more than 300,000 soldiers and civilians through battle and epidemics of cholera and smallpox. It was one of the largest wars anywhere in the world at the time, rivaling the Crimean War, and was the largest ever fought by Brazil.[43] It was a "total" war, one waged against civilians (mostly Guaraní Indians), a fact that gave the hostilities the tinge, if not the full obliterative reality, of ethnocide.

Though distant, the conflict was of broad international interest because of its racial and political dynamics as well as the huge territory that was up for grabs—some 14,244,934 hectares, an area about the size of New York State. It was quite exotic compared to European conflicts of the time: a war fought by Indians against blacks, by "nebulous republics" (Paraguay, Uruguay, Argentina), and directed by the only monarchy in South America, Brazil. There was the frisson of the omnipresence of women in the battlefields as combatants or camp followers, and guerrilla warfare that defied the norms of European contest. Though strangely primitive in some ways, the Paraguayan War was ahead of its time in the use of earth battlements against rifles and cannons and aerial recognizance (with balloons). Both technologies were used in the First World War some fifty years later.[44]

In discussions of the origins of the war, the usual *casus belli* of much of South America's history is invoked: imprecise boundaries, regional power disequilibria, a meddling international actor (in this case, Great Britain). Brazil craved free access to its cities on the Paraná River in Mato Grosso, which required sailing through areas controlled by Paraguay. The capture of the governor of Mato Grosso on the river was the precipitating incident of the conflict, followed by Paraguayan invasion of Mato Grosso. It was foolish for Paraguay with a population of 450,000 to take on a country with a population of ten million, but President Francisco López was given to grandiose impulses and had faith in his British munitions industries and his Guaraní guerrilla fighters' familiarity with wilds of the Mato Grosso outback; further, his standing army was four times the size of Brazil's.

Both sides framed the Paraguayan War as a crusade of civilization against barbarism. An assault on "power-mad" President López and his savage Guaraní Indians was one version,[45] and few statements encapsulate the vicious perception better than that of the Argentine statesman Domingo F. Sarmiento, whose *Facundo: Civilization and Barbarism* was one of the political literary icons of the nineteenth century: "It is providential that a tyrant caused all those Guaraní to die. It was necessary to purge the land from that human excrescence."[46] Paraguayan pamphlets viewed Pedro II as an incompetent poseur, the officers as buffoons, and Brazilians as reactionary and ruthless, wed to especially unsavory and unmodern forms of government (monarchy) and labor mobilization (slavery—which Paraguay had abolished in 1842). They ridiculed the Brazilian army as one of "black apes."[47]

Guerrilla warfare was the preferred indigenous practice, and López's Paraguayans did hold out in spite of staggering losses, with the campaign fought in its final two years by starving old men, women, and boys.[48] As Dionisio Cerqueira noted, "It was no fun fighting against so many children."[49] Sir Richard Burton, who seems to have been everywhere in the nineteenth century, in this case as British envoy to Rio, described the war in its grinding detail and despaired at the "thought of so much wasted heroism."[50] He documented the first military use of balloons to spy out the positions of the Paraguayans, who proceeded to burn soggy grass and hide under a mantle of smoke.

Arguments about territorial hegemony, frontier boundaries, and frail treaties are part of the story, but there does appear to have been a deeper economic and cultural clash. Paraguay was different from its surrounding states: its population was almost entirely composed of Guaraní Indians, the national language was Guaraní, and its political and relatively egalitarian economic institutions had been shaped by natives and the practices of Jesuit missionaries who had yearned to create a Christian socialist utopia along the lines of Tommaso Campanella's *City of the Sun*.[51] Some observers saw Paraguay as a kind of early socialist state, later overlain by a caudillo dictatorship that, though corrupt, provided many benefits for its citizens. The country was innovative: it built some of the first railways in South America, strung telephone wires, and constructed ports.[52] In addition, Paraguay was the first South American country to vaccinate for smallpox, no trivial thing when the population was mostly composed of highly susceptible indigenes.[53]

The incredible Paraguayan resolve during the war is sometimes explained in terms of racial characteristics: atavistic devotion to a pathetic leader, a passive acceptance of destiny, the residue of theocratic life, and so on.[54] What has fallen from sight is that Guaraní throughout the continent had been objects of relentless enslavement by Brazilians for hundreds of years.[55] Missions in South America reached their zenith in Paraguay, with more than 150,000 Guaraní residing in them by the end of the eighteenth century. These had been targets for slavers prior to the consolidation of the Paraguayan republic, and many had in fact fled from Brazilian territories. Atlantic importation of slaves to Brazil had ceased in 1850, and it is not unimaginable that the Guaraní population foresaw a dreadful future in Brazilian shackles and, at least this time, had the weapons (and vaccinations) to resist.

The lethality of the war has been the subject of some controversy: did almost 70 percent of the population die along with the caudillo López,[56] so that in the devastating irony of military historian, Tasso Fragoso, "it was really his country that died for him"?[57] While historical demographers have quibbled, the discovery of an 1870 census seems to support the astronomical death toll.[58]

The Brazilian military was profoundly affected by the war. First, the difficulty in mobilizing recruits turned what had been seen as a quick operation (a "slam dunk") into a protracted ordeal that was made more awful because of difficult terrain, poor training, worse logistics, tenuous transportation, desertions, and epidemics. Although the state and some convents sent about four thousand manumitted slaves as cannon fodder for the dreadful war, the bulk of the fighting men were the outcome of military dragnets, most of whom were likewise men of color.[59] The problems of impressments created an immense resentment against the Crown and exacerbated the racial and economic distance between Brazil's more prosperous classes gallivanting in Rio and those who actually defended and died for the *patria*.

The military had viewed the comportment of Pedro II's court with unbridled scorn for the civilian management of the war, the shameless profiteering, and the emperor's capering on his yacht in military drag. His son-in-law, Count d'Eu, theoretically in charge of much of the war, was both bellicose and baffled. His quasi-deafness was as much a metaphor of character as a real affliction, and even though he acquitted himself reasonably well, officers such as Benjamin Constant de Magalhães viewed the royal commanders with total contempt, a disdain that no doubt later fed the revolutionary impulses of the officer class.[60] Most of the management of the conflict fell to the Duke of Caxias, best known for his pitiless suppression of rural uprisings and especially for his abilities in prosecuting guerrilla warfare, having honed his skills in uprisings in Maranhão. Even he hadn't the stomach for the final days of hunting down López and his teenage sons in the swamps of the Pantanal* and, on the pretext of his health, returned to Rio de Janeiro to manage the country while Pedro II toured Europe.

Brazil got everything it wanted out of the war—the end of the caudillo Francisco López, 142,000 square kilometers of Paraguayan territory, and open navigation on the Río Paraguay—but it was a pyrrhic victory. First, the military was profoundly alienated by civilian behavior that they viewed as basically cynical, unpatriotic, and incompetent. The army increasingly favored abolition, in part as an outcome of the numbers of blacks in its ranks who had fought valiantly and helped win the Paraguayan War. The military also had a sense of the broader incompatibility of slavery with modern warfare. Internal military alliances for abolition arose after the Paraguay War and dovetailed with antislavery sentiments in other sectors of society.[61] While full emancipation was not granted until 1888, the death knell for both the monarchy and slavery came in 1886, when Isabel was politely informed by Floriano Peixoto (later a president of the First Republic) that the army would not hunt for runaway slaves any more.

At an ideological and nationalist level, the army (which had grown substantially during the war) saw itself as a different, new kind of institution. Rather than reinforcing old-fashioned European sovereigns and emphasizing the difference between the rulers and ruled, the army had incorporated men of different races and social classes from all regions of the country and had been able to unify, organize, and mobilize them for the national good: this reinforced a kind of mystical link between the army and the nation.[62] Recruits

* The Pantanal is an enormous wetland in what is now Mato Grosso and Bolivia.

rather than royalty became the incarnation of the *patria*, the embodiment of national honor. The military, in all imaginable skin colors, was put forward as an ideological and workable, practical model for a modern Brazilian polity.[63] The army saw a "nation in arms" as perhaps the only disinterested force for the construction of a new type of state and nation.

The military yearned for a new role in Brazilian politics, especially as the army itself became more allied with Republican politicians from the new merchant and coffee classes of São Paulo. These inclinations were inflamed by the resentments of the military old guard with patriotic credentials from the Paraguay War who were deprived of modest raises, reasonable incomes, and pensions by an intransigent, uninterested, and profligate court. The military now preferred to give orders rather than take them, to change governments from an absolutist monarchy whose power was invested in divine right to a republic whose legitimacy had broader sources in secular political rights.[64]

The powers of reason and science as necessary elements for progress in a modern polity were also key ideological tenets, as the military technocrats of the young tropical country compared the scientific advances of Germany, Great Britain, France, and the United States with the intellectual "torpor" of the tropical Iberian realms. Positivism was central in shaping the ethics of the young cadets who were to make up the new military and governing cadre. Among other charismatic professors in Praia Vermelha, Constant—himself a Paraguayan War veteran—was the principal advocate for Comtean ideas and imbued young cadets in the 1880s with this passion.[65]

Life and Letters at Praia Vermelha

Praia Vermelha was less of a military institution than an applied polytechnic school with a strong philosophical bent. With its focus on social philosophy and technical and military training, it provided a demanding curriculum for talented young men. Devoted to the applied sciences such as engineering and survey, such an institution allowed one to learn the skills that were defining technical and economic progress in Europe and the United States and mimicked to a certain degree the field training of the British Royal Geographical Society. In contrast to the national universities, which largely focused on the liberal arts and suitably genteel professions such as law, Praia Vermelha expected its students—clever and mostly poorer boys—to become the technocratic problem solvers of the regime and officers of the realm. It was perhaps not surprising, then, that egalitarian echoes of French Republicanism wafted through the halls and dormitories of the academy. As members of the precarious "middle" classes, whose access to systems of patronage in the last days of Pedro's court was marginal at best, and highly sensitive to the vast inequalities of Brazil's oligarchic slavocracy, these youths had political dreams inspired by revolutionary ideas. Brazil, with its monarch and its millions of slaves, seemed in the mid-1880s atavistic and vastly out of step with the rest of the world, including its own educated youth.

While the "cadet philosophers" shared the dreams of a new society, they mostly rejected radical goals even as Marxist revolutionary thought galvanized Europe. The emerging antimonarchist Brazilian elite preferred governance that did not depend on uprisings in the French model but rather a peaceful transfer from royalty and oligarchs to a modernizing, secular technocratic class. Given the elite and military's lack of interest in true structural change, the ideas and political philosophy of Comte's Positivism was a good fit for Praia Vermelha, whose young cadets were seeking a practical social philosophy that could serve as a map for their society. Comte, popularly known as the "father of sociology," has passed from fashion and now resides mostly in footnotes, overshadowed by others who built on his ideas, such as Marx and Darwin. But his influence on nineteenth-century thought was great in Latin America, and he prefigures many stage and evolutionary theories that appear later in the nineteenth and twentieth centuries in biology and cultural history.[66]

Comte's "third way" between rigid monarchy and redistributive revolution had the potential to keep the social structure (and its attendant wealth and privilege) intact, while changing the political apparatus to make it more responsive to economic change and to the rising urban entrepreneurial classes. It was not a liberal democratic philosophy, but it was reformist. Since Comte posited a kind of rational scientific priesthood as the vanguard of reform, it is no wonder that the young Brazilian cadets thrilled to his ideas, seeing in the Positivist program the contours of their future.[67]

Comte sought to integrate science with religion and revolutionary thought with socialist utopian ideas. While it would be impossible to summarize his ideas in detail here, two threads of his social analysis were especially important in the Brazilian variant of Positivism, indeed so important that they constitute the motto on the Brazilian flag: "Order and Progress."

First, Comte argued that predictable laws underpin the functioning of the natural world, and this can be revealed by objective scientific methods and observation—hence the use of the term "Positivist experimental science" to describe most laboratory methodologies even today. Comte extended this view by arguing that human history likewise followed the actions of deep structures, and these can be understood through an impartial analysis of society and social history, the arena of study that he called "sociology." It is "objective" observable nature rather than the issues of theory and final causes that concern Comte. In his sociology, Comte examined the structures of sociopolitical systems—the *order* or deep "laws" in his evolutionary schema—and saw human history involving the three stages of civilization: it begins in superstition, evolves to abstraction, and ends, according to him, in positive rationalism. The question of social dynamics—*progress*—is revealed once a science of society emerges and its inherent laws are unveiled. In Comte's view, deep processes in society were equivalent to "laws of nature": they were universals and acted everywhere the same. With deeper knowledge of these social laws, human beings could create a just society through proper guidance and interventions, rather as we might imagine that a missile trajectory can be adjusted through correct applications and control of thrust.

Comte's program would be anchored by a scientifically informed secular "priesthood"—a technocratic elite—with the intellectual and spiritual tools for the creation of an ideal society, an ethical culture. This elite would direct society forward through its selflessness, oriented by scientific methods and moral thought.

Imbued with such uplifting principles, excited by their role in the crucible of national identity, and clearly the only social segment able to incarnate Comte's technocratic vanguard, the army (and its cadets) felt "it was the Right and the Duty of the army to assure Brazil's destiny."[68] Comtean thought shaped the Brazilian military and governing institutions for most of the twentieth century and gave impetus and prestige to Brazil's well-developed technocratic cultures.

There is nothing to suggest that Manuel da Cunha realized that he was sending his impressionable son into the one of the most socially complex, racially diverse, intellectually demanding, and antimonarchist institutions in Brazil. It was the incubator for many radical leftist political activists, like Luis Prestes, whose battalions sought to stimulate local uprisings for land reform to rectify centuries of social injustice and inequalities of wealth in the Brazilian Northeast. It sheltered the half–Terena Indian (and Euclides da Cunha's companion) Cândido Rondon, who would become Brazil's first activist on behalf of indigenous rights and would travel with Teddy Roosevelt through the wilds of the western Amazon (da Cunha had hoped to join him on that expedition). It harbored men of letters like Alfredo Rangel and military historians like Tasso Fragoso. The builders of the cities of Brazil, like Lauro Müller, came from its ranks, as well as scores of governors, senators, and politicians.

It is in this strange military institution where Euclides da Cunha first comes into view in 1888, intense, neurasthenic, and like his friend Cândido Rondon, an impoverished, nerdy boy of uncertain race but strong ideas. He was writing adolescent poetry. He had begun his journalistic career in the school newspaper, was enthralled by the heroes of the French Revolution, and wrote sonnets on the most Jacobin of them: Danton, Robespierre, St. Juste. He longed for the grand gesture and insurrection with all his yearning heart. The cadets were in ferment, the city was rife with rumors. Their teachers and idols were at the very center of the machinery of change.

Da Cunha and this cohort were students in an institution subsidized by the emperor that seemed as much a penal colony as a training ground. Bad food and poor sanitation were hallmarks of Praia Vermelha. They had other complaints: their stipends were miserable, their situation seemed at least as much metaphor for the imperfect capacities of monarchy as anything their imaginations could devise.

Fired up by the revolutionary ideas of their teacher Constant, inflamed by the return from Europe of the Republican icon and pamphleteer José Lopes Trovão, the cadets were increasingly unruly. Wildly cheering Lopes Trovão's arrival from Europe from the ramparts of the institution in the rising November heat of Rio's summer, the cadets were hardly the obedient servants of Dom Pedro.

The commander of the school, General Clarindo de Queirós, was highly displeased at the breach of discipline and forbade any cadets from participating in the protests and antimonarchist events that were to coincide with Lopes Trovão's return. To avoid truancy of any kind, de Queirós planned a troop inspection by the minister of war, Tomás Coelho, on the sweltering Saturday afternoon of November 3. In a clever maneuver, de Queirós then postponed the review until the next day, thus assuring the presence of the disgruntled cadets throughout the weekend, which cannot have improved their temper.

Tomás Coelho, in the company of Senator Silveira Martins, visited Praia Vermelha to quell the Republican fervor of the unmanageable cadets. Due to the possibility of cloudbursts, Silveira Martins carried an umbrella as they examined the facility and cadets on the morning of November 4, 1888. The cadets, so the story goes, had all planned to break their swords and bayonets as a sign of their contempt for the monarchy, and to shout revolutionary slogans—at least this is what Arnaldo da Cunha would note in his diary after a dinner with da Cunha some years later.[69] At the crucial moment, however, they all presented arms obediently enough (although with a certain sloppiness noted by Coelho). All except for Euclides da Cunha, who, after a few tries, did not succeed in breaking his saber over his knee and instead threw it to the ground at the feet of Coelho. The general embarrassment of the moment was enhanced by a defensive posture taken by Senator Martins, who prepared to fend off any attack on the minister with a swordlike deployment of his umbrella. There is some suggestion by Rondon that it might have been a cruel practical joke, that the berating words uttered by da Cunha were really directed at his fellow cadets and not at the military reviewers.[70] Roberto Ventura, citing an article in the *Gazeta de Notícias*, indicated that da Cunha was demanding better treatment in terms of promotions.[71] Da Cunha was hustled off to the infirmary with a diagnosis of "overexcitement," and as General de Queirós and the dignitaries exited the parade ground, the companies burst into shouts of "Viva!" for Lopes Trovão. Unable to control the cadets and profoundly ashamed, de Queirós mounted his horse, clattered away, and returned only the next day.

None of da Cunha's classmates seemed to have called on him while in the infirmary, but da Cunha was enormously moved by a visit from the poet and medical doctor Francisco Castro, who told him, "Not all who surround you are incapable of seeing the grandeur of your act, or the generosity of your ambition. I just came to extend a hand in solidarity."[72] Meantime, da Cunha's father implored the monarch to merely expel the young man rather than subject him to military punishments.

While his adolescent cohort may not have been impressed, knowing Euclides as reclusive, unhappy, and maladroit, his act of rebellion catapulted him into political celebrity. The event caused a sensation, with the *Gazeta do Rio* carrying the debates about the meaning of the act in its pages, including commentary by the famous abolitionist Joaquim Nabuco (who felt that the government ought not to permit its military institutions to become hotbeds of insurgency) and by Minister Coelho. The Paulista papers, such as the *Província de São Paulo*, saw the incident as highly significant and indicative of antimonarchist sentiment within the military. The monarchist papers, on the other hand, viewed the outburst paternalistically: an apolitical eruption that was symptomatic of the end of term—too much late-night studying, poor diet, and anguish over exams.[73]

While the meaning of the act was parsed to its minute details, da Cunha was out of the military. By late November, through the good offices of Constant, he was sent down to São Paulo to work on the republican mouthpiece *A Província de São Paulo*, owned by Francisco Mesquita, who would become an influential patron, protector, and publisher of da Cunha throughout his life. The young man wrote scathing articles under name of Proudhom, after the French anar-

Figure 2.2. Cadet Euclides.

chist, for this newspaper. He would later return to Rio to complete a degree in civil engineering in late January 1889 and to prepare for revolution.

The monarchy was crumbling. On November 9, 1889, a bit more than one hundred years after the French Revolution and just a little more than a year after da Cunha's display, Benjamin Constant was empowered to develop the plan for the coup that was to occur a few days later, on November 15. The palace forces mutinied, and the army, under Floriano Peixoto, refused to obey the order of the prime minister. The immensely popular Manuel Deodoro da Fonseca was placed in power as a military government inaugurated the First Republic.

The iridescent warm fogs of the coast and of history flowed softly around Praia Vermelha, where the military trained its elite recruits in that distant time. Emperor Dom Pedro probably did not realize that one of history's conjunctures was unraveling around him and that his era was over.

Da Cunha's destiny was also imminent. Those who would continue to be central actors in his life surrounded him then: the minister of war, Benjamin Constant, Cândido Rondon, his intimate, and fellow novelist Alfredo Rangel, in whose villa in Manaus he would begin to elaborate his *Lost Paradise*. There were the young cadets who would establish the new Republic—and there they stood at that moment, galvanized (in both horror and amusement, or oblivious, like those in the back who saw nothing) as the young man threw down his sword in front of the minister and walked away, inscribing himself in the mythos of the Republic and, in an ironic inversion of Marx, appearing in history first as farce, later as hero.

The Republic began its bloodless revolt in the shambles of the monarchy, the end of slavery and conspiracies of the military clubs, in the impassioned idealism of the young cadets commanded by dreams of both freedom and control. It also began in the refusal of Euclides da Cunha to obey the imperial minister of war.

3

The Afterlife of Revolution

After the revolution, da Cunha was soon reintegrated into the ranks of the military and reinducted into the War College until the end of 1893, when he left as a lieutenant colonel. During the first months of 1890, he constantly visited the home of the revolutionary hero Solon Ribeiro to court his luscious daughter, Ana. Photos of Ana show a dark-haired girl with an elegant profile, sensual mouth, wavy hair, and Rubenesque physique. At that age she lacked the discernment to understand the precariousness of Euclides's military ambitions, and he, thoroughly inexperienced as well, had no insight into her either, as later events would show. By September they were married.

Their domestic routine, encompassed by the military circle of family and profession, remained roughly as it had been prior to their nuptials but with the frisson of family tension. Everywhere the discussions turned on politics of the Republic, which were proving to be as cravenly opportunistic as the politics of the Emperor had been. The spoils of the empire were divvied up among those who placed themselves most

Figure 3.1. Ana Emilia Solon da Cunha, Euclides's wife.

propitiously, who could most effectively work the lines of patronage of the Republic, and those, like da Cunha, who sustained the aura of committed republicanism. Solon Ribeiro, highly critical of an authoritarian government that held the reins of power more tightly than the monarchy and was at least as dedicated to patronage, soon caused domestic relations in the da Cunha household to become more strained. Euclides had thrown in his chits with Floriano Peixoto, while Ribeiro would resist Peixoto because of the latter's profoundly undemocratic instincts, the nature of the coup that thrust him into power, and Peixoto's strategic nurture of political fanatics.

In a world in flux, the crafty conspirator Peixoto had been waiting on the sidelines for his moment of opportunity. Elected as Manuel Deodoro da Fonseca's vice president, he found intrigue and absolutism much to his taste. In 1893, after the Rio stock market collapsed and President Deodoro's general fiscal and political mismanagement threatened the disintegration of the new Republic, Peixoto was catapulted into the presidency with his coterie of military Jacobins. Peixoto's notice fell on Euclides da Cunha due to the events prior to the revolution, his politically astute marriage, and his ardent Republicanism. Da Cunha would write this about Peixoto, the "Maréchal de Ferro" (Iron Marshal), and the plottings for the coup of November 3 that thrust him into power:

> The hero, who was an enigma to his contemporaries due to the obvious fact of being an eccentric, would later become an inscrutable problem for posterity because of the complete dearth of acts that could justify such great renown. He is one of those cases of a great man who didn't rise by absorbing, concentrating, and incarnating the yearnings and dispersed energies of a people, and then propelling these into national life. Rather, in our rapid shift to the new regime, he did not personify social forces but emerged as a completely new, distorting element of our destinies. . . .
>
> His house on Rio Comprido Street was the epicenter of resistance. One went there in broad daylight, there were no lookouts, and it completely lacked the precautions and anxieties with which a conspiracy romantically embellishes its dangers. The conspirators came prosaically, by street car. . . . You could see a salon, furnished only with a sofa, a few chairs, and two empty cabinets. There, with windows wide open, as though one were at the most licit of gatherings, they raved on about rebellion. Then suddenly like a cold shower, Marshal Floriano appeared with his characteristic mien of an eternal convalescent, his preoccupied gaze falling on one and all, but fixing on none. He sat down in his vague manner and, in the sudden silence, launched into a extended and detailed recital of the ailments that perpetually beset him. . . . This was the master of the Republic's pitiless Jacobins just prior to catapulting into dictatorship.[1]

In the time of Florianismo there were many opportunities for someone of da Cunha's training and capacities. Peixoto was cleaning house of loyalists of the previous regime and of those whose politics were not quite correct in the current milieu. Many old revolutionaries suddenly found themselves on the outs, and Solon Ribeiro was one of these, while da Cunha's War College companions as often as not found themselves in illustrious sinecures. Da Cunha himself received a message from the distinguished marshal, and so, attired in his uniform with saber at his side, he went to his interview with the ever more imposing Peixoto. Here is how he described this visit to his friend Lucio de Mendonça:

> I found him in his dining room, relaxed and in one of his more expansive moods. The oldest daughter, even at that early hour, was toiling away at her

sewing machine. . . . The great Dominator welcomed me—"There is an air of war about you. . . . You know, you needn't come in uniform. You are here as friends and not as soldiers." These were more or less his words. Now, my dear Lucio, get ready to prepare your most fulminating verses to vaunt my honest and total ineptness. The great guarantor of posts, referring to my recent graduation and enthusiasm for the Republic, announced that that I could choose a position for myself, since he himself felt he could not decide for me. . . . It is enough to tell you that we were in the midst of dispensing state governorships! . . . And I (at the time completely in thrall to Auguste Comte) declared innocently that I dreamed of a law for recent engineering graduates: a year of practical internship developing the Central Brazilian Railway. I won't tell you the rest. . . . When we departed, in the inert eyes of my interlocutor it was clearly written: "worthless." And yet I preened in an inexplicable satisfaction while descending the stairs at Itamaratí, then happily crossing the foyer below, I left embellishing I don't know how many dreams of my glorious future . . . a future I had just calamitously destroyed.[2]

So rather than a governorship of, say, Mato Grosso or a senatorial sinecure, da Cunha was shunted off to the Central Brazilian Railway—but not for long. The uprising of the pro-Monarchist navy in 1893 had stimulated harsh repression of any dissidents. Journalists, statesmen, any kind of sympathizers, including da Cunha's father-in-law, Solon Ribeiro, were clapped into prison. The new Republic was in civil war and under martial law. Rio was being regularly bombarded from the Bay of Guanabara by the rebel navy, and da Cunha as a military engineer was pressed into service building infrastructure from sewers to sniper towers.[3]

Solon Ribeiro, then a senator, had participated in oppositionist groups who resented Peixoto's coup d'état and was scheduled to be shot along with about two hundred others for his overt defiance of Peixoto's government. Da Cunha, deeply agitated when he got wind of the impending execution of his father-in-law, raced to Itamaratí for another interview with the Iron Marshal. He described it in this way:

The Marshal eyed me silently with that cold and tired look which we all now know by heart. Deep in his eyes, however, I glimpsed a malign and dangerous glimmer. I thought I was doomed. In spite of this, I rallied the last fragments of energy remaining to me: I told him I had raced across the entire city, and then added that I considered it to be impossible for me to live one more hour with the weight of such horror and dreaded suspicions. "Don't think, Maréchal," I said, turning to him, "that I came here as a banal supplicant for the life of my father-in-law. Let me be frank. So that you have no illusions about me, let me declare

> candidly that I don't recognize you as a leader, I'm not of your party, but I follow you because you safeguard the Republic that I also defend." Floriano gazed at me with a contracted yet still insolent expression. I thought I had erred in seeking him out. My anxiety was so great that I thought if he found a revolver to hand, I was lost. Then, suddenly, a monosyllable. I waited for the answer with a certain foreboding. His words would have been enigmatic for others, but for me they were perfectly clear: "Before your father had even imagined you" (the phrase he used was more pungent), "I was a friend of Solon's. You can leave."[4]

Da Cunha alternated between the life of a military engineer and that of a journalist in a time of serious political tensions between federalists, republicans, Jacobins and monarchists, when newspaper offices were burned, publishers were shot to death on commuter trains, and sabotage was rife. As a journalist, da Cunha urged that the political prisoners not be simply lynched or shot, as the rising passions of Peixoto's hardline Jacobins would have had it, but be tried under the "serenity of laws." Once again he had annoyed the authorities, and once again the army exiled him, this time to the mountains of Minas Gerais to work on building barracks far from convenient journalistic venues. Always ambivalent about army life, da Cunha began to doubt whether he could reconcile his "independent spirit with the discipline of the barracks and the caprices of the Republic."[5] Although he remained in the ranks for another year, he resolved to move to São Paulo for good, away from his wife's family, with whom he was in bitter conflict due to political differences.[6] Given his short temper and scholarly inclinations, it is hard to imagine him happily ensconced in the hearth of the bourgeois career military family, devoted to gossip and intrigues. Solon's periodic incarcerations and problems with the ruling military coterie didn't help. Surely his in-laws must have viewed him on some level as a collaborator with the more questionable elements of Peixoto's Republic.

Solon was later exiled to Brazil's various "Siberias"—outposts in Mato Grosso where there was an interesting probability that he might die from malaria, snakebite, or some other Amazonian affliction, taking him off the political stage once and for all. Later he would be reassigned to the garrison in Bahia, where he would preside, unsuccessfully over the second Canudos attack. Later he was sent to Belém, at the mouth of the Amazon.

Da Cunha was not embraced by the firebrands of the young Republic either, and his disillusionment with the new Jacobins of Peixoto was great. As he wrote to Ribeiro: "Here, I understand (and haven't even yet tried) that I'll get little or nothing from this political world basking in the bloody aura of Jacobins that view me with frank displeasure."[7] He wanted to have a direct role in the politics of the regime, but means of doing so through his military connections were increasingly closed to him. He would always remain a compliant dissident to the state, but in a way that constantly nourished his frustration, since he could never really rely on rewards or accolades to translate into any concrete security.

Ribeiro's household probably had plenty of commentary to make on his and Ana's conjugal life, and the political divergences certainly didn't help. Eventually, in a hurt fury, da Cunha broke off relations with the in-laws, asking for but one big favor: "that you no longer mention my name in the visitors' parlor."[8] Perhaps Republican politics had become too incendiary for what little domestic bliss he and Ana were to share.

Da Cunha decamped to his father's coffee farm, and although he was relieved for a while by physical labor, his own inclinations led him to a post with the superintendency for public works of the state of São Paulo. In his life as a functionary, he had ample time to read his Carlyle, the British historian Henry Thomas Buckle, Auguste Comte, Victor Hugo, Hippolyte Taine, and the epics of Walter Scott as he rode on horseback from work site to work site. Although he regularly romanticized rural life, especially when exhausted by political or bureaucratic contretemps, it was mostly too quiet and, in its way, too wearing. He and Ana also produced two children in this time: Solon, after his father-in-law, and Euclides II. His life seemed at an impasse.

Our Vendée

There was news from the north, from the Bahian outback: it seemed a strange monarchist- millenarian uprising was unfolding there. A small farming community whose main export was goat hides had somehow been thrust into the headlines and political anxieties of the Bahian elites. Local oligarchs saw the hamlet of Canudos as draining away their remaining labor force into a free peasantry, a kind of backland Utopia. Others saw the community as an attempt at Royalist restoration. Monarchic subversions and millenarian energies were exactly the atavistic elements that the new modernizing Republic longed to place well in its past. The hierarchy of the Catholic Church saw the lay leader, Antonio Conselheiro, as a challenger to ecclesiastic authority. The church had not been pleased by the public humiliation of some of its *sacerdotes* who had gone to "review" the orthodoxy of Canudos' religious practices.

The wandering ascetics who repaired churches and graveyards and preached in the remote villages of the outback were not particularly popular with the formal clergy. Antonio Conselheiro, the leader of Canudos, and his followers formed part of the vibrant folk Catholicism of the outback, where a lack of ecclesiastic presence over hundreds of years had given rise to lay preachers and their adherents, who integrated African and native practices, Catholic saints and liturgies into a complex spiritual amalgam. While sometimes these activities might be syncretic—a true blending of cultural elements—at other times indigenous and African rites and beliefs could coexist side by side with traditional Christian observances as a compartmentalized set of identities and practices. Conselheiro's survival as a figure of the folklore and mythology of the outback speaks to the durability of pilgrimages, the widespread distribution of holy sites, their cultural significance and historical resilience even today, and the backlands' ability to generate holy men, curers, and believers in the societies of the sertão.[9] Followers of these saintly men were called *beatos*, the blessed, and dedicated themselves to lives of religious devotion, itinerancy, and penitence. The roving preachers were also the news bearers and troubadours of the interior. They mobilized labor crews for repairing public works (such as wells) and religious infrastructure like churches and cemeteries. These wanderers, among them the leader of Canudos, provided sermons, spiritual counsel (hence *conselheiros*), and blessings. The lineage of their pilgrimages is usually placed in the medieval history of Iberian religious travels, but native and African millenarian traditions gave these travels more syncretically and culturally complex roots than has been generally recognized.[10]

In Rio, the fiscal collapse of the Brazilian economy did not help the political situation, and in such instability, the young Republic found Canudos a politically useful distraction. Local elites allied with the federal government and tried to impose order, setting off a set of events that would become a defining historical showdown.

The fateful trajectory had begun when some men from the village of Canudos went to collect some prepaid lumber for building their church. The timber was withheld, and local authorities feared that the disgruntled *canudenses* would bust up the port town and administrative seat of Juaziero and from there spread mayhem. Historian Robert Levine argues that this was a world of rumor and that there is little in the archives to explain why, suddenly, the Bahian militias were sent in. Trudging through the Sertão, the military finally arrived at the dilapidated village of Uruá, described as a rundown Indian post, and stayed for the night. Though physically unappealing, Uruá was the site of a regular Saturday market and well connected to the rest of the region, including Canudos, about seventy kilometers away. Three days after their arrival the 104 troops were greeted by Conselheiro's Catholic guard—a huge processional carrying a cross, religious banners, droning backland hymns and *kyrie eleisons*. They were armed with blunderbusses, agricultural implements, and wooden stakes. Da Cunha described this attacking force as "one of those penitential processions which the credulous backwoodsmen used to stage by way of propitiating heaven when the long summer brought the scourge of drought."[11] Mortality in this skirmish was low, but the symbolic impact of the *sertanejos*' triumph over the militia was incomparable. The soldiers eventually arrived back in Juazeiro in tatters, stumbling, completely defeated. As a matter of both image and larger governing capacity, the military's rout could not be left unavenged.

Without question, the most evocative descriptions of the Canudos battles are da Cunha's own words, and readers are referred to this classic of world literature. What follows is a quick summary to situate the upcoming chapters of this work.

After this first defeat at Uruá, a small military force, fewer than six hundred men under the direct command of Febrônio de Brito, was dispatched into the fray in an expedition organized by Solon Ribeiro, da Cunha's father-in-law. This force, however, encountered the guerrilla genius of Canudos, João Abbade, using the landscape and his *jagunços* to maximum advantage, and the guerrillas completely decimated de Brito's pathetic corpsmen. As a well-trained formal militia they kept in formation and so were unable to maneuver in the landscape. Abbade's snipers, perched in the rocks or the trees, were able to meld into the landscape and pick off the soldiers, and to move along goat trails and Indian paths back to Canudos at will. Soldiers were ambushed by handfuls of guerrillas who spooked their horses, which then raced into the dense thorn forests, where man and beast were horribly flayed, impaled on the landscape itself. There they remained, mummifying in the increasing drought, ghoulish sentinels saluting the next attackers. Skulls soon lined the main military routes to Canudos. African techniques of drop traps, spikes, and hidden entrances further confused the militia as Abbade's men crept to the flanks of the formations, then trapped the soldiers. The rebels even stole their food. The defeat of the second expedition was devastating.

This humiliating trouncing was followed by a crescendo of panic as the newspapers in the south fanned the Jacobin flames.[12] How could a ragtag bunch of *jagunços* defeat a modern army? It seemed possible only if weaponry and strategy were being supplied from monarchist supporters, eager for a restoration and inflamed by religious fanaticism. And perhaps Solon Ribeiro was at fault, a bit "unsound" in his loyalties, since he had been on the list of less desirables for some time. After this debacle Solon asked for deployment to Pará.[13]

Clearly what was required was someone who knew how to put down rebellions and whose political bona fides were impeccable. This was the military hero Antonio Moreira César, whose Jacobin sympathies and pitiless comportment in war were unassailable: he had crushed a secessionist uprising in Santa Caterina with implacable cruelty. In many ways considered the heir to Floriano Peixoto, Moreira César was the commander of preference for the third strike. His nickname was "Treme Terra": the earthshaker. He also had a more brutal moniker, "The Cut-throat," for his taste in summary executions of those who had participated in the Santa Caterina federalist revolt.

The third expedition set out in February 1897 with thirteen hundred men, sixteen million rounds of ammunition, and the latest in Krupp cannons. Uninformed of the terrain or the water availability, and flush with his success in the south, Moreira César and his battalions soon found themselves in an eerie landscape of ghost horses, buzzards, and Afro-Indian juju. Da Cunha described Moreira César as erratic, vain, and impulsive, suggesting that his character flaws might have been due to his epilepsy. Confident of his soldiers' abilities, Moreira César had no respect for what he incorrectly viewed as undisciplined brigands with little military acumen.

Moreira César and his soldiers simply walked into a guerrilla trap. Da Cunha described it this way: "Had not the enemy left the road clear for them up to now, failing to take advantage in the most favorable stretches to attack them? There was but one thing that worried them: what if they should find this rebel nest empty when they arrived? This likely disappointment produced an alarming thought: . . . their inglorious return without having fired a single cartridge."

Many, including Moreira César, would have all eternity to ponder their faulty assessments. The campaign proceeded with forced marches through the searing Sertão with the idea of "lunch in Canudos." The "Legal" forces—as the government military came to be known—set up the cannons and began bombarding the town, which essentially produced rubbled palisades that the attacking forces would have to traverse. Canudos "possessed the consistency and the treacherous flexibility of a huge net," wrote da Cunha. "In the somber story of cities taken by storm, this humble village must stand out as an extraordinary and tragic instance. Intact, it was very weak indeed. Reduced to rubble, it was redoubtable." Soldiers crawled through the wreckage, lost and disoriented, easy marks for snipers and urban guerrilla fighters of both sexes. Little streets became blocked-off cul-de-sacs, and the invaders became hapless targets. The problem wasn't getting into Canudos; "the difficult thing was to leave it."

Moreira Cesar called for a cavalry attack, and snipers in the church tower picked off the horses and riders as they waded chest deep into the Vasa Barris,

Figure 3.2. Euclides da Cunha, far right, in Bahia, awaiting departure to Canudos.

including the erratic commander of the charge, Treme Terra himself. Wheeling and refusing, once in the city, the horses were even worse at maneuvering in rubble of urban warfare. Cavalry was, after all, a military technique for open plains, not for guerrilla fighting in urban wreckage. It was not an assault but "a rash battering of a monstrous barricade." The remaining eight hundred men of the Republican army, withdrawing from the town, fled back through the Sertão. The inhabitants of Canudos went to the church plaza and "broke into a prolonged, shrill and deadly intentioned ululation"—a banshee sound of voodoo weirdness that followed the panicked battalions as they retreated and stumbled through a landscape littered with new corpses and the bleeding wounded, lacerated by the thorny brush and trampled by bolting horses. The confused and terrified troops were easily picked off by snipers positioned in boulder nests along the lines of retreat. Soldiers who saw their companions fall at their sides dropped their weapons and charged away from the infernal assaults that came from the front, from behind, and from the side, as though the landscape itself held the fleeing men in their sights. The troops in disarray had been largely drawn from local militias, and according to da Cunha, their inexplicable defeat soon took on a supernatural character.

Most of the combatants "had not so much as laid eyes on a single one of the enemy."

The *sertanejos* captured the artillery and thus effectively supplied Canudos with the most modern weaponry. The 16 million bullets aimed at the town had come to no avail. This expedition "had vanished utterly." The *jagunços* collected the bodies, placed the skulls facing the road at regular intervals, and festooned the thorn scrub with military regalia: "The barren, withered *caatinga* now blossomed forth with an extravagant flora: the bright red of officer stripes, the pale blue of the dolmans, the brilliant gleam of soldier straps, and the swaying stirrups."

Da Cunha followed the backland war and then wrote the tract that would change his destiny by framing Canudos (and by extension, the new Republic) in the idiom of the French Revolution. From the highlands of São Paulo, his battle cry *A Nossa Vendéia*—Our Vendée—was based on Victor Hugo's essay about Catholic royalist millenarian resisters to the French Revolution in Bretagne from 1793 to 1795, who were later massacred by Jacobins.[14]

A Nossa Vendéia was written before da Cunha had ever stepped foot in Bahia's backlands, but he had studied with tremendous detail the physical and biotic geography of the region and had been carefully guided in this by his friend the geographer Teodoro Sampaio.[15] His essay evoked the dangerousness of the landscape itself and the lethal tactics of guerrilla warfare in such an environment: "The soil of these outposts is covered, especially in the dry season, with a sparse and diminished vegetation; this, perhaps more than the fanatic hordes that follow Antonio Conselheiro, is the most serious foe of the Republican forces."

Da Cunha spoke of the "religious fanaticism catching hold of innocent and simple souls, and easily manipulated by the propagandists of the empire. The same barbaric courage, the same impractical terrain complement each other. . . . The same morbid heroism is as diffused and disordered as the impulsive chaos of the mesmerized." Likening the Brazilian military to the Roman legions pacifying the barbarians, he saw the civilizing Republic as a distant echo of the glories of the Pax Romana. "The Republic will triumph in this next test. We are condemned to civilization."[16] This "condemnation" would come to haunt him when he actually stepped off the train and into the landscape of his Iliad.

The insight and impact of this essay were great, and da Cunha's editor arranged for him to travel to Bahia during the final assault on Canudos as an aide-de-camp to General Bittencourt.

4

A *Quilombo* Called Canudos

An African Kraal

In the last days left to his "city of ruins" Euclides da Cunha gazed out over the lumpy hovels of the millenarian settlement of Canudos, his "mud-walled Troy." It was the second-largest city in Bahia, after the slave entrepôt of Salvador, but for him it had the appearance "midway between a warriors' camp and an African kraal." That was indeed more or less what this backland city was. Canudos was located at a sharp jog of the Vasa Barris River. It was unusually well watered for the backlands, a kind of oasis. It was encircled in a small mountain chain and, like many fugitive slave refuges, hard to get to, with ample possibilities for ambush and defense. Above the village rose Favela Mountain, which was later to give its name to impoverished urban squatter settlements—favelas—throughout Brazil.[1] Behind it was the village known as Poço da Cima (the upper springs). One watershed over was the settlement known as Mucambo: the name for communities of fugitive slaves. It seemed exquisitely geographically isolated: "the merciless climate, the periodic droughts, the

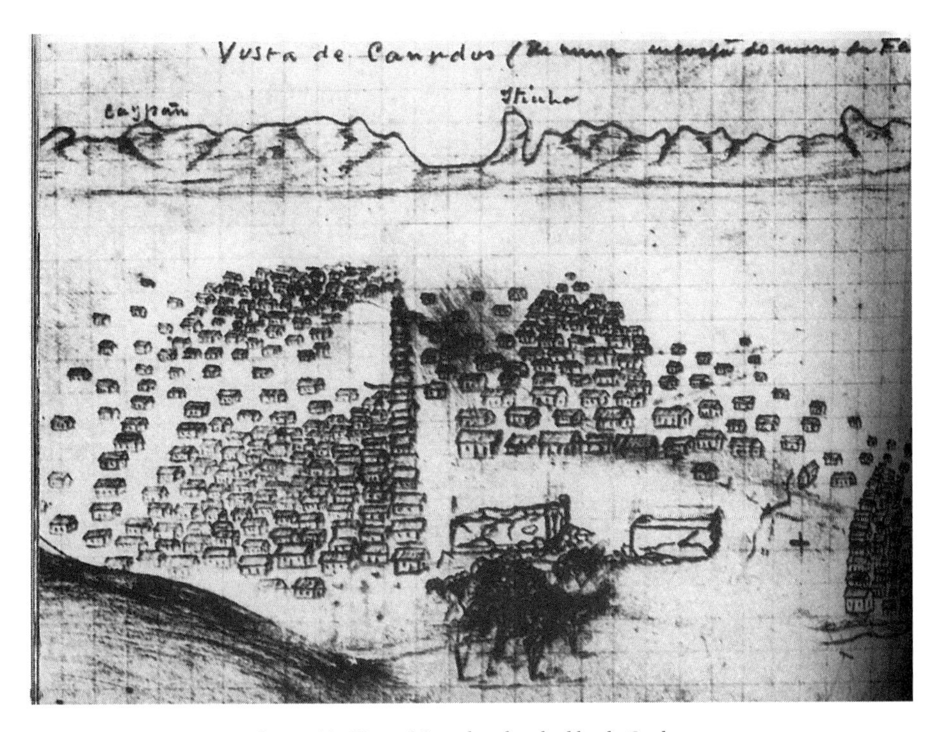

Figure 4.1. View of Canudos sketched by da Cunha.

rugged sterile soil of the barren mountain ranges lying isolated among the *araxas* of the central plateaus . . . this unprepossessing region, . . . this was the last refuge of the Tapuya."[2] As isolated as this place appeared, it was connected to all the backlands of the Northeast: Piauí, Sergipe, and Ceará. Once over the mountains, Canudos was linked to the gold and gem states of Goiás and Minas Gerais through the great artery of the Rio São Francisco. Canudos itself was at the center of a vast network of *quilombos*, Indian settlements, mission centers (*aldeias*), and hamlets, the backland circuits of market towns, trading posts, and pilgrimages. There were also routes of the great cattle drives, the *boiadas*, that assembled the beeves from the infinity of the Sertão and drove them to the coastal cities or to ports on the São Francisco. Canudos was at the intersection of landscape whose apparent stasis belied its deeper and unceasing movement, triggered by climate, spiritual impulse, markets, water, culture wars, and flight.

Climates and Catastrophe

The imaginary world of isolated and tranquil backland settlements, enclosed landscapes outside of history, is extremely inaccurate for the Bahian Sertão, where catastrophic El Niño droughts would unfold each decade with such severity that most of the population would have to migrate through the moribund countryside to places where water could be found, or else leave the backlands altogether.[3] As early as 1559–61, Jesuits recorded horrible droughts. Other chroniclers noted serious rain shortages in 1603, 1605–7, 1645, 1652, 1692. The eighteenth century had a severe E1 Niño in 1721–26, and the drought of 1777–78 killed seven-eighths of the cattle in the Northeast.[4]

The nineteenth century was also afflicted with devastating droughts in 1817, 1827–9, and 1858–60,[5] but the epochal years were 1877–79, 1888–91 (just prior to the period when Canudos was established by Antonio Conselheiro), and 1896–7, the years of the battles of Canudos. The young aristocratic naturalists Johann Baptist von Spix and Carl Friedrich von Martius, who traveled through this area during the drought of 1817, recorded the desperation of the fleeing populace (whose anxiety and dread they shared) huddled near slimy-green blackened springs or pits excavated in streambeds to imbibe alkali or pestilent water. Typhoid and other waterborne diseases were added to the afflictions that dogged these hordes migrating to the coast or one of the few remaining watered outposts. These fluids were so vile and toxic that the mules hauling the exiles' gear would drink it only when heavily laced with *rapadura*—molasses-laden brown sugar.[6] The drought in 1829 marked the first Northeastern migration to the Purús watershed six thousand miles to the west in the Amazon, which gives an idea of the immensity of the drought diasporas even this early on.[7] In 1858–60 the British consul reported drought so severe that "many families were willing to sell household slaves at any price."[8] The 1877 catastrophe would coincide with the dramatic colonization of the Amazon rubber forests. The drought of the 1880s was studied by da Cunha's friend and main scientific source on the Bahian backland, Téodoro Sampaio, who described the waters this way: "thick, turbid, with the reek of urine and the repugnant aspect of putrid liquids; none of us were pleased to drink this noxious fluid in spite of the thirst that tormented us in our arid and empty crossing".[9]

In the El Niño of 1877–78 (one of the worst of the nineteenth century) at least half a million people died, and hundreds of thousands of *flagelados*, drought refugees, migrated to the coastal cities or to the banks of the São Francisco, or gave themselves over to debt peonage in the Amazon rubber forests.[10] Epidemics of cholera, yellow fever, and smallpox increased the mortality of the starving hordes. The assets of the poor, their scrawny goats and little burros, also left their corpses on the parched land. The term *flagelados* aptly captures the existential aura of the backlanders: the scourged ones.

Once the refugees arrived in towns, their circumstances did not necessarily improve. They simply overwhelmed the local resources and were reviled or chased out, on to the next settlement, maybe to die en route. In other places they encountered state-sponsored programs to load them on boats and ship them to São Paulo, Maranhão, or Pará, a kind of national "middle passage" that delivered them into debt bondage. Certainly the horrible conditions on these *flagelado* boats echoed those of the Atlantic slave ships, although conditions may have been even worse since these people were not valuable commodities but starving "problems." The boats were often not seaworthy, and many *retirantes* (another term for drought refugees) died as overloaded vessels sank. Historian Gerald Greenfield, who has provided the most thorough study of the 1877 Niño and its politics, notes that the boats that transported slaves as part of the interregional market to São Paulo also carried drought victims. Mix-ups may well have occurred where free men fleeing the backlands accidentally entered slave markets.[11]

Another unfortunate destination was concentration camps mandated by Imperial policy. Rather than having migrants polluting and overrunning cities with a "pestilent presence" involving the disease, crime, and morality of the desperate, towns created detention areas outside their borders. They were known locally as *currais*—"corrals" with armed guards—where the *retirantes* lived rather less well than cattle, thrown together in hovels with virtually no sanitation. These camps were petri dishes of disease, and many died in their fetid confines. The *currais* continued to be used for the famished refugees in drought times until 1942,[12] when with the emerging knowledge of Nazi Holocaust camps, the similarities between groups of starving incarcerated undesirables—*sertanejos* and Jews—was too stark.

The New Middle Passage

The *retirantes* were not the only unwilling migrants. As the rains failed, so did many farming enterprises, and among those who owned slaves, their most "liquid asset" was their human chattel. While the importation of slaves had ceased in 1850, the rise of the coffee industry in southern Brazil in the 1870s created a ravenous demand for workers and animated a boom in unfree labor. Most vulnerable were the slaves of smaller enterprises in the interior of the Northeast, the cotton and castor growers living on the less productive margins of the Sertão. These were mostly native "creole" slaves, a population that actually considered itself Brazilian and had a sense of local identity in the *campo negro*—the black countryside.[13] The market for coffee slaves was surprisingly large: close to 100,000 slaves entered the Rio and Sao Paulo slave ports by boat from 1870 until 1888, according to the economic historian Roberto Slenes.[14] Robert Conrad, another analyst of the trade, estimates that yet another 100,000 may have moved more invisibly through internal overland routes, coffled together, walking more than a thousand kilometers.[15] At least 10,000 and possibly 20,000 slaves a year were exiled from homes and families in the Northeast from 1870 through the mid-1880s, the bulk from Bahia, with over 25 percent of the trade.[16]

On smaller plantations, relations between masters and slaves were intimate and mediated by an array of affinal and customary understandings of rights and obligations. The combination of changing financial and climatic conditions produced a rupture in both the economic possibilities of the interior and the implicit social pacts and moral economies between owner and slave. In a Pilate-like market process, boys and men were shunted off to slave dealers. Males constituted almost 70 percent of the trade, and 86 percent of this population was between eleven and forty years old.[17] Backland inhabitants suddenly found themselves swept up in a resurgence of slavery just when its end was in sight. Once again, the commerce broke up families and communities and fueled bitter resentment and rebellion among slaves and local indentured freemen. The shift to more extensive cattle production at the end of the nineteenth century may have reflected an increasingly intractable labor force, a scarcity of available *vaqueiros*, as much as the robust price for animals. The Niño droughts were a boon to water rich but labor-scarce plantations of coffee in the south and rubber forests to the north, and the commerce in humans that underpinned these thriving economies, but hideous for those captured in slavery's last snares.[18]

Brazil's internal slave markets help explain why women outnumbered men two to one in Canudos in da Cunha's day: it was the inverse of the regional slave markets for men. Canudos' importance as a safe haven for women who may have been abandoned, separated from their families by the slave commerce, widowed, or disgraced or for whatever reason, in need of protection, was widely noted. It was indeed a City of Women.

Journalists writing about the fourth expedition to Canudos were surprised by the omnipresence of women. Da Cunha would write, "Women, women, women—all the prisoners of war seemed to be women." They were everywhere; they did everything and were a strange subtext for the masculine theater of war. The Baron of Jeremoabo, the Sertão's most powerful landowner and politico and an indefatigable letter writer, wrote that in the last year of the existence of Canudos "it had 5,000 battle-ready men, not counting the women who also fight like fanatics."[19]

Peasants and Prophecy

In circumstances so devastating, so lethal, so simultaneously capricious and inevitable, one can understand the desire for miracles and messiahs, and relief from nature through the supernatural, one of the hallmarks of Northeastern rural culture.[20] Apocalypse was scarcely an abstraction. Yet the Sertão, seemingly dead at the end of drought, resurrected itself, flowered, and became rich with the rains—"nature in love with itself." This was a stock metaphor of Northeastern literature, from the simplest *cordel*[21] to da Cunha's *Os Sertões*, and fed the enduring myths of Northeastern resilience.

In a world so lacking in justice, so fickle and so deadly, the idea of utopias—intentional communities of fairness and virtue in life, rather than paradise in death—had a particularly strong hold in the Northeast. The utopian imagination was derived from Portuguese, Catholic, indigenous, and slave traditions. These included the reworking of medieval Portuguese Sebastianism, the story of the return of a warrior prince lost in battle against the infidels at Ceuta (Morocco) in 1578 to install—somewhere— a regime of righteousness and a new Jerusalem. This myth certainly resonated with the dreams and desires of the oppressed and has become an enduring metaphor in Brazilian politics.[22] The strands of loss, suffering, and utopian renewal also fit within the practices of folk prophecy and the Jesuitic utopianism of "Cities of the Sun" whose Christian communalism and syncretic practices stretched throughout the Bahian Sertão. Indigenous myths of "Lands without Evil" with roving messiahs and pilgrimages envisioned places of refuge and abundance.[23] Black syncretic beliefs and rites incorporated indigenous and Catholic spiritualism into religious practices with African roots; a complex millenarianism had always been part of the New World African cultural dynamic.[24]

The late nineteenth century produced scores of millenarian movements in native societies in Brazil and elsewhere.[25] Some included elements of prophecy, as Indian populations tried to adapt to a more pervasive integration into emergent modern economies and the definitive transformation of their ways of life.[26] In Mexico, the Tomochic millenarian uprising had striking parallels with that of Canudos, including charismatic folk prophets, indigenous populations (Tarahumara and Yaqui), valiant "bandit" fighters, and accusations of fanaticism and monarchism, and it likewise ended in a complete destruction of the town.[27] In the United States, the Ghost Dance religion that began in 1890 (and produced the massacre at Wounded Knee) reflected a similar phenomenon.[28] These revealed deep structural economic change, the rise of new power relations, and significant modifications in access to resources, traditional territories, and native landscapes of identity.

The threads of these different millenarian traditions—Catholic, native, and African—were significant parts of Canudos practices and legends. Modern historiography of the place reflects the rise of Marxist, subaltern, and moral economy approaches to rural studies that focus on deep inequalities and cultural mechanisms supporting precarious livelihoods in order to refute the perception of Canudos as a place of "religious fanaticism" and atavistic irrationalism.[29] Da Cunha's framing of *canudenses* as superstitious zealots reflects his (and the more general coastal) arguments about modern rationality versus primitivism. This attention to deeper rationalities is important but has tended to downplay the profound role of religious culture and local identities in fueling resistance.

Religious communities certainly reflected material circumstances but also built on their own powerful symbolic meanings. In Brazil, millenarian and *quilombo* communities had been hunted out since the sixteenth century not just for "irrationality" or "heathenism" but also because these communities siphoned away expensive (and necessary) labor and posed a profound social threat, since they embodied an alternative way of organizing life and livelihoods. These communities were also attacked for their symbolic energy and the concrete resistance that they materialized in their ritual places, agriculture, music, industries, and landscapes.[30] And in their possibilities of freedom and citizenship.

Rebellion in the Backlands?

Canudos was portrayed on the Brazilian coast as a primitive millenarian movement that yearned for the restoration of the monarchy. The people of Canudos did embody an anti-modernist, anti-statist, and anti-oligarchic movement that used religious terrain, language, and a lived "practical utopia" to frame its resistance to the existing political economy. Like almost everyone who had been a slave (or whose family members had been), the *sertanejos* practically deified Princess Isabel, whose May 13, 1888, decree ended chattel slavery in Brazil.[31] But that she was adored had less to do with the monarchy as a political institution than with her decree. The people of Canudos had no interest in the restoration of the monarchy or national ambitions, in spite of the frenzied accusations in the Jacobin press of São Paulo and Rio.[32] Their assessment of the questions of power was different.

In the *sertanejo* view, the Republic's "unholy" secular politics put backlanders in a state of spiritual peril by invalidating their ways of interpreting their circumstances and the supernatural means of intervening in them. The secularization of religious practices with its emphasis on civil rather than religious marriage, its deconsecration of processions, saints' days, and the like, and its belittling of the thousands of small religious and meaningful syncretic practices placed the *sertanejos* at the mercy of vengeful gods and helpless in the face of daily forms of oppression. Syncretic religion provided moral guidance for everyday life, political critique, and revenge. As the historian of native uprisings Eric Van Young suggests, "The lexicon and practices of popular piety provided the language of resistance at every turn. An intense and highly localized religious sensibility, its icons and symbols were the idiom and the arts of critique and resistance." And desperate consolation.[33]

The Baron of the Backlands

El Niño droughts emptied out the countryside—not only pathetic small-scale holdings but even substantial properties. The 1880 El Niños had been so devastating that animal stocks and backwoods populations had not recovered, and thus the well-watered flatlands in the bend of the Vasa Barris River were attractive to squatters, pilgrims, refugees, and "vagrants." Canudos was an abandoned ranch owned by one of the most powerful oligarchs of the region, the Baron of Jeremoabo, Cicero Dantas Martins, and seems to have been a *quilombo* even before it became a utopian community. Da Cunha wrote that the vicar of Cumbê, who had ministered in the Alto Sertão in 1876 and gone to Canudos, reported that clustered all around it was "an idle and suspect population, armed to the teeth." In a short space of time, beginning in 1893, "it would be transformed into the mud-walled Troy of the *jagunços*."[34]

Canudos emerged as an economic and cultural space of multiple ethnicities in part because it provided access to land exactly when the enormous traditional oligarchic d'Avila and Brito holdings made virtually everyone outside their clans a squatter. Other than the massive holdings of these families, only about 1–5 percent of the total area of Bahia was in formal tenure, and such lands were mostly on the "sugar" coast or near Salvador.[35] The reality was that the monopoly on land and access to it was the best way to coerce labor in the absence of slavery.

People may have been nominally free, but the land became captive. Brazilian slavery had provided provisioning grounds where subsistence and small surpluses were generated, but there was no guarantee of access to these plots after slavery's demise. Indeed, no institutional arrangements at all were made for the economic life of post-abolition slaves. In the United States, forty acres and a mule sometimes staked ex-slaves. Nothing was provided in Brazil. The social pacts, such as they were, between landowner and cultivator were changing profoundly, and alienation of traditional land rights was one such change. Almost overnight, a huge landless class was created in the backlands.

Access to land after abolition was mediated by the "coronels" through sharecropping, corvée, renting, and grazing shares, as well as wage work. Access to lands often required extra work on the owner's cash crops regardless of other agreements. This was known as *sujeição*—subjection—and the term aptly defined the social transactions.

Historian Monica Dantas has described the agrarian and social history of Itapicura, one watershed over from Canudos, a place that sent most of its population to the "rebel" city. Dantas reviews how small farms became more precarious after abolition: access to land was contracting, labor demands increased while rural oligarchs ignored their obligations, and the flimsy "safety nets" of backlands society were ignored or dissolved. Farmers were increasingly vulnerable, and many of the households had lost their men to the last pulse of slavery. Bolstering livelihoods became more problematic as enclosures claimed communal grazing lands and water and collecting rights became vigorously and viciously contested.[36]

The rural population became increasingly impoverished as the nineteenth century drew to a close. Dantas notes that numbers of small farmers were gradually removed from better parcels and lacked the financial or political wherewithal to purchase them or dispute removal. Small farmers and newly freed slaves were squeezed on several fronts. Their agriculture supplied the caravans the small markets and fed the estates, but they toiled largely unseen and were viewed as vagabond "surplus" labor. Elites saw this autonomous labor, where it was not under "subjection," as fundamentally criminal and described such workers as *vadios* (vagrants) or *jagunços*."[37] "From slavery to vagrancy" was one of the catch phrases of the Northeast at the time. New laws were enacted to press the landless or semi-landless "vagrants" into military service, to incarcerate them, and to use them as an unpaid work gangs.[38]

Canudos, where values of Christian (or syncretic) charity and communalism held sway, where free access to watered land was assured, became a compelling draw for backlanders facing the post-slavery economic regimes. This was especially so in times of intense drought, as the backlanders knew what might lie in store once they fled the land they had been working: refugee camps, peonage, exile, and death. As the droughts of 1894 and 1896–97 set populations once more in motion, landed small farmers might sell up, while the landless simply walked the routes of ancient *sertanejo* migration, following pilgrimage trails to the *brejos*, the wetlands of the high Sertão, to Canudos.

Formerly prosperous towns with thousands of people were sometimes reduced to fewer than a hundred. The settlement of Quemadas declined from 4,504 in 1892 to three occupied houses in 1897. Towns within a radius of two hundred miles lost more than half their populations to Canudos, and distant hamlets became virtually deserted.[39] By 1897, the time of the final assault, Canudos had exploded from a modest hamlet to a population of perhaps twenty-five thousand, the second largest town in Bahia.[40]

The massive migration to Canudos was disastrous for the *Coronais*. This letter to the Baron of Jeremoabo highlights landowner anxieties: "Antonio Conselheiro continues to be the reason that many people leave here and other places are now threatened with becoming depopulated. The exodus is widespread."[41] Departures on the scale described by the Baron's correspondents were alarming. Equally troubling was this note from his cousin, José Américo: "We'll soon see this Sertão confiscated by him [Conselheiro] and his people: he has more than 16,000, It's a miserable bunch; anyone who was a slave, all the criminal elements from all the provinces, there is not one human being among them . . . and I lack workers, I have only four drovers . . . and we're not doing well because of the lack of rain."[42]

Of Cattle, "Cannibals," and *Quilombos*

The Sertão of the Bahian backlands was largely a cattle landscape. The animals had been introduced in the early sixteenth century by the d'Avila family, along with the earliest slaves, and slowly moved from the coast to the banks of the Rio São Francisco to the most remote *currais*—cattle stations where the herds roamed, barely tended by black slaves, probably descended from the West African Hausa-Fulani people, whose livestock traditions and horsemanship from the Sahel were easily adapted to the arid Brazilian outback.[43] Though outwardly placid, the ranching Sertão was a landscape of violence. The backlands were militarized in invisible but lethal ways, as landlord vigilantes, bandits, slave, and Indian hunters, and *quilombo* militias roved the land enforcing their interests and taking their opportunities.

Cattle were so voracious and destructive that they were forbidden within fifty miles of the coast.[44] The takeover of vast areas by cattle herders monopolized land; indigenous livelihood systems—gardens, orchards, and hunting grounds—were trampled by serene bovine pillagers. Fellow travelers—invasive grass species that traveled to the New World as straw in the rough sleeping pallets of the slave ships and as fodder for horses and cattle—wreaked havoc in tandem.

Cattle were initially maintained under a slave system in the Sertão, but drovers and cowboys were part of a "slavery without overseers," similar to that of gold mining in Minas Gerais or Amazon forest collection, where close surveillance was impossible.[45] The *capataz*, or cattle boss, might travel with a couple of slaves and perhaps an Indian cowboy as well.[46] The *capataz* operated under a reward known as *sorte* or luck, where one of every four calves born could go to his herd to sell or breed as he preferred, an incentive structure that mimicked that of Sahelian and West African herders. This form of livestock administration was still in place in the late twentieth century in the savannas of the northern Brazilian state of Roraima and the grasslands of Marajó.[47]

This system involved both bondage and patronage, and while hierarchical, it required considerable flexibility to function in a world well beyond formal means of surveillance, where the rancher had unchallenged rights to administer justice within the confines of his holdings and to "manage" justice beyond them. In the immensity of the outback, where rustling by runaway slaves, Indians, bandits, and other ranchers was a likely possibility (and where a *fazendeiro*'s own *vaqueiros* might themselves be captured and resold, or might flee to a *quilombo*), the great stockowners created their own militias for control of people and resources on the cattle frontier. They were necessary for defending their honor, to punish transgressions, and as troops to fight the endless inter-oligarchic family feuds.[48] The reshuffling of ownership of lands, springs, and waterways after every major drought provided a constant stimulus to personal and political rancor among the elites.

The great landowners were permitted by law to maintain order however they chose to; they were authorized to provide *couto* or asylum for their militias. There *jagunços* could flee, were given ranchos, wives, and provisioning grounds, and were expected to serve and protect the family with whom they were allied. Disciplined by violence, this population was both admired and reviled by locals and outside observers and was the foundational social matrix for what has come to be called social banditry. Aware of all the water and all the paths through what seemed to be only trackless thickets of the *caatinga*,[49] the *jagunços* inhabited the complex intersection of racial, cultural, and economic politics at the margins of the great *sesmarial* landscapes.

In times of duress, and under the afflictions of the Northeastern droughts, these *jagunços* might easily engage in simple thievery as a well as "social" or "redistributive" banditry, though they were often seen as plain marauders, ambushing convoys and stealing cattle and women. They were more likely to engage in plunder when the precarious livestock and planting that formed their livelihood were undone by the lack of rain and when patrons abandoned the outback as they did during drought. *Jagunços* were "criminalized" when they slipped beyond the control of their masters but were idolized and feared when integrated into power structures. *Jagunços*, along with the wandering saints, were the most powerful figures to emerge from the humble soil of the outback with stories of bandit kings like Lampião, and his consort, Maria Bonita—Robin Hood figures to some in the Sertão.[50]

Those same *jagunços* became the defenders of Canudos and the clandestine heroes of da Cunha's tropical Iliad. They were, wrote da Cunha, "Conselheiro's best disciples . . . capable of loading their homicidal blunderbusses with the beads of their rosaries."[51] These *sertanejo jagunços* would later morph from Rebel Warriors to Defenders of the Nation when the terrain of battle shifted to the Amazon.

Indians and *Aldeias*

There were still many Indian groups in the Sertão at the end of the nineteenth century. The occupation of the interior of the Northeast proceeded only through extensive, centuries-long Indian wars, including a massive offensive against the coastal tribes, a campaign known as the Guerra dos Bárbaros—the Savage Wars—in the seventeenth century, and ceaseless guerrilla war elsewhere.[52] In the Northeast, these campaigns lasted well into the nineteenth and even twentieth centuries, in spite of the external imaginary of the "vanishing Indian."[53]

The coastal world of the Northeast and Bahia had been dominated by Tupí groups, who had been embroiled in the frontlines of Portuguese expansion and were sucked into the early phase of native slavery, where cane, cholera, and malaria took their toll. Just back of the Tupi terrains were the territories of the Karirí, of the Gê linguistic group and the kinfolk of da Cunha, whose lands extended from the São Francisco basin practically to São Luis in Maranhão, and paralleled the coastline roughly to Salvador. As tough semi-wild cattle pressed westward through the thorny woodlands, the masters of the beasts and slaves who ran them found a land of resistant forest fighters.[54]

Historian Bart Barickman has pointed out that the presence of natives remained stronger in the outback than is usually recognized. Historical ethnographers such as Hal Langfur and John Hemming have emphasized that the popular idea of the "vanishing Indian" obscures the reality of relentless resistance.[55] The region was deeply contested, and settlement throughout the Sertão was profoundly affected by the dynamics of both guerrilla and structured warfare.[56]

In this context, the widespread mission settlements—the *aldeias*—were complex, in some cases generating truly syncretic cultures based on native and Catholic spiritual sources, while in others the Indians were basically catechized slaves.[57] Indian lands of the *aldeias* were often simply expropriated by local elites. The Karirí, close to the slave-demanding sugar economy and the relentlessly expanding cattle systems, were unfortunately located at one of the most brutal intersections of labor demand and land dispossession on the continent.[58] As Karirí and other groups rustled or hunted livestock, they opened themselves up to enslavement and "Just" war. Landowners retaliated with *entradas*, forays to pacify the outback, and usually, after killing the men, sent the women and children to be sold on the coast; perhaps da Cunha's beautiful grandmother had been one of these.

Within the *aldeias*, natives practiced a syncretic version of Christianity—traditional native festival days would be celebrated with *missas de caboclo* (native masses), and native deities and myths could be worked into the catechism.[59] The autonomy and syncretic nature of Canudos, distant from the formal church, the Brazilian state, and their economies, made it a place of interest for indigenous populations, especially the Karirí.

The Karirí were insurrectionists. In the famous Revolt de Malês, when Muslim slaves sought to overthrow the Portuguese and establish a New World Caliphate in Salvador, Bahia, the Karirí expressed their solidarity with the rebels though they did not participate in the urban uprising.[60] At the time of the Canudos uprising, Karirí Indians from the *aldeia* of Natuba decamped entirely for Conselheiro's city, as did Tuxas tribals and Kaimbé from the *aldeia* of Souré. This was why there were three hundred Karirí Indians in Canudos, including their powerful shaman, dancing in full regalia before the final attack and dying there.[61] It is possible to imagine that with a slight twist into an alternative universe, with a different destiny for da Cunha's grandmother, Euclides himself might have been stamping war dances in the central plaza of Canudos as the Brazilian militias prepared their assaults. Certainly he had native cousins and kinfolk among the Karirí making their last stand at Canudos.

Quilombo do Canudos?

Canudos, like the "sanctuary state," the *quilombo* of Palmares, has a durable hold on the national mythology of Brazil. The resonance of Canudos lies in its utopian imagination and its ideas of resistance, citizenship, and autonomy. Canudos and Palmares were places of ethnogenesis: the re-creation of new cultures, roles, rules, and meanings of self in the face of immense, almost unimaginable processes of disintegration.[62] In these places, subjects became citizens. Such refuges or *quilombos*, and there were thousands, were, as the historian of slave uprisings, João Jose Reis has put it, sites of "the Invention of Liberty."

The term *quilombo*, as historian Stuart Schwartz points out, has a complex etymology that places it within a set of Angolan institutions, and indeed the most famous *quilombo*, Palmares, was known by its inhabitants as Angola Janga (little Angola). Schwartz, who bases his analysis on the African historical work of Joseph Miller and John Thornton,[63] focuses on the Jaga/Imbangala role in Angolan polity formation. The Jaga were mercenaries, lived on a war footing, and adopted children, usually war booty, from other groups into their ranks. Over time they created a highly multiethnic warrior cult that incorporated large numbers of strangers who lacked a common ancestry into processes of "ethnogenesis" through novel syncretic social forms and identities. *Quilombos* in Africa were organized through ritual affines and hierarchies rather than traditional lineage and clan obligation. These could unify ethnically diverse populations that had been torn from their lands, gods, and lineages.[64] They were reinvented in Brazil. Da Cunha was quite right to see Canudos as a "warrior camp."

What was important was that *quilombo* was becoming a synonym for *kingdom* or *polity*. Given the predominance of Bantu (Angolan and Congo) slaves in the Brazilian and especially Bahian slave economy, the large South American *quilombos* were understood by slaves as separate places with their own sovereignties, rules, and institutions. *Quilombos* were compelling sites of autonomy, cultural blending, and, despite relentless persecution, freedom in the face of slavery.[65] By no means could the entire suite of ritual practices be reactivated in Brazil, nor would *quilombo* leaders necessarily become Jaga, nor only male. Indeed, the foundational myth of Palmares has Aqualtone, an Angolan queen, as the initiator of the settlement: according to legend, she fled slavery with some retainers to the mountain retreat of the palm-covered Serra de Barriga, later to give birth to the Brazilian black cultural hero, the embodiment of black resistance, Zumbi.

There were other symbolic and ritual practices that could integrate diverse ethnicities into a functioning community. These could have derived from indigenous sources like the cooperative non-kin sodalities—ritual age groups—so characteristic of Gê societies. The Catholic Church also was able to meld different beliefs into an array of "double saints" who had covert meanings and valorized and comforted the fugitives from the vast diasporas that were shaping Brazil. Burial brotherhoods—*irmanidades*—along with labor groups, religious festivals, and patronage combined into a significant sub-rosa social glue within the idioms of Catholicism. Between holocaust and diaspora, natives and Africans had a great deal of social reconstruction to attend to and used whatever cultural means they could to achieve this.[66]

In Rebel Lands

Much of the history of nineteenth-century Brazil could be written through its uprisings. A listing of them reveals a combination of slave revolts, urban insurgencies, secessional movements, rebellions of political reform, and attempts at revolution. A listing does not by any means exhaust the number of revolts but gives a sense of the frequency, geographic sweep, and widespread nature of insurgency: Bahia, 1809, Pernambuco uprising, 1824, Cabanos revolt (Pernambuco), 1832–35; Rio de Janeiro, 1831; Recife, 1831; Cabanagem revolt in Amazonia, 1835–40; Salvador, Malês revolt, 1835; the Sabinada Revolt in Bahia, 1827–38, the Balaida revolt in Maranhão, 1838–40; the Farrapos War of Rio Grande do Sul, 1835–45; São Paulo and Minas Gerais, 1842; Praeira Revolution in Pernambuco, 1842; the Ganhadores strike in Bahia, 1854; Pernambuco, 1850; Viana, 1867; Quebra Quilos revolt, 1874–75; Vintem Revolt in Rio, 1880. As Schwartz puts it, "Many revolts produced *quilombos*, to the degree that fugitive rebels sought to avoid capture and punishment while cultivating in their new hideouts the seeds of future resistance movements."[67] Uprisings produced flight and new hidden polities.

Slave revolts whose purpose was to proclaim a new state, like the Muslim Malês who wanted a Brazilian caliphate, were relatively rare. Overtly "statist" projects, like the Malês, Cabanagem, or the Farrapos War (these last were stimulated by the examples of Caribbean revolutions, as we will see further on), which imagined revolution, stand in contrast to the far more general practice of creating spaces of cultural autonomy outside the labor relations and circumstances that defined the world of slavocracies. These resistant populations moved relentlessly into the interior and can be seen as Brazil's true colonizers of the vast continent. Flight and fight could be predicated on economic conditions, intrusions of the state and local powers into customary practices (like loss of grazing or collecting rights), resistance to military dragoons, changing social relations of production (alteration of the terms of work, the rise of capitalist forms of labor deployment, new forms of access to land, etc.), but mostly the social turmoil and evasion took place below the surface.[68] For each documented insurgency, hundreds of small acts of resistance and liberation occurred. As anthropologist James Scott argues, these "small arms" in class warfare were much more widespread and less noted but defended both dignity and survival.[69] When pushed too far, people "voted with their feet": they fled to *quilombos*.

Covert Communities in Plain Sight

Quilombos and *mocambos* were often understood as hideouts and seen as largely isolated from regional economic life. In some situations this was the case.[70] Recent research on *quilombos* and the communities that had their origins in them, however, is recasting our understanding of them. Some had tight yet covert linkages within the main economy, and others were more isolated and self-sufficient. Many scholars situate *quilombos* at the heart of both rural and urban provisioning, petty extraction, and healing. Firewood, fish, herbs, charcoal, manioc, greens, palm oils, fruits came from small holder and *quilombo* growing grounds and forests.[71]

Quilombos stretched from the *terreiro*—the urban safe houses and religious sanctuaries for ritual dancing and healing in the center of cities—to the suburbs and small satellite towns, short-term hideaways for religious practices as well as more permanent settlements.[72] *Quilombos* often faded into the background of the *campo negro*, the black countryside of free and semi-free tenants and sharecroppers.[73] These had relationships with networks of more distant communities and were often linked to plantations and ranches that even had *quilombos* within their own lands.[74]

Quilombos and communities derived from them were and still are a widespread form of settlement in the South American interior. They had varying degrees of autonomy, linkages to the regional economy, their own forms of governance, often collective forms of land holding, and relatively egalitarian distribution of goods. These kinds of communities were likely to be structured along what James Scott has called the "Moral Economies," with rights to subsistence and access to livelihoods.[75] *Jagunços* and *vaqueiros* of Canudos as well as other *sertanejos* would certainly have known about, had economic and kinship ties with, spent time in (if for nothing else to celebrate various festivals and dancing),[76] and had perhaps even been members of *quilombo* communities. *Quilombos* were often clandestine "settlement networks" that involved many communities, as was the case with some famous *quilombos* like Campo Grande, Ambrosius, and Calinda. These might more usefully be thought of as covert "counties" rather than individualized villages. A cattle station might well overlap with the *quilombo* universe, though it was a ranch outpost. What has fallen out of the analysis of the backlands is that *quilombos* were far more numerous and extensive than millenarian settlements and constituted a covert and very widespread system of territorial occupation whose histories, like Canudos, resides in runaways, resistance, economic change and climatic catastrophes.

The Secret Life of Liberty

Secrecy was key to the survival of *quilombos*. They were embedded in networks and circuits of information, goods, and people (potential sources of betrayal) and existed in uneasy relation to both native and national communities. Such communities remained known but unseen until some event triggered an attack.

How prevalent and secret *quilombos* were is suggested by a landmark piece of modern legislation. In 1988, the Brazilian constitution recognized the land rights of traditional populations based on their historical territories. Not exactly overnight, but very soon thereafter, *quilombos* and their inhabitants "came out of the dark," and scholars and land officals began to develop a much more profound understanding of how widespread these communities had been. They were everywhere, in national parks, in cities, in suburbs, in mountains, at frontiers . . . That they existed and continued to exist speaks to extraordinary powers of defense and resilience. A national mapping exercise of existing communities formed from *quilombos* (*communidades remanescentes*) revealed, unsurprisingly, that Bahia had the highest number of these types of communities. Enormous *quilombo* communities such as Kabinda still exist in Goiás. More than four hundred "residual" communities can still be found in Pará,[77] while more than a thousand exist in Bahia.[78] The site of Brazil's space program on the island of Alcântara near São Luis in Maranhão was overlain on a *quilombo* community.[79]

Beyond the historical ethnographic work that revealed the extent of *quilombo* presence, the ubiquity of fugitive communities can be also read in place-names. These include such obvious toponyms as Morro de Quilombo, Riacho Mocambo, Mucambo (which was not far from Canudos, and where in fact Conselheiro had repaired a church),[80] and various other place-names that include "Quilombo da" followed by some other name. African names such Novo Angola, Cabinda, Calunga, Calabar, or Caiene das Criollos ("Cayenne of the Creoles"—Cayenne, the capital of French Guiana, having abolished slavery), and many others derived from Bantu, Yoruba, and other West and Central African dialects litter the landscapes of the north, northeast, and central west of Brazil. In addition, certain saints especially associated with Africans or natives provide clues to their early origins: St. Benedict, Our Lady of Rosario, Our Lady Aparecida, and St. Anthony are particularly important in the African diaspora. That local elites saw Canudos more as a *quilombo* than a religious settlement is suggested by a sneering relative of the Baron of Jeremoabo: "He [Conselheiro] is more powerful than Napoleon! . . . Today I intend to naturalize myself as an African."[81] Others noted that Canudos was full of the *gente do 13*, black ex-slaves liberated by Isabel's May 13 dictum.

Language of Threat

The *jagunços* described by da Cunha as "indolent and armed to the teeth" could have been inhabitants of a Canudos *quilombo* that predated the community established by Conselheiro. As da Cunha put it, "Even before the arrival of Conselheiro, this obscure hamlet, . . . like the majority of the unknown villages in our backlands, contained the germs of disorder and crime."[82] With terms like *immoral*, *thieving*, and *dangerous*, the language of threat had in the past placed *quilombos* at the center of an intense and anxious official discourse of political security and criminality, requiring destruction of such communities. In addition to this idiom of the "other," da Cunha framed Canudos's necessary annihilation in the modern language of historical laws and the Spencerian logic of biological progress: the populations of Canudos "had not evolved."[83]

Canudos and Total War

Was Canudos a *quilombo*? Technically, if the term is limited to its narrow usage as "slave refuge," no, since the settlement grew and was obliterated after abolition. It may have been in earlier moments, as da Cunha noted; the Baron of Jemoabo's correspondents certainly classified it as a slave, ex-slave, and criminal refuge.[84] But if the term can be amplified into its broader meaning, a place of socioeconomic and political autonomy—and this is the way the term is widely used in modern Brazil—then Canudos takes on the functional, symbolic meanings and practices of a *quilombo*. With the debates about territory, identity, and land rights that have emerged since the 1988 constitution, which recognized traditional land rights and ethnic territories, a more complex historiography of *quilombo* communities easily encompasses Canudos. The term *quilombola* or *quilombo* is now an umbrella for a diversity of pathways, ethnic routes, and roots of settlements in ongoing legal battles to ratify historical holdings.[85] The "lineages of liberty" took many forms, including the occupation of abandoned agricultural settlements. Other lands, *terras do Santo*, were technically "owned" by a saint but occupied by Afro-Brazilian communities who made their living there. In addition, refugees from rural violence formed communities of safety and livelihood.[86] Following current usage, Canudos would certainly fall into the *quilombo* category, and in all likelihood, had it not been destroyed, it might well have received title to its land under the current legislation, especially given its renown.

Canudos had other cultural features that also marked its *quilombo* lineage. First, as I have noted, Canudos, like *quilombos*, had numerous syncretic and folk elements, rites, and spiritual practices that served to unify ethnically and culturally populations and gave social meaning to people who were for the most part dispossessed and disposable. Canudos clearly placed autonomy, folk religion, noncoercive labor relations, and communalism in opposition to the practices of the secular state, conventional Catholicism, and the oligarchic economies of labor coercion.[87] It was fundamentally a withdrawal from these national institutions and a symbolic and practical, rather than revolutionary, challenge to them.

The usual backland villages and missions were also culturally syncretic, but Canudos had elements that made it much more *quilombo*-like. These pertain to the *quilombo* as politically autonomous and as a *defended* settlement. Historians and rural sociologists such as Flavio dos Santos Gomes, Flavio Moura, and Victor Salles in their magisterial studies of *quilombos* have sketched some of the defensive features of *quilombos*, all of which characterized Canudos to an extraordinary degree. First, *quilombos* used the environment as the first line of defense. They were typically hard and exhausting to get to, well hidden, while access routes permitted a great deal of surveillance. Canudos, da Cunha notes, was difficult to make out in the landscape, a "monstrous urb" looking like a jumble of earthquake ruins as one approached it from the north and almost invisible if one came from the direction of the coast. The journalists and writers who covered the final expedition all noted to their dismay how grueling it was to get within striking distance.[88]

The difficulty of access was matched by the ease of surveillance. An active intelligence network was key to the survival of *quilombos*. Canudos was extremely well informed about the movements of the Brazilian military, whether this was through "secret agents," partisans, or children who ran messages, or through its sniper nests and lookouts hidden among jumbled boulders. Canudos clearly had sympathizers in every village, and many in adjacent valleys had kinfolk living there.[89]

Flavio dos Santos Gomes also underlines the how important guerrilla tactics were for the survival of *quilombos*. It is possible that the great insurgent warriors of Canudos developed their backland military skills not simply in the herding of animals and the daily violence of the Sertão but in the longer historical defense of these refuges. Much of da Cunha's most agile descriptions are devoted to what a perfect landscape Canudos had for clandestine war. Few places in Bahia were more blessed with tricky terrain for misleading guides, getting companies to run up against a mountain wall to their doom, or providing easy ambush than the environs of Canudos. Indeed, *Os Sertões* at its heart is a kind of paean to the brilliance of guerrilla strategy in a complex and difficult landscape where the *sertanejo* military imagination (and deadly practice) shines. Da Cunha's posture shifts from derogation to admiration of such inventive warfare.

Canudos had to have significant defenses and had them in abundance, including the most unlikely in this land of drought: "in this lonely region where mountaintops merge with the high tablelands, they had precisely selected

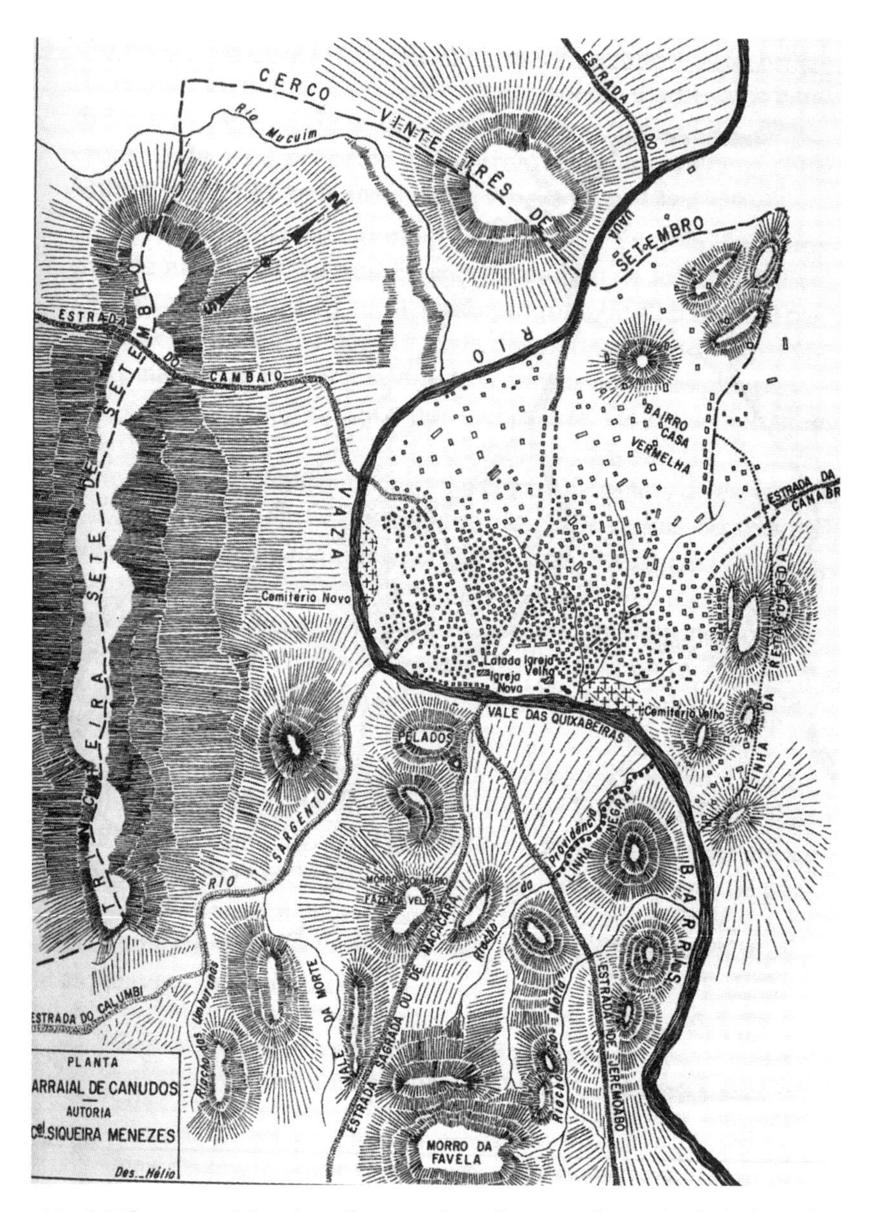

Map 2. Military map of Canudos with surrounding military installations for the final assault. Note the moatlike Vasa Barris River.

the spot that came nearest to being one enormous moat."[90] To cross the Vasa Barris into Canudos meant to take on snipers from every direction, including the church tower. Horses and men were hampered by the river flow, and their slow, clumsy movements made them simple targets even for imprecise weaponry.

Multiple entry and exits were also part of the *quilombo* strategy, and people easily slipped in and out of the village, much to the endless annoyance of the final expedition. Other defensive techniques like drop traps, slip nooses, snipers, poisoned stakes along the path, and juju such as skull-lined roads to scare intruders (this was used to great effect at the third and fourth expeditions) were widely used.

These were the internal military features of *quilombos*, but to hostile outsiders the key element of *quilombos* was a need for their total obliteration. *Quilombos* had been sites of "necessary annihilation" since the 1500s, for literally hundreds of years and thousands of expeditions.[91] Once discovered, *quilombos* were objects of total war where all military resources were used to the maximum and elimination was the goal. This reality defined the *quilombo* history that stretched from Palmares to Canudos. Canudos was the most artillery-heavy war in Brazilian history. Every person (women and children too) was ultimately described as a *jagunço* and considered to be a combatant.

Canudos was the largest and most symbolically important *quilombo* since the kingdom of Palmares. Its hold on the Brazilian imagination as a place of resistance, social reimagining, and economic justice remains transformative. It was not for nothing that the authoritarian military dictatorship of 1964–85 flooded the ruins of the rebel city for a dam in an (unsuccessful) attempt to expunge its memory and symbolic resonance.

This was the social landscape that awaited da Cunha.

5

Mud-Walled Jerusalem, Mud-Walled Troy

Euclides da Cunha, sharply attired and armed with the fashionable theories of his day about race, history, and landscape, was prepared to see an epic contest between Civilization and Barbary. He spent six weeks in Bahia, from late August to October, much of it well behind the front lines, and he was worried that he had arrived so late that he would not see battle. He knew little beyond what his friend Sampaio had suggested to him of the Sertão history and his own imaginary of the French Vendée uprising transposed to the Bahian outback.

As da Cunha made his queasy way up the coast with the São Paulo militia, he might not yet have known that another severe drought, what we now know as El Niño, was searing the backlands, parching the lost corpses of the earlier Canudos expeditions into unburied mummies. True to da Cunha's grisly description of a soldier "at his ease" splayed under an *umbazeiro* tree, his desiccated charger wedged for eternity in the vise of a nearby stone outcrop, long mane wafting on the slightest breeze, the Sertão had become a world of spectral inhabitants. Local folklore was confusing the dead, the soon to be dead, and those merely transiting through in the long years of the drought. The avian icon of Canudos was the *urubú*, the large, carrion-eating buzzard, a macabre trickster who in the popular songs and poems of the time kept asking the government to send more troops, as he complained that his palate had become so refined on his steady diet of lieutenants, captains, and generals.[1]

Da Cunha was with the São Paulo battalions of the fourth expedition, part of an army of eight thousand men who dug in very carefully, watched their supply lines, and concentrated on siege tactics and bombardments.[2] They positioned themselves in the early summer of 1897 and suffered significant losses through the ensuing ruses and ambushes of the *sertanejos*. The descriptions of the military dynamics that unfolded in the environs of Canudos have the flavor of deep interdigitation. The *jagunços* prowled around the "legal" Brazilian forces all the time.[3] Care had been taken this time around to include far fewer Northeasterners than in previous expeditions. Local army men were more likely to see in the faces of the *sertanejos* a sorrowful reflection of themselves, since the average fighting man was from the same social stratum as the backland guerrilla and perhaps might have had kin, belief, and history in common with him, and at the critical moment might desist through some secret solidarity. Instead, *gaúchos* from southern Brazil and battalions from the Amazon (including from Amazonas Commander Cândido Rondon) formed the bulk of the forces. *Bombachas*, the loose pantaloons that were the hallmark of the *gaúcho*, were as much in evidence as military uniforms.

Canudos itself, said da Cunha, "was, appropriately enough, surrounded by a girdle of mountains. It was a parenthesis, a hiatus. It was a vacuum. It did not exist. Having crossed that cordon of mountains, no one sinned anymore. An astounding miracle was accomplished, and time was turned back for a number of centuries. As one came down the slopes and caught sight of that enormous bandits' den that was huddled there, he might imagine that some obscure and bloody drama from the Stone Age was taking place. The actors on one side and the other bore on their countenances the indelible imprint of many races."[4]

It had taken months to dig in, with ambushes and epidemics making things worse. Tenuous supply lines and constant harassment kept rations short. The troops at the front, under the command of General Artur Oscar, were famished, and the drought-plagued land mostly provided only pestilent waters. A smallpox epidemic raged among the Republican forces. As da Cunha put it "One thousand mules were worth ten thousand heroes." The government forces were repeatedly enticed into traps that captured entire companies, as well ruses that picked off individual desperate soldiers. One lure was a *jagunço* with a neck bell just like that of domestic goats, used so the animals could be found in the outback. Ravenous troops hungry for roasted kid set out in hunt of this enticing dinner, only to discover that they were in fact the prey, and so met their end. The narrative of *Os Sertões* is alive with such *sertanejo* ingenuity, so clever and lethal.

Da Cunha was quite ill during his entire stay but relied on military doctors and avidly interviewed anyone back from the front lines, prisoners of war as well as hapless children who ended up in the crossfire. He himself would adopt a little *jagunçinho* and take him back to his home.[5] The Canudos region hardly lacked for orphans, or for that matter, child warriors.

Da Cunha wrote the two armies into reflections of each other. The army tents had been quickly covered by fronds of local vegetation and took on the primitive aspect of Canudos itself, like something shaken out of the earth. The military forces, the "legal" forces, and the rebels were buying food from each other, exchanging jokes, living in the same sort of hovels, dying of the same diseases, thirsting, all in close proximity.

The siege, as da Cunha describes it, initially took on a comradely form, with men in the front lines calling some name at the Canudos rubble, and getting a response. An amiable conversation would ensue—about families, conditions, and the like. Longing backland songs would be sung. Obscene jokes were told to the hilarious enjoyment of those on both sides. When a divergence in opinion occurred, "insults couched in a forceful argot" would be punctuated with bullets. But "the rebel settlement little by little succumbed to the slow process of strangulation," while the newly arrived artillery from São Paulo longed for the "feverish convulsions of battle."[6]

Da Cunha's own words are the most moving among those produced the cadre of journalists who, against their own initial posture, were overwhelmed by an increasingly unbearable reality: the guerrilla fighters were genius at military strategy, and there seemed not to be a coward among them.[7] The women, the wounded, the aged, and the ill remained stalwart. As da Cunha put it: "The *sertanejos* had inverted the psychology of normal warfare: their resistance was stiffened by reversals, and they were strengthened by hunger and hardened by defeat."[8]

A Land outside History

Da Cunha had arrived at Canudos during the final days. The siege by then had been going on for months. All inhabitants of the town were now criminalized and called *jagunços*, even the women and children; all were targets. Although there were still a few lines to sneak supplies in and out of the blockaded town

Figure 5.1. Huddled women prisoners from Canudos.

and its wells still must have worked to some degree, the Vasa Barris was drying up, and the mortality from famine and waterborne disease was growing. Conselheiro was taken to his next life not by a bullet but by dysentery. Incessant bombardment and sniping were the norm, even as the inhabitants of the town sang their hymns and the landscape rumbled with their prayers. When on September 25 the final line of the siege was put into place, the town was doomed.

Once the siege lines were set, a white flag appeared, and the most wounded and the women were taken out of the town that was about to become a crematorium. With his camera João Bastos captured the huddled, horrified women and children whose next step involved abuses of all kinds and a "next life" in some indentured situation among families or brothels in Salvador. Bastos's affecting images would soon become famous as photographic documentary of the War of Canudos.

The prisoners came in, and as da Cunha put it, "there were few whites or *negros* among them. . . . The unmistakable family likeness pointed to the perfect fusion of the three races. And around them all were the victors, separate and disparate, protiform types, the white man, the black, the *cafuso*, the mulatto with all gradations of coloring."[9] The backlands, according to the environmental theories of his time, had stabilized a racial mixture. The "pure form," the racially consolidated *sertanejos*, were what da Cunha would call "the bedrock of our race."[10] His earlier assertions about polygenic theories of racial origin and darker people as a different species from superior whites evaporated and never appeared again in his writing. But what happened now, to the horrified gaze of the journalists was the *degola*: throat cutting and beheadings. For the men it was "off with their heads, out with their guts."[11] Da Cunha's Sertão becomes a "cry of protest, somber as the bloodstains it reflects."[12] The hacking up of prisoners in front of their families, in front of indifferent military commanders who mustered not the slightest reproach, completely traumatized da Cunha. In the final days, as the brutality became even more vicious, he became impossibly more distressed.

Da Cunha still remains a powerful witness to the horrifying concluding events. His sympathy and revulsions oscillated between the *sertanejo* fighters and the Brazilian militia as his own identities as coastal elite, military aide, and backland half-breed warred within him. The final assault no longer vaunts the brisk reportage of battalions lined up and Krupps cannons firing away, but rather an elegiac and anxious lament of endless fratricide. Racial triumphalism, the premise so arrogantly articulated in the early part of *Os Sertões*, has ceased to mean anything. Canudos does not fit into any theory of history and civilization that da Cunha had advanced before he took the boat to Bahia. There is no racial stigma that underpins the epic brutality on the Vasa Barris, only the mark of Cain.

The final battle involved hand-to-hand combat, in each of the 5,200 houses that made up the "monstrous urbs" of Canudos. Many had stayed, according to da Cunha, because they expected that the dead Conselheiro would return with battalions of angels and the armies of St. Sebastian to crush the Republican forces. The beleaguered inhabitants were in a true end time, so perhaps they believed anything was possible. What all the reporters saw was unrelenting battle waged behind a "breastwork of corpses" and what eventually became an enormous pyre, raising the nauseating reek of rotting and roasting

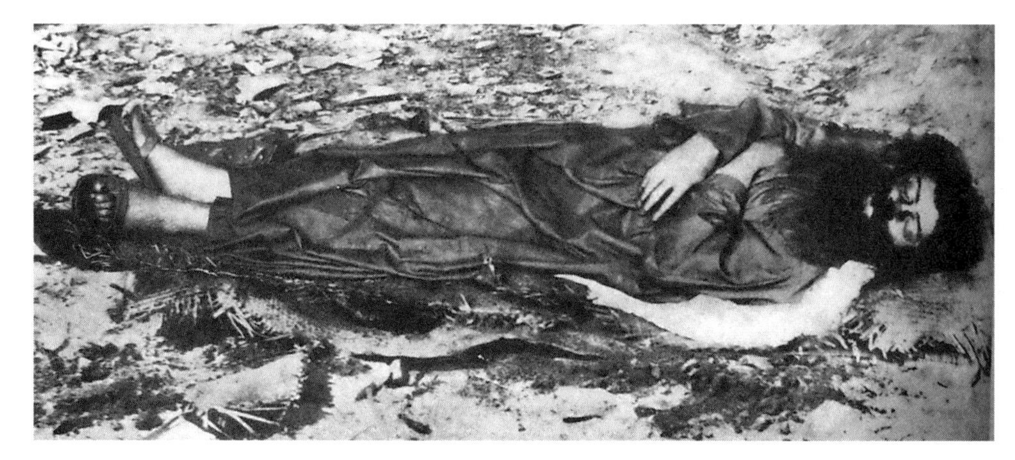

Figure 5.2. Antônio Conselheiro with his head still attached to his body.

human flesh. "The terrible exploits are veiled in obscurity for all time. . . . There before them, a tangible reality was a trench of the dead, plastered with blood and running with pus. It was something beyond their wildest imaginings."[13]

And as da Cunha notes, "Canudos did not surrender." The town was utterly destroyed.

Da Cunha had stated in the first paragraphs of his masterpiece that the triumph of the civilized over the more barbarous races was an inexorable historical process. However, he charges, "the campaign . . . was, in the integral sense of the word, a crime, and as such, to be denounced."[14] "Civilization," his ambiguous term, fundamentally shifts its meaning between the beginning and the end of the book.

Conselheiro was not the only supreme leader of the Battles of the Backlands to be sacrificed to the gods. As the Canudos forces enjoyed their victory parade through the streets of Rio amid cheering crowds with General Bittencourt and the president of the Republic, Campos Sales, waving and delighted, an assassin who had Sales in his sights leaped with an open knife. Bittencourt threw himself in front of the president and died of his wounds. It wasn't a *degola*, but it was a public death nonetheless.

Da Cunha returned to his father's farm, gravely ill and heartsick. Today we might describe him as having posttraumatic stress disorder. He eventually improved enough, however, to decamp to Rio Pardo, where he repaired a bridge and wrote his masterpiece under the patronage and intellectual support of his close friend Francisco Escobar. In a zinc-roofed hut, he produced the book that catapulted him into history. It seduced readers by beginning with fashionable clichés, but, like the *sertanejos*, da Cunha led them through Brazil's own heart of darkness, its mysterious geography and unknown backlands, to ambush them in Canudos.

PART 2

The Scramble for the Amazon

6

In the Times of Scrambles in the Land of the Amazons

Scrambles in the Old World and New

Africa was partitioned by the European powers after the Berlin Congress of 1885. Much has been made of this meeting of the diplomatic powers when lines were drawn on maps, deals were made, and the "Scramble for Africa" formalized what had been a rather disorganized process of reconnaissance and imperial annexation. Five European nations—France, England, Germany, Belgium, and Italy—carved up Africa for their uses, under the pretext of advancing the three Cs: Commerce, Civilization, and Christianity. They managed to assign themselves some thirty mostly tropical, largely unexplored colonies and protectorates, with sovereignty over more than a hundred million people. These terrains proffered the gold, gems, oils, plantations, and *Langdorfia* latex, the vine whose sap vied with Amazonian *Heveas* and Peruvian *Castillas* in the global rubber market..

Far less attention has been paid to the Scramble for the Amazon. This reflected the reality that the contests had been going on in slow motion for centuries. The Iberian ex-colonies

Figure 6.1. The Baron of Rio Branco.

continued in their internecine squabbles. French, English, and Dutch colonies sat on the Guyana periphery of the boom lands of the Caribbean, inching down toward Brazil. Hemispheric aspirants like the United States lurked, schemed, and filibustered around the edges of the great basins of the Orinoco and Amazon. Overall, South American colonies had been low-value chits in the larger dynamics of European diplomatic exchanges, and the attention of European powers was blurred by earlier failed colonialisms—their own (all had made earlier attempts at capturing the Amazon)[1]—as well as their disdain for the residue of a rickety imperium, in the case of Brazil, and republics run by temperamental caudillos on the ruins of the Spanish empire. There was a hodgepodge of vague treaties and old claims that could be activated in the interstices of the vast unknown territories. Tropical geopolitics, mostly languid, periodically exploded in clashes in swamps and forests far removed from the gilded government halls where the fates of these places were often decided.

One of the observers of the politics of the African Scramble was an obscure Brazilian diplomat who before long would make an indelible mark on the geography and politics of the tropics; indeed, in the Western Hemisphere he "made" its map. This was the Baron of Rio Branco, who would become well known to many of the adepts of tropical "Scrambles" as one of its deftest practitioners.

The Baron of Rio Branco

The Baron of Rio Branco, José Maria da Silva Paranhos, is acclaimed as the most skilled diplomat in Brazil's history. He, like Euclides, had come from fairly modest circumstances and a Bahian family. His father, the Viscount of Rio Branco, had risen through the ranks of imperial favor and ended with a sinecure as a Senator of Empire and the title of Viscount for his diplomatic service during the Paraguayan War, when both large areas of Paraguay and Bolivia became Brazilian, a "preboard" for later scrambles. After this war, in part because of the distinguished role black soldiers had played in it, he argued for the "free birth law"—that is, children of slaves born after 1871 were to be free at age twenty-one; the legislation itself is known as the Rio Branco Act.

The Viscount's son, an exceptional young scholar, had been a bohemian in his youth and, unlike the austere Euclides, a boon companion in the serenades, practical jokes, brothel visits, and drinking bouts of the elite *jeunesse dorée*. He was also a dedicated and careful historian and published two books on Brazilian military and diplomatic history before he was thirty.[2] In 1869 it had been "arranged" for him to be the deputy to the Imperial Parliament from Mato Grosso—that redoubt of military posts, nonvoting Indian tribes, and a few gold-mining camps was an all-purpose political trump card. But at that time he had neither wealth nor temperament for Brazil's intrigue-saturated political life. After six years, Rio Branco requested foreign posting in Brazil's diplomatic corps. His first placement was in 1876 as consul general in Liverpool, the hub of Britain's imperial commerce, where he could observe and monitor the flow of global commodities, which increasingly included Brazilian rubber and coffee. He was thus especially well placed from the beginning of his career to appreciate the commercial significance of Amazonia, a reality that often escaped his Rio-based colleagues. He would continue his diplomatic career in Paris (which, predictably, he adored), with postings later to Washington and, in the critical years of the African Scramble, Berlin. During the adjudication over French Guiana, he was usefully deployed to Berne. His responsibilities placed him in the most vibrant political circles in Europe during the fever of its active imperialism, and near the best archives of Europe. He was to witness the diplomacy of the great powers in the Scramble for Africa, and with it the mechanisms through which northern European dominance of the Tropics was achieved: the charter companies, the colonization schemes, the imperial proconsuls, the "back deals," purchases, and covert alliances that were to define this period of colonial expansion.[3]

To observe the complexity of the diplomatic relations associated with the African Scramble was useful schooling for deciphering strategy in large-scale territorial disputes, and in Rio Branco's time in Europe he watched the masters of late nineteenth-century international diplomacy and intrigue at work: Salisbury, Disraeli, Chamberlain, Bismarck, Leopold, Brazza, Gladstone, Rhodes, and, banker to Empires, the financier genius de Rothschild. It certainly must have struck Rio Branco how little was known of the resources of the African continent compared to South America, where resource reconnaissance had been a regular feature of colonization and conquest for the better part of four hundred years, although reports remained submerged in the archives of Iberian crowns and Vatican libraries.[4]

Rio Branco's own researches on the Paraguayan War and other controversies of the Southern Cone educated him about boundary disputes and treaty histories in his subcontinent and the interplay of military, diplomatic, geographic, and ideological strategies. Certainly the European experience of shaping Africa was of extraordinary interest, but there was also much to be learned from the neighbor to the north, the United States, a country that had during the nineteenth century managed to defeat the British in the War of 1812 and extended its political reach by annexing swaths of Mexican lands, colonizing northern terrains, buying out huge portions of European possessions, and effectively occupying parts of the Caribbean and Central America. The Spanish-American War had also given the United States a foothold in Asia with the annexation of the Philippines. This idea of continental-scale hegemony was naturally of tremendous interest to Rio Branco. The North American deployments blended formal and informal settlement, conquest, purchase, and diplomacy as different circumstances required and was highly instructive both as a model to emulate and as a warning. When Rio Branco took up the post of foreign minister in 1902, US territorial enterprise in South America was not just "on the horizon" but already actively inside Amazonia.

For the Baron, one clear lesson of the Scramble for Africa was the crucial importance of establishing and demarcating frontiers. For Rio Branco's Amazonian ventures, there were specific sets of treaties that shaped his tropical diplomacy, and ideas embodied in them that he used to great advantage to remake the outline of Brazil through the "lawfare" and the statecraft of negotiation and adjudication courts.

The international arbitrations involved review of earlier treaties, while cultural idioms about the places were advanced along with historical maps. These diplomatic exercises were infused with narratives of "merit" as well as *lebensraum*: "our language, our customs, our tastes" designed to work off the competing nationalist/imperial ideologies. Precedent was useful but not definitive: maps and documents were often inconclusive. The malleability of these processes was Rio Branco's greatest insight and underpinned his phenomenal success.

At the time Rio Branco began to engage Brazilian frontiers, exploration sciences and transportation technologies were rapidly changing, integrating "the South" into international commodity markets and European knowledge systems, at exactly the moment when the earlier apparatus of boundaries—the colonial treaties—was crumbling. All this made Amazonia an ideal place for a Scramble.

In Amazon Country

The Iberian approach to scientific and geographic knowledge about its colonies (and especially the Amazon) emphasized secrecy. Amazonia had been a separate Portuguese colony until the early nineteenth century, and its transit and exports existed in a different realm from those of southern Brazil. The scientific silence of the crown was matched by that of the Jesuits, whose rich archives remained largely hidden from academic or political scrutiny.[5] Although quite a bit was known about the New World tropics, the formal literatures and reports were embargoed, and most travelers other than the ecclesiastics were largely illiterate. This left the region full of tempting rumors and, as da Cunha would discover, lying maps. The great South American interior was as obscure as darkest Africa for most European powers. The charts of Amazonia's contours and waterways were examples of creative cartography and imaginative copying.[6] The place was not exactly a *terra nullius* (there were multiple claims on these territories) but rather a land more or less "undiscovered"—*no descoberto*—in the sense of not yet unveiled, with colonial boundaries, native nations, imaginary places, and ecclesiastic settlements framing and overlapping this cartographic void.

The heart of Amazonia appeared as an amorphous space on the charts, largely taken up by the *Pays des Amazones*: Amazon country. Some gold trickled out of it, and precious drugs, dyes, balsams and perfumes constituted the elite Amazonian exports, while ship provisions of rice, turtle oil, dried fish, and a kind of nauseating confit of manatee supplied the sailing trade. Things were changing rapidly in Amazon Country by the nineteenth century, all of which would shape the contours of the Scramble: the emergence scientific and broader imperial interests in terms of both "enlightenment" knowledge and resource espionage; better land survey techniques, training for "geography militant,"[7] and the revamping of transportation systems, from wind and rowing vessels to the modern steamship. Nineteenth-century Amazonia resided at the intersection of ambitious dreams of empires and new republics; it was embedded in old mercantile commodity circuits as well as an emergent globalized capitalism that relied on tropical raw materials for machinery, medicine, and industrial outputs for new markets of mass consumption.[8] Scores of travelers spread out over its terrains, as I (and da Cunha) discuss in much more detail in later chapters.

Geography Militant

While elements of the Amazonian lowlands were of economic interest (rubber, cacau, drugs, and gold) and certainly the place was scientifically compelling, the economic news and traveler exploits hardly captured the geographical complexity or the accuracy required by states to actually claim these areas. Information on the bends in the rivers and their actual names were often the purview of geographers sponsored by the British Royal Geographical Society like William Chandless in the upper Amazonian tributaries, Robert Schomburgk in the Guianas, and Percy Fawcett on the Rio Verde (all contested border areas), as well as armies of lesser known military surveyors, unglamorous and mostly unsung, who carried out the technical business of survey exploration. Latin American military schools like Praia Vermelha were incubators for home-grown surveyors such as da Cunha and Acrean rebel José Plácido de Castro. The fate of huge territories often hung on anonymous scribblings and sketch maps. An expedition with a decent (or at least convincing) cartographer might well achieve more than armies and fortresses.[9] The extent to which these played out in boundary events is eloquently and humorously spelled out by historian Graham Burnett's *Masters of All They Surveyed*, a paean to the lonely, poorly rewarded, but highly effective craft of colonial survey.[10]

The controls on navigation and international trade that had prevailed on the interior rivers for centuries had kept outside explorers and espionage at bay in Amazonia until the middle of the nineteenth century. There was also the perennial and irksome problem of finding and keeping rowers—a subject of continuous complaint in the pre-steam era. The most august of travelers could find themselves marooned. Even the ruling magistrate of the Amazon,

Francisco Mendonça Furtado (brother to the crown's foreign minister the Marquis of Pombal), was once ditched, along with the rest of his entourage, by his rowers off the coast of Marajó.[11] By the mid and late nineteenth century, better survey and steamships were transforming the possibilities of travel. Steamers could carry much more cargo and bypassed the need for the typically recalcitrant rowers. As steamships increasingly plied the region, so did scores of adventurers, scientists, and bureaucrats.

The opening of the Amazon to international commerce in the 1850s made issues of boundaries, sovereignties, and information all the more urgent. A "Great Game" was afoot in the Amazon, one that fed off earlier global competitions and modern aspirations, and there were no real frontier boundaries in place.

Uncertain Dominions

The political geography that the Baron and Euclides ultimately confronted had long, contentious, and illustrious roots anchored in ambiguous treaties and mythic and fantastical landmarks: the Lake of Xarayos at the western boundary of Brazil, the centipedelike Lake El Dorado in the Guianas, the lake of Rogualgoalo in the headwaters of the upper Amazon. Rivers had names in native languages but were baptized by their "discoverers," and multiple uses of the same names for different places (like Rio Maria or Rio Vermelho) compounded the confusion. A casual shift in a river course or a place-name by a cartographer meant that tens of thousands of hectares could change nationality with the flick of a pen, and often did.[12] The discovery of a valuable resource or an emerging strategic position could stir up dormant resentments and animate conflicts on the sleepy frontiers of ultra-peripheries, suddenly subjecting them to the scrutiny and meddling of global powers. The confusing and inventive features of cartographies was reflected in the ambiguities of formal treaties that were meant to divide up and administrate the "unexplored territories" of the "Lands of the Amazons," where even today the maps are a mess.

If there was a single truism that prevailed in tropical territorial diplomacy, it was that treaties were made to be broken. It is not my purpose to detail the politics of the successive pacts that recast all Amazon territories among the nations who claimed them, but the Brazilian case was especially important. Brazil's boundaries involved over twelve thousand kilometers from the Rio de la Plata north to the Amazon basin in some of the remotest and least formally documented parts of the planet, where the DNA of competing territorial claims could involve extinct empires, ecclesiastical lands, the assertions of ex-colonies, distant republics, new utopias, financial speculations, and modern imperia. Rio Branco, perhaps better than any scholar of his generation, knew very well the significance of the archaic treaties and, perhaps more to the point, how to defeat them.

Slow-Motion Scramble: An Island Called Brazil

To grasp the dynamics of the Amazon Scramble, it is useful to understand the general framework treaties, some of the localities, the forms of state legibility (such as new mapping exercises) as well as the ideologies invoked as forms of claiming, and the specific talents and strategies used by Rio Branco. He was building on a long Brazilian history of recasting formal boundaries. Four main treaties had initially structured Brazilian claims, and Rio Branco overrode them all.

The first, the Treaty of Tordesillas, the Papal Bull of 1494, divided South America between Spain and Portugal almost immediately after the return of Columbus along a meridian 370 leagues (about 1770 kms) to the west of the Cape Verde Islands. Lands to the east of the line would belong to the Portuguese crown; those to the west to Spain. Since measuring longitude was a dicey enterprise and the interior path of the Tordesillas line was a matter of speculation, it was more a vague instrument of policy than a barrier to action.

Almost as soon as Tordesillas was decreed, there were substantial modifications to its location, mainly due to Portuguese and Spanish concerns about their valuable Indian Ocean holdings rather than the amorphous New World lands. In the space of forty years, the cartographers and treaties had moved the Tordesillas line more than seven hundred miles to the east and then back. The fluidity of abstract boundaries before longitude could be adequately measured, coupled with the most tenuous forms of administrative occupation, would ultimately privilege "citizens" on the ground as final arbiters of national territory. The Tordesillas line, it should be noted, was negotiated for boundaries between *Iberian* crowns. The French, the Dutch, the Irish, and the English were not signatories and habitually ignored it.

The Brazilian and Spanish colonies were linked by their shared rivers: The Amazon River joined what today are Peru, Ecuador, and Colombia with Brazil. The Paraguay/Parana/La Plata drainage connected Bolivia, Paraguay, Argentina, and Uruguay with Brazil to the east. The explorers from São Paulo plied this immense Paraguay basin northward in sailing vessels moving on the relentless Antarctic winds that blow from the pole to the tropics for part of the year, the *monsões*, from whence we get the English term *monsoons*. Later the term *monsões* came to be applied to the adventurers, laborers, and traders who routinely plied and explored the western territories of the Parana basin. With a combination of sailing and rowing it was possible to move from

São Paulo to the center of the continent in Mato Grosso. To return, the currents carried one down the Paraguay River to the La Plata, a route that had connected silver from the Potosí mines for untaxed contraband for Atlantic embarkation. It was the great interior integrator, as well as its perennially contested boundary.

A map from 1600 clearly shows an "Island of Brazil," with the substantial Lake of Xayaros linking the Amazon and Paraguay Rivers. By 1601, Antonio de Herrera in his *General History of Castillian Activities in the Islands and Countries of the Ocean Sea* (derived in its essentials from Alvar Nuñez Cabeza de Vaca's *Shipwrecks and Commentaries*) had provided the most widely known assessment: "There is a huge lake they call the Xayaros formed of many rivers that arise from the heights of the Andes. They form the river Paranaguazu and with this said lake, communicate with the Amazonas."[13] The Lago de Xarayos, named after the highly civilized and soon extinct Indian group, was seen as a pivot between the two great drainages and the source of both Río Paraguay and the Amazon, thus suggesting that Brazil might incarnate itself from the Platine estuary to the mouth of the Amazon as a giant, continental-scale island, a sort of New World Australia. From the beginning of the 1600s, the possibility of a connection between the Paraguayan and Amazon drainages was an inspiration for territorial expansion beyond the Tordesillas line. To transform indistinct frontiers into the outline of a new nation required discursive and geographic imagination. While the myth of El Dorado as reality and metaphor stimulated explorations and expeditions over the vast interior, the spatial conception that nourished a Brazil with a large interior was "Island Brazil," a geographical image that ultimately galvanized colonial diplomacy.

As the seventeenth century wore on, the diligent mapmakers moved the Brazilian "Tordesillas" limits farther west, where the link to Amazonia corresponded increasingly to the Madeira and Paraguay River basins,[14] capturing, at least by the map, millions of hectares, most of what was called the "Pays de Amazones" and occupied by large indigenous nations. This cartographic nimbleness was matched by an equal agility within the landscape itself. A number of expeditions moved from Mato Grosso through the missions in the Moxos plains and then on to the Rio Madeira, taking the measure of the terrain and populations. These forays may well have been secret, but soon they were a major clandestine circuit for the spoils of gold mining and native slaving.[15] The route was probably far more widely known in pre-Columbian times via the connections from the extensive indigenous Moxos plains to the Xingu River polities, so the rumored integration of the watershed was probably informed by earlier trade and travel routes that would certainly have been active in sixteenth-century expeditions of Cabeza de Vaca and Sebastian Cabot, early speculators on such links.[16]

Inventing the Interior: Island Brazil

The invention of the "interior" of Brazil by its mostly coastal population was made more appetizing by the discovery of rich veins of gold in the early eighteenth century in the Guaporé basin, in what is today known as Mato Grosso, and the territories of the Captaincy of Goyáz, between the Sao Francisco River and the Araguaia-Tocantins. Gold needed slaves to find and to mine it, so whenever the luminous metal appeared, captive labor was soon applied to it, later followed by freelancers of various kinds. The mameluco* *bandeirantes* from São Paulo rapidly colonized the area with their mining slaves, areas far to the west of the Tordesillas line. Though small in numbers, once beyond the gold rush town of Cuiabá, gold strikes—*garimpos*—were soon sprinkled along the riparian landscapes, giving an impression of greater numbers and settlement than actually existed. The region also became a refuge for large numbers of *quilombos* as mining slaves fled up the tributaries. By linking the mines of Cuiabá by river travel on the Guaporé and the Madeira, then into the main Amazon channel to the port of Belém, or sailing on the the Paraguay/Paraná and connecting the savanna goldfields of Goyaz to these rivers via rough overland mule routes,[17] the Portuguese of the eighteenth century indeed possessed a diffuse El Dorado in the midst of the huge tropical interior, bounded by the Amazon and Paraguayan river systems.

The town of Cuiabá was quickly founded in 1727, and the capital of the new Captaincy of Mato Grosso, Vila Bela, was established on the banks of the Guaporé two decades later, as the state rapidly interceded in order to politically validate the territorial claims of the freelancers of the Portuguese empire. To the west were no lowland Spanish colonies to speak of, only the missions of the Moxos and Chiquitania, which were independent although under crown protection.[18] These mission territories in what is modern Bolivia had functioned as a "bulwark state" between Portuguese gold and the Spanish silver mines of the Andes.[19] The missions became an increasingly militarized buffer zone at the interface between Spanish and Portuguese colonial territorial interests and, at least as importantly, protected native labor from the chattel demands of *encomienda* and *mita* on the Spanish side and slavery on the Brazilian. These communities, formerly ultra-backwaters, trading in cattle hides, cacau, herbal teas, and tree oils, found themselves on the front lines of imperial contest by the time of the Treaty of Madrid in 1750.

The architect of this treaty and the strategies that underpinned it was

* Offspring of natives and Portuguese.

the Brazilian-born Portuguese overseas secretary Alexandre Gusmão (1695–1753), whose negotiations emphasized actual territorial occupation. Historical geographer Jaime Cortesão has emphasized four main reasons that Gusmão's idea of Brazil as an "island"—bounded to the west by the Paraguay and to the north by the Amazon basin—had such appeal and infused early territorial ideologies.[20] First, if the tropical "northwest passage" could be found from the Paraguay River through the giant swamp today known as the Pantanal into the Amazon territories, an extraordinary linkage of massive river routes was possible and could be controlled by Brazil, an issue of great geopolitical interest.

Next, the idea of an "island" would make sovereignty over Brazil's interior and the transfer of Amazon lands to Brazil geographically logical and would significantly undermine any claims Spain made to the lands to the east of the Paraguay and south of the Amazon River, to which it was in fact entitled under the original provisions of the Tordesillas line. In administrative terms, it would help unite the Portuguese colonies, since the Amazon estuary had been separately administered since 1623, and through this 1750 integration, it would enhance the southern seat of power.

Then, instead of endless squabbles over the location of the imaginary grid line of the forty-sixth meridian, the Tordesillas line, natural geographic formations—in this case enormous rivers—would become general territorial markers, something visible and obvious to all. This question of gigantic rivers would not resolve everything, as we shall see, but it was a step forward.

The Treaty of Madrid (1750)

The stimulus for the Treaty of Madrid lay in the messy territory of multiple claims between the Jesuits, Brazil, and Argentina. Portugal began establishing significant colonies in Argentinean lands, such as the Colonia of Sacramento, in areas obviously west of the Tordesillas line and at the heart of some of the richest Jesuit settlements. The extensive grasslands were coveted by Brazilian cattlemen, but the region had had a complex history of invasion and skirmish. In this context of highly contested boundaries to the west and to the north, the first Brazilian scientific boundary commission (1729), composed of the two "mathematical clerics" (*padres matemáticos*"—Diogo Soares and Domingos Capassi) was initiated and coordinated by Alexandre Gusmão, an essential first step in the definitive structuring of the boundary.[21] This enterprise involved using the astronomical innovations in cartography pioneered by the French Royal geographer Guillaume Lisle (1675–1726) to generate greater precision in geographical mapping. The maps by the padres were meant to accurately document the extent of Luso occupation and the physical location of the rivers and landmarks. The Spanish crown, however, in a practice that it (and its ex-colonies) seemed to often repeat, launched no counterpart to the Portuguese survey.

The specific Portuguese intentions with the 1750 treaty involved three territorial outcomes: to win Spanish recognition of the terrains occupied by Portuguese/Brazilians in Rio Grande do Sul (thus extending Brazilian coastal control to the south); to recognize Portugal's claim to the eastern shores of the Paraná, Paraguay, Guaporé, and Madeira Rivers, thus securing formal control of the mining districts of Mato Grosso; and to attain sovereignty over the Amazon basin, not just the Atlantic estuary (as defined by the Tordesillas line). These consolidated "island" Brazil. In return, Portugal would cede the Sacramento Colônia, at the mouth of La Plata River (Brazil had other river routes to the Río Paraguay), and, of special interest to Spain, the Philippines, key to the highly lucrative Asian silver trade. Those islands were technically Portuguese by virtue of the Asian projection of the Tordesillas line onto the other side of the globe.[22]

This treaty had broader implications for South American territorial history, however. First and foremost, the treaty largely eliminated the Tordesillas line as a demarcation between the New World holdings of the Spanish and Portuguese empires. The annulment of this foundational division was a pathbreaking precedent and served to open up the Amazon to many contenders, since the territorial "rights" of Spain no longer counted in substantial territories and Brazil barely had a presence there. Next, the treaty ratified the use of the Roman law concept of *uti possedetis*—he who has, keeps. The use of this principle in border negotiations would inform the substance of South American disputes for the next 150 years. Further, the borders were to follow clearly discernible landmarks, in this case rivers, that at least made the borders far more obvious—but whether a given river was in fact the indicated border became a problem in other frontier cases. A final element was the use of boundary surveys to consolidate the shifts in territorial dominion.

The Madrid Treaty included a clause that underpinned Amazonian contention later: a straight parallel line would be drawn from the middle of the Rio Madeira (in this case taken to be just above the extensive falls) to the (unknown) headwaters of the Rio Javari at 6.53 degrees south latitude; this would constitute the western boundary between the Portuguese and Spanish colonies in their (unexplored) northern reaches. This clause would inflame the Scramble a century later, as this principle was ratified in the Treaty of San Idelfonso.

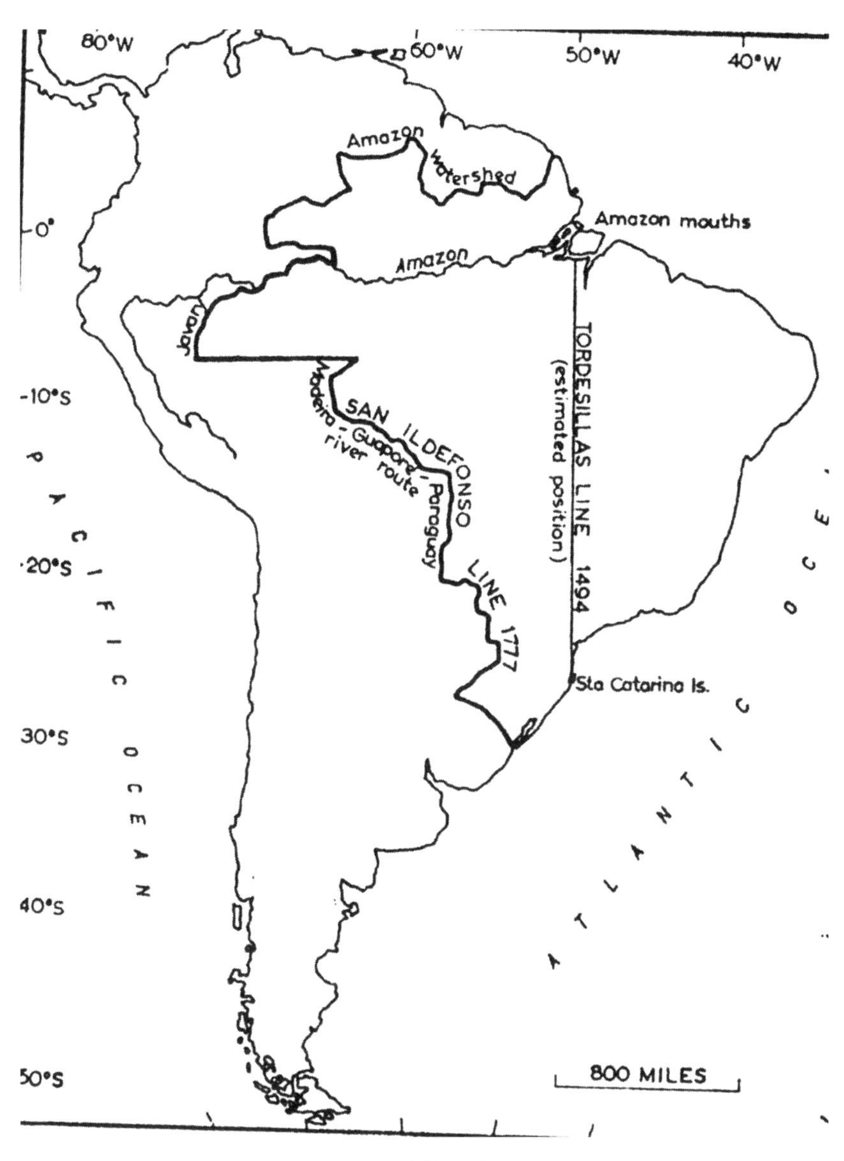

Map 3. Idelfonso line.

The Treaty of San Idelfonso (1777)

The territorial lines indicated by Treaty of Madrid and the Treaty of Idelfonso were meant to be demarcated by joint boundary commissions. This task was somewhat completed in the La Plata areas, but the circumstances of the northern boundaries—hostile Indians, the problems of surveying in the dense forests and impassable swamplands, and general recalcitrance on both sides—meant that the western Amazonian holdings were actually never surveyed and thus remained vague and contestable. Earlier colonial disputes had a long afterlife that reverberated into the Brazilian Republic's Amazonian controversies, as the "Amazon Country" morphed into the "realm of rubber."

The line of the Idelfonso Treaty is shown in Map 2. The map also reveals the extraordinary shift in territory from the Tordesillas to the Idelfonso Treaty. This for all practical purposes remained the border for next ninety years. The Idelfonso line would become a key boundary in the subsequent conflicts that embroiled Peru, Bolivia, and Brazil, as we shall see.

Ayacucho: A Treaty of Friendship with Limits

The Ayacucho Treaty also featured prominently in the Scrambles of the Western Amazon. Initially promulgated as a "Treaty of Friendship and Navigation" between Brazil and Bolivia in 1867, the treaty revisited the 1777 Idelfonso line, but under the tensions and anxiety of the region during the Paraguayan War. The threat of isolating Bolivia's Amazon holdings (and thus making them economically worthless) as Paraguay fell to the Brazilians made free navigation on the Paraguay River a priority. Bolivia's Amazonian boundary treaty of 1867 referred to the famous line "from the Madeira to the headwaters of Javari."[23] The key point of contention was where exactly the line between the Madeira and the Javari was supposed to lie. There were several options depending where one chose to start on the Madeira—its midpoint (this was favored by Peru), the Beni (the position preferred by Bolivia), or an upper tributary like the Itambari (the position preferred by Brazil). The next point was that the "headwaters of the Javari," were still unexplored: they were thought to extend roughly to 10.20 south latitude, but no one really knew. The Ayacucho Treaty took an oblique line from the Beni River to the Javari at 6 degrees south, to the true headwaters of the Javari. This maintained the precious Acre territory in Bolivian domain but transferred some 97,000 square kilometers to Brazil. This agreement was hashed out by the Viscount of Rio Branco, the Baron's father.

Bolivia lost a great deal of land in this treaty but still maintained sovereignty over the territory of Acre and five important ports of the Paraguay River, plus navigation rights on the Paraguay and the Amazon.[24] In addition, exclusive navigation rights above the falls of San Antonio and the development rights to and use of roads around the falls of the Madeira could valorize Bolivia's lowlands, which had languished after the mission system was abandoned. There were, however, many explorers involved in "resources review," whom we will meet in the next chapters. Bolivia agreed to emphasize the principle of *uti possedetis* over the region in this Treaty of Friendship and Navigation, a position it would come to regret.

This exercise in diplomacy initially seemed to have navigational rather than territorial consequence, but by the 1880s Acre was becoming one of the most commercially desirable areas on earth, and Bolivians, Peruvians, and Brazilians were dusting off their treaties and reviewing their maps. A key question on which the Scramble would hinge involved historical jurisdiction over this increasingly contested region and whether in fact Bolivia actually had the territorial rights over the realms below the Idelfonso line. Peru contested Bolivia's claim by arguing that the extinct Viceroyalty of Peru had jurisdiction over all South American *audiencias*, while Bolivia had asserted its rights through the authority of the vanished *audiencia* of Charcas.* Bolivia rejected the extension of Spanish rights over territories after the end of the Spanish empire. Since Charcas had dissolved into several states including Bolivia, Bolivia's territories were, in this line of reasoning, immune from Peruvian territorial assertions derived from its earlier imperial administrative position. Deciphering and recasting the rights of phantom administrations become yet another element of da Cunha's ideological task, since the complexities of the Peru-Bolivia claims would fall into adjudication in 1902, with Brazil as a highly interested party, since Brazil had bought out Bolivia's land claims as part of a later boundary agreement that I discuss in more detail in chapters 9 and 10. The conflict was to be mediated by the president of Argentina.[25]

It was in this context of increasing ambiguity of borders that Rio Branco managed his first boundary "conflagration." It wasn't in the Amazon but rather in the Misiones, contested between Brazil and Argentina. The frontier had a festering set of treaty disagreements, many of which would be echoed in his later diplomatic and military skirmishes in Amazonia, and where the intellectual, international and political connections developed in this conflict would serve him well in the Scramble.

The Misiones Apprenticeship

In 1893, Rio Branco first displayed his flair for diplomacy as the chief of party as Brazil pressed its case against Argentina for the control of the Misiones area, some 33,669 km^2. Initially occupied by Jesuits and thus outside the jurisdiction of the crown of either Spain and Portugal until the end of the eighteenth century, this terrain had become highly contested once the Jesuits

* The *audiencia* of Charcas extended from southern Bolivia through Argentina.

were expelled from Brazil in 1781. Vulnerable to ceaseless slaving raids by Brazilians, the former mission Indians rose up in rebellion, but were ultimately chased across La Plata River, functionally ceding the lands, among the richest of the entire mission system, to the Brazilian nationals, who happily settled on the well-tended ranches and farms. The Paraguayan War of the 1860s further complicated the claims as Argentina, Uruguay, and Brazil disputed each other's territorial authority, ultimately to be resolved through mediation.[26]

The Misiones adjudication relied on US president Grover Cleveland as its plenipotentate. Rio Branco went to Washington, DC, and became integrated into its diplomatic circles. He also went regularly to New York to confer with professor of international law John Bassett Moore of Columbia University, who would become a key figure in the Amazon Scrambles. Moore was one of the great practitioners in an emerging international culture of adjudication and global law, and he became a judge in the International Court at The Hague. Moore was a frequent consultant to Brazil's foreign ministry and was Rio Branco's legal "muse." He helped establish the modern uses of the Monroe Doctrine and Pan Americanism by both the United States and Brazil.[27] A lively public intellectual on international diplomacy, Moore wrote for *Harper's* and the *New York Times*; today there is a stamp that commemorates him.

The case of Misiones, one of the most complex nodes of territorial dispute, involved a long history of overlapping claims, mapping mistakes, and historical drama. Eventually it was a diplomatic triumph for Brazil, even though Brazil's precedence over Argentina was initially difficult to imagine. Rio Branco's residence in the United States, his affability, and a widening circle of "friends in Washington" certainly cannot have escaped President Cleveland's notice. But the hallmark of Rio Branco's work was his painstaking scholarship in historical reviews of chronicles, maps, pacts, and history of settlement. There was also no substitute for knowing the actual geography: one of the key river boundaries in Misiones proved to have been misidentified, and the earlier territorial demarcation by binational committees had not been carried out, opening the way for Rio Branco's unique historical style of scientific "lawfare."

The agility of the arguments and the depth of the research in the adjudications projected Brazil strongly into the international arena and helped forge Rio Branco's relations with United States, which would stand him in good stead in the Purús controversies and probably cemented his decision to ultimately ally with North America rather than Europe in many larger diplomatic questions.[28] Rio Branco's success with the Misiones would have been achievement enough, because this was a rich area, contested and well known, well mapped for centuries. Rio Branco would point out that virtually the entire populace was composed of Brazilians, even though this was a relatively recent development, brought about by the tumultuous expulsion of Argentineans and Jesuits and the pitiless suppression of native Guaraní rebellions.[29] His successful strategy would become a template for later conflicts. *How* Brazilians got there, *when* they arrived, and *who* they were (citizens or not) were of little interest; *that* they were Brazilian was key.

Recipes for Territorial Rights

Rio Branco used a set of techniques in this initial foray that later defined his diplomatic strategy in Amazonia. First, he challenged claims based on the inherent vagueness of the "founding treaties" such as the spectacularly contestable Treaty of Tordesillas of 1494 but also, later, the equally questionable lines of the Idelfonso, and in Amapá, the ambiguity about the location of Rio Vincent Pinzon, and the endless sequences of boundary revisions. The fact that no one knew where anything was, that the maps were as likely to document imaginary terrain as real places, helped his cases a great deal. That such treaties often ended up favoring Brazil had everything to do with the emphasis on Brazilian de facto occupation, versus the de jure constructions that other South American legalists favored. Rio Branco was ready to reinvent tradition and recast treaties, relying on a relaxed view of diplomatic precedent when it served him. While treaty boundaries may have been open to question, modern survey lines were less so, and so he insisted on rapid demarcations in his adjudications. Next, Rio Branco used his historical scholarship extremely pragmatically, and he generally chose to make his arguments by focusing on current polities rather than on precedents developed by earlier colonial administrations. By insisting on superseding all previous claims and surveying the contested territories at once, Rio Branco was able to definitively consolidate Brazil's boundaries using the politics of existing states rather than archaic polities and extinct empires.

Third, when possible, "discovery" was itself an important component of claim. "New lands and new stars" was a catchphrase for the early Portuguese imperial exploration. This involved finding "new lands" and describing celestial means to navigate to them—that is, the navigational "software" of exploration to situate routes and places. These were a hallmark of Luso imperialism. Rio Branco was successful in his careful use of Brazil's modalities of claiming, including early Luso-Brazilian expeditionary data, which were much more precise and "scientific" than comparable imperial enterprises of other European states due to Portugal's deep navigational history and technologies.[30] While planting crosses and saying to uncomprehending populations that the interior of the continent belonged to a European ruler was helpful in its way, the actual determination of location and the extension of the colonial "footprint" were key elements in this form of "conquest," and the Portuguese and their master pilots were its best practitioners.[31] Thus, the 1638 expedition by Pedro Teixeira to Quito, with rowers and astrolabes, to claim the Amazonian terrains two thousand kilometers up the Amazon River was precise in siting locations and placing markers, especially compared to the earlier careening excursion down the Amazon by Francisco Orellana (1541–42), where distances had been basically counted in days.

But mere "discovery" and geolocation would not, on their own, carry the day in the nineteenth century. Rio Branco's fourth guiding principle, inspired by his father and Alexandre Gusmão, was that physical occupation meant sovereignty. De facto settlement would, in his view, be more powerful on that big continent than all the lawyers and lying maps that could be generated in national capitals. Centuries of relentless movement through the interior by Brazilian *bandeirantes*, residents of religious settlements, and runaway slaves, along with an active colonization policy, had pushed Portuguese boundaries almost to the Andes. The cattle, peasants, adventurers, and fugitives pressed westward with their own expansive epics in the *sertões* and *selvas*. These were then capitalized upon by the Brazilian state.[32] Rio Branco knew how to make the most of popular history, while the Spanish republics remained wedded to royal and colonial legalisms based on de jure possession of essentially unknown areas.

Travel histories, constellations, and old treaties notwithstanding, it was warm bodies rather than arid arguments that increasingly prevailed when it came to negotiations. The shift to the Roman legal concept of *uti possedetis* in the mid-eighteenth century became more critical after the expulsion of Catholic religious orders in 1767, because they left a vacuum in their wake, a kind of *terra nullius* in the middle of the continent that stretched from Paraguay to the Orinoco, exactly between the least known and most contested terrains of Brazil and the former Spanish colonies. Also, Roman law carried with it the appropriate luster of empire.

Another element of Rio Branco's arsenal was bilateral negotiation. His observations of the African Scramble had revealed how very quickly things could degrade once multiple alliances or disagreements occurred on small points. The Brazilian experience with the triple alliance with Argentina and Uruguay after the Paraguayan War was in itself a cautionary tale.[33] Rio Branco admired the clarity of the relatively simple binational agreements worked out in the US imperial pattern.

Varying cultural idioms about the places were also invoked along with maps and treaties. Thus the ways that "territorial rights" could be woven together involved physical, juridical, and cartographic assessment but also deployed ideas about localities, people, and place that infused these exercises with narratives of "merit," "nationalness," folklore, "taming," and other idioms of nationalist sentimentality designed to show adjudicators why such places truly belonged culturally to one country rather than another: "to speak our language and have our customs."[34] These kinds of materials were central to Rio Branco's style of statecraft and much preferred over battles in remote places where fighting conditions were awful.

There were two final practical elements to Rio Branco's Scramble strategy as the adjudications moved forward. The first was not to let "demilitarized" or modus vivendi status linger too long, since the neutralized territory rapidly became a realm of open access for any variety of territorial schemes, as was the case in Amapá (which languished in this status for almost fifty years), and the second was to move demarcation teams to the regions as quickly as possible, rapidly funneling their results into the mediation and treaty language. With deployment of these tools, the relatively nonconflictual Brazilian Amazon boundary treaties with Venezuela (1905), the Netherlands (1906), and Colombia (1907) were speedily resolved.

The competition over Amazonia was out of control by the 1890s. The rubber economy was exploding, and there were enormous gold strikes in the Guianas. Armies trudged from one swamp to another. The western zones were ruled by the great rubber barons, the Fitzcarraldo brothers on the Ucayali and Madre de Dios, the Casa Arana on the Putumayo, and the Suarez brothers on the upper Madeira. Guyana goldfields were full of escaped slaves, Caribbean miners, and adventurers of all stripes. Utopias and independent republics were being proclaimed from Acre to Amapá. Entrepreneurs, speculators, financiers, and developers swarmed. The river traffic burgeoned. The African Scramble and US expansion in North America (to the west and the south) and its Caribbean and Central American filibusters (including Nicaragua and, just north of Amazonia, Panama) provided templates for what might unfold if no attention was paid to Brazil's northern frontiers. All Amazon "borders" were fluid and very loosely specified at the end of the nineteenth century.

There were numerous territorial disputes in play in the Amazon, involving as actors and adjudicators Brazil, Bolivia, Peru, Ecuador, Colombia, Paraguay, Surinam, British Guyana, French Guiana, Argentina, the United States, Italy, France, Great Britain, Belgium, the Netherlands, Russia, and Switzerland.[35] In this context, it is obvious why Baron Rio Branco took frontier demarcation as his central charge.[36]

The Misiones experience served as a suitable apprenticeship for his entrance into the Amazon Scramble, where the stakes, the terrain, and the adversaries would be far greater. Having triumphed in the south, Rio Branco had to take on the land of Cabo Norte, the area now known as the state of Amapá, a place that had been in conflict for four hundred years, with the last fifty having been more or less a territorial free–for-all. His adversary in this case would be among the most powerful countries in the world, France.

7

Imperialisms, Revolutions, and Resolutions in the Caribbean Amazon

Unquiet on the Northern Front

Rio Branco's first foray into Amazonian diplomacy involved a controversy that had been brewing for close to four hundred years. The "Wild Coast" or "Lands of the Cabo Norte," later known as the Franco-Brazilian "Território Contestado" or "Contesté"—now known as the state of Amapá—had been a complex scramble. Located in the Caribbean Amazon, it was a region at the intersection of native, African fugitive, and European colonial boundaries and New World imperial rivalries. It was also a region of habitual insurgency. Brazil's adversary, France, was among the most powerful nations on the planet in the eighteenth and nineteenth centuries, indeed so powerful that it had caused the Portuguese monarch to flee his own land and set up shop in his unruly tropical colony.

The Caribbean Amazon can be usefully understood as bounded to the north by the Orinoco, to the south by the Amazon, to the east by the Atlantic, with its currents, economies, and imperialisms linking to the Caribbean, the Atlantic beyond, and points north and east. The ambiguous interior boundary was framed by its runaway slave communities and native settlements. The Caribbean Amazon was connected by trade and escape to global circuits of commodities (including slaves), insurgencies, and ideas. The integration of the Guianas into both Amazon and Caribbean processes has been largely overlooked, in part because most studies have followed Northern European colony outlines rather than the implications of non-European territorial configurations and how malleable the colonial frontiers actually were.[1] Rio Branco was facing a region rife with failed European experiments and imperial ambitions, fugitive cultures, and rampant malaria built on the ruins of one of the New World's great tropical civilizations.

To describe all the complexity of this region is more than this short chapter can aspire to do. However, the defining processes that shaped the Contestado can usefully be divided into several loose periods: its complex pre-Columbian history; the early consolidation of Portuguese hegemony on the Amazon, the emergence of active European colonies in the Caribbean Amazon (with their massive and rebellious slave populations); Enlightenment experiments; the rise of regional revolutions; and the final politics of adjudication. The region was, if nothing else, a testimony to the limitations of European imperialisms in peripheries that today seem lost to the world but that figured in European diplomacy and conflict for four hundred years. Caribbean currents, trans-Atlantic storms, and Amazonian forces shaped this region's ecologies, polities, and utopias.

A Glance Backward

For scholars of Amazonian pre-Columbian history, the Contestado territory—today known as the Brazilian state of Amapá—is of considerable interest. It is a place of antiquity of human occupation in the New World, part of the sophisticated estuary and north coastal cultures of Amazonia, and includes sites of raised fields, ceramic masterpieces, exquisite funerary art, and other evidence of highly developed civilizations. It has an extraordinary pre-Columbian astronomical observatory, a two-thousand-year-old "tropical Stonehenge" of some 127 upright stones arrayed in circular and other orientations, for marking solstices and equinoxes.[2]

The Guianas were connected to the Caribbean by pre-Columbian diasporas and trade routes, and to the Amazon interior by commercial, military, and migration routes through the Orinoco, the Rio Negro, the Essequibo, and down the Parú, Jarí, and Trombetas to the Amazon River itself. Trading states like the Carib and some Arawak polities moved precious commodities

Map 4. The Caribbean Amazon.

Table 1. Main Amazonian Scramble territories, 1900–1909.

Place	**Litigants**	**Treaty/adjudication and date**	**Area (km²)**
Amapá	France-Brazil	Swiss adjudication, 1900	260,000 to Brazil
Acre	Bolivia-Brazil	Treaty of Petrópolis, 1903	191,000 to Brazil, 3,200 to Bolivia
British Guyana	Great Britain–Brazil	Victor Emmanuel, 1906	5,000 to Britain, 3,000 to Brazil
Peru/Brazil	Peru-Brazil	Velarde–Rio Branco, 1909	40,404 to Brazil, 38,850 to Peru
Peru/Bolivia	Peru-Bolivia	Argentinean adjudication, 1909	56,980 to Bolivia, 85,470 to Peru

including jade, goldwork, ceramics, dogs, salt, and drugs through labyrinthine inland waterways, along the coast, and to the islands of the high seas.[3] The populations of the coastal estuaries were among the most complex, populous, and developed settlements in the New World at the time of the arrival of the Europeans.[4] While the Indian wars did not figure as prominently in European occupation of this area as elsewhere in the New World, the famously recalcitrant locals perhaps took their inspiration from earlier Arawak resistance and guerrilla strategy.[5] What is clear is that the "landscapes of memory"—the travel routes of the natives—also included extensive food forests able to support populations in ways largely invisible to Europeans and thus made autonomous forest polities a central feature of the modern history of the Caribbean Amazon. The muddy mangroves and swamp forests abhorred by colonists were both a defense and a larder to the natives, and later helped support fugitive slaves.[6]

The first European "discovery" of the region was by Vicente Pinzón, who had traveled with Columbus as the captain of the *Niña* and in 1499 landed in what is now Pernambuco, in the Brazilian Northeast, a few months before Brazil was officially "discovered" in 1500 by Pedro Alvares Cabral. Pinzón turned north, sailed for about fifty leagues to the Amazon estuary, and then continued on up the coast. The Oyapoque River was said to have been discovered by him and was later baptized with his name, as was another huge watercourse, the Araguarí, right below Cabo Norte, a river that would become, as von Humboldt described it "celebrated in the annals of diplomatic disputes."[7]

This confusion over river names was later to become a flashpoint in Franco-Brazilian territorial battles. The problem was that thirty rivers drain to the Atlantic between the Amazon and the Oyapoque, and many were baptized as the Vicente Pinzón.

Military conquest of the Wild Coast was very difficult for reasons of geography, guerrilla warfare, and disease.[8] Successful engagement required alliances with native polities who mediated trade, travel, subsistence, and security. Historical ethnographer Neil Whitehead has shown in detail the complexity of the indigenous interactions and how European involvement (and trade goods) enhanced both trading polities and lineage societies by reinforcing the power of their elites, increasing both competition and cooperation among them, and inflaming regional indigenous politics.[9] In addition, the Caribbean Amazon was at a complex intersection of a "Great Game" between France, England, the Netherlands, Spain, Portugal, native Amazonian nations, and the *quilombo* settlements and polities over the region. Runaways, castaways, mutineers, pirates, and privateers also plied the wild coast with their own economies, politics, and ambitions.[10] It was, in spite of its seeming remoteness, very cosmopolitan, a strangely transnational space. Daniel Defoe's novel *Robinson Crusoe* (1719) takes place on an island off the Wild Coast and provides a sense of just how international the region was: Crusoe, a British planter, also a Brazilian landowner and slaveholder, is shipwrecked. His tale involves outwitting native groups (cannibal Caribs), international powers (Spain, Portugal), mutineers, and pirates; in the end British technical skill and moral virtue naturally triumph over European and local barbarianisms.

The modern Brazilian state of Amapá, derived from Contestado lands, probably evokes no image at all. Its main exports today are manganese ore and the *Euterpe* palm fruits (açai) that are processed into popular health products. Those interested in extractive reserves and tropical conservation note that Amapá, like Acre and Amazonas, is one of Amazonia's "eco-states." It is also the site of some of the most well-known Amazonian *quilombos*: Mazagão Velho and the very beautiful lake-based Curiaú. The history of Contestado remains mostly a footnote, but even though today it may be seen as a hyperperiphery, it was a political chit in European dynastic politics for hundreds of years, and a significant part of the Amazon Scramble for several reasons. It was a large area; it had resources, most especially gold and latexes. Its boundary politics were famously confused, as was its ethnic makeup. It also marked an evolution in Rio Branco's diplomacy: "informal colonialism" carried out by slaves fleeing Brazil could be construed as a migration of "Brazilian citizens," and there was direct use of prominent scientists in the imperial practices on both sides. This adjudication also marked the use of US hemispheric politics (the Monroe Doctrine) in Rio Branco's American public relations and negotiations. These tactics would be deployed again in the Western Amazon. Just how complex the region was is suggested by the convoluted diplomatic history and relentlessly shifting boundaries of a place that was essentially a no-man's-land. In this sense too, the experience was a useful preparation for the diplomatic challenges in the Western Amazon.

Sovereignty by Decree versus Boots on the Ground

The Amazon channel of the seventeenth century was a kind of immense free-for-all, a *terra nullius* open to adventurers as well as royal charters. European decrees would grant sovereignty to various regions in the Amazon Caribbean with a pointed lack of concern for (or ignorance of) competing claims. For the Portuguese, the central ambition was to consolidate their hegemony, since in principle the Tordesillas Treaty gave them rights to much of the Amazon estuary and control of the gateway to the vast interior, even though the riverbanks were subject to persistent incursions and settlements by Ireland, England, France, and Holland, none of which were parties to that pact.

In 1605, Henry IV proclaimed that French territories extended in a swath from the Orinoco to the Amazon (about 1,000 kilometers), in spite of a profusion of Dutch, Spanish, Portuguese, and English settlements along the coast and interior watersheds. With a French colony on the island of São Luis in Maranhão (475 kilometers to the south of Belém), this decree was meant to place the coasts to the south and the north of the Amazon estuary under the aegis of the French crown and thus gave control of a huge territory as well as Amazon navigation to France. By 1613, Maranhão-based General Daniel La Ravadière and battalions of Tupinamba Indians set out to materialize this edict by establishing French military settlements on the Amazon. In their absence, São Luis was attacked by Portuguese forces, and after continuous assaults, by 1615 the French had been expelled from Brazilian territories below the Amazon, mostly taking refuge in their Guiana outposts.[11]

While La Ravadière was being routed, King James of England deeded British rights from the Amazon to the Essequibo River (about 780 kilometers apart) to his agents Thomas Challoner and John Rovenson. These two appealed to financier Robert North (hence the North Cape, or Cabo Norte), whose influence extended to many of Britain's wealthiest nobles, who were pleased to take on this interesting mercantile adventure under the name of the charter company, the North Company. King James was not alone in his interest in charter companies. There had been many Dutch incursions into tropical America (and North America, for that matter), and by 1621 the Dutch West Indies charter company was formed. Its main concerns were in the south Atlantic in northeastern Brazil, but the colony of Suriname, or Dutch Guiana, was being developed on the Wild Coast with very loose boundaries. Dutch settlements existed on the north shore of the Amazon from the Jarí River to the estuary, at the mouths of the southern tributaries of the Xingu and the Tapajos, and the Guiana coast at the Essequibo River, where the Netherlands appeared to also threaten the Spanish claims on the upper Orinoco. It was here in the Wild Coast colony that many Dutch plantation owners (with their slaves and capital) resettled after defeat by the Portuguese on the northeastern coast of Brazil in 1654.[12]

The Portuguese crown was concerned that the largely unexplored territories above the Amazon would be sucked into northern European empires and parceled up like the Caribbean and North America. Portuguese colonial governors in Brazil had been distracted by the protracted war with the Dutch over Pernambuco and the rebel slave state of Palmares. Brazil was too big and Amazonia too different to manage under the existing ruling structure, In response to what seemed an emerging territorial catastrophe, in 1621 Philip IV decreed that Portuguese Amazonia should be administered as distinct captaincies—Maranhão, Grão Pará, and Ceará—and as a separate colony, since winds and currents drove toward the Northern Hemisphere and made it easier to reach the Amazon from Lisbon than from Salvador. The initial strategy of Pedro Teixeira, the first commander of the Amazonian captaincies of Grão Pará and Maranhão, was aggressive military action to eradicate foreign forts and colonies on the Amazon channel. While Teixeira's modern fame accrues to his remarkable voyage up the Amazon to Quito and back with 47 canoes, 1,200 black and Indian militia, and 120 armed military men who pounded in markers, claiming the lands they passed through in the name of the Portuguese empire, his renown at the time followed from his obliteration of Dutch and English settlements along the main channel. In this process he captured the commander of the Dutch West Indian Company, Nikolas Oudan, and James Purcell of the North Company. Purcell, undaunted, after being released from Iberian custody returned to the Amazon the following year (1628) to make a go of it with tobacco plantations and active native trade at the fort of Torrego, on the north side of the river roughly opposite the mouth of the Xingu. This establishment once again was crushed by Teixeira, and after these events North Company and the Dutch West Indies Company ceased their efforts on the Amazon channel, retreating to outposts on the Wild Coast.[13]

After reasserting control over the Amazon channel, the Portuguese began expanding sovereignty to the north, proclaiming a new captaincy in 1637, that of Cabo Norte, a territory stretching from the Vicente Pinzón River (in this case the Oyapoque) about 420 kilometers to the north of the Amazon, and stimulated missionary colonization on the lower Amazon and the estuary as a proxy for actual Portuguese settlement.[14]

European patent letters still abounded, because the Caribbean trade had become prosperous and the economic potentialities of the region had mythic allure. Brazil, a colony itself, was seen as having limited local powers over the continental immensity, especially the mangrove coast of the Guianas. The French king, Louis XIV, indifferent to Portuguese concerns, granted rights to the charter Compagnie Equinoctal to lands extending from the Amazon to the Orinoco and, on the west, to the confluence of the Rio Branco with the Rio Negro. In 1676, after a triumph over the Dutch in the Guianas, lands were assigned by royal decree from the Amazon to the Island of Trinidad to the new charter company, Compagnie Cap de Nord. France's northern colonial boundaries were often contested, but by 1690 the French were definitively installed in Cayenne, while Belém and the lower Amazon was clearly under the aegis of the Portuguese crown. Between them lay the lands of the Cabo Norte, the Contesté/Contestado with no other European enterprise in between, a seething *terra nullius* with multiple overlapping territorial assertions.

The Luso-Brazilian Strategy: The Socio-ecology of Survival

Luso-colonialism had used a mestizoized set of "tropical hands" as its agents, entrepreneurs, and proconsuls in Africa since the mid-1400s. They were culturally and through intermarriage probably more adapted to the tropics and benefited from the malarial resistance of their African progenitors, coupled with European immunities. In Amazonia, these colonialists forged alliances through marriage, mistresses, and clanship with Amerindians and West African slaves. This genetic mixing would, over time, enhance the resistance to malaria and yellow fever, reducing the mortality of the offspring of these Luso-colonials compared to newly arrived Europeans who died like flies, and native Americans—confronting both African maladies and European diseases to which they had limited immunity—whose populations crashed.[15]

This "marriage policy" also helped ensure continuity in political liaisons. These affinal relations "institutionalized" connections from colonials to locals and provided access to the knowledge systems necessary for security, trading economies, and traversing the complex regional and social geography of the Amazon estuaries and tributaries for acquiring backland products.[16] It also provided a complex interstitial space for mixed-blood ascension and autonomy. There were also ecclesiastic proxies—not a major element of colonial strategy on the part of the Protestant Dutch and the British, but a regular feature of Luso imperial practice in Amazonia and elsewhere.[17] Finally, a decentralized administrative structure—the Amazon captaincies—that could respond rapidly and decisively with local militias (rather than foreign armies) made the Portuguese far less "outsiders" and more agile than those arriving with their patent letters and mercenaries from European monarchs. The early Luso occupations in Amazonia had a lot to do with embeddedness and strategic uses of military, spiritual, and sexual conquest. In any case, other European empires were busy elsewhere. French Saint Domingue (now Haiti) was the wealthiest sugar colony; the Dutch had the lucrative Indonesian spice islands, and the British had their own valuable tropical islands and American colonies, far more manageable than a poor settlement perched on the edge of a massive jungled continent. These countries were also deeply involved in North American hemispheric colonial politics and European intrigues in ways that Portugal was not. Manhattan, after all, was ceded to the British by the Dutch in exchange for Suriname in 1674.

A French Imperial Outpost in the Caribbean Amazon

The French presence in Amazonian South America is not widely known, although today the Ariane rockets launched from French Guiana lift into the tropical heavens carrying about half of the world's commercial satellites. French Guiana's capital, Cayenne (previously known as Mocambo), was claimed in 1676; its name bespeaks a Creole spice residing in the back of kitchen shelves rather than an outpost of a powerful tropical empire. Coffee lovers may have read of the biopiracy that introduced the plants of that privileged drink into Brazil, today the world's largest producer.

In the eighteenth century coffee was highly coveted, and its international trade was largely monopolized by the Dutch. It was first grown in the New World in Suriname, but soon administrators in Cayenne were experimenting with its cultivation in its governor's distinguished collection and acclimatization botanical garden, La Gabrielle.[18] Sale of seeds and cuttings was forbidden in order to maintain the valuable coffee monopoly.[19] In 1737 Francisco de Melo Palheta was sent on a diplomatic mission to Cayenne. Brazilian historical lore has it that Palheta's secret and perhaps true mission was to obtain coffee cultivars from France's La Gabrielle. The method involved a passionate romance between the Brazilian diplomat and Madame d'Orvilliers, wife of the governor of French Guiana. As he left his amour, Palheta bore the contraband coffee plants tucked into the charming bouquet he took home to Brazil as a romantic memento, and thus he introduced the commodity that would define the Brazilian agricultural economy for much of the nineteenth and twentieth centuries.[20] It was quite a memorable affair—one certainly not forgotten by those who had wished to maintain the coffee monopoly. And then there is Papillon, the escapee from Devil's Island (when French Guiana became a penal colony), whose memoir was later made into a film starring 1960s heartthrob Steve McQueen. But these disjointed anecdotes hardly capture the nature of its history and how seriously and how long France yearned to claim the Contestado to consolidate a major New World presence. By the beginning of the eighteenth century, formal treaties began to specifically address the sovereignty issues, which were as murky as the mangroves and just as treacherous. This diplomatic backdrop with its highly mobile boundaries framed the "on ground" ventures, most specifically the diasporas from Brazilian and Guiana's slaveries into the Contestado landscapes.

Diplomacy and the Wild Coast

The foundational treaty for this region is that of Utrecht (1713), which placed the boundary of French Guiana at the "Vicente Pinzón," in this case the Oyapoque, at latitude 3.8 north of the equator and about 420 kilometers from the Amazon. The treaty gave Brazil both sides of the channel on the lower Amazon and preserved its control over navigation through the estuary. It also limited French commerce of and with natives below the boundary line. Nothing was said about the western limits, since these still were mediated by the Tordesillas line (until 1750) and in principle under Spanish control, and in any case the interior remained largely a cipher.[21] Utrecht was part of negotiations in dynastic conflicts (the War of Spanish Succession) among the European crowns including France, Great Britain, Austria, and the Low Countries as well as Portugal. Predictably, there were problems.

The central problem was, which river was the "Vincente Pinzón"? The Oyapoque? This northern version of the Pinzón—as noted, about 420 kilometers to the north of the Amazon—was the preferred Brazilian boundary. On the other hand, the French argued that Yapoc (the maps have more than twenty different spellings) was a generic term for "island," and the only place with a large enough island to merit such a name was Marajó, the Switzerland-sized island in the Amazon estuary. Through this linguistic manipulation, the magistrate of the French colony, the Marquis of Ferrolle, would argue that the French claim stopped at the banks of Amazon. Others argued that Oyapoque actually referred generically to "river," and there were numerous rivers that paralleled the Amazon. These included the Calçoene, the Caraporis, the Rio Amapá, the Rio Cunani, and most critically the Araguarí at 1.1 latitude, about 120 kilometers north of the main channel, just below Cabo Norte, which technically flowed into the Amazon estuary. Geographical luminaries like the royal cartographer Guillaume Lisle, de la Condamine and even Pedro Teixeira had placed the "Vicente Pinzón" below the Cabo Norte, giving that ubiquitous name to the Araguarí River, what we might take as the "southern version" of the "Vincente Pinzón." Even von Humboldt weighed in, placing the "Vincente Pinzón" at the Calçoene River, which, at latitude 2.3, was more or less the midpoint between rivers we now call the Oyapoque and Araguarí.[22] Quite a bit of terrain was at stake depending on the boundary: some 261,588 square kilometers, an area about half the size of California.

The Rio Araguarí was one of the largest rivers that paralleled the Amazon, so travelers on both the Brazilian and French sides often took it to be the political boundary line between the two powers—as in fact periodically it was. Treaty after treaty followed in an attempt to keep up with the social history that was racing ahead of the stumbling diplomacy, with maps being recopied at a furious pace. These exercises in "geography militant" included French geographers' intentionally inserting false latitudes for some rivers.[23] Who in those backlands knew what documents were being prepared in the European courts? What locals did know was that presence and powers of any European state were tenuous in those watery forests.

The Treaty of Madrid (1750) gave Portugal control over Amazonian terrains to the west of the former Tordesillas line and was mostly concerned with demarcation between Portugal and Spain in the Western Amazon, leaving the interior boundaries between states in its former domain undefined. By introducing the idea of *uti possedetis*, the treaty radically tested the balance of regional powers. Most European occupation hugged the coast, and the ability to penetrate further into the interior was limited by fugitive communities and native settlements. The Madrid Treaty, however, nourished the impulse of Enlightenment leaders in both France and Portugal to enhance and consolidate their territorial presence through what would turn out to be disastrous colonization schemes, as we shall see shortly.

The French Revolution with its abolition of slavery and then the rise of Napoleon in French governing circles also had significant effects in Contestado politics. In the unexecuted Treaty of Paris (1797), the French regime rejected the "northern" Pinzón (the Oyapoque) in favor of one "middle" version of the Pinzón at the Calçoene River. This boundary line, located about 2.3 degrees north, stretched back to the source of the river and continued the line back to the Rio Branco. It would have annexed a huge swath of Brazilian terrain to French territory. The next treaty, that of Badajoz in 1801, stated that the boundary line was the Araguarí, whose boundary would stretch to its source of that river and then back to the Branco. A few months later the Treaty of Spain delimited the Carapanatuba River, a third of a degree north of the equator, which empties slightly above Macapá, as the true boundary. This river also is an Amazon tributary and would have involved the arbitration of navigation rights. The Treaty of Amiens of 1802 brought the boundary line back to the Araguarí. All these treaties lapsed with the Treaty of Fontainebleau (1807), which dismembered Portugal and sent its monarch into exile.[24]

After Napoleon's armies invaded Portugal, forcing the Portuguese king to flee to Brazil, the colony took military action in 1808 to avenge the ignominious exile of its monarch: Brazil captured Cayenne and held the boundary up to Suriname at the Maroni River at 5.7 latitude, giving Brazil control of more than 630 kilometers of the Wild Coast for eight years, also reintroducing slavery.[25] At the Congress of Vienna in 1815, the Prince Regent of Portugal and Brazil reasserted the boundaries of the Treaty of Utrecht, which set the frontier of the two countries at the Oyapoque. Both parties were to name a joint commission to survey the area, but this exercise, as usual, was never carried out. This produced a continuing flux of settlement and armed skirmishes, and indeed several circumstances, like runaway slaves and army deserters (or evaders), were making the region more complicated. In 1841 the two countries agreed to neutralize the area, to abstain from colonization and military exercises until something could be "worked out."[26] The formal treaties and their locations are indicated in table 2, which shows how frequently frontiers of the Contestado shifted and gives a sense of the territory at play. It would take more than fifty years to resolve the final boundaries, during which regional tensions only increased. The constant drafting of treaties was one means of claiming: France and Brazil (and later Belgium) dreamed of formal colonies in the Contestado. And others—those who had been slaves, military deserters, detribalized natives—also had their own dreams of polities in those amorphous, aquatic lands.

The Black Amazon

In eastern Amazonia, African imports and Indian slaves were a highly coerced labor force of more than 90 percent of the population. The number of slaves sold into all the Guianas and Amazonia was not trivial—over half a million Africans (542,000) found themselves in the mangroves, estuaries, forests, *varzeas*, and flooded savannas of the eastern Amazon and the Guianas: 144,000 in Amazonia, 294,000 (almost as many as to the entire United States) in Suriname, 72,000 in British Guyana. French Guiana imported only

Table 2. Migration of French-Brazil boundary lines: treaties, negotiations, events.

Name of treaty, negotiation, event	Date	Boundary	Degrees north	Favored	Other comments
Utrecht	1713	[ambiguous]		France	River names confused
		Japoc	4.14		
		Amapá	2.7		
Madrid/Idelfonso	1777	Araguarí	1.15	France	
Paris	1797	Calçoene	2.3	France	
Badajoz	1801	Araguarí	1.15	France	
Madrid	1801	Carapanatuba	0.10	France	France achieves sovereignty over north bank of Amazon
Amiens	1802	Araguarí	1.15	France	
Brazil invades Cayenne	1808	Oyapoque	4.14	Brazil	Retaliation for French occupation of Portugal
Paris	1814	Araguarí	1.15	France	
Vienna	1815	Oyapoque	4.13	Brazil	Interior boundaries indicated at 2.24 N and longitude 55.39
Neutralization of territory	1841				
Adjudication	1856	Calçoene	2.3	France	Compromise between Araguarí and Oyapoque
Rio de Janeiro	1897				Agreement to seek Swiss arbitration
Final arbitration	1900	Oyapoque	4.14	Brazil	

31,000 slaves, a trickle when compared to almost a million slaves who disembarked in the French Antilles.[27] Recent African arrivals were in general more inclined to flight to *quilombos*, and the activities other than sugar production assigned to slaves—gathering, hunting, ranching, fishing, and rowing—were extremely difficult to supervise. Runaway workers became a chronic problem. Not surprisingly, areas thought to be free zones (like the area north of the Araguarí) became preferred refuges, as well Cunani and Calçoene Rivers farther north.[28]

By the late eighteenth century, the *mocambos do índios*, *mocambos do negros*, and *mocambos dos índios e negros* appeared in the official literatures as slaves, deserters, and the discontented, black, red, and white, fled into the maze of estuary islands, up the *igarapés*, on the banks of hidden lakes, to the headwaters of tributaries like the Trombetas, lurked near plantations, or evaporated into forests of the Cabo Norte.[29] In addition, the triumph of the Saramaka maroons of Suriname in 1762 after decades of revolt produced yet another "nation" pressing against the boundaries of European sovereignty from the interior.

What the Dutch called *negeropstands*—black rebellions—were rife throughout the region: Berbice (1762), Demerara (1772),[30] Cottica, and the Coromantee (1794) uprisings were explosions of more common "everyday forms of resistance." The level of violence in these slave systems were as extreme as the dissipation of their masters, as many observers at the time noted. Survival often required flight.[31] The anxieties of the French about an invasion from the African Boni at the headwaters of the Oyapoque were quite real.[32] Indeed, insurgencies in the Caribbean Amazon were often nourished by the hope that black allies from the free nations of the interior would assist local uprisings.[33] Abolition of slavery in France (1794) made the Contestado lands of singular interest, and reports of the time show large numbers of *quilombos* interacting with towns, plantations, and markets. They were so prevalent near Macapá and along the Araguarí that there was talk of bringing in native militia, the Mundurucu, to roust the interior fugitive settlements.[34] The French and Portuguese crowns had agreed to return runaways to their owners, since flight was becoming a general problem; extradition agreements applied to criminals hunted for crimes like cattle rustling, slaves "stealing" themselves, insurgency, military desertion, and murder, as many fever deaths were blamed on poisoning by African and Amerindian sorcerers, and indeed there was quite a bit of poisoning.[35]

Fugitives to the lands of the Cabo Norte relied heavily on its rich extractive resources—palm products, latexes, nuts, game, fish, and medicinals from the extensive forests and mangroves—amply described by Aublet and a legacy of earlier Amerindian occupants. These biotic treasures were the survival resources for the runaway communities, who learned from natives or, as detribalized natives, already knew how to use these assets that were invisible to outsiders and also brisk items of regional trade. There were agricultural plantations of manioc, sweet potatoes, and rice, usually destroyed by the militias,[36] but the difficulty of eradicating the fugitive communities suggests that they employed a broad range of decentralized supporting resources well beyond agricultural cropping,[37] as well as constant replenishment the *quilombos* by new runaways. Military reports complained bitterly about the how rapidly *quilombos* reconstituted themselves.

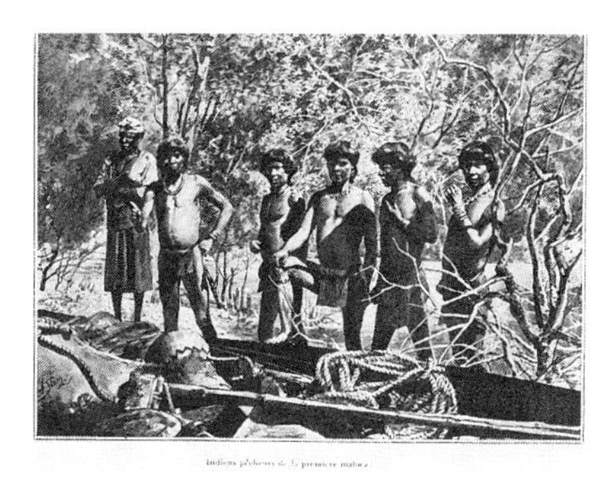

Figure 7.1. Syncretic cultures of the Caribbean Amazon: African and native boatmen.

The syncretic cultural complex extended not just to landscape and lifeways but to lifeblood itself. Studies of blood types comparing phenotypically black populations on the upper Oyapoque with adjacent "isolated" native populations showed blood chemistries more typical of native communities than of African blood types, a result also found in *quilombos* of Amapá, where more than 50 percent of the blood characteristics reflected native contributions, a ratio much higher than elsewhere in Brazil. Significantly, the Duffy marker, a West African genetic mutation for resistance to *vivax* malaria, was also widespread. These hybridized populations were able to withstand malarial ravages, while the white colonists died in droves.[38]

The secret settlements of fugitives, whether black or native slaves or army deserters, followed patterns seen throughout Latin America of slaves fleeing to develop autonomous communities in the interior. Figure 7.1 shows the complex cultural interface of natives and blacks in the depths of the Caribbean Amazon. These communities were successful, many surviving up to the present day. But they were a problem for European colonization and were not seen as a solution or alternative. The European colonial imaginary for the Contestado had its own images of state-led utopias.

The Amazon Experiments of the Enlightenment

The political and economic importance of New World colonies increased markedly in the eighteenth century, and the Amazonian holdings needed consolidation. The French and Portuguese foreign ministers who oversaw the Contestado from the lofty perch of continental courts of the mid-eighteenth century—Étienne François, Duc Choiseul (1719–82), and Sebastião José de Carvalho e Melo, Marquis of Pombal (1699–1785)—seemed almost mirrors of each other. They had home intrigues, but it was the overseas politics that increasingly marked their regimes. Both were unusually influential as Enlightenment politicians in the final period of Europe's absolutist monarchies. They were able but stubborn administrators, interested in managerial and trade reforms. Both were vehemently anti-Jesuit: Pombal suppressed the order in Portugal and its colonies in 1759, while Choiseul eliminated the order in France in 1775.[39] Both had spectacular colonist failures in the Caribbean Amazon.

Both foreign ministers were well aware of the implications of the 1750 Treaty of Madrid for their uncertain borderlands: it eliminated Spanish control of the Amazon interior and emphasized de facto possession. This stimulated Pombal and his half-brother Francisco Mendonça Furtado, magistrate over the Portuguese Amazon, to embark on a frenzy of fortress building at strategic points, including the immense and imposing citadel at Macapá on the north channel entrance of the Amazon. The treaty indicated nothing about "interior" frontiers, and thus these were now up for grabs. Choiseul and Pombal paralleled each other in their views of colonial policy. First, they yearned to make their Amazon peripheries as economically exciting as the French colony of Saint Domingue, the jewel of the "sugar Caribbean," or the Brazilian Northeast. Both areas were generating generous returns for their colonial masters, and understandably, the metropoles hoped to transform the undefined Amazon lands into profitable enterprises. Next, tropical colonization seemed amenable to rational and scientific approaches to these men of the Enlightenment, instead of the ad hoc and messy procedures that had thus far prevailed. Both invested in "acclimatization" gardens, botanical exploration and plant collection.[40]

Choiseul sent a protégé of the Jussieu botanical clan, Jean Baptiste Aublet (1720–1778), a pharmacist and the first formal ethnobotanist of Amazonia to study local economic botany, plant adaptation, and agronomy as a support system not only for local colonials but also for the Antilles.[41] As with Louisiana, the idea was to use mainland Guiana as subsistence support for the productive sugar islands, rather than importing preserved beef from Ireland or wheat from France. In addition, French royal botanist and biopirate (and archrival of Aublet) Henri Poivre sought to use the region to acclimatize various plants of the spice trade and thus to break the Dutch monopoly on precious essences. To this end he secretly "acquired" trees of cloves, cinnamon, nutmeg, mangosteen, lychee, and candlenut to develop high-value exports in addition to sugarcane. These were planted out on the grounds of the governor's mansion, La Gabrielle.[42] Aublet, an unsung Amazon hero of New World natural history, produced foundational taxonomic works in ethnobotany and even introduced de la Condamine to *Hevea* latex. Aublet himself was taken with another kind of latex, *Parahancornia amapá*. This tree, known as *amapá* (which gives the Brazilian state its name), produces a latex used as milk, as well as a rumored antisyphilitic on which Aublet hoped to make his fortune. Aublet died shortly after returning to Paris, however, and his collections were acquired by both Jean-Jacques Rousseau and Kew Gardens.[43] Aublet's live collections eventually were spirited to the botanical garden in Belém.

Both Choiseul and Pombal sought to develop the labor mix that would transform their torpid Amazon colonies into economic dynamos through a better labor configuration or a better type of laborer. Choiseul had eliminated slavery in France but had little interest in doing so in the colonies, yet he was concerned about the loyalty of slaves in international conflicts, no doubt partly because of the persistent rebel wars between black fugitives and Dutch militia in Suriname, just to the north of his colony, and the regularity of insurgencies throughout the region. The Caribbean Amazon was very unsettled, with the Dutch to the north, the Portuguese to the south, and fugitive polities of "Revolted Negroes" and "Palenque Nations," as the maps had it, in the interior. Choiseul had lost Canada to the British, and it was that victory that inclined him toward white colonists as a military strategy for protection against incursions into the Antilles. In Guiana he hoped for a tropical version of the temperate-zone model of yeoman colonization *à la canadienne*. Colonists would provide not only security but the opportunity for an Enlightenment community on the balmy shores of the equatorial Atlantic. Choiseul and his agents were maintaining a useful network of spies throughout the region, and his interests in colonization were both military and economic.[44] Choiseul and Pombal were both interested in white colonization for strategic reasons, due to the requirements of *uti possedetis* that inhered in the Madrid Treaty (whites would have a "reliable" nationality, unlike Indians or slaves) and concern whether slaves (who were chattel, not citizens) would be suitably loyal in this realm of uncertain sovereignty. Choiseul, who brooded over his loss of Canada, took his defeat there as an object lesson and embarked on what was to become a disastrous colonial venture.

The Alsatian Amazon, or Colonizing in the Land of Candide[45]

Choiseul had been very impressed by a handbook written by a French Guiana planter named Brûletout de Préfontaine. A precursor to what would become an influential colonial genre during the nineteenth century, Préfontaine's book extolled the individualistic agrarian frontier life and the rich soils and natural resources of French Guiana in his *Maison rustique a l'usage des habitans de la partie de la France equinoxial*. His 1763 book was so enthusiastically embraced in Paris that Choiseul believed that Guiana might yet become the ideal yeoman colony. Further, as Choiseul noted, "peopled with whites, the Guyane, well placed on the winds for the French islands of the Gulf of Mexico, is favorably situated to provide them with assistance and even to act offensively against the British islands."[46] To this end, Choiseul made plans for a large white settler colony on the Kourou River, near Cayenne. While many tropical hands, including Governor d'Orvilliers, counseled against this (and resigned over the program), Choiseul designated Étienne François Turgot to organize the expedition, and soon extravagant promises of tropical agrarian utopia and two-year stipends had inflamed the imaginations of large numbers of marginalized French provincials, mostly rustic blonds from the Alsace and Rhineland.[47] The future governor Victor Pierre Malouet described the motley bunch this way: "It was a deplorable spectacle, even for one of my experience, to see this crowd of imbeciles, among whom, in addition to agricultural workers, were merchants, families of artisans, a crowd of civil and military servants and finally a troop of clowns and musicians who were to provide amusement for the new colony."[48] In 1764–65 some twelve to fourteen thousand hapless Alsatian colonists would find themselves in the coastal lowlands in the rainy season with no colonial preparation on site, only rotted provisions, and soon they began to die like flies, albeit in the ghoulish manner of Edgar Allan Poe's *Masque of the Red Death*. Absent productive activities for colonists, and "in order to prevent melancholy"—a malady at least as likely to undermine the colony as malaria, so it was thought—the administrator, Jean Baptiste Mathieu Chanvalon, encouraged a lively round of parties, intrigues, and distractions (clowns and musicians did come in handy). These frivolities, however fun, did little to check the malaria and yellow fever, the problems of the putrid provisions, and the insalubrious life in swampy tents during the rainy season. Few colonists ever even left the squalid settlement of Kourou to take up the cultivation of their lands.[49] While estimates vary, of the twelve to fourteen thousand settlers at least ten thousand died quite quickly.[50] Of the remainder, about nine hundred stayed, impoverished and ill, and a few hundred returned to France. The episode established this part of Amazonia as a white man's tomb and reinforced slavery as the model for tropical colonization, and with it the cultures of slave flight, everyday forms of resistance, and rebellion.[51]

Yet other enterprises were discussed. In 1776 the Baron de Bessner, inspired by the successes of the Jesuits in Paraguay, suggested that the ecclesiastics (banned from French and Portuguese holdings) could be transferred back from Europe to create new colonies by rounding up 100,000 Indians and 20,000 Bush Negroes from their sanctuaries in the interior, thus reviving the Jesuitic utopia of Paraguay in the Wild Coast.[52] This scheme was mercifully

Figure 7.2. The banished in Kourou. Some are burying comrades, while others dance with the Mulatresses.

nipped in the bud, but it at least recognized the viability of the hybrid black and native settlements in the interior.

Morocco in Amapá

Pombal was aware of the French failure but had a different approach to the labor question, one not based on white yeoman colonization but thoroughly rooted in tropical laborers. He viewed the stagnation of the Amazon colony as the outcome of three central problems: first, the monopolization of native workers by ecclesiastics; next, the lack of slaves (and credit to buy them) so that even with a perfect climatic conditions, sugar cane and tree crops like cacau languished for lack of hands; finally, white colonists and overseers' lack of familiarity with frontier conditions and slavery.[53]

Pombal tended toward an activist state that would subsidize the purchase of slaves via the charter company of Grão Pará and Maranhão, so that agroindustries and infrastructure would be supplied with the needed workers. The Directorates that he created would banish the religious orders and replace ecclesiastic missions with a secular system. Indians who had toiled on clerical lands were to become administrated wage laborers for rent to settlers too poor to purchase slaves. "Directors" of mission villages—now labor depots—would receive a generous 17 percent of the value of the goods produced by their wards, while assuring the crown its mandatory 10 percent. These two means—ramped-up slavery and Indian rental—would, in Pombal's view, loosen the regional labor constraint. Natives from the Directorates were also assigned onerous jobs in civil construction, woodcutting expeditions, rowing, and state provisioning. This policy had its local commentary when the native rowers, transporting the regional magistrate, Francisco Mendonça Furtado (half-brother to Pombal), near Belém, jumped overboard, swimming to shore and abandoning him to estuary waters (he barely survived).[54] The arduous nature of these tasks and the general abusive treatment of the natives, plus the fear of disease, often led to Indians to desert Directorate villages and flee to the interior, sometimes into the multiethnic native *quilombos* or *mocambos*.[55]

Pombal's zeal to secularize the mission system not only reflected his vehement anticlerical sentiments but also fed a strategy to decisively transform the widespread ecclesiastic settlements (technically under Vatican sovereignty) into communities that were formally Portuguese, and thus evidence of de facto (and Luso) occupation. Pombal and Mendonça Furtado's correspondence on questions of settlement reveals just how anti-Jesuit anxieties dovetailed with Pombal's imperial territorial necessities to show de facto occupation.[56] Following the Portuguese model in these matters, Pombal also urged intermarriage in the colonies, the "portuguesation" of place-names, and the end of the *lingua franca* in favor of Portuguese as a means of creating a Luso colonial identity. (The *lingua franca* was, if anything, a pan-national/ethnic creole language, used throughout all of Amazonia by the religious orders and not limited to Portuguese domains.) Pombal's polices were supposed to reinforce Portuguese presence in the face of the requirements of *uti possedetis*.[57] French policy in Guiana, on the other hand, favored racial segregation, eagerly awaited the white settlers who were at always at the heart of its colonial policy, was less insistent on language, was still very Catholic, and was oriented in its race relations by the far less assimilationist Code Noir. Pombal's approach had enormous impacts on the native population of Brazil, largely because indigenes in Directorate villages were basically worked to death and exposed to diseases, which contributed to the processes of cultural dissolution of native traditions and languages. This approach also created a class of proletarians and a peasantry, some of whom would eventually end up in Belém and help fuel the Cabanagem revolt.[58]

Pombal had general policies, but he also had in mind a specific model of a sort of colonist utopia: he sought to transfer the frontier town of Mazagão, a Portuguese outpost in Ceuta, Morocco, to Amapá.[59] The colony would be a slavocratic military outpost that if successful might define a new form of occupation. Pombal was certainly aware of the catastrophic Kourou experiment, but in his colonial exercise the state would intervene to smooth out the transition while the colony, accustomed to remote militarized border conditions, embarked on sugar and rice production with its own slaves. Transporting the entire community of some twelve hundred citizens with ample state and urban planning to their new tropical home seemed the epitome of rationality, especially when compared the deadly chaos of Choiseul's colonization. But various problems ensued, not the least being disease, lack of experience with floodplain agriculture and forestry, and the adaptation issues inherent in the differences between a life in arid Morocco and one on the equator with upwards of three meters of rain per year. The predictable catastrophe ensued, with most of the whites dying in fevered penury, as the black slaves slipped away into nearby *mocambos* or created their own, turning Mazagão, meant to be a Portuguese frontier outpost, into a black and mixed-blood polity.[60]

The problem was that Europeans without immunities simply died from multiple bouts of fevers, usually *vivax* malaria but also yellow fever. The continuing mortality made it seem that the only viable colonization had to be based on Africanized (or African) laborers with their sickle cell and Duffy resistance to both *falciprum* and *vivax* malaria. Indeed, the most successful small farmer colonizations were the autonomous runaway settlements and the new ethnically complex indigenous communities that persisted in the shadow of the imperial colonizing efforts. Knowledge of the native pharmacopeia also helped. Quassia bark, named after a maroon medicine man, was one of these,[61] but the ethnopharmacology of the region seems to have been rich in fever remedies.[62]

Both as symbol and as concrete alternatives, the maroon communities were but one inspiration for revolt. Throughout the Caribbean, in Jamaica as well as Suriname, independent maroon polities had made deals with imperial powers that respected their autonomy as independent states. The invention of freedom, as historian João José Reis has put it, had ceased to be an abstract idea as the region entered the age of revolution.

Revolutions and the Wild Coast

In the Caribbean Amazon, the realities of worlds of autonomous *quilombo* and indigenous communities influenced ideas about political possibilities.[63] The violence that inhered in the extensive slave communities of the Caribbean and Brazil was constantly stimulating flight. These refugees took advantage of regional macropolitics. The ambiguous boundaries, shifting European sovereignties, and creation of neutralized zones produced an open political "commons" where both the terrain and the diplomatic politics favored confusion and jurisdictional paralysis, which provided excellent conditions for runaways, military fugitives, and refugees from, or instigators of, revolutions. The colonies were also transformed by formal revolutionary ideologies from Europe, elsewhere in the revolutionary Caribbean, and even France's deported revolutionaries—those moderates of the 18 Fructidor uprising who were meting out their last fevered days in tropical banishment.[64]

The slave economies of the Eastern Amazon were integrated into Caribbean circuits of goods, people, and ideas through winds and currents, as well as the clandestine flow of knowledge and gossip in the subaltern worlds of slavery and resistance to it.[65] Local revolts were stimulated in different ways by the autonomous black polities of Suriname, the French and Saint Domingue revolutions, as well as the rise of Bolivarian abolitionist politics that contributed to "almost revolutions" and decolonization throughout the Spanish Amazon. Indeed, from the Orinoco to the Amazon channel, the Caribbean Amazon might well be understood with its maroon states, *quilombos*, recalcitrant natives, renegades, *bandoleros*, and disenfranchised Creoles—a place in fairly constant rebellion throughout the late eighteenth century—as the one of the most revolutionary places in the hemisphere. Slave uprisings and runaways were reliable features of slavocracies and nourished continual low-intensity warfare, As the eighteenth century wore on, there were new and palpable threats. News of revolution and the rights of man flew through the New World. In 1790 French Guiana, like Saint Domingue and Guadeloupe, had an uprising. The rebellion, a small affair, occurred in the Approugue Valley. Famously, on hearing that the French revolutionary pronouncements of the rights of men declared them all equal, 162 slaves went to their master saying they had been declared free in France and wanted to take advantage of their liberty. Similar events occurred elsewhere.[66] Some rebels were captured and killed by a black militia whose members were promised their freedom in exchange for repressing that of others. But what it meant to be a *citoyen* could not simply be met with repression in the Amazon, since workers could withdraw into the forests or the maroon world. This separate *quilombo* reality, along with the limited institutional and military capacities of the government, probably defused the more radical revolutionary impulses: one needn't overthrow the existing state; one could construct a new polity away from it. Simple labor withdrawal was widespread in the Caribbean Amazon.[67] On news of the revolution in French Guiana, Brazil sent five ships to the mouth of the Approugue River to avert any "ideological contamination" by the revolutionary state. Von Humboldt was denied entrance into Portuguese Amazonia from his sojourns on the Orinoco partly because of his revolutionary sympathies and the more general concern about espionage.[68]

The attempted revolution in French Guiana was of only minimal interest in the métropole because the colony with its 10,000 blacks, some Galabi Indians, and 1,300 whites had little economic importance and this uprising was nothing compared to the explosive situations elsewhere in the Antilles. The insurrection was not entirely devoid of bloodshed, embellished as it was with the torture and grisly killings of the ringleaders, but it was hardly the violence-saturated revolt of Saint Domingue or Guadeloupe, and the formal complaints about the revolt tended to grouse about black insolence and indolence rather than murderousness.[69] The revolt was expressed more through everyday resistance—labor withdrawal from slavery into small-scale farming and local exchange networks of products like cacau. Indeed, colonial officials could fume and rage all they liked that no one was picking the cotton or cooking the cane, but there was little they could do.

There were other slave uprisings on the edge of the Caribbean: the Coro rebellion on the Venezuelan coast, whose instigator had spent time in Curaçāo, had been a leader of insurgent runaways from that island, and knew the Haitian rebel leaders. There was the *bandolero* resistance from the cattle estates and cacau groves in eastern Venezuela.[70] Elsewhere in Brazil, especially in slave-dense Bahia, rebellion unfolded constantly.

In 1794 France formally put an end to slavery in its colonies, at least for a time, but that did not mean it wanted its colonies to be unprofitable . French Jacobins sent out a taskmaster to oversee French Guiana—M. Cointet (a nephew of the revolutionary Danton), who yearned to abolish access to black subsistence plots and get the *citoyens* back to work on export crops as part of their patriotic duty. Cointet spent two years at his tropical post, but the "peaceful revolution" seemed to only infuriate him. If *citoyens* could not count on freedom through revolutionary means, it could be achieved through flight—a stark difference between the islands and the mainland.[71]

The French revolutionary state began to rethink its relation to its Wild

Coast outpost, recasting it as a penal colony. More than 190 "unsound" revolutionary deputies, along with a few hundred other undesirables (royalists, priests, and a few regular convicts) were exiled to French Guiana in 1797, after the coup of 18 Fructidor. Death from recurrent malarial fevers associated with the tropical prison were so certain that the place became known as the Dry Guillotine (la Guillotine Sèche).[72] The memoirs of Fructidor exiles reported mortalities of more than 90 percent in the communities where they passed their final days. They complained bitterly of the complete indifference of the local black population to entreaties for labor. The colonial administrator apparently had told the local population that the exiles were royalists in favor of slavery, so perhaps the ex-slaves' general diffidence was especially pronounced where the banished were concerned.[73]

French Guiana remained the preferred venue for incarcerating political prisoners until well into the twentieth century. The 1848 revolutionaries—more than twenty thousand of them—also found their final resting places in the feverish coast.[74] The colony was among the largest and most important of the nineteenth-century penal colonies and perhaps was the inspiration for Franz Kafka's famous short story "In the Penal Colony." When steam engines came into widespread use in the mid-nineteenth century, criminals formerly assigned to prison galleys of the Mediterranean were sent to the tropical gulag as a form of colonialism on the cheap, as in Australia. Prisoner/colonists became chain gangs of colonial agricultural laborers, but unlike Australia's salubrious colony, Guiana was lethal: virtually all its eighty thousand inmates are lost to history.[75]

The end of slavery in French Guiana, the winds of revolution, and the presence of independent autonomous communities in the interior of the northern Amazon had extraordinary impacts on slaveholding in northern Brazil. News of revolutions was carried by boatmen, slaves themselves, gossip, and all the routes through which information flowed. Abolition in French Guiana galvanized slaves along the Amazon.[76] The lands of the Cabo Norte had become a "transnational" space in the sense of a no-man's-land, or at least *sem rei* and *sem lei* (without king or formal law). Many fleeing slaves moved up the Araguarí, while the western boundaries of French Guiana were still undetermined and possibly extended to the Rio Negro, so it was reasonable to flee past the cataracts to the headwaters of rivers like the Trombetas to the forests and savannas of the Guiana shield, where interaction with the maroon polities was possible.[77]

French Guianan slave owners, unhappy with this turn of affairs, sought refuge in Pará, many taking up residence in Bragança, to the east of Belém, and some even into the prime sugar (and slave) zones of the Mojú River.[78]

Other things moved besides people. Two of the Guianan migrants, Michel Grenouiller and Jacques Sohut, had been valued botanists at La Gabrielle and eventually were contracted to direct the newly created botanical garden in Belém. They managed to smuggle out clandestine plant matter from La Gabrielle to plant and test in Belém, including the important cane cultivar Tahiti (now known as Cayenne), which came to dominate Brazil's sugar production.[79]

Freedom's Short Reprieve

Slavery was reintroduced into French Guiana when Napoleon rescinded abolition in 1804.[80] By then the Contestado was populated with numerous free refuge settlements. This state of affairs changed rapidly in response to the exile of the Portuguese crown to Brazil. French Guiana was captured in a joint British-Portuguese assault and the entire region, from the Maroni River to the Amazon, was held from 1809 to 1817; slavery was reintroduced by Brazil, and consolidated just as its hold had been loosening. An escape hatch had crashed shut.

The Caribbean revolutions did, however have effects in Brazil. As historians João José Reis and Flávio dos Santos Gomes have noted, insurgents of the many nineteenth-century Brazilian revolts could be found wearing medals with Dessaline, the revolutionary governor of Saint Domingue, around their necks, and the revolution in Saint Domingue seemed to trigger the "Tailors' Revolt" in Bahia. Radical revolutionary proselytizers wandered through the Contestado, and there was general landowner anxiety about slave uprisings.[81] The strongest impact of these revolutions and emerging free settlements may have been in the Amazon itself, and in the "almost" revolution of the Cabanagem (1835–40), an uprising that resulted in the death of about a fifth of the Amazon population, in which rebels did take hold of the state apparatus, echoing, albeit abortively, the decolonial movements of the Haitian revolutionary Toussaint and those of Bolívar elsewhere in South America.

The Cabanagem Uprising

The Cabanagem revolt reflected schisms within elites who were positioned uneasily above a vast laboring class of detribalized Indians, free whites, free persons of color, Brazilian- and African-born rural and urban slaves who had plenty to be resentful about, especially at a time when the liberation movements had already resulted in abolition in the Spanish empire and in the Caribbean, and slave-free lands of the Cabo Norte existing just beyond Brazil's boundaries. Local Amazonian elites wanted regional presidents (or governors) chosen from members of their society rather than some meddling deputy selected by the recently exiled Portuguese Corte in Rio. This was an issue of particular importance for the prideful Amazonian ruling class, since it had been a separate colony and not an adjunct of the south until the Portu guese monarch had fled.

Local Amazonian commercial interests and resentful politicos had had considerable independence and fomented anti-Portuguese sentiment that exploded in 1835 with the murder of some twenty Portuguese merchants and businessmen. As happened on many occasions during the early nineteenth century, what began as claims for local sovereignty by local grandees soon morphed into deeper rebellions and quests for regional autonomy. Many of the elite Amazonian revolutionaries changed their surnames from the Portuguese to Amazonian toponyms like Tocantins, Tapajos, Marajó to make their political alliances and regional identities obvious. The imperial representative, cowed by the early dissidence, offered amnesty to the local elite insurgents and was able to be sworn in as governor, but the rebels soon regrouped and took over the city of Belém.[82]

Within this rivalry between traditional powers and governors selected by the crown, there were inevitably, given the zeitgeist, calls for the abolition of slavery. As the rebellion moved down the ranks, into the streets, and up the rivers, increasingly there were calls for social revolution.[83] Subsequent uprisings soon took over plantations and ranches. Marajó, the great island ranching economy, was captured by its slaves. As insurgency expanded out from Belém and into Amazonian tributaries, not only were scores being settled, but the violence took on dimensions of intertribal warfare as well, as "colonial tribes" like the Mundurukú faced off the relentlessly insurgent Mura. Amid the uproar, to evade massacres, to flee slavery or incarceration, many fled up to the headwaters of Amazon tributaries and into the ambiguous sovereignty and anonymity of the Cabo Norte.[84] Alfred Russel Wallace, like other later travelers, would report that many of the black boatmen he employed had had important roles in the failed revolution.[85]

As slaves fled from the lands of one colonial power to those of another, extradition patrols crossed into each other's territories, often snagging other slaves in addition to their own fugitives, or, in the case of *quilombos*, rounding up entire populations and dragging them back across frontiers. These forays predictably had the effect of maintaining intercolonial tensions at a high pitch. "Property" anxieties were compounded by deeper political worries: labor withdrawal, autonomous communities, rebellion, and revolution made the Cabo Norte zone very unsettled, a realm of seething low-intensity class, abolitionist, and international conflict. In 1841, in response to the continuing military incursions and the ambiguity of sovereignty on both sides, the Contestado was "neutralized."

In 1848, as an outcome of revolutionary movements in Europe, slavery was once again abolished in French Guiana, again stimulating waves of Brazilian runaways fleeing to the lands to the north of the Amazon. This prospect was made more interesting in 1855 by the discovery of gold on the Approugue River by a Brazilian fugitive from Ouro Preto in Brazil's mining state of Minas Gerais, who comes down through history known only as Paulinho.[86]

Amazon Klondike and Cunani: The Utopia from Below

Dreams of El Dorado had gilded the very first imaginings of this part of the Amazon by Walter Raleigh. Now the possibility of instant wealth attracted thousands of people into the Contestado, sucking labor from local agriculture, attracting hands from elsewhere in the Antilles—Guadeloupe, Martinique, Suriname (the Saramaka—from the black polity on that river—came as panners but mainly as canoeists and river traders),[87] Brazil, and the United States—as well as coolie labor brought in to substitute for slaves. The gold rush engaged everyone from small-scale clerks to the speculators and financiers on the Parisian Bourse.[88] Mining involved formal and informal economies that operated more or less in tandem. These mines were fantastically rich: from 1857 to 1861 the Cayenne registered more than 2,600 kilograms per year of gold, and in the 1890s Guiana's exports hovered between 2,600 and 3,000 kilos annually.[89] Strikes occurred all through the Guiana headwaters, and the region also had considerable amounts of rubber, which it also began to transact.

The "internationalization" of the neutralized zone was becoming more pronounced, and in the midst of the deadly serious business of territorial control, what seemed to be an odd filibuster unfolded on the Rio Cunani, where residents declared themselves an independent republic. The Rio Cunani was part of the Calçoene district, where gold was discovered by an escaped African slave, Clement Tamba. Trajano Benitez, a small local trader who had been a runaway slave from the cacau district of Cametá on the main Amazon channel, proclaimed the "independent Guianan Republic of Cunani" in 1885. The "republic" lasted until 1891. The Cunani settlement was basically a *quilombo* that sought to legitimate itself as a state with international recognition. With just six hundred people scattered along the river, whom Arthur Reis described as "runaway slaves, criminals, deserters who had contact with a few river traders, the *regatões*, the little town claimed the entire demilitarized

Figure 7.3. Striking it rich: rubber barons were not alone in their taste for gramophones and opera. Note also the African stool.

zone from the Oiyapoc to the Araguary."[90] French geographer Jules Gros was named Cunani's president, while his colleague and fellow explorer M. Guignes became his chief.

While today such tropical "utopias" have an air of absurdity, this practice of filibustering was well honed by the end of the nineteenth century and had its adepts in icons like Sam Houston in the United States and, as we will see later, Luis Galvão in Acre, so its strategy as a means of annexation by a more powerful state was novel but not unheard of. The map of the self-proclaimed republic covered most of the demilitarized zone, a highly distressing bit of cartography from the Brazilian point of view, especially because the "republic" wanted to base itself on French legislation and the national language would be French. As described by their colonial advocate, geographer Henri Coudreau, the populace was munching baguettes and drinking wine, and the charming village squares reflected Gallic taste, not to mention the absence of slavery (Brazil wouldn't abolish slavery for another three years). The Cunani flag displayed the French tricolor in the upper left in a field of forest green. The republic had its own stamps and currency. The Cunanians hoped to be absorbed into the French overseas system or else preferred their complete independence.[91]

A French *Géographe* Militant

Even more upsetting, at ideological and diplomatic levels, were the reconnaissance expeditions of Henri Coudreau (1859–99). Coudreau was a explorer of a decidedly romantic bent, one of the cadres of "geography militant" adventurers who engaged in general colonial resource reconnaissance, geographical survey, basic linguistics (a kind of explorers' "Berlitz" set of travel phrases), and quick colonial anthropology. He strove to equal the mileage and distinction of the travels of another French geographer, Jules Crevaux, who eventually met his end in the Gran Chaco of Bolivia/Paraguay, eaten by Toba Indians, at least according to other members of that expedition.[92] Coudreau traveled widely through the eastern part of the Amazon basin (with Crevaux's guide, a local maroon named Apatú) producing the usual "Travels to . . ." colonial adventure books, a durable genre that crammed bookshops and best-seller lists of the late nineteenth century (and still does today). Coudreau spent more time than most of such travelers among native populations[93] and was very taken with indigenous cultures.

Impressed by Coudreau's extensive local knowledge, the governor of Guiana employed him to reconnoiter the Contestado from the Oiyapoc to the Amapazinho River. His books *Les français en Amazonie* and *France equinoxiale* described local natural resources and regional history, less infused with French colonial triumphalism than with descriptions of how local former slaves throughout the region had received various Brazilian expeditions "à coups de fusil"—with rifle fire and resentment. Coudreau was exceedingly optimistic about the prospects for French colonialism, most especially in the interior grasslands, the "prairies of the Contesté," where the "recidivists" then installed on Devil's Island could become colonists of the interior. Rather

Figure 7.4. Henri Coudreau: intrepid explorer and French colonialist.

than send these "pariahs to what is in fact a slaughterhouse" in the lethal coastal climate, Coudreau thought that France should transport them to an agricultural colony in the interior, which would satisfy both humanitarian and economic needs *à l'Australien*. Further, "the French presence at the mouth of the Amazon would not be without interest to *La Patrie*." Coudreau's use of the term for temperate zone grasslands, *prairie*, was evocative of the salubrious, successful North American colonization of the Great Plains rather than the more accurate tropical term *savane* with its overtones of the malarial grasslands of doom and pestilent mangroves of the coast. Coudreau helpfully proposed that immigration companies then moving millions from Europe to North America and southern Brazil might be hired to assist in transporting new colonials to the (expanding) French Amazon.[94]

By invoking the principle of *uti possedetis*, and the flight from slavery, Coudreau coupled his colonialism to utopian Cunani, a maroon society, a place that could be one of the "schools for a future socialism" complete with the charms of a recovered Eden: "Cunani, noble river on whose banks Brazilian slaves in flight from their masters came to find their freedom." Here in this village would rise "the city of the future."[95]

Coudreau's point was that the Cunanians "never liked Brazil (this is explained by their previous status as slaves) and are very much in favor of France. Their republic *avant la lettre* has existed for twenty-five years, they have sent petitions to the French government four times requesting administrators. In the meantime, England has tried to establish a naval station there, and the US sent a war ship to study the hydrology. . . . Portugal sent emissaries there that tried to set up a principality that would be secretly supported by the house of Bragança. But the Cunanians want either to be French or to be free. . . . One more Republic is on the World Map!"[96]

Coudreau continued his exposition of regional development by focusing on polytechnic schools embracing both native knowledge and European practical sciences. Ahead of his time, Coudreau viewed the region as especially important for extractive goods of rubber, cacau, and Brazil nuts; in short, he viewed Cunani as an economic powerhouse, not even counting the gold in the alluvium. Coudreau ended his treatise with "The current leaders and all the population are disposed to attach Cunani to France." His final comments extol with high praises the graces, caresses, and kindness of Indian women. Speaking of the broader existential exhaustion of his voyages, he yearned to live in the depths of the wilderness with two or three "Femmes Indiennes."[97] Perhaps unsurprisingly, his wife Odile began to travel with him and was there when he contracted fever and died in 1899 on the Trombetas River, a place famous both for its malaria and for its maroon settlements. Odile Coudreau later became a celebrated *exploratrice* in her own right and continued her expeditions until 1903.

Coudreau had carried out reconnaissance for governor of Pará (and da Cunha's schoolmate) Lauro Sodré on the Amazon tributaries of the Xingu and the Tocantins-Araguaia in the 1880s and had mapped much of the estuary, along with the rivers of Marajó Island and its autonomous communities. It would be difficult to imagine anyone in the European world who knew the Contestado better, and he, a quite glamorous figure, was famous in European geographic and political circles. But the pro-France positions and extensive colonization programs Coudreau was advocating for the Contestado were extremely worrisome for Brazil, as was the gold rush. The huge number of Francophone migrants—the population of the Calçoene alone had exploded by more than ten thousand people—was a clear threat to Brazil's purported sovereignty.[98] The doctrine of *uti possedetis* that had worked so effectively for Brazil in other boundary disputes might backfire in this case.

"All the rivers seem to be auriferous"

With more gold strikes, the bucolic days of the Cunani were soon over. Gold seemed to be everywhere: nugget-bearing rivers that transected the Contestado were suddenly lined with commercial houses, and *regatões*, small commercial boats, plied the rivers with trinkets and necessities, while the region seemed only to belch up gold in every side stream. The Calçoene region, however, was a becoming a flashpoint: French vessels increasingly plied the river, and British boats loaded with foodstuffs used for ballast and traded for rubber and gold dust. In 1895, an Englishman and an American, Mr. Hargreaves and Mr. Cerdemn, resolved to start an agricultural colony with seventy men transported on the *Gazelle* (they were turned back).[99] Both Brazil and French Guiana were up in arms, but what was clear is that bit by bit the region as a whole was becoming "frenchified." The Calçoene, once home to Brazilian fugitives, now teemed with populations from every French colony and indentured labor from Asia. The entire Contestado seemed to be changing from a place of Afro-Brazilian refuge and rebellion into an outpost of the French *métropole*. Adventurers (and adventuresses) from all over flocked to the new Amazon Klondike. A French Guianan regional magistrate, Eugenio Voissien, was placed in charge of Calçoene, as most of its gold went out through Cayenne, and he began to prohibit the entrance of Brazilian mining in this zone.[100]

The Brazilian response was to create an administrative triumvirate headed up by Viega Cabral, Gonçalves Tocantins, and Antonio Coelho on the Amapá River, essentially in defiance of the neutralization orders. They then placed

Figure 7.5. Women of the Golden Placers.

a customs station on the Calçoene River, where they taxed materials coming from Cayenne at 10 percent. The response in French Guiana was to appoint another magistrate of their choosing, Trajano Gonzales, who was, after all, a black former Brazilian, to administer the French interests on the Calçoene, and indeed gave him the title Capitão-Governador do Amapá. This title infuriated the triumvirate, who took Trajano prisoner. This news was accompanied by insinuations that Cabral had been bilking miners and embargoing entrance of the French into the Calçoene. This spurred the governor of Guiana into action. He sent Captain Lunier and a crew of soldiers on the warship *Bengali* to investigate the matter. When they arrived on the Amapá River, a pitched gun battle ensued, Lunier and four of his men died, fourteen Brazilians were killed, and the local town was sacked and burned. The survivors loaded the *Bengali* and steamed back toward Cayenne. This bit of Amazonian chaos—a French military incursion into what was clearly Brazilian territory, capture of a French magistrate, death of a military captain, sacking of a city, and so on—made the volatility of the situation clear.[101]

Faced with this distressing configuration, in 1893 the foreign minister of Brazil, Carlos de Carvalho, sent two powerful secret agents into what was certain to be an acrimonious fray. Rio Branco, still a diplomatic functionary based in Europe, was deployed to the French archives to excavate the region's geographic and diplomatic history. He was preparing for what Coudreau called—apparently without irony—"a savage tournament of historical geography." The other strategy was to send a master of the tropics to review the place. Swiss zoologist Emilio Goeldi, the director of the Pará Museum and an eminent scientist in European circles (he'd been a protégé of Ernst Haeckel), was sent into the Contestado and especially to the Cunani under the mantle of scientific collection.[102] Coudreau was prominent in European scientific circles, and his assertions needed a counterweight.

The intersection between science and espionage in Amazonia surely has no better exemplar than the Amapá expedition of Goeldi. In fact his archaeological finds in 1895 of caves filled with gorgeous funerary urns, true masterpieces, was one result, and these can still be seen in Belém's Museu Goeldi today. But it would appear that there was another object of this foray: a devastating colonial report.

Goeldi's report stated the following:

> The Contestado territory is, and I can affirm this in the most positive way, inhabited in most of its area by Brazilians. Without exception Brazilians occupy the area from the Amapá River to the south, as they do the Cunani, the Oiyapoc. The language used is Portuguese, the way of life, the customs, the manners, all this is just like Pará because they are almost all Paraense. In those areas there are maybe a half-dozen foreigners, of these a few are creoles from Cayenne. The only place where this is inverted is on the Rio Calçoene, where creoles from Cayenne, Martinique and Guadeloupe, and from southern France are decidedly the prepon-

Figure 7.6. Emilio Goeldi: melding scientific exploration with espionage.

derant number. France maintains constant relations with this river directly and via Cayenne and Martinique.

But in the spirit of any arbitration, it is not enough to know whether the people living in the Contestado are of Brazilian origin. What is equally or perhaps more important is whether they *want* to be Brazilians. In this I can affirm because what I saw and convinced me is that they in fact want to belong to Brazil and not to France. What Mr. Henri Coudreau wrote about their sympathies for France were the grossest lies. Many people in Cunani were indignant when I read them sections of the book by Coudreau. . . .

> As to the value of the Contestado, I don't want to underplay my firm conviction that the lower areas (Cabo Norte and Amapá) will not be worthwhile for a long time. On the other hand, the northern regions are beautiful and worthy of discussion from the Cunani to the Oiyapoc. It's not for nothing that France was content to take the northern part from the Calçoene area and above. That way France would keep the best section and we would have the worst. Cutting this gordian knot in half would absolutely not be convenient for Brazil. The division should be *everything or nothing*. If the Contestado is limited to only the Amapá part, it is not worth the trouble and time to fight over it. But the north of the Contestado is as good as the southern part is bad: litigation is completely justified and Brazil should look after its legitimate rights.[103]

Goeldi clearly had his terms of reference, to explore who occupied the region and what citizenship they preferred. This question goes to the deepest irony in these proceedings: formally many of the runaways and ex-slaves would not have "citizenship" in Brazil, since criminals and military deserters would have no rights, so it is difficult to imagine the wholehearted desire of these refugees to be Brazilian. Other issues in Goeldi's brief were the charges raised by the French about what transpired in various incidents: Had the Brazilians prohibited the French on the upper reaches of the Calçoene? Goeldi argued no, it was a generalized prohibition. He noted that Brazil should "pay attention to the north": "The proximity of creoles in the Calçoene so close to the Brazilian agent Cabral in Amapá would be a constant source of complications. . . . In a straight line the distance between the two sections is of a few hours by land, by boat the distance is less than a day's travel. A certain Lourenço Beixamar of Marajo, a person who very much sympathizes with the Cayenne Creoles . . . has opened a rough road between the Calçoene to the Amapá River, with the obvious goal of preparing a route to attack Cabral from behind."[104]

The region was about to explode. The little skirmishes were playing out on top of a huge economic surge, not just of gold but of rubber. King Leopold in 1884 had thought the Contestado might be a useful Belgian colony.[105]

Mocambos, Maps, and Amapá

In 1898, France and Brazil agreed to arbitration by the Swiss government. Goeldi, a Swiss citizen who had in fact suggested the Swiss Federative Council as mediators, became both an adviser to the state of Pará and an informal consultant to the Swiss arbitration judges, after he had left Brazil for Switzerland on the pretext of illness. The arbitration required a mere eight months to assemble the arguments, a short period for a region that had been in such contest, and another eight months of rebuttal.

The position of the French was that the boundary to which they adhered, the "Vicente Pinzón," was the Araguarí, which flowed more or less parallel to the Amazon and slightly to the south of the Cabo Norte. The interior boundary would extend from a parallel following the Amazon to the thalweg of the Rio Branco River. This was the boundary suggested by Coudreau, over which he asserted France had diplomatic and historical rights. Brazil, on the other hand, contended that the Oiyapoc, some hundreds of kilometers to the north, was the Vicente Pinzón and the western boundary was the border was that with Suriname.

In the adjudication, it was clear that more was better. The French delegation, led by M. Legrand, former minister to the Netherlands, Gabriel Marcel of the French National Library, and various former governors of French Guiana and ministers to Venezuela, presented their three volumes and thirty-five maps. Rio Branco, on the other hand, acting as a special minister, presented three atlases encompassing four hundred years and two hundred gorgeous facsimile maps, three volumes of formal documents and letters pertaining to the area, and the massive study of the region, *Oiyapoc and l'Amazone*, by Joaquim Caetano da Silva, a tome that reviewed the history of these debates in detail with Rio Branco's own annotations. Rio Branco also emphasized the importance of the Monroe Doctrine and the necessity of limiting European spheres of influence in the Western Hemisphere, a feature of the adjudication that was thoroughly noted and widely commended in US policy circles. John Bassett Moore wrote a very approving analysis of the adjudication for the *New York Times*, highlighting this element of the diplomacy, in which, given his close relation to the Brazilian delegation, he most likely had a hand.[106]

The presence of Goeldi in Switzerland was also certainly useful in swaying the adjudicators, especially because as a zoologist and museum director, he was viewed as a more reliable scientific observer than the hot-blooded Coudreau, whose reviews were casually written, impassioned, and novelistic compared to the dryer styles of the museum director and Rio Branco. What was clear, in a bizarre twist of the doctrine of *uti possedetis*, was that the Contestado had been occupied for a long time by Brazilian runaways, and they had "our (Brazilian) ways, our language, and our customs" and thus captured enormous new territories (261,588 km^2) for the *patria* that many of them had rejected. The Pará Museum was renamed the Emilio Goeldi Museum.

Rio Branco was a hero: he had taken two of the thorniest boundary adjudications and transformed them into Brazilian triumph. Two years later, in 1902, Rio Branco would become Brazil's foreign minister. He would be taking on a series of scrambles in the western Amazon that were as volatile as the gold mines of the Contestado. There he would confront hemispheric aspirants and international stalking horses in a place that was immensely richer and more populous: the Purús watershed.

8

"American Amazon"?

Colonizations and Speculations

Neighbors to the North

When the American gunboat *Wilmington* steamed through the Amazon in 1899 with a secret US-Bolivian treaty aimed at "Americanizing" the Acrean territory, the Brazilians, though outraged, were not exactly surprised. The antecedents to the *Wilmington* affair—and the revolutionary response to it—lay in a set of long-held schemes, conceptions, and explorations during the previous decades reflecting a US attitude toward the Amazon that Brazilians, and especially Amazonians, found suspect. As far back as the 1850s, the US Confederacy had had dreams for the colonization of Amazonia, and scientists sponsored by America's top scientific institutions (Harvard Museum, US Naval Observatory, the Smithsonian) floated down the Amazon in support of this agenda. In a later decade, Americans began to develop plans for a "New World Liberia." North Americans with entrepreneurial ambitions for the region never seemed to be lacking, with many "up-country" schemes emerging in the 1870 and 1880s. The United States, in dealing with its southern neighbors, had taken a stance of protecting its citizens and other interests "from the Halls of Montezuma to the shores of Tripoli." In events of the day, where Venezuela—backed by the United States—was pitted against the British over control of the Orinoco estuary, and where a vigorous American colony was planned,[1] more spirited implementation of the Monroe Doctrine might have changed the last phrase of the Marine hymn from "the shores of Tripoli" to "the Orinoco sea."[2]

An American Slave State in the Amazon

The Baron of Rio Branco, an assiduous historian, was well aware of long-standing American interest in Amazonia, due to his own time in the United States, his long friendship with John Bassett Moore, no mean historical analyst of American diplomacy in his own right, and US forays during the imperial period, when Rio Branco's father was foreign minister. Matthew Fontaine Maury, Maury's brother-in-law William Lewis Herndon, Harvard Museum director Louis Agassiz, and their ally the Brazilian statesman Tavares Bastos, had to convinced Emperor Pedro II of the virtues of allowing ships from any nation to travel on the Amazon and of letting Americans settle there in large numbers.[3]

Maury himself never stepped foot in the Amazon but deeply affected American, and especially Confederate, ambitions for the region. Herndon and Agassiz became icons of Amazonian exploration, producing accounts of their travel that are still regarded as classics. American and British readers today are largely unaware of the imperial context of their travels, though Brazilian analysts have not hesitated to place these writers squarely in the literature of empire and territorial conquest. To many in Brazil, they were New World "Stanleys": precursors to imperial acquisition.[4]

Maury, a dedicated Confederate, was an eminent scientist and, like Agassiz, was head of an important institution, the US Naval Observatory, whose equivalent today might be NASA. Both were men of science and believed profoundly in God's design, in scientific racism, and in the virtues and the necessity of American colonization of the Amazon. In Maury's view, this was the best way to develop the riches lying fallow under the louche energies of Brazilian dominion. The mechanisms to achieve this change would include free trade, open navigation, steam travel, and American entrepreneurial spirit in the form of colonization.[5] And slave labor. For Maury, black labor and white management would be key to transforming this immense region.

Born in Tennessee to a prestigious but downwardly mobile Virginian family, Maury was a brilliant autodidact. He wrote what many considered to be the foundational work in oceanography, one of the seminal practical maritime texts of the nineteenth century, *The Physical Geography of the Seas*, a tome credited with expanding American maritime dominance at midcentury and founding the new scientific discipline of oceanography. Maury's technique involved promoting widespread observations from various fleets, whalers, and merchant ships on position, water temperature, prevailing winds, pressure, and other elements of interest and having them sent to him at the Naval Observatory. He then assembled this mass of data into a system of maritime maps of wind and water flow. This strategy of information collection was rather like today's Audubon Society Christmas Census[6] or a kind of wiki *avant la lettre*. By coordinating thousands of disparate observations, Maury was to see surprising connections in practical navigations and to collate immense amounts of observational knowledge into substantive maritime charts, creating navigational tools so powerful that he was described as a kind of Confederate Newton.[7] He passionately believed that the physical phenomena he was observing were manifestations of divine intelligence and godly design. Lovingly and poetically he described the movement of winds and waters, the subtle machinery that directed the globe, as the handiwork of the "Architect of Creation." Even for the times this was not a particularly trenchant analysis in the realms of physics or astronomy; nonetheless his prestige and popular reach were great.

Maury was one of the most decorated American men of science of the nineteenth century and received numerous awards and accolades in Europe.[8] He was an initiator of the American Association for the Advancement of Science and helped found what later came to be known as the Virginia Institute of Technology. Maury was well placed within Virginian and Washington politics, and in his role as the top navy scientist he was intimately connected with the US ambitions and international diplomacy of seafaring, steamship development, and river trade. He was powerful enough, and in the 1850s the navy was important enough, that he could mobilize the national resources to send his brother-in-law Herndon along with Lardner Gibbon on an expedition that would produce one of the durable Amazon travel classics: *Exploration of the Valley of the Amazons*.[9] As Maury outlined it in his letter of instruction to Herndon, the expedition was to "prepare the way for that chain of events" so that soon the region would be understood "as an American Colony."[10] This reconnaissance was meant to provide the empirical foundation for US colonization of the Amazon: to, as he put it: "revolutionize, republicanize and Anglo-Saxonize that valley."[11]

Maury's interest in Amazonia and Confederate imperialism had many sources: his professional work on the flow of currents, his yearnings for a

"Southern Manifest Destiny," and his preference for certain theories about environmental determinism and racial hierarchy. All of these coalesced in a vision of a "Confederacy in the Tropics" that would reach from Virginia into the Amazon, and a broader hemispheric division between American slave and nonslave economies.[12] The politics of Maury place him centrally in the Southern faction of the Young America movement in the United States.

The Young America movement had several strands. One was a loose cultural and critical literary association whose purpose was to develop a distinctly "American," postcolonial, New World voice. Among this group were icons of the American canon Emerson, Melville, and Hawthorne.[13] Others were radical populists in favor of agrarian reform, involved in the free soil movement, and supporting the various 1848 uprisings. But beyond cultural nationalism and the socialist sympathies of one part of the Young Americans was another that took its inspiration from a more imperial and politically conservative cast. These Young Americans were engaged in a deep political re-imagination of US political sensibilities and global role. This involved a shift from the centralized governance, isolationist, yeoman trajectory of the postrevolutionary Jeffersonian period to the more imperial thought and laissez-faire economy that inhered in Jacksonian democracy.[14] The important Southern segment of this movement was inflamed by a fiery nationalism, Manifest Destiny, and international interventionist politics, and was also profoundly pro-business.[15]

"Our Sweet Sea"

Maury's argument for linkages between America and Amazonia took its inspiration from his study of wind and currents, from which he concluded that a log released at the mouth of the Amazon would float through the Caribbean (that "American Mediterranean" and "our sweet sea") past the Mississippi delta through the Gulf of Mexico and the Florida Straits. . Thus, like many other rivers of the Southern states, the Amazon fed into the Caribbean. In his view, it had two estuaries: the first where it poured into the Atlantic and the second, its "true estuary," where it deposited its sediments in the seas off the southern Gulf Coast. This was the logic that folded Amazonia into North American hegemony. Oceanographically, Maury said, "that river basin (the Amazon) is closer to us than to Rio and puts . . . the mouth of that river within the Florida pass and as much under our control as is the mouth of the Mississippi." The implications of this are noted by da Cunha in his fragment where he calls the Amazon "the least patriotic of rivers."[16] Da Cunha's attention to this detail suggests that he was familiar with (and nervous about) Maury's analysis, and indeed he worried about Maury's positions a great deal.[17]

In Maury's view, in earlier times civilizations had emerged from discrete watersheds like the Tigris or the Yellow River. But now, he believed, large multi-river basins would become the great cultural and economic integrators.[18] Amazonia was seen as part of an "American Mediterranean," with the colonization and commerce of systems of watersheds (including the giant waterways like the Mississippi, Orinoco, and Amazon and Central American rivers such as the Motagua, Patuca, and Cocos) mastered by a "New Rome" based in the southern United States—New Orleans or Norfolk.[19] The relationships of the winds and tides meant that "ships sailing from the mouth of the Amazon for whatever port of the world are forced to our very doors by the southeast and northeast trade winds: New York is the half-way house between Pará and Europe."[20] For Maury, the ocean currents mingled not only the waters and sediments of Amazonia with those of North America but also their destinies.

Maury viewed the tropics as a cauldron that powered the earth's marine machinery, but he also thought the tropics would fulfill other divine purposes.[21] The linking of American Manifest Destiny to God's glorious ocean devices had several implications. By midcentury, the Southern politicians more or less understood that slave economies would enjoy no further territorial expansion in North America. Excluded from the northern Great Plains and the West, Southern slavocrats shifted their gaze to the tropics. "Southern Manifest Destiny" was a kind of inverse of Jefferson's imperial dreams. Instead of the Jeffersonian "Empires of Liberty," the war cry of some Southern secessionists was "Imperial Republics of Slavery." As the conservative journal *DeBow's Review* would put it in 1849: "We must meet our Destiny, a Manifest Destiny over all of Mexico, South America, the West Indies."[22] Some antebellum Southerners, such as soon-to-be Confederate president Jefferson Davis, already viewed the Gulf of Mexico as Confederate territory. Others, like Mississippi governor John Quitman, a veteran of the annexation of Texas and the Mexican-American wars, turned his gaze to the tropical terrains, full of squabbling caudillos, proto-revolutions, native peoples, and freebooters of all kinds, and saw a Central and South America that could be disciplined and developed as part of a new American "Confederacy."[23]

A Confederate tropical Manifest Destiny would be beneficial in many ways. Maury, like many other Southerners, feared a Malthusian crisis in a South overrun with black slaves, leading to problems of race war and miscegenation.

Since slave systems could not expand on the North American continent, they needed a dumping ground for "excess" population.[24] The Amazon would be the salvation of American slavery:

> The Amazon valley is to be the safety valve for our Southern States. When they become over populated with slaves, the Africa slave trade will cease and they [Southerners] will send these slaves to the Amazon, just as the Miss. valley has been the escape valve for the slaves of the Northern now free states, so will the Amazon valley be to that of the Miss. . . . It would be relieving our own country of the slaves, it would be hastening the time of our deliverance and it would be putting off indefinitely the horrors of that war of the races which, without an escape is surely to come upon us. . . . It is becoming a matter of faith among leading southern men that the time is rapidly approaching when in order to prevent this war of the races and its horrors, they will in self defense be compelled to conquer parts of Mexico and Central America and make slave territory of that—and that is now free.[25]

If Amazonia was colonized, the tensions between the Northern and Southern states would be significantly reduced, and "the Union would be saved!"[26] Maury walked a subtle line vis-à-vis the larger international slavery question as well: "Shall Amazonia be supplied with this class from the U. States or from Africa? In the former it will be the transfer of the place of servitude but the making of no new slaves. In the latter it will be the making of slaves of freemen and adding greatly to the number of slaves in the world."[27]

Beyond the ideological and territorial ambitions lurked economic concerns. Southern cotton soils were becoming depleted, and Maury came to believe that "the only remaining cotton country . . . is to be found on the southern tributaries of the Amazon."[28] Maury more or less envisioned in one of his more rapturous passages the entire basin devoted to cotton production.[29] With the British demand for the crop accelerating, Maury felt that British self-interest, given their immense dependence on the cotton industry and in spite of their aggressive abolitionist politics, would permit them to cast a blind eye to the way the commodity was being produced.

In a stressed Southern economy beset with depleted soils, the loss of the lucrative river trade to the new railroads, and a dim future for the institutions of Southern life in the face of an emergent industrializing North American economy, the rejuvenating energies of Amazonian colonization would provide rescue.[30] This move to the tropics, coupled with the commercial and entrepreneurial spirit of the United States, would transform the Amazon valley in the same way Confederates (and their slaves) had remade the Mississippi from a wild place into Dixie, a prosperous "Land of Cotton." The Amazon with its regular climate and rains, would, in the view of Maury, become the breadbasket of the world, a cornucopia of the most varied products.[31]

"How fortunate the Amazon is empty"

But should Amazonia be peopled with, in Maury's words, "an imbecile and indolent people"?[32] The answer for Maury was clearly no—"the sort of labor necessary to the extensive cultivation of cotton plants is compulsory labor." "Looking into the future," Maury wrote, "I have seen an African slave populations of America clustered around the border of this 'Mediterranean sea.'"[33]

Maury had strong environmental determinist views on race. He was a devotee of Arnold Guyot, a protégé of Agassiz, who had trained with him in Neuchâtel, Switzerland. Agassiz had arranged for Guyot to join him at Harvard, although Guyot eventually took a post at Princeton. Guyot was a glacial geologist and a meteorologist; in fact he ultimately became the founder of the US Weather Bureau. Guyot was a creationist whose views on environment and racial hierarchies echoed those of Agassiz and were outlined in Guyot's book *The Earth and Man* and somewhat more rabidly in later life in his *Creation, or The Biblical Cosmogony in the Light of Modern Science*.[34] The prominence of these three men, Agassiz, Guyot, and Maury, who headed major scientific institutions of their day, speaks to the legitimacy and power afforded to scientific racism.

The Earth and Man contained rousing arguments about Northern intellectual acuity, European imperialism, and a very specific justification for imperial occupation of the tropics. This statement inspired Maury enough to quote it in his letter to Herndon: "It is reserved for the European race not only to exhibit the most perfect phase of Human Civilization but to impress that Civilization on other races of the World." More to the point, "the progress of the Negro would never develop from within, but by necessity be imposed from without." Informed by these ideas on the racial superiority and the fashionable environmental determinism of the day, Maury would argue that "this [Amazonia] is a place for slaves. The European and Indian have been battling with these forests for 300 years and not left the merest mark. If someday its vegetation is tamed, if one day its soil is reclaimed from the forest, its wild animal and reptiles subdued by the by the plow and axe, it will have been done by the African. This is the land of parrots and monkeys and only the African is up the task which man must realize there."[35] While the brawn would be black, the technical and sophisticated knowledge would remain the domain of their white masters.[36]

With abolition, there would be four million slaves suddenly loosed into the American scene. It would be far better to take white Americans and their slaves en masse and, as had happened in the Mississippi, have them people a new place with a fruitful system until it reached it its full productivity. These views required some empirical reconnaissance, and it was Maury's kinsman Herndon and midshipman Gibbon who were charged with the task.

Maury's imperial position was that of Henry Walter Bates: "everything in Amazonia remains to be done."[37] Amazonia was, said other explorers like Charles Mansfield, "a boundless waste."[38] Mansfield would put it this way: "Like dogs in a manger they themselves don't use the land and inhibit others from doing so. What a monstrous insanity to guarantee with treaties the possession of such lands to the Iberians." This view implicitly expressed the fashionable imperial ideas of Emmerich de Vattel, the Swiss jurist of international law whose views on sovereignty and "natural law" suggested if a country were not effectively occupying its lands, or held more than it would use or cultivate, it should not oppose itself to others able to do so. Even liberal polities in Europe that had earlier frowned on the idea of empire were finding it politically much more to their taste by the mid-nineteenth century, and Vattel provided useful rationales.[39]

The tropes of emptiness, primitivity, and incapacity were hardly new in the annals of tropical claiming, but the nineteenth-century North American interests in the Amazon Scramble animated several types of logics: economic interests melded to divine right, merit bonded to destiny, chosen people (within the racial hierarchies), preferred political system (republic), and the virtues of free trade (central to the economic theory of the time), as well as in the larger political imagination involving righteous dominion and heavenly purpose. "How fortunate it was that the Amazon was empty," wrote Maury, "since then it could be populated by North American slaves."[40]

Maury's Instructions

Maury's letter to Herndon of November 13, 1850, was wild about the possibilities of colonization. In Maury's opinion, which later echoed throughout Herndon's tome, opening the river to free trade would soon induce a flood of colonists and their slaves from the United States, and with steamboats and open navigation a vibrant economy would emerge—"it would be regarded for all practical purposes as an American colony." Maury admonished Herndon not to let on to officials that he was reviewing Amazonia for its possibilities for Confederate colonization.[41] Instead, Maury emphasized that Herndon should forge friendships with governments and interests on the upper Amazon—Peru, Bolivia, Ecuador—because if these countries cemented regional navigation rights with Brazil, a means of "free trade" with external trading partners (the United States) might indeed be possible. By 1853, when Maury published his collection of essays on Amazon colonization, *The Amazon and the Atlantic Slopes of South America*, the strategy that he preferred was clear: develop treaties and alliances with countries in the upper Amazon that had won open navigation rights in the Amazon, and through these stalking horses the United States could acquire access to the whole channel.[42]

Maury's letter of instruction to Herndon urges exploration of "familiarity": Could one grow Southern crops like cotton, sugar, rice, tobacco? Was there any coal? Did the Amazon cut its banks as the Mississippi did? The exotica Herndon was to note were not very exotic: chinchona bark for quinine and reports on the rubber industry. Herndon himself remarked that that the Amazon itself was really just like the Mississippi at high flood.[43] Maury was at pains to reiterate that the Amazon and the Mississippi were more or less analogues and commercial complements of each other. He relied heavily on Herndon's reports and repeated that though the river itself might be enormous, a jungly Amazon of swamps and snakes, the mighty Mississippi with its bayous and water moccasins was not so different. The Amazon would not be beyond the scope of Southern skills and practices. Rather justifying territorial occupation based on the taming of "ecological otherness," the claims here were based on *similarity*, an unusual characterization for tropical colonial ambitions. They were wedded to the congruence of their economic systems (slavery), their markets, and the potential new outlets for American products.

Herndon was obedient to the wishes of his kinsman. His travelogue of 1853 refers constantly to the similarities of pasture, of fruits (though he does note the exceptional deliciousness of the native *guanabana* and of course cacau), of cuisine and how local foods could substitute for North American staples. He comments on what he sees as the excellent fertility of the soils. He notes approvingly that Richmond wheat could be purchased all the way from the Andes to the mouth of the Amazon. While there is plenty of the usual Amazon tropicalia (turtle eggs, close calls with alligators, irritating bugs, some rough rapids, etc.), Herndon's account has less heavy breathing than the usual "darkest Amazon" narratives: for him, the natives are basically not so bad, though they could benefit from military colonies and compulsory work; Herndon's rowers seem relatively tractable compared to those mentioned in other reports, including that of his compatriot Gibbon, who languishes on the Beni and fears abandonment by his guides. All the officialdom that Herndon meets yearns for American know-how. The weather is pleasant, insects are not great but not impossible, the place is healthy, and the potential for native products both collected and cultivated is much greater than Herndon had realized, although he emphasizes possibilities of coffee and chocolate, already powerhouses of the global commodity economy. Herndon is at pains to describe the general approval he finds everywhere for US colonization and free trade. The implicit argument is that if any foreign power were to have colonies there, by far the most "pre-adapted" would be Southern slavocrats and their chattel, who had done it all before. Maury hinted to Herndon to look for large areas to acquire on Peru's upper Huallaga (today a major coca-producing zone), what Maury would describe as the "New Tennessee."[44]

Herndon's account amply fulfilled the instructions and desires of his brother-in-law for the trip: possibilities for colonization, relative ease of travel (he mostly stayed with the Catholic missions), the widely positive view of Americans. He was aided enormously by a terrific "fixer," his Peruvian traveling companion Ijurra, who mediated, translated, and procured for Herndon's travels. Herndon's trip could be summed up this way: "I presume that the Brazilian government would impose no obstacles to the settlement of this country by any of the citizens of the United States who would choose to go there and carry their slaves: and I know the thinking people on the Amazon would be glad to see them."[45]

Herndon's book became a best seller. The US Navy alone published and distributed ten thousand copies, but Herndon's account was just one of a two-volume travel narrative. Herndon's companion, Midshipman Lardner Gibbon, had taken different routes and informed on other parts of the basin: Herndon went through Peru into Brazil, while Gibbon went to Brazil via Bolivia, through the Mamoré branch into the Madeira. Gibbon's memoir was somewhat lost in the flurry of publicity that attended the Herndon narrative. Indeed many modern editions of the *Travels* completely omit Gibbon. He was younger (in his early twenties) and less connected, traveled the tougher route over the Bolivian Andes, and saw a different Amazon. His report was sometimes dismissed as juvenile, perhaps due to his appreciation of spirit and beauty in women and horseflesh (with ample and admiring descriptions of both). He spent time amongst muleteers and traders and had a good deal of rough travel—over the Andes and through the inundated Llanos de Moxos. He ascribed his safety and survival on these treks to his mule Rosie, whom he sadly had to return to the muleteers once his travels took on a more aquatic turn as he moved farther into the flooded lowlands.[46] The most daunting part of his (or anyone's) journey from Bolivia on to the main channel of the Amazon was getting through the almost 200 kilometers of rapids and falls on the Madeira River, the graveyard of many explorers and ambitions. The "Devil's Cauldron" of these rapids separated the rich Bolivian rubber forests from the Amazon River access to Atlantic markets.[47] Gibbon's careful notes on the rapids and his claim that whatever difficulties existed, the long-term benefits of getting around systems of falls, whirlpools, and the like would far outweigh its costs later struck a chord for many seeking the "main chance" in the Amazon.[48]

Gibbon was a better ethnographer by far than Herndon, who was mostly interested in native people as coerced labor or objects of annihilation: "This seems to be their destiny. Civilization must advance though it tread on the neck of the savage, or even trample him out of existence."[49] For Gibbon "the industrial, agricultural and manufacturing people of this country are principally among the aborigines," and he goes into a recitation of methods of smelting, jewelry making, weaving, planting, brewing, brick making. The cultivation of multicolored native tree cottons and the richness of the dye plants are duly noted. The mineral exchanges between the high and lowlands and the placer mines of gold are commented upon and their value calculated. He reports that among the Chiquitano Indians there was great love of music making and instrument manufacture (taught by the Jesuits) and aptitude in reading, mathematics, and languages. This was hardly a savage realm. Gibbon was in the lands of the mission cultures of the upper Amazon and in the ghost empire of the great pre-Columbian societies of the Moxos.[50] Though the Jesuits had been gone for most of a century, the vibrant syncretic culture built on the ruins of an earlier civilization remained.[51] Regarding the Yacaré, Gibbon said,"there are two characteristic of the Indian we particularly notice: his honesty and his truthfulness." Gibbon's view of natives is a Rousseauian counterpoint of the noble savage to Herdon's savage brutes.

Gibbon paid close attention to diseases: one hundred Indians die of smallpox in the Bolivian town of Trinidad while he is there; the Brazilian garrison has some kind of languishing affliction, or "Fort Fever" (perhaps beriberi). Gibbon himself has malaria. He notes runaway slaves on the Bolivian side of the Madeira (said to number two thousand—an enormous number at the time, one that hardly boded well for the new immigrant slave–based production yearned for by Maury), the village of Borba, composed almost entirely of blacks, and the free black militias in Mato Grosso.[52] He provides detailed information on forts, economic activities of the most varied types, military men, and equipment because he was in a position to actually see them. Gibbon's report reflects the social milieu where he traveled, closer to the realities of Amazonian economies, and was dependent on the goodwill of locals. While he was at times a guest at missions, he often slept rough, in farmhouses and muleteer taverns or in the open. Unlike Herndon, he had no interlocutor.

For later adventurers and explorers, it was Gibbon's work that was the more useful guide, not the popular Herndon account. Euclides da Cunha would laud Gibbon's cartography as one of the few authentic efforts in this land of lying maps, and coincidentally the one that best corresponded with Brazilian positions.[53] Maury wrote no letter telling Gibbon what to look for. In terms of measurement and observation, Gibbon's survey was more precise and his judgment less clouded by an external agenda. The more reliable technical content of the travels lies within Gibbon's volume.[54]

Gibbon's narrative, in contrast to Herndon's, is not a story of the primitives and yokels yearning for American salvation. He describes a dinner party in La Paz where the lovely hostess engages him in a lively conversation about politics: "She expressed approval of the American people but not some of their actions. . . . She asked me to explain to her the meaning of all the articles she saw in the La Paz newspapers on the subject of Cuba. Turning suddenly, she looked up and said 'what are you doing here Senhor Gibbon, do you want Bolivia also?'"

The answer, although Gibbon did not know it then, was yes.

Tropical Dixies

Soon after the travels of Herndon and Gibbon, and the agitated promotion of colonization by Maury, both the United States and Brazil would be involved in their respective convulsive and hideously lethal wars (the Civil War for the United States, the Paraguay War for Brazil). Maury's career in the navy ended with the Civil War, and he fled to the Confederacy and later to England to avoid arrest. He maintained his commitment to Southern slavocractic immigration and ended up promoting and helping (not especially well) to manage colonization programs for Confederates to Mexico. Maury had met Austrian emperor Maximilian, who had decorated him in 1858 for his scientific work. That monarch became the ruler—"protector"—of Mexico. In 1866, Maury became Maximilian's imperial commissioner of immigration and dreamed of installing a "New Virginia" in the settlement of Carlota along the road between Veracruz and Mexico City.[55] Maximilian's reign depended on the French troops of Napoleon III to control native rebellion against him. When that army left, Maximilian was eventually overthrown by Mexican revolutionary leader Benito Juárez. After Maximilian's death by firing squad, Maury would return to the United States under the Union amnesty agreements to teach physics at the Virginia Military Institute.

Amazonian colonization faded for a time as geopolitical project, but Maury's ideas did stimulate a migration of Southerners who preferred immigration to Brazil, where the "peculiar institution" still thrived, to the problems of reconstruction in the United States. While they were inspired and influenced by Maury, their move was not part of a broader national strategy but rather a personal response to abolition and the defeat of the Confederacy.[56] If there was a geopolitical project, it pertained to Southern Protestant evangelization in the papist Brazilian empire. Some authors credit the arrival of Southern Protestants as central to the establishment of Protestant religious sects within Brazil.[57] Other sites of interest for Confederate colonies were in the Orinoco (a possible toehold for annexation of parts of Venezuela).[58] Relatively large Confederate colonies were founded in Belize, Honduras, and Mexico as well as Brazil, and it is estimated that more than ten thousand Confederates migrated out of the United States.[59]

Southern migrants were also encouraged by the 1850s writings of Maury's contemporary Colonel Lansford Hastings.[60] Hastings had had an active life as a colonizer and dreamed of emulating Sam Houston by wrenching land from Mexico, proclaiming it an independent republic, and later having it annexed by the United States. To encourage westward migration, he had a brisk sideline churning out books on routes for immigrants into California.[61] His prestige in this arena declined drastically due to the unfortunate "Hastings cutoff" through the Sierra Nevada that was used by the Donner party, who, stranded by snows in the Sierra Nevada, ended up in cannibalism.[62] This understandably undermined confidence in his set of North American schemes, but he was undaunted and redirected his frontier ambitions to the Amazon.

Hastings went to Brazil, carried out some preliminary assessments, promptly wrote his "Immigrants Guide to Brazil," and organized a colony in Santarém, a town at the mouth of the Tapajos.[63] "The Amazon," Hastings noted, "reminds us of the Mississippi." The colony itself was not so successful, and an account of the travails the migrants endured to arrive in the humid little entrepôt reads like a melodramatic novel with extortion, shipwrecks, mutiny, and onboard epidemics.[64] A few families endured and were quite successful, and as elsewhere in Brazil, Americans were considered innovators in agriculture. One of the first rubber plantations was established there with some twenty thousand trees by descendants of the American colonizers. It was this early plantation experience that later, unfortunately, convinced Henry Ford to establish his plantations and company towns near Santarém.[65]

Many English-speaking tourists and scientists washed up on the doorsteps of the Santarém Confederates and enjoyed their hospitality.[66] The Anglophone enclave at Santarém attracted adventurers of all types, including an Englishman, Henry Wickham, his wife Violet, and their four children, who resided there, struggling economically for several years. In the 1870s Wickham devised the biopiracy that would ultimately unravel the Amazon rubber economy when he shipped out some seventy thousand seeds to Britain's Royal Gardens at Kew.[67] Other migrants, including some Confederate military men, joined Latin American armed forces. John Randolph Tucker, a rear admiral in the Confederate Army, was invited to join the Peruvian navy with a few hand-picked Confederate officers. Later Tucker was appointed president of the Peruvian Hydrographical Commission of the Amazon, which surveyed the upper Amazon from the ports at Iquitos and upper Ucayali tributaries. Tucker and his cohort of Confederates were responsible for naming the Ucayali port of Leticia at the intersection of Peru, Colombia, and Brazil. It was named after President John Tyler's granddaughter, the first person to raise the Confederate flag.[68] James Orton, traveling under the auspices of the Smithsonian in 1867, enjoyed meeting the Confederate crew on the Ucayali as he traveled around the upper Amazon, noting natives, economies, shipping, and the value of exports in a less famous but more economically quantitative reconnaissance of the region.[69]

The impact of Maury and Herndon on Amazonian enterprises and entrepreneurs was palpable in imaginary travels as well as concrete tropical ventures. In light of Confederate colonies on the Amazon, the young Mark Twain's yearning to take a steamboat to the Amazon and become a coca entrepreneur seems not so far-fetched. He had read Herndon's book and, learning about coca cultivation, dreamed of introducing this substance to the world at large.[70] Others, alive to the colonization discussion but concerned about a different dimension of the slavery question—"the problems of the Free Negro"—began to dream of colonies in the Amazon, not for masters but for freed American slaves.

New World "Liberias"

On December 3, 1861, in his address to Congress, President Lincoln asked that steps be taken for colonization of slaves liberated in the confiscation of property "used for insurrectionary purposes," as they were now essentially wards of the state. "Steps should be taken for colonization . . . at some place or places and climate congenial to them. It might be well too, to consider whether free colored people already in the United States could not, in so far as individuals may desire, be included in such colonization. . . . To carry out the plan of colonization may involve the acquiring of territory and the appropriation of money."[71] Congress gave the executive the power to begin to explore this state-sponsored colonization—in essence deportation of the emerging class of ex-slaves.[72] Lincoln had, after all called for a "colony of freed Negroes in Central America and provinces in Nuevo Grenada," on the northern coasts of South America, a place now known as Panama. Lincoln was to support "New World Liberias" in five major public addresses, including two State of the Union speeches, and in the preliminary Emancipation Proclamation.[73] Among vigorous advocates of this position was Washington newspaper man and congressman Francis Blaire, Lincoln's informal adviser, founder of the Republican Party, and major negotiator with the Confederacy. Blair had been in favor of simply annexing Central America, arguing that "the door is now open in Central America to receive the enfranchised colored race born amongst us."[74]

It fell to William Seward, the secretary of state, to address this question in substantive ways. Seward was not fainthearted when it came to acquiring huge expanses of territory: he later purchased Alaska—some 1,500,000 square kilometers—from the Russians in 1867.

The ideologies of environment and race at the time suggested that tropical venues would be the most appropriate places for the newly liberated black and mestizo population to settle, and to this end in 1862 Seward approached the ministries of countries with tropical colonies—Britain, France, and Holland—about the feasibility of such immigration. Seward's note of inquiry indicated that many free blacks indeed wished to emigrate if they were given certain guarantees by the United States and the nations to which they would go.[75]

The European powers were not happy about the prospect of a sudden onslaught into their colonies of free blacks who would maintain their US citizenship. Partly this resistance stemmed from generalized racist views of the intellectual and laboring qualities of blacks, and as a rule, if migrants were to be had, white immigrants were preferred. The other problem was that the relocation of large numbers of US citizens would enhance the potential and pretexts for US incursions. The number of former slaves in the United States was among the highest of any nation in the hemisphere, and most of the New World colonies and young republics were having enough trouble with their own domestic insurrections and racial conflicts without a deluge of more or less indigent black expatriates from the United States. Depending on how the Civil War went, taking in liberated or free blacks as colonists might generate later questions about harboring fugitives and "stolen property," possible pretexts for invading colonies and countries distant from the protective European military apparatus.[76] Predictably, the Central American countries registered complaints about the unilateral nature of the wished-for colonization and, given the racist ideologies of the day, about the caliber and kind of colonist.[77] Seward responded to such concerns with assurances that consent of the colonized country would always be sought.

Lincoln and Seward had pressed Kansas senator Samuel Clarke Pomeroy into service to review possible colonization sites in northern South America and Central America for a new "Deep South." Pomeroy is best known today for his political bribery trial and later for his chairmanship of the Santa Fe Railroad, but before his election to the Senate he had been the financial agent for an emigrant aid society in Massachusetts. Responding to the agenda proposed by Lincoln and Seward, Pomeroy leaped into the fray after a brief tropical inspection. He produced a pamphlet, *Information for Persons Proposing to Join the Free Colored Colony of Central America*, that was widely circulated, much to the irritation of the Central American republics, since Pomeroy's plan involved the annexation of the Colombia province of Chiriqui on the northern rim of South America, later known as Panama. Seward, responding to outrage from Foreign Minister Molina, who represented Nicaragua, Costa Rica, and Honduras, asserted that Pomeroy was merely acting as an agent, had no negotiating powers, and really was just carrying out a survey. This, however, was not the impression the pamphlet gave: it included departure dates from Baltimore, baggage rates, and it featured the usual hyperventilating prose of the land salesman and speculator, describing the area as extremely productive, able to produce temperate and tropical crops, a "semi-tropical California," and a place of unparalleled healthfulness.[78] In light of the staggering mortality that attended the construction of the Panama Canal, this last was an especially dubious claim.[79]

The Union triumphed at Antietam, and the Preliminary Emancipation Proclamation was declared on September 22, 1862. It appeared that indeed there would be soon be four million free blacks released in a destabilized, rancorous South and a war-weary Union. But the war was not over yet, and it was not the time to alienate allies, so officials muted US ambitions for colonization in the Latin America.

Deporting large numbers of "undesirables"—political prisoners, debtors, criminals, vagrants—was not an unusual practice in the mid-nineteenth century. The tropics were famous for their gulags: Botany Bay in Australia is probably best known, and as we have seen, France shipped off revolutionaries from the 1848 revolts to French Guiana. Banishing desperate, insurgent, or incarcerated populations into peonage in the tropics was an increasingly normal practice as the century wore on and slavery was gradually eliminated: tropical commodity extraction and commercial agriculture clamored for labor. Prison gangs increasingly featured as a new form of coerced labor.[80]

Amazon Liberia?

One person with keen interest in Lincoln's vision was the US minister to Brazil, General James Watson Webb. Famous as a bon vivant and ladies' man, Webb had owned newspapers (the *Courier* and the *Enquirer*) and railroads and later enjoyed many diplomatic posts, including those to Austria and Turkey. Webb was somewhat of a theorist of comparative slavery, having produced a small book on the topic: *Slavery and Its Tendencies.*[81] He reflected a strand within the abolitionist wing of the Democratic Party that thought the solution to the inevitable problems of emancipation lay in tropical resettlement of ex-slaves. Webb viewed black colonization as an alternative to what otherwise would develop into socially undesirable miscegenation and "inevitable" race war.[82]

Following this view of the nation's needs, and with Webb's understanding of black adaptation to the tropics, the unusual features of Brazilian slavery, and its free men of color, Webb advanced a set of proposals to and negotiating points he hoped to raise with Emperor Pedro II.: (1) the colonization should be cheap; (2) liberation need not be immediate, nor should servitude linger; (3) immigrants would take up an "apprenticeship" in the colony; (4) colonization costs should be paid for from the products of the apprenticeship; (5) colonists would ultimately end their political connection with the United States because Brazilian society, more tolerant of people of color, would provide more possibilities of advancement.[83] In short, Webb was arguing for transforming slaves freed in the United States into indentured workers in Brazil. He suggested that the emancipated slaves would be entrusted to a joint stock colonization company (headed, naturally, by himself), which would then resettle them in Brazil. Thus former slaves would be transformed from personal to corporate chattel until they paid off their settlement costs.

Webb framed these arguments in his dispatch of May 1862 for Seward by using a comparative analysis of characteristics of slaves, societies, and political economies. The analysis showed how Brazil's labor difficulties and US racial problems would be solved through the creation of a free black colony in the Amazon. Webb's reasoning went like this: (1) free labor was scarce; (2) slaves were scarce and expensive (due to the prohibition of slave imports after 1850); (3) the demands of the burgeoning economies of southern Brazil were draining labor from the northern provinces, where workers were increasingly needed. Webb believed that US ex-slaves were more docile and hardworking than "the fierce, warlike and intellectual" Africans who made up the Brazilian slave population, who were "ready for insurrection and capable of extensive conspiracies to effect their liberation." Webb felt that Brazil's problems with black insurgency could be resolved via a huge influx of American ex-slaves. The belligerent Afro-Brazilian insurgencies would be undermined through the calming cultural impact of thousands of "docile" American ex-slaves and the flooding of Brazil with another form of labor.

Webb's solution argued for immediate expatriation of US blacks to Amazonia in order to "render Brazil the richest among kingdoms of the Earth." As he put it: "The African slave trade can never again supply the Negro labor alone suited to the region, and white labor is quite out of the question." Webb proposed that the United States should initially pay for the transportation of former slaves and North American black freemen to the Amazon, where Brazil would supply the lands, about one hundred acres per colonist. The costs inherent in the immigration (transportation, land costs, etc.) could be defrayed by the income generated by the products of the "apprenticeship" of several years (up to ten). After a time, black immigrants could take up Brazilian citizenship with the rights that accrued to free men in the Brazilian empire.[84] Webb urged the rapid adoption of this model due to the prejudice that prevailed within the United States. "The US," said Webb, would be "blessed by his [African American] absence and the riddance of a curse which has well nigh destroyed her."[85]

Webb's ambitious plan was deflected by Seward, whose mild response emphasized a decision to resolve the US slavery question within the nation—a policy turnaround from the previous postures of the administration and the program Seward himself had outlined earlier. Further, these extravagant plans, coupled with machinations of US entrepreneurs in countries of the upper Amazon, Peru, Bolivia, and Ecuador, contributed to a certain coolness and obstructionism in Brazilian diplomacy toward an Amazon-American colony, whether of slavocrats or free blacks.[86] At a time of significant racial problems and uprisings in Brazil, the idea of the infusion of a huge population of "free" blacks was alarming, especially since racial ideas of the day suggested that they hampered modern development. In any case, this particular American ambition for the Amazon was tabled, as the United States was caught up in reconstruction and assassinations while Brazil and its allies devoted themselves to the crushing of Paraguay.

These "American Amazon" schemes did not bear much fruit, but the development of such programs occurred at the highest levels within the US government, were promoted by its most august scientific institutions, and were widely publicized at the time. While the purpose was to "offshore" US racial problems through state-mediated programs justified by Confederate Manifest Destiny or a New World Liberia, the next iterations were based on adventurers and entrepreneurs and carried no virtuous colonial social gloss, but rather incarnated the sparer lines of resource imperialism and speculations.

American Adventurers and the Republic of Poets

In moving west, Americans had developed several ways of taking "open" or contested lands and profiting from them. One technique was to grab territory, subdivide it, and sell the lands outright. Another way was to build infrastructure—roads, ports, and railways—and control the revenues from them. One could also sell infrastructure derivatives—for example, the right to build or the right of commerce on rivers. Another practice was to build infrastructure, negotiate rights to their adjacent lands and speculate on these. These risky possibilities with potential for large returns all required raising capital.

The Amazon was alive with adventurers. In the 1860s and 70s, the US engineer, Mexican revolutionary sympathizer, and Bolivian Amazon surveyor George Earl Church began to plan major railroads to link Bolivia with the Amazon ports on the Madeira, the precursor of the construction of the disastrous Madeira-Mamoré rail line. Today Church's name survives only as a footnote in Amazonian studies, but he was as ubiquitous an imperial explorer-entrepreneur as his friend the celebrated imperial botanist and biopirate Clements Markham, or the widely traveled Sir Richard Burton, diplomat, ethnographer, libertine, and translator of the *Kama Sutra*, *Epigrams on Priapus*, and *The Arabian Nights* as well as the Portuguese masterpiece *The Lusiads*. These men were regularly at the intersection of journalism, exploration, and uprisings, with developed tastes in history and ethnography. Markham and Church knew each other from their ramblings in the upper Amazon and were close friends. Markham was Church's literary executor and promoted him for the post of vice president of the Royal Geographical Society.[87]

Church engineered railroads and infrastructure surveys in Argentina, Costa Rica, and the Bolivian, Peruvian, Ecuadoran, and Brazilian Amazon.[88] He had considerable impact on Amazonian business speculations. His field research, newspaper articles, negotiations, and programs were meant to launch him from his career as a civil engineer, populist revolutionary, and journalist into the lofty realms of an imperial capitalist. Like that of Lardner Gibbon, Church's experience in the upper Amazon was widely used and cited, and the would-be explorers of the Britain Royal Geographical Society pored over his pamphlet *Desiderata in Exploration*.[89]

In 1868, with the new 1867 Ayacucho Treaty boundaries in hand, Church negotiated a concession to canalize the Madeira-Mamoré Falls or to construct a railway around them with the Brazilian government. This negotiation package included navigation rights on the Bolivian Amazon affluents and the right to exact toll and freight charges for twenty-five years through his Bolivian joint venture, the National Bolivian Navigation Company, incorporated in New York. Because the Madeira River is in fact in Brazilian lands, Church's proposition required the consent of Brazil. He was able to convince the Brazilian government to give him the Madeira concession directly, and he integrated these rights into a second corporation, the Madeira-Mamoré Railway Company, along with mineral and land rights adjacent to the route of the railroad, a concession of some 560 square kilometers. The persuasive Church raised some six million pounds—a huge sum, almost 600 million in current pounds[90]—in bonds from London venture capitalists despite Bolivia's very dubious reputation in international lending circles. This was quite a feat, since the Bolivian commercial agents Peto, Betts and Co. had trundled around Europe trying to mortgage all of Bolivia's resources. Church described what was on offer this way: "The whole of Bolivia—animal, mining, financial, and almost spiritual—exclusive rights of international communication and river navigation—Bolivia inside out, present and future, Bolivia down to the base of the Andes."[91] In some financial circles this freehanded approach to mortgaging national territory was considered questionable and the deals unenforceable. Church's survey experience in the region and the fact that he had worked as a formal analyst for several upper Amazon governments (Peru, Ecuador, Brazil, Bolivia, and who knows what other clandestine partners), as well as his long association with Markham, gave him a legitimacy that was lacking in others hawking upper Amazon real estate.[92]

It certainly helped that gold had recently been discovered in the Caupolicán district just to the south of the Acre, an area to be served by the proposed railroad. Church's Bolivian company bought the rail rights from Church's Brazilian company for twenty thousand pounds in cash, a very large sum at the time, and certainly a useful one in any era. While some might have viewed this as a kind of self-dealing, it did consolidate the bundle of rights into one company. Church began his enterprise with subcontracts to a Philadelphia construction firm, P. T. Collins, which sent some 750 American laborers, 200 Bolivian Indians, and 200 Ceará migrants to the site where they actually built tracks and ran a train (whose main engine they named Colonel Church).[93] The business itself collapsed in a complex multinational cloud of litigation involving Bolivia coastal *compradors* dismayed at the possible deflection of trade from their Pacific venues to Atlantic ones, various British interests, and bribery scandals involving the Bolivian president, along with the usual disease and strife in construction projects in the Amazon. When everything fell through—a fairly regular destiny for Madeira-Mamoré ambitions—the transportation development rights ultimately were resold to King Leopold II of Belgium.[94]

"An East India at our very doors!"

The Bolivians continued their own ambitions for settlement and opened negotiations with yet another American, Azanel Piper. Armed with Chandless's maps, Piper reportedly had also partially explored the 1,700-mile length of the Purús.[95] In the first real attempt at a charter company in the upper Amazon, the Colonization Company of California incorporated in San Francisco in 1870, in the hope that the vision of yet another frontier would attract the pioneers and gold rush magnates at the edge of the Pacific. Unlike Church, whose enterprise focused on infrastructure development and navigation rights, Piper preferred to speculate directly on the land and the minerals. He obtained two enormous parcels for his concession. The first was 230,000 square kilometers, which he agreed to colonize with Americans and Europeans over a period of twenty-five years. The colonization company would have rights over all territory not formally registered under Bolivian law ("vacant and uninhabited lands") and to all Siriono or other "nomadic" Indian lands from the Madeira to the Rio Grande (the river just outside of the modern city of Santa Cruz). The company would have the rights to emit its own currency and develop its own banking system, as well as exclusive navigation rights on the Purús, Juruá, and Madeira.[96] The Company could levy taxes and develop infrastructure. In short, it would have all the prerogatives of a state. Immigrants would produce spices, fruits, rubber—the usual plentitude—and have access to the tractable and able labor provided by the local settled Indians, whose virtues were so nicely evoked by Gibbon.[97]

The real lure was the Caupolicán area, to which Piper's company had exclusive territorial and mineral rights for fifty years, an area deemed by Church and other observers to be extremely rich in alluvial gold deposits, which could be exploited by the new techniques elaborated in California gold fields and silver mines. On top of that were coal, cobalt, copper, tin, salt, and diamonds. Bolivia's riches were meant to echo California's frontier as an upper Amazon El Dorado. Piper explicitly compared Bolivian riches to those of California, envisioning a robust agrarian economy supplying the mines. The territory would stretch from the Madeira (the Iteñez and Itambari) to the headwaters of the Javari. These were the "boundary" lines of the 1777 San Idelfonso Treaty. The land grant to Piper would encompass the same areas as the Bolivian Syndicate Charter.

The venture was too speculative and very expensive; lacking broader inter-

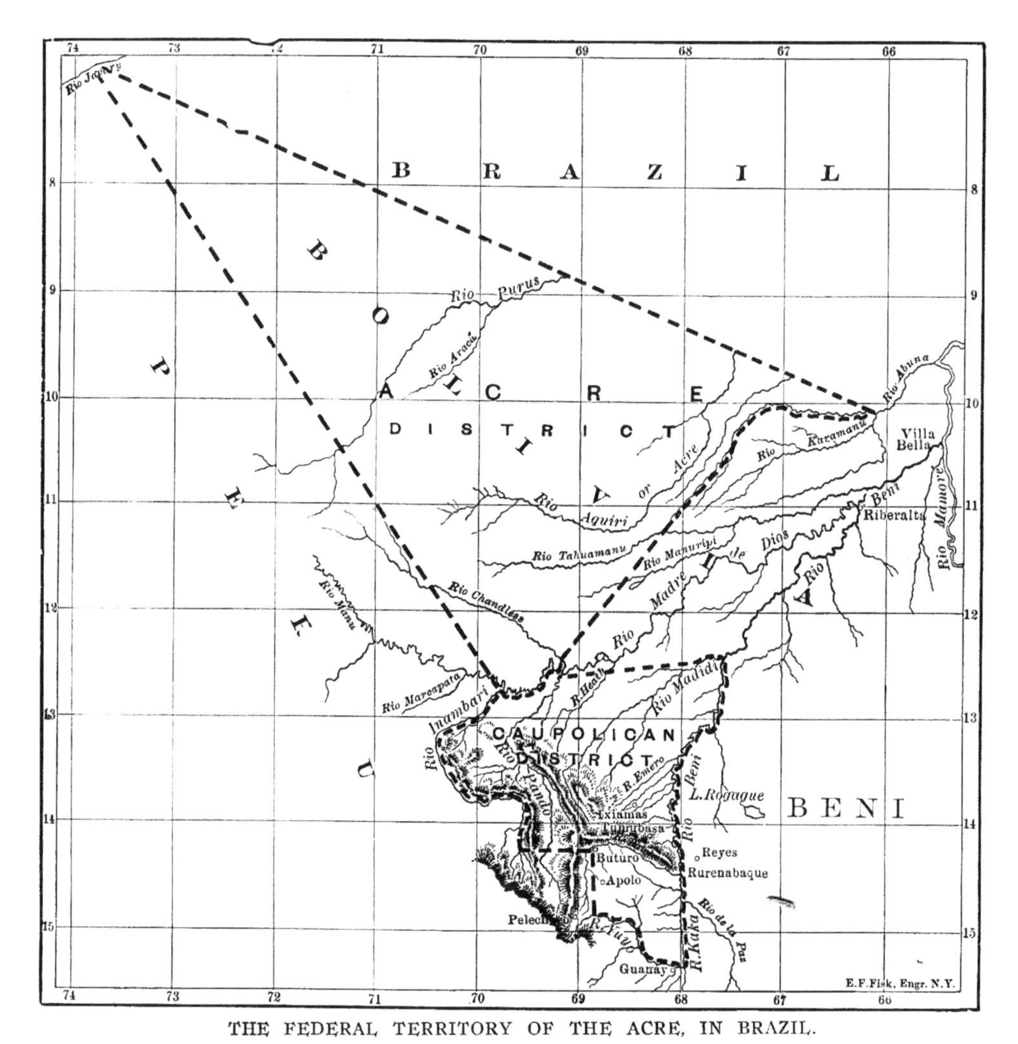

Map 5. Map of the Azanel Piper Charter.

national financing, it never got off the ground. But upper Amazon schemes were in the air, because the value of rubber was soaring and explorers of many stripes were making assiduous resources assessments, including one Sir James Conway whom we will meet in the next chapter.

The Amazonian populations in the 1870s were increasing rapidly. El Niño droughts were driving the northeastern *sertanejos* deeper into peonage and farther up the tributaries, where significant income could be generated from the rubber, quinine, and cacau collected and shipped down the rivers and out the great Amazon channel to emerging markets for these coveted tropical

Figure 8.1. Llamas loaded with rubber for a trip over the Andes to Pacific ports.

products. International finance capital of many sorts was now active in the Amazon: British, Belgian, and French banking houses were supplying trade goods for the tappers and credit for building the cities that were beginning to explode. By the end of the 1890s more than 100,000 people, mostly Brazilians, were in the Acre basin alone, generating some 17 million dollars in export value there, a modern equivalent value of some half a billion dollars, an immense sum at the time.

In addition, this place, as Church described it, contained "unsurpassed auriferous wealth." Yellow (metal), red (Indian labor), and black (rubber) gold were all to be found in the litigious zone. And if no way was found to move material through the waterways, there was always future transport over the Andes with llamas.

9

Wall Street, Rebels, and Rio Branco

Preboard for Syndicates: *Wilmington* and the Republic of Poets

In 1899, a new customs house was installed at Puerto Acre[1] at the behest of Bolivian foreign minister and plenipotentate José Paravicini. This customs house taxed the revenues of the rich Acre basin above its confluence with the Purús, as well as the rubber from other watersheds such as the Orton, and even some rubber from the Madre de Dios. The levy caused a three-million-dollar loss to the state of Amazonas, highly displeasing to the Manaus businessmen and the state government, not to mention the *seringalistas*—rubber barons and traders obliged to pay it.[2] Paravicini had some other policies for Acrean development—education of girls, for example—but the fundamental appeal of the territory for the fiscally precarious Bolivian government was the potential revenue from a booming rubber economy. The Acreans had different a different view: a few months later the customs house was a broken ruin. The *seringueiros* were in a guerrilla war with the Bolivians, ambush by ambush.

In this light, the travels of the gunboat *Wilmington* to the upper Amazon in March 1899 was cause for considerable alarm. The entrance of a US warship to the rivers of the most valuable forests on the planet without authorization was perturbing, and it suggested that a more complex Scramble was coming into play. It did not bode well that the Americans were more than two thousand miles into the Amazon basin at an especially robust imperial moment for the United States. After all, this was the country that had annexed much of the North American continent, an expanse that previously been in the hands of French, English, Spanish, and Indigenous nations. The Americans had seized about a third of Mexico, had filibustered around Central America (including William Walker's slave polity in Nicaragua), and were negotiating for the construction rights for a canal and with it the annexation of a Colombian province that would soon become the client state of Panama.[3] All this with the ink barely dry on the Paris Peace Treaty that ended 1898 Spanish-American War, which had delivered Spain's great island colonies—Cuba, Puerto Rico, Guam, and the Philippines—into the hands of the United States. It did not help that the *Wilmington*'s captain, Chapman Todd, had neglected to get formal permission for his Amazon itinerary.

It was in these treacherous seas of clandestine war and territorial anxieties that Todd and the *Wilmington* had to navigate when he arrived in Belém. At a formal reception he stated he stated he was on "a mission of friendship and solidarity with Brazil." He further framed the nature of his friendship in this way: "Remember: there is room enough for everyone in this great country. I am going up the Amazonas. . . . I am going to make a report and I am sure it will be gratifying to this great country."[4] One passenger on the *Wilmington* was the US consul to Brazil, K. Kennedy, and in Belém at the time of the arrival of the gunboat were Bolivian foreign minister Paravicini, fresh from the Bolivian frontier; Guillerme Uhthoff, Paravacini's aide; and the customs officer of the Bolivian frontier, Ladslau Ibarra.

The *Wilmington* was to take a US-Bolivian pact to the president of the United States, William McKinley, signed by Pavacini, Kennedy, and Bolivian vice consul Luiz Truco. The accord dealt with an "agreement" over the Acre River basin, still technically Bolivian but in militant dispute by Brazilian *sertanejo/seringeiro* guerrillas. Bolivia had already sent an expedition to the Acre basin to back up its claims with armed men. The "Little Agreement" between the United States and Bolivia, as it came to be known in Amazonian circles, involved seven provisions that would make Brazilian blood run cold.

The first article of that document established that the United States would, through diplomatic means, raise with Brazil the question of recognition of Bolivia's rights to the entire watersheds of the Iaco, Acre, and Purús in light of the limits established in 1867 by the Treaty of Ayacucho. Article 2 was that the United States would furnish the necessary arms should a war break out between Brazil and Bolivia. The third article established that the United States would pressure Brazil and Bolivia to name a commission to establish the international boundaries of the of the Juruá and Javari Rivers. The fourth clause stipulated that Bolivia would enjoy free navigation rights on all the tributaries of the Amazon and duty-free status at the customs houses of Belém and Manaus for products coming into and going out of Bolivia. The United States would defend these rights. The fifth condition was that in view of the previous clause, Bolivia would discount tariffs on US products coming into Bolivia by 50 percent and reduce tariffs on rubber going to any US port by 25 percent for ten years. The last two clauses of the Bolivia-US pact stipulated the disposition of responsibilities in case of war: Bolivia would denounce the Ayacucho Treaty in accord with the United States and establish that the new territorial boundary would pass through the mouth of the Acre River, and the remaining territories (those between the mouth of the Acre River and the existing occupation) would be handed to the United States in freehold. The United States would pay the war expenses and receive the rents from Bolivian customs houses as future payment for these military outlays.[5]

This was quite a list, with unpleasant implications for Brazil. First, it was obviously a set of agreements about prosecuting a war and distributing its costs and benefits, including a generous territorial giveaway to the United States. Second, the remaining clauses were trade and transport agreements that would make the rubber holdings and other resources in the far western Amazon very economically alluring.

A Spanish Subversive

Spanish journalist, adventurer, and café revolutionary Luis Galvão,[6] whose job it was to report on the doings of illustrious visitors in Belém, portal to the Amazon, noted the unusual confluence of dignitaries and, as a good journalist, began his interviews. Thoroughly charming, and sharing the same tongue as the Bolivians, he was soon able to supplement his income as a writer for the newspaper *Provincia do Pará* with another job, one with the Bolivian legation in Pará. It was through this connection, its access to diplomatic papers, and the revelations of his drinking buddy Ibarra, who was upset by what he viewed as disloyalty on the part of Paravacini (who Ibarra thought was trading away his country), that Galvão discovered that the United States and Bolivia were in negotiations over the Acre territories. The first months of 1899 had shown the Bolivians that any control over the Acre was going to be very difficult because of the number of Brazilian settlements and *seringalista* displeasure at taxation and annoyance at Bolivian rule. The Bolivians, for their part, were seeking "diplomatic support" from the United States in the face of an increasingly insurgent population. The *Wilmington* then proceeded to Iquitos, the capital of the Ucayali and center of the latex trade in the northwestern tributaries, where the passengers were visiting with Peruvian rebels who dreamed of controlling the Ucayali and its rubber economy, at least according to the some press reports.[7] Perhaps some larger US regional strategy was afoot.

Galvão's revelations were a major journalistic coup that triggered infuriated denunciations at all levels of Brazilian society, international outcry, and tropical revolution in Acre when a copy of the pact between the US and Bolivia was printed in the *Provincia do Pará* on June 6. The region and nation exploded. The Brazilian statesman, international jurist, and famous abolitionist Rui Barbosa entered the fray, reminding fellow Brazilians about US activities vis-à-vis Cuba and Hawaii and cautioning that history could repeat itself.[8] Others, such as the journalist and historian João Lucio de Azevedo, placed the episode firmly in the realm of economic imperialism with direct references to the European ventures in Africa, situating the Amazonian case within the political economics of the African Scramble.[9] The voyage of the *Wilmington* was an object of formal protest by the Brazilian legation headed by Joaquim Francisco de Assis-Brasil in Washington to Secretary of State John Hay. Hay apologized to Assis-Brasil, asserting that Todd had had no intension of violating the laws of Brazil, but also took the stern posture that in fact Todd was the wronged party and the population of the Amazon had been discourteous to an American official.[10] The public outcry was such that the first Bolivia-US agreement about the Acre territory became a dead letter, but it galvanized the Acrean rebels. Meantime, Luis Galvão made his way to Acre.

The Acrean revolution was directly related to the *Wilmington* episode and the dangerous agenda revealed by Galvão. The insurgents used the agreement point by point to further incite local rebels.[11] On Bastille Day, July 14, with support from the state of Amazonas and its intrigue-minded (and open handed) governor, José Cardoso Ramalho Junior, the Acrean *seringueiros* rose up and expelled the Bolivian delegation based in Puerto Acre. The Acreans feared complete dispossession of their holdings and had mobilized to forestall any local Bolivian action.[12] This event ushered in a new republic.

The Independent State of Acre was proclaimed with the Spanish adventurer and journalist Luis Galvão as its president. With his usual dramatic flair, Galvão proceeded to issue decrees and proclamations, baptizing the new revolutionary state as the "Republic of Poets." What the uprising unleashed, frivolous though it seemed and gilded with the romance of rebellion, was serious in the extreme. Bolivia began a military invasion of the region and its infant republic.

Bolivian president José Manuel Pando knew the Amazon well, having been incarcerated there for revolutionary activities in his youth, and had been part of a boundary survey team in 1891, along with Belgian astronomer Luis Cruls, then head of the Brazilian Astronomy Institute. Pando was aware that the Purús, Iaco, and Juruá were full of Brazilians in territory that he thought Bolivian under the 1867 Treaty of Ayacucho. Also the Suárez brothers, Nicolás and Gregorio, the masters of latex production in the Bolivian lowlands, were applying political pressure, since lands that should have been under their aegis as lords of the Bolivian Amazon were claimed at every bend in the river by dark-skinned debt-enslaved Northeastern *sertanejo* migrants.[13] In order to secure territory that had defied Bolivian sovereignty, Pando sent another military expedition into the Acre. This was an arduous mission: the company traveled from the Mamoré to the Orton River, north into the Acre valley, and overran the *seringais*—the rubber estates—that were central to the rebel base. In this instance Bolivia defeated the insurgents. Galvão was secreted to Manaus and paid off with seven thousand pounds, allegedly for some sensitive information in his possession, probably pertaining to the role of the Amazonas governor in supporting the revolution.[14] The "Emperor of the Amazon"—Galvão—vanished after this adventure (and with a tidy sum) from the pages of history.

Bolivian flags unfurled over the Acre valley, survey teams were organized, and the Acre River was opened to international navigation. Control of the region was still very much open to question, given incessant guerrilla skirmishes. La Paz sent a plenipotentate and extraordinary minister to the United States to request US intervention in the pending problems with Brazil, perhaps waving the *Wilmington* Pact.[15] Olinto de Magalhães, the chancellor of Brazil, held his own in the flurry of diplomatic communiqués but urged Assis-Brasil to impede insofar as possible any international repression against the rebels, even though Brazil formally did not recognize their sovereignty over the Acre. At least not yet.

Magalhães had worked with Rio Branco in Washington on the Misiones arbitration, so he was not a novice in this arena, but the situation was rapidly deteriorating, and Brazil's president, Manuel Ferraz de Campos Sales, did not really have the measure of what was unfolding in the Amazon forests. The United States was the main buyer for Brazil's coffee, and Campos Sales, a native of São Paulo, was aligned with the mercantile interests of his home state and thus loath to alienate North American commerce through what he viewed as some silly Amazonian incident, an insignificant boundary squabble. Further, previous chancellors, including Magalhães, had been compliant about the oblique line of the Ayacucho Treaty even as the region rang with Brazilian surnames and the twang of the *sertanejo* Portuguese. These earlier diplomats had upheld Bolivian rights over these vast forests inhabited by Brazilians, an unfortunate precedent give the economic value of the lands. In the meantime, Bolivian customs houses had been raised on the Iaco, Juruá, upper Purús, and Acre Rivers.

From Washington, Assis-Brasil alerted Magalhães that Bolivia was seeking business alliances through the agency of Charles R. Flint of Export Lumber. Flint was one of the great capitalists of the Gilded Age, a great consolidator of companies, and indeed his conglomerate of adding-machine companies would eventually become IBM. He had developed a major merchant bank and most centrally had merged a number of enterprises into the company known as US Rubber. He was also well known as a munitions dealer.[16] Flint's efforts on Wall Street could carry great weight, and his business coteries were comfortable with international ventures. They had consolidated extractive economies in Central America (including chicle latex used in Chiclets and Beemens chewing gum) and were central to Wall Street speculations in Panama, including the latex industry there.[17] And then there was Felix Aramayo, the Bolivian diplomat who was exploring other forms and sources of financing in Europe.

Outsourced Colonialism: Charter Companies

Bolivia's agent in London, Felix Avelino Aramayo, was especially active with and very clear on the virtues of charter companies. With Aramayo's proximity to London banking houses, the ubiquity of charter companies in Britain's empire, and his penchant for international deals, the possibilities of "charter" colonialism captured Aramayo's enthusiasm as a Bolivian development option. Bolivia had little domestic capital, and it seemed quite likely that the zone would become expensively militarized if other strategies of getting the revenues (and reducing the costs of governing) did not prevail.[18] Aramayo decided to approach the Belgians, the masters of colonialism in ultra-peripheries.

Aramayo hoped for funds from various Belgian consortia and overseas banking enterprises, a move that made sense since Leopold's imperial team had successfully developed African charter companies for the "Free Congo State," had flourished in the humid tropics, and were used to operating in the rubber trade. Indeed, as Leopold's private secretary Carton de Wiart noted in his 1901 trip to the Amazon, "there were still many chapters that remained to be written which he [Leopold] might or might not realize in South America."[19] At the time, Belgian capital was ranked third in total investment in Brazil, after that of Britain and Germany.[20]

The charter company was one of the mechanisms, along with direct colonization or military conquest, for the extension of imperial economic ambition. Charter companies can be seen as a type of "outsourcing" of colonial management, with a subject nation providing concessions for resources in exchange for royalties. The companies would raise international capital from bankers, states, and royal houses and would arrange for initial funding in the form of loans and military and diplomatic guarantees from sovereign nations. Among the most famous or notorious charter companies of the nineteenth century were those initiated by King Leopold for the Congo Free State (and the largest source of rubber outside the New World) and funded by his Banque Africaine.[21] Cecil Rhodes's British South Africa Company had been the means of annexing the goldfields of the Transvaal. The French West Africa Company, the British East India Company, and the Dutch East India Company were mercantile ventures that prefigured imperial territorial acquisition.

The virtue of charter companies was that they served both the interests of private capital (a proven set of resources, a range of financing arrangements in both the public and private realms) and guaranteed territorial hegemony through the long-term lease arrangements with the subject state (typically twenty-five but often fifty to one hundred years),[22] backed up by military powers of the imperial state. The advantages to the colonial state were access to overseas territories and management of resources and markets without the costs, complications, and resentments of direct imperial administration. Charter companies were a sort of "starter" colonialism and could become stalking horses for colonial territorial annexation. Bolivia's economic and political weakness, made that country an obvious entry point for any state with colonial ambitions in Amazonia. As Colonel Church had wryly noted, the whole place was more or less for sale.

The mechanism of the charter company was alluring since not only did it permit initial military security through company mercenaries, but should there be a larger menace—whether political (a revolution, let's say) or military threat—the nation where the company was incorporated could choose to intervene on the behalf of its beleaguered citizens. A typical case was that of Panama, where in 1903 an uprising resulted in US intervention and its annexation. Another was the invasion of the Transvaal by England when British gold miners clashed with Boer farmers over their lands, triggering the First Boer War. Da Cunha was to make much of the guerrilla tactics of the Boer War and drew close analogies with the triumph of the Boers over Rhodes and

Britain, emphasizing how the resistance of the Acrean *seringueiros* to Bolivian Syndicate echoed the South African situation.

Aramayo saw in charter companies great relevance for the economic exploitation to the Bolivian Amazon. An entrepreneur alive to the potentials of private capital bolstered by state securities, he and his family had already acquired extensive mining interests in Bolivia. Aramayo was well integrated into the inner circles of President Pando's regime, and as a financier, he was sensitive to the economic and imperial possibilities provided by London's financial "City."[23] And the Bolivian state was broke. Aramayo, not surprisingly, looked to the Continent, to Belgium.

The Charter and the Brazilian Congo?

Belgian companies held large concessions throughout the southern tributaries of the Amazon. The cattle operation Cibils produced beef extracts (basically Bovril) used for invalids and for military campaigns from its large herds in Mato Grosso near São Luis de Caceres, a spread of some 750,000 hectares. Large Belgian concessions existed on the Rio Juruéna, a tributary of the Madeira in Mato Grosso. The Belgian Banque Africaine had financed 11,000 hectares of rubber estates on islands in the Amazon estuary and lower Amapá. Belgian interests had purchased from a former president of Bolivia, Adolfo Ballivian, rubber forests on the Abunã, and when George Church's companies failed, they bought monopoly rights to steam navigation on the Guaporé and the right to construct a railway around the San Antonio falls, rights that were later sold to the American entrepreneur Percival Farquar.[24] Leopold himself had kept an eye on Amazonian resources ever since he had offered to take over the "neutralized zone" in the Amapá and turn it into an independent state along the lines of the Congo Free State. The king had maintained significant interest in the Araguaia-Xingu region, where he hoped to use development of railways to the São Francisco River and steamships to colonize and later expropriate the terrain, as he had done in the Congo. Belgian "shock troops of empire," field managers and surveyors who had cut their teeth on Belgian African enterprises, could be found in numerous enterprises in Brazil. The same resources—rubber, animals, and metals—were in play, as well as a "cannibal" population that had to be subdued. The Amazon as a mirror image of the Congo was noted widely at the time. Thus the monarch lurked on the edges of the Acre deal, although ultimately he declined.[25]

While success eluded Aramayo with the Belgians, on June 11, 1901, Aramayo and Frederick Willingford Withridge, financial agent for the Vanderbilt family, signed a contract that would lease 207,200 square kilometers of Bolivian Acre territories for a period of thirty years through the creation of a charter company, the Bolivian Syndicate. Another 25,900 square kilometers in the Caupolicán were also leased but on quite different terms. There all mining rights for fifty years belonged to the company, and it had the right to *purchase* the entire territory. These were exactly the same territories that Azanel Piper had yearned to colonize (and on similar terms), but these lands now would come under the aegis and management of the new corporation. Sir James Conroy—later the director of the Bolivian Syndicate—had spent some time in these regions and saw in them an "Amazonian Transvaal." It may well have been that the Acre, through extremely valuable on its own, helpfully diverted attention away from the Caupolicán, the golden jewel in the crown of the Syndicate, and thus permitted more complete possession of, as Church put it, "these auriferous lands."

The members of the Bolivian Syndicate included the great financial powers of Wall Street and American business: the US Rubber Company, Metropolitan Life, Export Lumber, Frederick Whitridge , Emlin Roosevelt (cousin of the soon to be president Teddy Roosevelt), the Central Trust, Vermilya and Co., the banking firm of Morton, Bliss and Company, the British bankers Brown, Shipley and Co., and clandestinely, through an illegitimate son, the House of Rothschild.[26] Several members of the Syndicate were players in Panama as well: Flint, Morgan, the House of Rothschild. International territorial/infrastructure deals and speculations were exceedingly appealing to these Gilded Age magnates.

Baron Rio Branco, still in charge of the Berlin legation in 1901, was extremely distressed at the sudden European interest in the upper Amazonian territories. Charter companies, as evidenced by the African Scramble, were a dangerous precedent and would ignite further insurgency, rather in the same way that Rhodes's Transvaal Charter had exploded in the ongoing Boer War. In the midst of guerrilla war, the region's future was up for grabs.

A Simple Contract

The charter was straightforward. The Syndicate, based in New York, would have fiscal administration of the Acrean territory, with exclusive rights for thirty years to levy taxes and exact customs, tolls, and land rents in conformity with Bolivian law. It had the right to use force, to maintain an army to defend its rivers and privileges. No less than 60 percent of the profits on the initial capital of 500,000 pounds sterling would go to Bolivia, and the rest to the charter company. The company could buy any land in Acre not claimed by others under Bolivian law, and all mineral rights would belong to it. The

Syndicate would construct or subcontract the development of roads, ports, telegraphs, and other crucial infrastructure, had rights of free navigation and could, at its discretion, grant navigation concessions. It would study ways of uniting the Acre river systems by rail with the rivers Orton and Madre de Dios. Both parties—Bolivia and the Syndicate—would maintain a regional manager to facilitate communication. More sinister were the articles that indicated that the company would respect local property rights as long as they had been registered according to Bolivian laws. Since there was some question about sovereignty, many concessions had been conferred by Brazilian rather than Bolivian authorities, and most were not registered at all since people had simply squatted on enormous territories in these remote outposts. The usual means of enforcement of property rights and other legal niceties in the Acre was the law of the .44 (Winchester), which typically carried little documentation. This meant that the majority of Brazilian holdings would be technically invalid and their owners dispossessed.

The charter's second concession, the Caupolicán district, was to be exempt from all taxes for fifty years. The syndicate could tax and place tolls on the river and had exclusive title to all the lands once they were surveyed. The Caupolicán was perhaps as compelling to some of the syndicate members as was the Black Gold of rubber. The Caupolicán would, then, become a charter colony at the intersection of Bolivia, Brazil, and Peru.

The Bolivians caviled about the nature of the charter, but Syndicate lands were to become an American colony, in fact if not in name. When President Pando advanced the proposal for a vote within his government, he was explicit about the "diplomatic help" that would accrue from the neighbor to the north. General Pando understood that occupation of Amazon forests and swamps was a questionable military option for Bolivia, but he hoped that US resolve in defending the "interests" of its citizens—as it had in various filibusters and when it sided with Venezuela (and its US agents) against Great Britain over occupation of the mouth of the Orinoco—would provide a useful precedent for intervention in the upper Amazon. Pando himself would later lead a doomed expedition to the Amazon, abandoning his presidency while he personally led battalions to the jungles.

Secretary of State Hays even wrote to his consul, George Bridgman, in La Paz with the list of American Syndicate members and, in a completely "unofficial manner," expressed his hope for Bridgman's enthusiastic assistance and support in their enterprise.[27] Hays' informal imprimatur and the list of the "lions of Wall Street" seemed to ensure US government commitments to its powerful citizens. At the same time that the Bolivian Syndicate was under discussion in the Chamber of Deputies in La Paz, a Belgian charter under the name L'Africaine formed to develop ports on the Paraguay and railway concessions from Santa Cruz to Potosí was also approved. Bolivia was giving away a great many sovereign rights.

Assis-Brasil met with Secretary Hay and began to form a line of argument that would hold special power in the Acrean case. In consultation with John Bassett Moore, whose diplomacy hewed closely to the Monroe Doctrine,[28] Assis-Brasil argued that "while today it might be an Anglo-American colony under the pretext of a charter company, next year it might be a purely British, purely Belgian or French colony, or who knows who else." The arguments that were about to made in the general diplomatic campaign against the Syndicate were best articulated by Rio Branco:

> If the British enter into this business with the Americans, it is natural that interests and commercial rivalries would lead other European countries to take advantage of this opportunity to bury the Monroe Doctrine. Up to today it has been a means of scaring off Europeans and it has served us well on many occasions, notably the bitter experiences with France over the limits of Guiana. The fact is if we appear to disagree with or have a conflict of interest with the United States, who until now have seemed our firm allies, this will do us considerable damage, weakening our position substantively in the eyes of Europe. If by chance the US invites European governments to exploit the lands of South America by imposing complete freedom on the Amazon, it will be difficult for them to refuse the invitation.

Rio Branco closed his note with a chilling coda: "I swear that a half-dozen ambitious men in La Paz or New York will not succeed in awakening the covetousness with which the US in other times regarded Amazonia, which caused so much perturbation and trepidation in more than one cabinet during the empire."[29]

The Baron had the previous US escapades in Amazonia clearly in mind.

Revolution in the Tropics

The Bolivian Syndicate, with its highly internationalized set of partners, set off the next revolution in Acre and a suite of diplomatic and military events that took Brazil into war with Bolivia and Peru. The Acreans, under revolutionary leader Plácido de Castro (a former schoolmate of Euclides), led an armed insurrection. Given the Syndicate's similarities with African charters, the Acrean rebels were soon compared to the "Homeric insurgents" of the Boer War. Others, of a more capitalist bent and inspired by the Syndicate experience, were eager to extend the possibilities of charter companies. Auguste Plane, the chargé for French commercial missions, wondered after his trip to the Amazon why his nation's capitalists couldn't organize for a similar enterprise.[30]

The threat of external military intervention was very real to the Brazilians and the Peruvians alike. Both countries strongly protested the Syndicate, since Peru believed the entire area under contest belonged to it, along with a substantial chunk of Brazilian territory, as a consequence of the Idelfonso line ceded to the Viceroyalty of Peru in 1777. The Syndicate, with the powerful military potential of the US behind it would make any other claim almost impossible to prosecute. Chile was preparing to provide gunboats to quell the Acrean uprising. Other alliances pro and con were coalescing in covert ways. The Brazilians registered pointed complaints through diplomatic channels that given the length of the leases, the company lands were a de facto American colony; that the presence of an army so close to undemarcated borders was a cause of immense concern; that the Brazilian populations within these areas had no safeguards of person or property; that Bolivia had unilaterally transferred shipping rights on Brazilian waters.[31]

When Syndicate director Frederick Whitbridge and Andean explorer and field director of the enterprise Sir Martin Conway arrived in Berlin with a letters of support from the British foreign minister Joseph Chamberlain, Rio Branco intervened strongly. The Syndicate had come to request monies from German bankers, not because they lacked capital but because they preferred a wider diversity of investors and a greater range of allies. Germany's imperial prestige added to that of Britain would have further legitimized the Acre enterprise as an international colony in the heart of South America. Rio Branco expressed in the clearest terms to German foreign secretary Manfred von Richthoven that the boundaries were in contest and asked him to advise his national bankers of the risk of lending to the Bolivian charter. Rio Branco imposed on Edouard de Rothschild to use his influence as well. Assis-Brasil meantime had had an acerbic visit with Secretary of State Hay, who telegrammed his consul Charles Page Bryan in Rio, "Let it be known to whom it may concern, that it would be very disagreeable to this government that the interests of innocent Americans suffer in the Acrean Question."[32] Thus, while expressing a considered position of neutrality in the boundary question between two sovereign states, the behind-the-scenes posture that the United States and Bryan took was that US Acrean interests would be "vigilantly watched and resolutely protected."[33] This was the "diplomatic support" that Bolivia yearned for.

Rio Branco was sworn into his post as foreign minister in December 1901 in the midst of the Amazonian volatility, and he began to completely revise the stance and policy of earlier administrations with his usual originality.[34]

The War at the End of the World

Once the Acreans discovered the details of the agreements of the Bolivian Syndicate, the charter company of which they would now be subjects, they prepared their rebellion. Their revolt was easily justifiable after the arrival of Lino Romero, designated representative of President Pando. A more profoundly divisive and incompetent figure for a region as politically delicate as the Acre is difficult to imagine. Romero arrived in April 1902 and immediately emptied out the administrative positions of the previous Bolivians. This move would alienate his local countrymen, who had been relatively obliging to the Brazilians, thus calming resentments over earlier battles in the way that those on frontiers often made accommodations to the "enemy" due to their shared circumstances. Romero proceeded to populate these posts with his cronies in the capricious manner typical of petty caudillos. He further required that all residents register their lands six months after his arrival and pay a fee to do so. This was a lucrative sideline, with possibilities for acquiring the lands of holdouts, as well as a useful arena for corruption, but for the Brazilians it seemed the first step in their dispossession.

The Bolivians lacked military manpower in Acre, and those Bolivians who were there were functionaries rather than fighting men. The battalions in place were suffering from beriberi and malaria. But rumors were flying, and Romero had irritated widely. The Acreans prepared an uprising. A revolutionary junta, formed by the members of the leading commercial houses and the larger estate owners, began to organize men and materiel and to smuggle arms into the backwaters. Plácido de Castro was placed at their head. Although he was making his fortune as a surveyor, and although as a gaucho he preferred a caudillo style of governing, he was taken up in the revolutionary possibilities and rhetoric of the moment. It certainly helped that several of the estate surveyors, including Plácido himself, formed its strike force, since they knew the terrain, had tropical backwoods skills, and had the maps.[35]

The Brazilians yearned to attack on Bastille Day as they had when led by Galvão, but they were not quite ready and so chose August 8, 1902—symbolic in its own way, as it was the Independence Day of Bolivia. And so on that day, before Bolivian officials had even gotten up, Plácido and his rebels arrived. "It's a bit early for the party, don't you think?" quipped the sleepy superintendent. Plácido replied: "It's no party, Commandant, it's revolution!"[36] He

Figure 9.1. Plácido de Castro (on white horse) and his revolutionaries.

and his rebels took the capital, Xapuri, a town whose revolutionary bona fides would last into the end of the twentieth century as the center of the rubber tappers' movement, the home of labor organizer and environmentalist Chico Mendes and the site of his assassination. With the capture of Xapuri, Plácido declared the Independent State of Acre with the usual florid manifesto. This revolution was carried out with the intention of annexation of the independent republic by Brazil.[37]

The battles were not yet over, and Pando himself chose to lead battalions down into the Amazon a few months later, in January 1903. Defeating the guerrillas in the absence of absolutely overwhelming force could hardly be done in a place as physically complex as the upper Amazon, where, as da Cunha noted, those most formally trained would be its least capable soldiers. Military men from the Andes found the thousand-kilometer slog through the Orton and Madre de Dios watersheds grueling, and troops deserted (and died) by the score. As with the Paraguay War and the battles of Canudos, these tropical conflicts involved combating the environments and disease as much as the enemies. While *force majeur* could ultimately prevail, the locals were implacable, and the formal armies, if they triumphed, did so at great cost. Bolivia had no large army; the mortality of Bolivian combatants in the

Acre was such that recruitment was quite difficult and desertion frequent. Nonetheless, General Pando, who clearly preferred to resolve territorial questions by military campaigns rather than palace politics, had lurched off to the jungles for the last campaign.[38]

The Acreans continued their guerrilla war. Rio Branco was well versed in the details of this particular conflagration as well as its diplomatic antecedents, thanks to his time in the imperial court when his father was foreign minister and the Ayacucho Treaty had been signed. A series of private letters from Assis-Brasil had kept Rio Branco abreast of the Acre imbroglio. The Baron, Assis-Brasil, and John Bassett Moore had decided on a different means of resolution of the Acre question, even though Brazilian battalions had their orders and were moving toward the border of Mato Grosso and into Manaus, and the lands claimed by Bolivia were in the hands of Acrean rebels. Brazil's position was helped by more general problems in the imperial world. Germany had declined to participate in this conflict, and Britain, eyeing the state of affairs in South Africa, seemed little interested in yet another guerrilla war.

There were three key elements that underpinned the Rio Branco approach in the Acre and echoed his general diplomatic principles. First, the Treaties of Madrid, Idelfonso, and Ayacucho would have no standing in this controversy. Those pacts pertained to New World terrains of the extinct Spanish and Portuguese empires. While the agreements might provide useful guidance for modern independent nations, in Rio Branco's view they had no legal sway in modern republics. In any case, Idelfonso had never been ratified by Portugal, in part because of the lack of a boundary survey and the fiery enmity of Iberian rulers held for one another at the time. Thus traditional treaty claims not signed by the *current* republics had no validity. Next, *uti possedetis de facto* and not *de jure* would be the most powerful claim for disputed territory. And finally, Rio Branco would negotiate only bilaterally. Three-way diplomacy would require that Brazil give up significant territorial concessions, which it was not interested in doing with Bolivia or Peru. The boundary questions of Peru and Bolivia would be taken up separately.

At issue, as we have seen, was the legality of Bolivia's right to transact terrains with undefined boundaries. As the region had become more contentious and militarized, Bolivia and Peru agreed to submit to arbitration by the president of Argentina, José Figueroa Alcorta, and in December 1902 an agreement was signed in La Paz. Bolivia would be the stalking horse for Brazil's interests, with the Bolivian position argued in da Cunha's *Peru versus Bolivia*, while Rio Branco would negotiate directly and in sequence with Bolivia and then with Peru. In the meantime Brazil had to devise a way to thwart the Syndicate.

Members of the Bolivian Syndicate may have been the lords of finance, but the concessions were worthless if they were inaccessible. Brazil had opened the Amazon to free navigation, but its nineteenth-century proclamations were limited to the main channel and regular ports of call, such as Tabatinga and Manaus. Free navigation rights did not apply to the tributaries of the Acre territory, such as the Purús and the Juruá. Open navigation there was a courtesy enabled by Brazil, which now prohibited it for goods moving to and from Bolivia. The river was shut down. The United States advanced a mild protest about this state of affairs in January 1903, noting the hardship this caused North American business in Bolivia, and urged rescinding this prohibition. Navigation remained closed. England, France, and Germany protested vigorously, largely because their commercial houses and agents were howling, as this embargo disrupted the flow of rubber and their upriver trade. Despite the uproar, for the time being the Purús and all its affluents were off limits. This was quite a dramatic step.

The presence of charter companies like the Bolivian Syndicate set a dangerous precedent; as Assis-Brasil had suggested, they might be American today but European tomorrow. So the better strategy, from the Brazilian viewpoint, was to enforce the Monroe Doctrine and let European imperialisms (and their charter companies) infest Asia and Africa. These arguments were advanced by John Bassett Moore, who was advising the Brazilians and a firm advocate of the application of the Monroe Doctrine whenever new European claims appeared in the Western Hemisphere.[39] The Syndicate, after all, included Europeans (the British and, covertly, the Anglo-French illegitimate offspring of de Rothschild),[40] and though owned by Wall Street magnates, the concessions could be sold, or controlling shares could shift to international agents. This form of economic imperialism could lead to the foreign partitioning of the interior of the continent, a possibility that would cause anxiety in Washington as well as South American circles. Shifting sovereignties over the malarial coasts and gold diggings might not have mattered much in the Guyanas, but incursions into the heart of the vastly lucrative rubber economy was a completely different set of politics.

Finally, the Bolivian Syndicate was basically a corporation elaborated by some of Wall Street's most seasoned speculators. Given this, it should be possible to acquire its concessions. After all, substantial areas of the United States—Louisiana and Alaska, for example—had been bought from sovereign states. If Bolivia (or at least those in power who stood to gain from the transaction) could be placated and the Wall Street speculators appeased, the Acre could end up in Brazilian hands without a costly war. With this possibility in mind, and with Assis-Brasil discreetly hinting to Secretary Hay that this might be the solution, Rio Branco authorized the House of Rothschild to negotiate on Brazil's behalf with the Syndicate. If the financiers were justly compensated, there was no reason for the United States to intervene in Acre. Bolivia was too poor to pay off the Syndicate, but Brazil could guarantee loans that would, and also paid two million pounds to Bolivia.[41] Whitridge began negotiating in February 1903, and by March, an initial payment of 550,000 pounds flew into the Syndicate bank account.[42] The satisfied financiers, who never actually did anything but get a buyout and whose field directors never got closer than Manaus, cheerily counted their cash. Once the transaction was completed, Brazil immediately opened the Purús to international trade. The details of the treaty needed to be worked out, though, and Peru was becoming less tractable.

After much negotiation, the Treaty of Petrópolis required that Brazil pay two million pounds to Bolivia and the Bolivian Syndicate in exchange for 191,000 square kilometers of Acre territory. This boundary along the Rio Verde would be surveyed by a member of the British Royal Geographical Society, in this case the always self-promoting Percy Fawcett.[43] Bolivia received in exchange 3,000 square kilometers—a sliver of Brazilian territory between the Abunã and Madeira Rivers—and the commitment to build a rail that would bypass the more than 200 kilometers of Madeira rapids. This was the infamous Madeira Railway, whose virtues were described by Church and where he had such interminable (and bankrupting) difficulties. The new rail enterprise would hew to Church's trajectory, including catastrophic levels of disease, worker unrest, and the ruin of another financial empire, that of American magnate and infrastructure developer Percival Farquar.[44] Farquar finished the railroad just as the rubber industry breathed its last, making rail transport irrelevant in the western Amazon. Bolivia and its boundary were still in contest, however. With the Treaty of Petrópolis in November 1903, a modus vivendi and demilitarized zone were put into place while the boundary was surveyed.

The treaty took the Brazil-Bolivia conflict out of play, but the pact was powerless over the northerly battles raging on the upper Purús between Peruvian *caucheiros* and Brazilian tappers. Peru moved into place, asserting its Idelfonso rights, which included all the claims of Bolivia and terrains from the Andean flanks almost to the mouth of the Madeira on the Amazon. This was da Cunha's litigious zone.

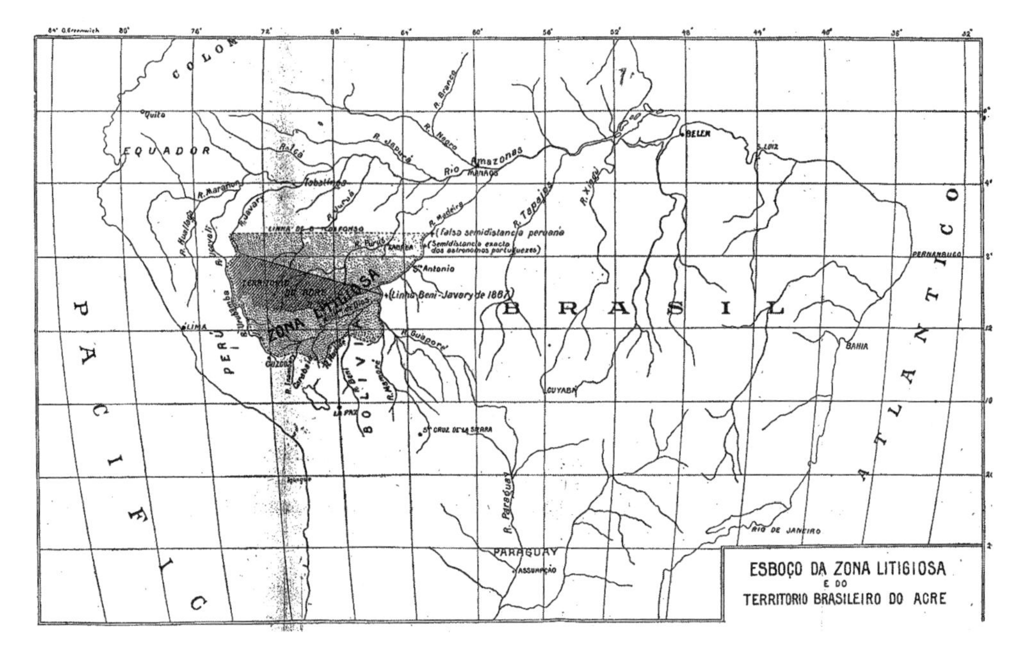

Map 6. Litigious zone.

10

Peru, Purús, Brazil

A Simple Dichotomy . . .

Peru's *caucheiros* recognized the virtues of the Purús. The upper reaches of the Madre de Dios and the Ucayali had had significant stands of the latex known as *caucho*, from the genus *Castilla*, but output was declining because to get the latex the trees had to be cut down. The terrains of the upper Juruá and Purús were an untapped frontier and incarnated a loose economic boundary between the production systems associated with *Hevea*—rubber or *seringa*—and those of *Castilla*, or *caucho*. A simple dichotomy can be made about the forms of extraction of latex and labor. *Caucho* extraction required killing the tree, so exploiting it was a nomadic but highly profitable proposition. The latex of one *caucho* tree could equal the returns of a year of tapping rubber. *Hevea* trees, on the other hand, were tapped every few days by a sedentary population.[1] *Caucho* was largely worked by native labor under varying forms of coercion. *Hevea* trees were tapped by migrants in more or less stable settlements under many forms of labor deployment, including debt peonage. In the upper Purús and the Juruá these two socio-environmental systems encountered each other, with the ricochets of their gunshots echoing from the remotest *tambo*—tapper hut—to international capitals and the highest realms of diplomacy.

This frontier between *caucho* and *Hevea* rubber is often taken as a biogeographic boundary, but analysts like Jacques Huber, director of the Museu Goeldi and rubber specialist during the boom; Harvard ethnobotanist Richard Schultes; and USDA rubber specialist Russell Seibert have noted that the ranges of *H. brasiliense* and *Castilla* often overlapped throughout the basin.[2] The encounter of *Hevea*s and *caucho* was really a clash of political ecologies, one that would ultimately be expressed in an international boundary. While "Acre fino"—fine Pará rubber—was taken from *H. brasiliensis*, the high prices in the global markets in the 1890s and 1900s made many gums less integrated into the earlier trade like *Castilla* and *balata* (*Manilkara*), economically viable and sent adventurers into the most far-flung Amazonian backwaters. Other "weak heveas"[3] like *H. lutea*, *H. viridis*, and *H. guyanensis*, and *H. benthamia* occurred widely alongside *Castilla* and were also in use, especially on the Ucayali and Putumayo. They fetched the lowest prices and were often used as an additive (or adulterant) for other latexes. But the *Hevea*s in the eastern Amazon were being overtapped,[4] and news of the extensive *Hevea* forests of the Purús began to galvanize the trade.

Overexploitation of rubbers of all kinds had certainly affected the trees—whether *Castilla* or *Hevea*—driving their collectors from the eastern Amazon and western Peruvian jungles to clash at the remote reaches of the Purús and Juruá, but there were more complex political economies that underpinned the movements of latex seekers into the headwaters. These included the monopolization of *caucho* in the upper Amazon: the houses of Arana, Suarez, Dias, and Fitzcarraldo had consolidated their empires and expelled smaller operators. There were also changes in labor regimes in response to stratospheric latex prices and accelerating global demand. These forces were all pushing the less powerful enterprises out of the terrains of the great masters of the regional rivers and into the Purús watershed.

Geographically, the great western *caucho* empires extended in a swath from the Colombia-Venezuela frontier to the *llanos* of Bolivia. The Casa Arana dominated in the Guaviare, Putumayo, and lower Ucayali and the Fitzcarraldos (and a number of their lesser vassals) claimed the mid- and upper Ucayali and Madre de Dios watersheds, while the Suárez brothers dominated Bení-Madeira.[5] These powerful "über barons" formed a bulwark of latex monopolies that was actively expelling many smaller or less compliant operators from their domains, leaving eastward movement into Brazil as the only option.

Figures 10.1a and b. Two *über-caucheiros* of the Upper Amazon: Arana and Suárez.

One of these fugitives was the former henchmen of the Fitzcarraldos, Carlos Scharff, who took over the upper Purús after Delfim Fitzcarraldo was killed by Indians. Why Scharff appeared on the scene is more complex than the clichéd narrative of rapacious *caucheiros* slashing the latex trees into oblivion, although that was certainly one part of the story.

The powerful economic incentives provided by the increasingly feverish rubber demand made almost any kind of latex extraction a going concern, but the key problem was labor organization, a question of broader concern in industrial and agroindustrial enterprises at the time. Julio Arana, one of the most powerful "Caucho Kings," was a person of modern sensibilities (he much admired Henry Ford) when it came to the organization of the firm. He had begun to radically rethink *caucho* and wanted to modernize the extraction in a number of ways, including labor control, by "rationalizing" production, changing administration, and labor deployment. Traditionally, native men trekked into the forest for hunting parties, ritual practices, raiding their enemies, and capturing women, so gathering *caucho* to return it as tribute to tribal chiefs in exchange for steel and manufactured goods fit relatively well into existing cultural practices.[6] Part-time collection allowed indigenes to engage other necessary activities (like agriculture) and kept tribal structures of authority intact; it may well have augmented the powers of chiefs at the intersection between indigenous labor deployment for latex extraction and the circuits of exchange for manufacture.[7]

Arana rejected this system as inefficient. Latex gathering shifted from ad hoc to obligatory collection mediated by violent coercion of natives, although this was not exactly a new invention on his part. Terror was required to control populations that could move freely through the forest, so an un-

Figure 10.2. Armed militias at Carlos Scharff's caucho post.

imaginably more vicious system of control was installed, enforced by Arana's Barbadian overseers—harsh white/mestizo section chiefs who were paid on commission for the latex—and his indigenous child warriors who hunted down fugitives and procured new captives. These power relations were brutal in the extreme and caused huge dislocations as well as mortality—Bora and Huitoto Indians were moved around within the Putumayo, and at least thirty thousand died in the process.[8] Large native populations were relocated by the *caucho* bosses. Amahuaca, Piro, Campa, Shipibo, and Witoto, for example were taken into the Madre de Dios.[9] This "blood rubber" and terror slavery narrative of the Arana reign on the Putumayo eventually dominated the understanding of labor relations in all the latex economies.[10]

The differences between the *caucho* and *Hevea* economies are explored in greater detail in chapter 14; indeed production processes and labor relations were quite different in these two economies, and this difference would be a key theme in da Cunha's narratives about the past and future of the upper Amazon.

Fernando Santos-Granero points out that when labor practices began to threaten national interests, especially as the territorial conflict with Brazil heated up, the Peruvian state tried to regulate these labor regimes but was at the same time using freelance *caucheiros* as well as powerful "Caucho kings" to expand and consolidate Peru's tropical territories. The efforts of the Casa Arana to consolidate holdings on the Putumayo transferred what had been Colombian lands to Peruvian sovereignty. Arana had added significant territory to Peru by expelling Colombian *caucho* barons during Colombia's War of a Thousand Days, a time when no Colombian military was likely to appear. This kind of "accidental imperialism" preceded formal state claims. Peruvian magistrates in Iquitos were thus inclined to overlook the matter of native slavery, given the geopolitical stakes.[11] Julio Arana would eventually argue that it was geopolitics, not production practices, that lay at the heart of the denunciations against his Peruvian Rubber Company.[12] Roger Casement, who led the international denunciations against Arana, noted the extensive degree of Peruvian military protection received by the agents on the Putumayo.[13]

Outcry over the brutality associated with caucho extraction was beginning to rise in the press. The "terror slavery" that mimicked that of the Congo was causing international problems, as journalists and adventurers roamed in the backwaters without Casa Arana sanction and various religious orders complained of the exploitation of Indians. There were rumors that the British threatened to boycott Peruvian rubber.[14]

Another Great Game in the Amazon

The Brazil-Bolivia conflict had certainly concentrated Peruvian attention on forests of the upper Amazon, while the Peruvian *caucho* barons effectively argued that they were the agents of Amazon "Peruvianization," since it was their (tormented) labor force that was integrating the trackless forests into the Peruvian nation.[15] The Purús watershed, some 2,800 kilometers long (and navigable for 2,600) was a flashpoint: the Acre revolution had unfolded on its southern tributaries while warfare with Peru was exploding on its main channel.

Peru's claims on Amazonia, outcomes of the Madrid and Idelfonso Treaties, were vast. What was at stake was, as da Cunha put it, "the largest territory that had ever been contested between two nations, some 720,000 km^2 in one of the very least known parts of the planet."[16] If Peruvian claims held, it would place towns like Lábrea, which at the time had greater export-import accounts than Manaus, under the control of Lima. The upper Purús was a nexus where conflicting colonial documents and maps did little to defuse the clamor of the countries over rights and where ambiguous boundary histories provided useful options for extending national territories. Peru was moving from peripheral status in Brazilian quests for territoriality to the center of its imperial anxieties. Not only did Peru's claims place an enormous area of very valuable Amazonian lands in contest, but this conflict was the last of the Brazilian Scramble. Once it was finalized, whatever the configuration, Brazil's immense border would be completed, and the outlines of the modern nation defined.

Echos of Idelfonso and Louisiana

The official Peruvian position was that Peru's lands extended to the historical lines established by the 1777 Treaty of San Idelfonso, from the midpoint of the Madeira following the river course down to its headwaters, and thus Bolivia had no legal right to transfer to Brazil what it thought were "its" terrains. Peruvian diplomats maintained that it was indeed unfortunate that Brazil had paid all that money to buy off the Bolivian Syndicate and the Bolivian government, but the "purchase" of these lands, on its own, did not guarantee territorial claims.

The only treaty of the modern era that addressed the boundaries between Brazil and Peru was that of 1851, one of the usual "commerce and navigation" treaties. Article VII of that treaty, as da Cunha was at pains to point out[17] and as John Bassett Moore would summarize in his pamphlet on the topic,[18] stated: "In order to prevent disputes respecting the frontier, the High contracting parties, The Empire of Brazil and the Republic of Peru are regulated by the principle of *uti possedetis*: consequently they acknowledge the frontier at Tabatinga, thence a line north to the River Japurá, and to the south to the River Javari. A mixed commission will decide the frontier according to the principle of *uti possedetis* and will arrange for the exchange of territories as may seem proper for fixing the most natural and convenient boundaries."[19] Although Peru was adamant in its assertion of the 1777 Idelfonso lines in the Purús controversies, in 1851 it had in fact accepted a quite different principle.

The principle of *uti possedetis*, as already noted, had been recognized by several nineteenth-century boundary treaties, including that between Brazil and Venezuela (1852), the convention between Brazil and Paraguay (1856), and that between Brazil and the Argentine confederation (1857), and indeed the contretemps with the French in Amapá was also ultimately decided, as we have seen, on the basis of Brazilian occupation. Brazil and Bolivia had invoked this principle over the Acre territory in the 1903 Treaty of Petrópolis. Peru, however, hotly disputed all accords that were not in compliance with its claims based on the Idelfonso treaty. Diplomatic telegrams crackled between

Lima and Rio. Meanwhile, the *varadouros*—the forest trails—from the Ucayali filled up with *caucheiros*, enslaved Indians, and their masters roving into the upper tributaries of the Juruá and Purús. If possession was nine-tenths of the law (and indeed it was the entire substance of *uti possedetis*), then there had to be boots on the ground and Indians hacking away at trees.

Brazil had occupied the southern banks of the Amazon and its main tributaries, especially the Purús and the Juruá, and did so well into their upper courses. Given this physical occupation, Moore, employed by the Brazilian legation to shape the issue in light of international jurisprudence, argued the legality of Brazil's claims on the basis of early nineteenth-century US frontier precedents, injecting a novel set of legal principles into Amazonian frontier politics. Moore invoked the position of President James Monroe and his plenipotentate William Pickney concerning the boundaries of Louisiana, that were set forth in a letter to Pedro Cevallos, the Spanish minister of state. The legal substance of this position would later structure what would become the US doctrine of Manifest Destiny. Following the Monroe/Pickney argument about the Mississippi drainage, Moore asserted that Brazil's sovereignty over the Amazon River implied "deep sovereignty" over the hinterlands of the Amazon's tributaries, especially since most of the upper watersheds of the Juruá and Purús were occupied by Brazilians. Next, Moore argued, "a settlement is entitled not only to the lands actually inhabited, but to all those which may be needed for its security"[20]—an especially important point given the guerrilla war unfolding on the Purús, but one that, given the enormous size of the hinterlands of these rivers, justified territitorial claims over huge watersheds. Thus, "effective occupation" as required by *uti possedetis* was not the only form of claim on the upper Amazon forests; "deep sovereignty" over tributary watersheds and the broader needs of national defense were others. Using this reasoning, Moore folded the Amazon into the logics of acquisitive US territorial jurisprudence, consolidating this with *uti possedetis* of Roman law. These complemented the more traditional ideologies of "discovery" as claim, since the presence of Brazilians preceded that of the Peruvians. These more theoretical positions of Rio Branco–Moore diplomacy were further buttressed by the reality that Acre had been a purchase, just like Louisiana.

The Ground War on the *Caucho* Frontier

The clash between Peru and Brazil was due in part to the complexities of their historical treaty politics, but there were material pressures as well. The *caucho* system had led to local extinctions of the *Castillas* and a gradual eastward shifting of *caucho* hunters in search of the trees, even as the soaring demand for *Hevea* rubber kept stimulating upriver western migration from Brazil. Eventually these two migratory flows clashed in the upper Purús and Juruá.

In the late 1890s and early 1900s, Peruvian rubber *patrones* such as Carlos Scharff began to move large numbers of indigenous workers from Peru into the Juruá and Purús watersheds of Brazil to avoid the having their native workers hunted out and incorporated into the Casa Arana (or Fitzcarraldo) monopolies and to avoid possible state intervention in their activities as rumors and newspaper denunciations of *caucho* slavery became more widespread.[21]

The upper Ucayali and Madre de Dios had land connections—*varadouros*—to the upper Juruá and Purús, products of some of the historic trading routes of the Piros Indians and possibly even some ancient Incan supply routes. Leopoldo Collazos, a foreman of the Fitzcarraldos, is often described as the first Peruvian into the region, although others credit Delfim Fitzcarraldo.[22] Collazos arrived in 1899, established an outpost, then descended to Manaus and returned the next year with merchandise—and with a "secret agent" from Peru, Manuel Pablo Villanueva, who would later become Peruvian envoy in Manaus. Villanueva, clearly on the authorization of Lima, or Pedro Portillo, the magistrate in Iquitos, hoisted the Peruvian flag at the settlement of Cátiana at more or less the old boundary line of the Madeira-Javari where it crossed the Purús, and proceeded to open a customs house. As had happened in Acre when Bolivians erected a taxing entrepôt at Puerto Alfonso, the locals rose up in defiance. Oton Bacelar, the ostensible master of the Cátiana settlement, challenged the imposts with *seringueiro* rebellion. Collazos and Villaneuva quickly decamped to Iquitos, where they began to make military preparations with the garrison there for the conquest of the upper Purús.

At the end of 1901, Carlos Scharff had traveled from the Ucayali to the Juruá and into the upper Purús watershed with several hundred Indians. Historian Leandro Tocantins has described Scharff and his modus operandi this way: "Insinuating, ambitious and violent, the ideal characteristics for success in that savage environment, he soon obtained the power and respect among the local populations."[23] Scharff used the Peruvian military based in Iquitos to mediate his local quarrels, rather than the bands of thugs that more powerful river masters typically commanded. Rubber barons of all types and nationalities were using nationalist arguments to advance and ensure their personal economic claims, even as states used them and their workers as materializations of national authority in remote outposts.[24]

Scharff is interesting on another level. He used the native "tribute model" extensively in his estates and seems to have depended on the discipline main-

Figure 10.3. Barbadians and enslaved Indians.

tained by powerful chiefs.[25] These might be indigenous peons or populations governed under their tribal structure, as was the case with Venâncio Campa, who mobilized his clansmen for *caucho* extraction and himself became a kind of *patron*. Da Cunha would note the large plantation managed by Chief Venâncio. Scharff ultimately had more than two thousand "civilized" and native laborers on the upper tributaries of the Purús.[26]

Scharff had an ambiguous role in the Peruvian Scramble. He makes appearances as a backwoodsman in the remoter tributaries of the upper Amazon, but he seems in many ways to have been an agent provocateur mobilized to bring attention to Peruvian presence in the region, to manufacture conflict, and to draw in the military from Iquitos into the Purús. Whether this was just his character or an assigned charge is not known, but he was a consistently a point man on the Peruvian side, apparently able to command military backup whenever he needed it. Scharff liked to display a diploma that named him the governor of the upper Purús, incited local Peruvians not to pay the customs fees to Brazilian officials, and embarked on vicious *correrías*—hunts of native peoples. In many cases the indigenous people were in a kind of frontier "live and let live" relation with the *seringueiros*, who were for the most part tethered to their tapping routes and typically did not hunt Indians. These relatively cautious intercultural relations had underpinned the early successes of Brazilian occupation and were described in the history of places such as

Sena Madureira's *Gazeta do Purús* as well as in da Cunha's laudatory remarks about Manoel Urbano (see chapter 15, with a picture of Urbano).[27] Those who were tapping trees alone in the depths of the forest certainly did not want to upset relations with natives who could pick them off at will. Peruvian *caucheiros* traveling in large armed bands of twenty men or more, however, were happy to stir up trouble if it meant they could capture a new labor force and concubines.

One observer of the dynamics between *Hevea* tappers, Indians, and *caucheiros* was the earnest amateur ethnographer Alcot Lange, who provides an unusually detailed eyewitness account of this destabilization.[28] Lange, a wealthy young New York adventurer and gifted photographer, spent time—really the best term would be hung out—on a tributary of the Javarí and was a participant-observer when the durable relations of the cannibal Mangerona tribe and the traditional rubber estate based on *H. brasiliense* were caught up in battles triggered by incoming *caucheiros* hunting for labor and trees. Many of the enslaved natives died miserably in transit in the pestilent entrepôt of Remate de Males on the Javari. To evade this fate, Lange explains, the Mangeronas embarked on an array of defensive maneuvers, such as ambushes or poisoning creeks to make *caucheiros* ill, and he recounts a battle in which every *caucheiro* was killed—a necessary action for avoiding a later, obliterating reveng—and eventually eaten.[29]

Scharff was unreliable in his economic obligations—taking latex and not furnishing goods in return, or not delivering latex to his suppliers as promised—and local Brazilians threatened to expel him. He urged the Peruvian authorities to send in troops, and on June 23, 1903, twenty soldiers and two sergeants appeared at the mouth of the Rio Chandless, along with Iquitos police chief Jorge Barreto. They began to set up another customs post. A similar action occurred on the Amônea in the Juruá watershed, part of a coordinated effort to assert Peruvian presence on the upper stretches of these rubber rivers. Apparently the powers in Lima had reread the 1851 treaty and embarked on a policy of occupation.[30] Small steamers from Peru flying the Brazilian flag changed them to the Peruvian colors as they passed over what Peru understood as the boundary line, then made their way up tributaries of the main rivers, interrupting local commercial monopolies, defying Brazilian custom houses, and moving materiel (untaxed!) through the *varadouros* into the Ucayali. Rio Branco had two wars on his hands in the Purús watershed: the battles over Acre territory and guerrilla conflict on the upper tributaries with Peru. There were no formal troops fighting for Brazil.

If Plácido de Castro was the hero of the Brazil-Bolivian battles, his counterpart in the Peruvian clashes was José Ferreira de Araújo. Ferreira was a

Cearense migrant who had left for the Amazon in 1872 to meet up with a relative who had migrated there to escape the Northeastern drought of 1845. His long history in the region gave him great legitimacy in the ensuing conflicts. Infuriated by Peruvian actions, the *seringueiros*, under the leadership of Ferreira de Araújo, organized resistance, so on June 25 Barreto and his soldiers woke up to find themselves surrounded by two hundred Brazilian tappers. After a skirmish, the Brazilians captured the outpost on the Chandless; some fifteen Peruvians (most of the local garrison) died, and Barreto and Scharff were imprisoned in Manaus. Both were later released, probably to the regret of the locals. Scharff returned to Iquitos.

Colonel Pedro Portillo, the commander at Iquitos, and Scharff made plans together to retake the customs house at the mouth of the Chandless. Scharff, with a small contingent of thirty military men, two lieutenants, and a militia of some three hundred *caucheiros*, traversed the *varadouros*, and arrived on the upper Purús at Curanjá. A delegate from the Peruvian forces demanded that Ferreira leave the post. This emissary was promptly taken prisoner. A rescue party was sent out, and it also fell into the clutches of Brazil. Peruvians rallied to retake their post at the mouth of Chandless, and constant battles ensued as the Peruvian force moved down the Purús destroying *barracões* (trading posts), wooding stations, and domiciles as they passed. Finally they set up their position at Sta. Rosa, billeting sixty men there and fifty-eight across the river at an entrepôt called Funil, which da Cunha would describe in detail. Ferreira de Araújo and some 270 tappers resisted this incursion. The battle was on, and depending on the nationality of one's source, it was a triumph either for Brazil or for Peru. Fifty-eight Brazilians were killed, but the Peruvians fled back to the Ucayali. When Ferreira's men overran Funil, the Brazilians killed their earlier prisoners.

This incident became a source of festering bitterness between the Brazilians and Peruvians on the mixed reconnaissance commission, as da Cunha made amply clear in a private communiqué to Rio Branco. The battle had ended on April 1, 1904, but news of it reached Rio only at the end of the month. Guerrilla war raged in the upper Purús and Juruá; Brazil closed the upper Amazon to trade, especially munitions destined for Peru, and once again caused fury as it interfered with international commerce. The embargo of Peru, the disruption of the flows of rubber, and the mounting military forces on both sides made some kind averting action imperative for maintaining globalized trade.

By May Hernán Velarde, foreign minister of Peru, met with Rio Branco, and they began to hammer out the dimensions of a possible neutralized zone, since Iquitos was arming for war. Llamas and mules loaded with Mannlicher

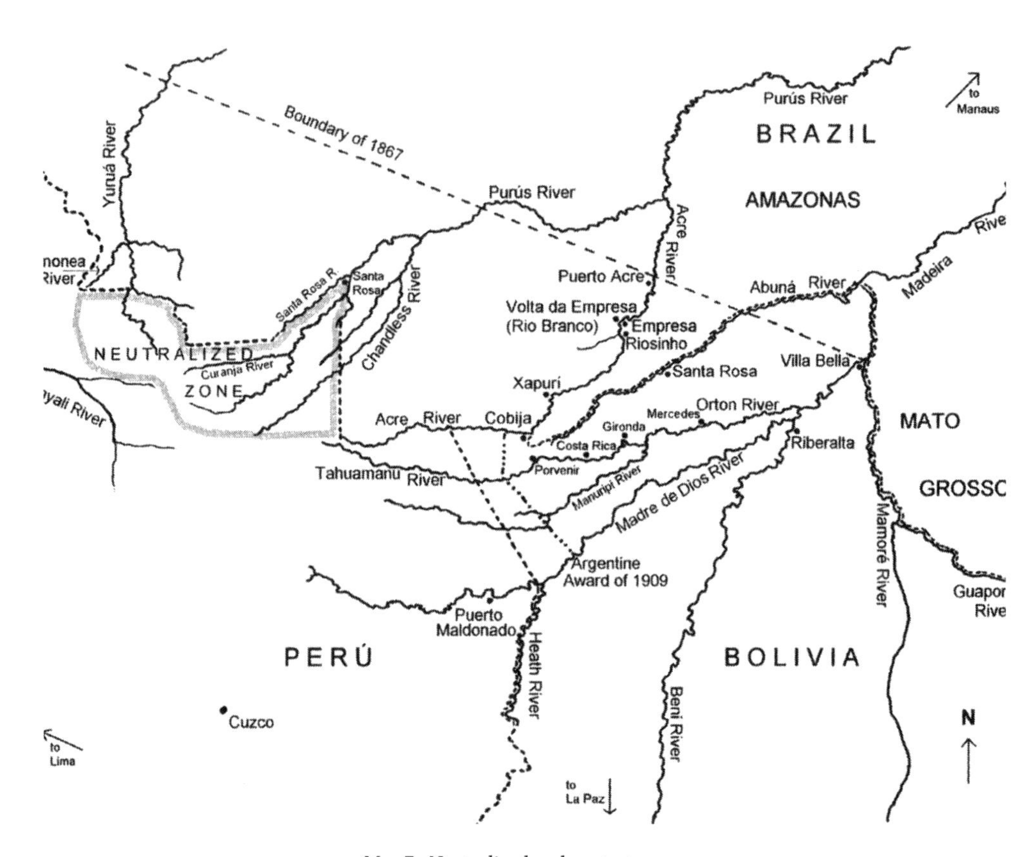

Map 7. Neutralized and contest zone.

rifles staggered over the Andes to offloading points on the Urubamba to float munitions down the Ucayali to Iquitos. The garrisons in Manaus were on high alert. Each country, in a frenzy, imagined a major invasion taking out a key Amazon city (either Iquitos or Manaus), although neither had the wherewithal to do it. Brazil's battleships were hardly capable of making the trip from Rio to Manaus. Peru had purchased gunboats in Britain but could not move them into place because of the Amazon embargo. In a letter to da Gama, da Cunha likened the saber rattling to an impotent "Offenbachian duel—each brandishing their weapons in futile postures, an absurd, lethal conflict."[31]

In light of these tensions, the treaty ambiguities, and guerrilla war throughout the region, it was crucial to demilitarize the area while Brazil and Peru resolved the boundary issue. There were complex machinations, since what would be the "neutralized zone" carried within it an implicit position about the total terrains that could be negotiated. Peru's Velarde preferred that the entire watersheds of the Purús and Juruá become neutralized, taking in the Idelfonso claims. Rio Branco continuously diminished the geographic areas under discussion. In the end, a *modus vivendi* signed in July 1904 included in a "neutralized zone" on the Purús in the area above the Manoel Urbano River. The document stipulated joint governance in several areas with customs houses, and provided for a binational reconnaissance survey for the Juruá and Purús to supply data on the nationality of the occupants and other ancillary details, since both nations had little information about the place in contest other than that it was generating rivers of money. In the meantime, Rio Branco had won ratification of a secret nonaggression treaty with Ecuador, which would hold while hostilities with Peru continued and would keep the area from exploding.

Friends to the North

Peru was trying to pressure the United States to intervene diplomatically in the case. The Peruvian legation argued that Peruvians were threatened with invasion and framed their position largely through the narrative of a second round of Brazilian imperial acquisitions in upper Amazonia. From the Brazilian perspective, sensitivities about the syndicate (a US enterprise with another country had been thwarted by Brazil, after all) and Theodore Roosevelt's interventionist bent suggested that a public relations effort had to be mounted to sway US opinion, since Peruvian diplomats were giving interviews to American newspapers emphasizing Brazil's insatiable Amazonian imperialism, a matter of no small economic importance. In this environment, the pamphlet prepared by Moore at the behest of Brazil, *Brazil and Peru Boundary Question*, explained the antecedents of the situation and the thinking that underpinned the posture of Brazil. This was sent to American diplomats, journalists, and other opinion makers to defuse anti-Brazilian sentiment.

This was the context that prevailed that May when da Cunha wrote his four articles on Peru. In August he would meet Baron Rio Branco, and by the end of the year he would set off to the Amazon.

11

Euclides and the Baron

A Labyrinth . . .

Riding out to construction sites and administering the building of public works was tedious and exhausting for da Cunha. It satisfied neither his creative nor economic yearnings: he was beset with dissatisfactions in all his callings—military, engineering, farming, journalism. Any luster he might have enjoyed from association with General Solon Ribeiro's military career had dissipated in subsequent intrigues, along with Solon's dubious performance in the second Canudos campaign. His final days in obscure outposts hardly spoke of a surfeit of confidence in his trustworthiness. In a time of nepotism and connections, while others thrived through their revolutionary associations, da Cunha's carried little benefit.

Furthermore, da Cunha had managed to alienate a succession of regimes. His caustic positions on corruption and cronyism and his moderate views on activities like summary executions during the Jacobin frenzy had made him some obdurate foes. While *Os Sertões* was a literary sensation, its description of military incompetence and, pitiless extermination of modest peasants, ex-slaves, and the last remnants of some regional indigenous groups could hardly be construed as complimentary. He had broken with most of his confreres from the Praia Vermelha as they rose on the wings of patronage to ever more august sinecures.

Da Cunha had hoped to teach engineering and sciences at the new São Paulo Polytechnic, and begged his friends to intercede for him. He had, as usual, managed to alienate the key person—the incoming rector—in one of his newspaper articles, closing off that possibility. Working his father's farm had its allurements and fed his nostalgia for the simple rural life while subsidizing his household costs. But da Cunha's desire for the rural was for the imaginary terrain of his childhood, not the adult anxieties of a marginal coffee enterprise. In addition, things were not going particularly well with the increasingly miserable Ana. The long correspondence with his father was often deeply affectionate, but the career instability and the emotional volatility of Euclides worried the elder da Cunha.

The dead end of the army forced da Cunha to return to the dreary tasks of surveying and engineering. His correspondence over these years is one long litany of complaint, even when he could get some reading done on trains and muleback. Plan checking and supervising construction held less and less attraction, and his youthful idealism about this type of public service had waned in the tedium of the day to day. By 1904, many letters to his literary confreres began with variations on "in the few minutes I could snatch from my over worked obligations . . ." Machado de Assis received letters of humorous protest: "I'm completely caught up in a labyrinth of . . . sewers!"[1] In April of that year, the São Paulo economy slumped. When da Cunha's superiors informed him of a cut in salaries, he simply had had enough. In one of his furies, he insulted his supervisor and left his engineering job with the Santos sanitation department with the usual drama of bitter words and slammed doors.[2] Nothing by way of new employment was in sight. A few days later he sold the future rights to *Os Sertões*, including translation rights, for the relatively trivial sum of one conto and eight hundred milréis, about two months' salary.[3] The volume would eventually go through more than sixty printings and was widely translated.

He had been elected to the Brazilian Academy of Letters in 1903, and that carried glory but no stipends. By mid-1904, the unemployed da Cunha was mulling over various projects: perhaps a book on the Duque of Caxias, the repressor of countless black and peasant uprisings in the Northeast who was hailed as the military genius of the Paraguayan War. Or should he write on the Naval Rebellion? or a biography of Emperor João IV?[4]

Da Cunha was strongly supported in his writing and in his politics by Júlio Mesquita, the editor of the *O Estado de São Paulo*, who had even tried (unsuccessfully) to launch him into a political life in São Paulo. With no prospects after his contretemps with the czars of the Santos sewer system, da Cunha began again to write his columns for the *O Estado* and another paper, *País*. Their subjects were the tensions heating up in the upper Amazon and the conflict with Peru. As he had with "Nossa Vendéia," he was about to capture the mood of the time.

The Geopolitics of Euclides da Cunha

Between May 14 and 29, da Cunha wrote four major articles on Amazonian geopolitics "Contrasts and Comparisons," "Inevitable Conflict," "Against the *Caucheiros*," "Between Madeira and the Javari"[5]—a lot of Amazoniana for someone who had barely mentioned the region before.[6] These with other essays on geopolitics were gathered in *Contrasts e confrontos*, a collection published after he became part of the Rio Branco team.[7] As with "Nossa Vendéia," he was angling intellectually and politically for a role in the unfolding conflict.

As tensions had escalated between Peru and Brazil throughout the spring of 1904, as we have seen. Rio Branco resolved to send a battalion of three hundred men to this Amazon frontier, and he signed a secret nonaggression pact with Ecuador in order to consolidate upper Amazonian allies on an increasing volatile and militarized frontier. Ecuador had a simmering resentment with Peru over Amazonian boundaries,[8] based in Ecuador's claimed right to sell to a British charter company what Peru had deemed national territory. Using colonial maps, Peru argued that the administrative claims of the Viceroyalty superseded those of an *audiencia* and had won the day. Peru was also deploying this precedent in the case with Bolivia, but Acre was still a fresh and unpleasant object lesson in the limits of Spanish imperial organization on the current frontier and the dangers of taking on the numerous and riled-up Brazilian tappers. Bolivia's enormous loss of territory after the Acrean War was a somber reminder of the potential costs of defeat.

There were worries in Peru that the conflicts on the Purús and the Juruá could spill over into the nearby Ucayali basin and potentially result in the capture of the Amazon city of Iquitos, the only Peruvian port (and Peruvian navy base) that connected to the Amazon and the Atlantic. Brazilians worried that Manaus might fall to the Peruvian navy. Under terms of the Treaty of Petrópolis, the Peruvian government opted for demilitarization and reconnaissance with Brazil while the debate over Bolivian boundary rights continued. It was in this uneasy chess game of incipient war that da Cunha entered the fray.

Da Cunha had yet to set foot in Amazonia, but he read well: for his grounding in regional history he pored over *Tesouro descoberto* by the great eighteenth-century Jesuit João Daniel, who had languished for twenty years in an Inquisition jail. Da Cunha also scrutinized the work of the usual throng of naturalists: Alexander von Humboldt, Henry Walter Bates, Alfred Russel Wallace, Agassiz, Johann von Spix, and Carl Friedrich von Martius as well as Brazilian politicians like Tavares Bastos. There were the works of the British explorer William Chandless regarding travels up the Purús and Juruá Rivers: Da Cunha's Amazon itinerary would retrace his route. Euclides, again, was captured by Brazil's unknown interior, guerrilla conflict, and national identity.

From Nationalism to Imperialisms

In 1904 Brazil was in the midst of a major modern pulse of globalization as its rubber monopoly fed insatiable international industrial demand and its coffee exports soared. The nation was also thrilling to its own stylish cosmopolitanism, with flourishing transnational artistic, literary, and political cultures and a new sense of itself as a modern hemispheric power.[9] Rio was being rebuilt with airy boulevards inspired by the monumentalism and visual drama of Georges-Eugène Haussman's Paris. The fashionable shops and cafés of Rua Ouvidor burbled with the accents of numerous languages, with topics focusing at least as much on international politics and economics as on romantic affairs.[10] Fancy hotels served whiskey, champagne, and tropical ice creams on outside terraces, a delightful innovation.

Brazil was also emerging into its own *internationalism* in complex political and diplomatic interactions with numerous states. From 1890 to 1910 it was involved in major border revisions involving diplomats from Argentina, Bolivia, Peru, Colombia, Venezuela, the Netherlands, Switzerland, Italy, the United States, Britain, and France. With Rio Branco at the helm, Brazil was active in sponsoring pan-American conferences, making sure its pavilions were visited at the world's fairs, and angling for a seat on the International Court of the Hague. Brazil inaugurated a foreign service with embassies and ambassadors, not just legations, and began seeing itself as a diplomatic player in modern global affairs.[11]

Da Cunha's writing marked him as a member of the assemblage of thinkers on Brazil's international relations and the larger questions of imperialism and civilizations, especially addressing state formation in Russia, the United States, and South Africa, places with the most obvious parallels to Brazil. His perspective had a durable influence on the military geopolitical ideologies of the Amazon during the twentieth century and on theorists on Lusophone imperialism like Gilberto Freyre.[12]

For all his seeming social ineptness, da Cunha had a profound sensibility of political zeitgeist. His writings of May 1904 show him situating himself for the next round of his political and artistic life. His newspaper articles bear the stamp of his unusual blending of geography and history for evoking a national character, showing how these underpinned political cultures, destinies, nations, and emergent civilizations. It is useful to review the themes and content of these short articles written in that unhappy May, because they presage the ideas he later expanded in his political writing on Amazonia: the dialectics of place and national character, racial politics, guerrilla warfare, and the contrasting natures and destinies of Brazil and Peru's Amazon economic ecologies. These articles help explain da Cunha's rise into Rio Branco's closest diplomatic circle and why later his relations with the Peruvians on the joint reconnaissance commission were so rancorous.

It is often assumed that Rio Branco's interest in da Cunha was due to the prominence of *Os Sertões*. Although the Baron adored writers and artists and catapulted many of them into positions of great political importance,[13] he was not sentimental or dazzled by artistry. Rio Branco's Amazon statecraft required good historical geography and cartography. In each of Rio Branco's adjudications and negotiations in Amazonia, the lead ideologue was carefully chosen with the style of the opponents and adjudicators clearly in mind. Da Cunha, who knew something about Northeasterners, guerrilla warfare, and environment, was a quick study on Amazon history and, an additional plus, knew field survey techniques.

Contrasts and Comparisons between the Madeira and the Javari

The series of articles that da Cunha published that May build upon one another, developing ideologies of national character that embody his "post-Canudos" racial positions and how these would play out in the new arena of Amazonia. They no longer embrace the notions of social Darwinism on the superiority of the "civilized races"; they are no longer environmentally deterministic in the mechanically reflexive way of his earlier writings,[14] but they do remain wedded to the effects of place in the creation of national identities, cultures, and economies through interactive processes that today we might call "co-evolution" or a kind of "socio-environmentalism." Da Cunha's view of the interaction of locality, culture, race, and livelihoods is more subtle than his earlier positions on the "march of history" and comes closer to the modern geographic subdiscipline of political ecology. Da Cunha explores national history and imperial trajectories as well as the implications of the converging and tumultuous Amazon frontiers for Peru and Brazil.

"Contrasts and Comparisons" is the first of these essays:

> Anyone tracking the itinerary of von Humboldt amongst the mountains and peoples of Peru notes an interesting parallelism: People and landscape mirror each other. History there seems like a vulgar plagiarism of Nature. It is noted in all periods: by the archaeologist viewing the bas-reliefs of the crumbling temples, the geologists describing the strata that fold up in sharp escarpments, and the colonial chroniclers and their moving narratives of the conquistadors. We see in these jumbled contrasts how the social facts imitate inorganic realities, reappearing, reproducing, and reasserting themselves between two extremes: on the one, the Andes and civilization of the Incas; on the other, the Peru of earthquakes and of edicts.
>
> As one travels from the crumpled and demolished coastal lands, confined to the trembling filigree of seismic faults and their periodic cataclysms, and then ascends the imposing immobility of the Andean Cordillera held in place by a rigid skeleton of dolomite, one moves from a rebellious and febrile republic, periodically agitated by the irritable weakness of its caudillos, to the wreckage of a patriarchal empire propped up by an inflexible theocracy and caste system. . . .
>
> One can't disguise these differences and these identities. They are there in strip of the littoral with its sterile dunes and in the untamed fierce *montaña*[15] with its impressive forests. A hazy, almost mythical past is counterposed to an undefined and deplorable present: they face each other, they repulse each other.

Da Cunha goes on to describe the triumphs of the Inca civilization: sanctuaries worked from living rock, the extensive irrigation systems, excellent roads, architecture, and monuments to defy cataclysm:

> From Cajamarca to Cuzco there is probably not a kilometer where a small pyramid, an obelisk, a pillar, a bit of ruined portico, a plinth of granite with etched bas-relief, a phalanx of monoliths, or a caryatid of bluish porphyry does not evoke that extraordinary race—one that was innocent of iron but that chiseled rock with delicate bronze tools and created monumental sculpture from blocks of mountains. . . . [The Inca also] sought the eternal glacial springs, captured and directed them, sometimes by adjusting the slopes, other times by punching

> subterranean channels through mountains or even—a detail that considerably pushes back the beginnings of modern hydrology—connecting the flow from one mountain to the next in numberless conduits. Finally, in places where they couldn't find solid bedrock on which to erect their monuments they invented a kind of gigantic polygonal apparatus to support them. The Inca created an architecture to resist catastrophe. . . . But they did not foresee the calamity of the arrival of the sixteenth-century Spaniard.

Imbued with what sociologist Gilberto Freyre has called the mystique of the engineer,[16] da Cunha clearly was enchanted by the technocratic, constructive, and constructing Inca: "That strong and peaceful race which gave pride of place to its agricultural inspectors, its engineers who opened roads and canals, to the architects who raised their temples—this civilization was undone by treachery and the brutality of Spain."

Comparing this edifying Inca civilization to the disintegration that attended the conquistadors: "Their history became an obscene copy of those temblors that in minutes turned cities to rubble." He contrasts the unity of the native race, its sure and ordered methods, with the disorder and chaos brought by the large-scale pillage presided over by "all the races of the world." The "Peruvian," he points out, "even more than the 'Brazilian,' is an ethnographic fiction."

Da Cunha describes the racial diversity of the Chinese, blacks, Spaniards and how all were bent on plunder:

> They snatched the gold, the silver, the nitrates, the guano, the mummies, and the stones from the temples. . . . They hacked at the coasts and the islands, they clawed the flanks of the mountains, they profaned the funeral pyramids and sacked the tombs and the *huacas** of the ancients, prizes at times more valuable than the richest mines: it is enough to note that that with a fifth of the gold from but one of these, the town of Trujillo was built. . . .
>
> Nothing characterizes better the parasitism, the contempt for traditions, the lack of solidarity, and the unstable energies. . . . The past is only treasure for plunder . . . to be used in a strictly utilitarian manner. Appropriating at random, the flimsy and turbulent society tangibly expresses in its debasement, its contrasts to the dead Civilization: it is reflected in the miserable huts built chaotically on the massive foundations of monuments, or even of cities: we cite only Huamachuco, built with block ripped from temples, a dismal collection of

* A *huaca* is a ritual or sacred site, a monumental place in Inca culture.

old facades, huddled and humbled rearrangements of the former metropolitan grandeur, broken up for hovels with sagging roofs and thick walls—a melancholy architecture of ruins.

Da Cunha goes on to argue that Peruvian imperial political culture was equally as impulsive, sterile, and agitated by the constant intrigues, the battles, the endless proclamations, the dangerous presidential successions done *à americana*—with revolvers.

In these differences, one almost senses that the Peruvian incursions into the remote frontiers of the upper Juruá are a rebuff, a yearning to abandon that narrow sliver of land where a Nation whose Inca origins presaged the loftiest destinies is buried, hemmed in between the vastest of seas and highest peaks in lands convulsed by the disharmony that has so decreed its misfortune. . . After one has ascended, then crossed over the second chain of the Andes there unfolds a stable nature—one devoid of catastrophes and ruins, one with its creative forces still intact, awaiting only the prodigious force of labor. This tropical hinterland offers, through the discipline of toil in prolonged and mindful efforts, and in its durable intimations of its own natural harmony, a refuge from the turmoil that was always denied Peruvians in their other lands. It incarnates possibilities that could awaken the best aspects of their character, aspects that today are displaced by a kind of belligerence that is both a weakness and an anachronism. But this setting might further jeopardize them, creating yet greater misfortune, if in a novel way it breathed further life into the pillage of the *caucheiros*, who, in their devastation of the great forests, reanimate a long, ancient apprenticeship in calamity.

This essay sets the context for his continuing themes and future arguments: Peruvian society is balanced between a glorious past, a febrile, increasingly militaristic and precarious present, with an economy based on roving plunder, and perhaps an illustrious future in Amazonia, where a harmonious interaction between human beings and nature might emerge if not infected with the (national) parasitic taste for pillage. Da Cunha posits a rupture between people and place as the defining characteristic of modern Peruvian civilization.

In "Inevitable Conflict," da Cunha elaborated his ideas that "nomadism" inhibits interaction with environment,[17] positing a kind of taming by places that produces an integration of races and productive political ecology. He expands the idea that the larger economic and historical imperatives of Peru are fundamentally Amazonian and sets the stage for the epic conflict between Peruvian and Brazilian empires that he sees unfolding in the Purús:

> The Peruvian incursions (into Amazônia) do not merely assert the greed of a few adventurers maddened by the ambition those extremely rich rubber forests arouse in them. It is graver: they are the material expression of a historical impulse bursting with irresistible purpose. These forays are not merely determined by the unstable and dispersive social energies of that South American republic, one most despoiled by caudillo politics, but by inviolable physical laws". . . .
>
> Really, to contemplate Peru through the prodigious vision of a von Humboldt or the limpid intelligence of C. Weiner[18] is to understand immediately that the destiny of Peru, with its constrained swath of land stretching from Arica to Trujillo, immured between the Pacific and the Andes, oscillates between two inflexible extremes: either the complete extinction of their nationality as it becomes overrun by an enormous immigrant population, one that embraces every modality of temperament from the industrious German to the half-enslaved coolie, or a heroic unfolding toward the future, a daring engagement with Amazonia, a national salvation embodied in the rush to the headwaters of the Purús, achieving in one glorious venture both access to the Atlantic and a more fecund setting for expressing Peru's national energies. . . . There is no escaping this dilemma.

Like Brazil, Amazonia could hold for Peru its national destiny and identity. Euclides emphasizes culture—in this case a predatory one—more than race per se as the molding force of national character, an idea he will revisit elsewhere in his writing. As in his first essay, da Cunha reviewed the devastation of various types of economy based on mining everything from silver to guano to graves. "And yet in this febrile and parasitic activity, Peruvians are not alone in their predilections for plunder," he says: "any street in Lima displays the most varied ethnographic gallery on earth." He reviews these racial types and mixtures:

> Among these four principal types (blacks whites, Asians and natives) are uncountable forms of mestizos: from the mulattos of all bloods, *zambos* and *cafusos*, to the Indian half-breeds, to the *cholos* who remind us of our own *caboclos*, and those fascinating Chinese-cholos in whose faces are found the lineages of practically all the races. These also find their alienated expression in the pillage of history and landscape. . . . In these disordered energies is resumed the inevitable conflict of temperaments. Life unfolds there without connections, without the discipline that derives from that harmony of forces so necessary for production, and that extinguishes the dispersive impulse: instead, it is a society without qualities, with no defined traditions nor incisive national trace, one shot through with recent migrants who profit and abandon. This political disorder mirrors the

> upheaval of natural forces that periodically convulse the landscape. Peru is a place of a perennially incipient national culture.
>
> It hasn't escaped Peruvian statesmen that this play of contrasts derives in large part from a society that lives contingently and seems merely warehoused on the stretch of land along the Pacific. Peruvians come from that unfortunate place where barren earth conceals infrequent riches that diminish day by day and are never restored, and gives the unproductive labor of locating such wealth the sorry elements of pillage, . . . a restless mingling all the races but not yet a people.
>
> In [Amazonia], so other, there is redemption in the work and the stability of farming for those people with the best elements for a higher historical apprenticeship. But even as their hopes emerge, they are thwarted by the Cordillera. . . . In fact, the Pacific Ocean, even as they scrape out the Canal in Nicaragua, will have little impact on the national progress. Peru's true sea is the Atlantic, and its obligatory egress is through the Purús.
>
> Many of its best statesmen knew this. The expansion into the Amazon was viewed as an elementary task in their struggles for survival.

Da Cunha then describes the several failed efforts to build rail lines over the Andes and points out how "the great tributary of the Amazon . . . has oriented for some time the administration of that republic. . . . Since 1859 and Faustino Maldonado[19] . . . successive expeditions were launched to the east, impelled by some self-abnegating souls to those remote places in an extraordinary intuition of the true interests of their country. These precedents reveal, in the disturbances that swirl today throughout that zone, a meaning much more complex than a few melees with rubber tappers." In terms of the larger conflict, he argues that unlike the Paraguay War, where the caprices of a caudillo were stopped only by a protracted guerrilla war with terrible slaughter, a military encounter with a nation seeking its future would be profoundly unpredictable.

This essay, "Inevitable Conflict," was compelling on several fronts, especially because the populations' composition provided racial material for a contrast between the "nationness" of Peru and that of Brazil. Peru, like Brazil, was awash with ethnic mixtures of every kind but, as da Cunha argues, unconsolidated. This was an important point since Peru was (and is still) often perceived as more racially uniform than the famously multihued Brazilian polity. Given the pervasive influence of racialist ideas in scientific anthropology, especially the concept of mestizoization as racial degeneration, the similar starting point permits an almost experimental analysis of two multiracial societies and the evolution of differing tropical "civilizations." Peru, shaped by a political economy of pillage, generated a society that remained fundamentally unstable, uncohesive, nomadically immigrant, and incoherent, a place of "perennially incipient national culture," and in the sociocultural hierarchies of the time, less developed.[20] This contrasts to the Brazilian melding of race and cultures into a "national civilization." The essential conflict for da Cunha is between the effect of plunder versus that of production on national character, histories, and ecologies. These become moral landscapes as he assesses the two nations' Amazon ambitions, a theme he developed in more detail after he traveled on the Purús. Both Brazil and Peru, he suggests, would have their destinies inscribed in Amazonia. The real question was whether nomadic plunder or an emergent civilization would prevail.

There was another subtext. Da Cunha was arguing against the Brazilian policies of "whitening," which proposed European immigration as the key to Brazil's development, as highly miscegenated populations were viewed as obstacles to progress.[21] Da Cunha's statements reflected his post-Canudos thinking about the nature of race, national identity, and the virtue of racial complexity—the "bedrock" of his Brazilian "race."

"Against the *Caucheiros*," his next article, considers whether Brazil should send troops to the western border to prevent possible incursions into Brazil's territory by Peru in spite of the modus vivendi. The essay is a treatise on guerrilla warfare and presents it as the most effective, *modern* form of combat. Da Cunha's experience at Canudos, where guerrilla forces had held off much larger, better armed forces, had changed his understanding of tropical campaigns. The obvious effectiveness of Acre's transplanted *sertanejos* against the regular forces of Bolivia were evidence for the efficacy of homegrown, locally adapted militias. Citing the Boer War as another example, he argues that the South African campaign owed its triumph to informal militaries on the modern battlefield: "The speed and violence of gunfire impose a dispersion completely opposite to the contrivances of halts and formal maneuvers . . . soldiers without the habit of thinking for themselves become befuddled, disjointed, and useless, because the greater their formal adherence to ranks, the less their individual ability to act." Clearly distressed by Brazil's dispatch of formal battalions to Manaus, he describes the new battlefield in Amazonia:

> Those who await us are not organized troops but the *caucheiros*—alert and furtive: barely organized in their small boats, dispersed about in their light canoes, maneuvering quickly, alone, in the shadows and in wakes of the floating logs, attacking suddenly from the flowery borders of the *igapós*,* disappearing,

* Flooded forests.

imperceptibly swallowed up in the endless bayous [*parana-mirims*],* where they conceal themselves in a fretwork of fronds and branches, vanishing in the infinite curves and uncountable channels that characterize the fascinating hydrology of the southwestern tributaries of the Amazon.

The material image of a campaign there would be that of thrashing around in labyrinthine backwaters. Our strategists will not have the relatively simple task of fighting the enemy: the real problem, perhaps the impossible one, will be to locate him. Politicians delude themselves if they imagine that the mere appearance of a few regiments in the enormous tract of lands between the Purús and Juruá is enough to protect the locals and to impede trespass through an undetermined, unmarked frontier. Compact battalions, trapped as much by preconceptions and the habits of rectilinear combat will be more useless the more disciplined and tied they are to coordinated formations. The better, the more impeccable their military organization, the worse their ability to adjust themselves to the theater of operations: to face the terror of rapid clashes or to flee the treacherous ambushes. . . .

Besides, the necessary forces to repel an invasion are already there, agile and acclimated, in the irregular troops of Acre, composed of the fearless *sertanejos* of the Northeastern States, who have been transforming Amazonia for the last twenty years. They make up the truly modern army. . . .

We can entrust ourselves to these diminutive titans, their galvanizing competence, toughened and seasoned by broiling suns; they are accustomed to rapid and decisive deliberations and dazzling tactics—they improvise in combat with the same spontaneity with which humorous rhymes leap from their lips in the *folguedos*.†

This is hardly the da Cunha of "Nossa Vendéia" and the early pages of *Os Sertões* predicting the inevitable victory of history's merciless project of racial triumphalism and a murderous modernism. His view has shifted to the spontaneous, quick-witted autochthonous "organic intellectuals" who, far more than the structured coastal army, now constitute the most *modern* and most effective protection for the fatherland.

These "diminutive titans" are the subject of da Cunha's last essay on the Peruvian conflict, "Between the Madeira and the Javari." Central to his argument is the idea that the divergence between the nurturing husbandry of the region's Brazilian northeastern migrants and the callow pillage of the

* Sluices.
† These are Northeast festivals that often include a kind of rhyming and insult competition. Today's improvisational hip-hop competitions are analogues.

Peruvian *caucheiros* is a clash of "tropical civilizations." The question of Brazil's manifest destiny is raised and compared with the US history of western territorial expansion. The space between the Madeira and the Javari was now the great zone of conflict, where both the Treaty of Madrid and the Treaty of Idelfonso drew the territorial lines that Brazil was now about to defy. Rather than the land of emptiness, da Cunha's Amazonia is an inhabited tropical realm. The essay outlines a history of Brazilian settlement that he will later embellish to justify the Brazilian possession of western Amazonia.

> There is no region in all of Brazil that has experienced the vertiginous progress of that most remote stretch of Amazonia where neither the devotion of the Carmelites nor the engrossing semi-commercial, semi-evangelistic activities of the Jesuits could prevail. A little more than thirty years ago the place was desolation itself. What one knew of the place went little beyond the disheartening lines of Father João Daniel in his imaginative *Discovered Treasure*: "Between the Javari and the Madeira, a distance of more than 200 leagues, there are no settlements of whites, Tapuyas nor missions." This eighteenth-century adage would be repeated in 1866 by Tavares Bastos.[22] "Amazônia is an aspiration: leaving the suburbs of Pará you penetrate only wildness."
>
> Since then, nothing has explained the oblivion of that territory.

Da Cunha begins his comparison between Brazil and the United States by describing the different approaches to their frontiers:

> In the US it was the reckless wave of population finally breaking on the shores of the Far West, while for Brazil, for most of its history, it was the actions of the *bandeirantes* that for so long defined our regional claims, traveling the monsoons up the rivers, conquering the torrents, and defying the treacheries of the great tropical rivers of the Paraná and the Amazon. Yet these travels incarnate a unifying historical purpose that will become well understood only when the national spirit is robust enough to write the epic of these expeditions to the Heart of Brazil.
>
> Our two main lines of penetration to the outback were those from São Paulo and Pará, both of which converged in Cuiabá. There, strangely, they positioned themselves from the outset as though seeking every possible impediment, loitering in the western limits of territories that could have been easily traversed without extraordinary efforts.
>
> . . . The Purús and the Juruá are, after the Paraguay and Amazonas, the most navigable rivers of the continent. After descending from the eastern heights of the last Andean foothills, where the rapids surge and their only cataracts thunder,

the waters flow in a soft declivity that even the most precise instruments cannot always distinguish.

The waters adjust to this rare surface uniformity, so expressively evident in the most casual contemplation of a map. They pour into great parallel courses that run between the Madeira and the Javari and slowly drain the soft regions that stretch out from the Bolivian plains, where a harmonious nature veils the opulence of an incomparable flora in the labyrinths of its bayous.

But, oddly, no one sought these immensities. The nation—Brazil—that secured possession of the lands at the headwaters of the Rio Branco, the Rio Negro, the Solimões, and the Guaporé with palisades and battlements at the old Portuguese fortresses at San Joaquin, Marabitanas, Tabatinga, and Príncipe da Beira—these four enormous strongholds challenged our traditional rival, Spain—But our metrópole completely avoided, indeed fled from, those distant western tracts of territory, at least until someone else revealed them for us. In 1851 it was Count Castelenau and the lieutenant of the North American navy, Frank Maury.[23]

It was a revelation. The discovery coincided with a renaissance of national ambition. In the press, the robust practical spirit of de Sousa Franco allied itself with the dazzling intelligence of Francisco Otaviano in irresistible propaganda to open Amazonian trade up to all flags. . . . This policy was supported by the lucid analysis of Agassiz, the studies of Bates, the observations of Brunet, the works of Silva Coutinho, Costa Azevedo, and Soares Pinto until the civilizing decree of December 6, 1866,* was announced.[24]

In this massive extension of new horizons, one tenacious explorer, the Britisher William Chandless, unexpectedly mapped out the guidelines of a singular ambition: to discover a passage from Acre River to the Madre de Dios—the old challenge of the link between the basins of the Amazon and the Paraguay. He didn't merely resolve this question, but indeed transcended it. This enterprise took him to the remotest stretches where the two great rivers mix their waters in their farthest sources. Impressed by the natural marvels around him, he transformed himself in to a pioneer of bounding ambition . . . meanwhile, the efforts of an intrepid backwoodsman, Manoel Urbano de Conceição, an unsung hero (as are the majority of our true outback conquerors), scouted up the networks of streams and backwaters to finally reveal the headwaters of the tributaries of the Purús. He prepared a large part of the lands for the rapid and intense settlement for another master of the wilds, Colonel Rodrigues Labre. . . .

It was a magnificent transformation. Shortly after, successive waves of immigrants reprised the exciting tumult of the arrivals of the eighteenth century.

* This announced the free navigation of the Amazon.

The latex of rubber trees, cacao, copaiba, and all sorts of vegetal oils substituted for the gold and diamonds that had nourished the earlier unbounded ambitions of the *bandeirantes*.

The land, until then the domain of nomadic Indians, in less than ten years had more than six thousand souls and was linked by way of the Fluvial Company of the Amazon from the most remote outposts of Sepatini and Huitanãa to Manaus. The initial trajectory of 1,014 miles had to be extended to take in the larger tributaries that went all the way up to Ituxi in Acre. And then finally to a city, the real city of Lábrea. Lábrea! In its converging energies it was the integrating molecule of civilization, rising unexpectedly from the vast, wild solitude. It brought to the headwaters a character quite divergent from that of our other backwoods settlements: it had the improving aspect of two journals, the *Purús* and the *Labrense*, the sumptuous luxury of a sought-after theater; it had schools, paved and straight streets. These successes, which form the most uplifting chapters of our current history, are also the most moving examples of the application of transforming principles to society. . . . Really, what manifests itself there is the natural selection of the righteous.

For these pioneers to invest in the unknown, the simple craving for riches is not enough; one needs more than anything a will, a perseverance, a stoic courage joined to a privileged physical constitution. These places are today, in the midst of our national feebleness, the theater of a vital competition. Alfredo Marc found on the margins of the Juruá some Parisians—authentic Parisians!—trading the charms of the boulevards for the arduous exploitation of a bounteous rubber estate. From our jungles emerge all the valiant, converging there from all quarters of the earth. But outstripping them in number, in vitality, in a better organic adaptation to the climate, and with that debonair and fearless style with which they have always faced danger, are our admirable, indomitable *caboclos*, the backwoodsmen from the northeast who have transformed and molded the foreign adventurers, imposing on them our language, our customs, and in the end our destiny, and so extending in their energies the dominant component of our nationality.

. . . Returning to the earlier parallel, the Yankees, after pausing for years at the base of the Rockies, enthusiastically traversed them, attracted by the rich mines of California. This is exactly the moment when we advanced all the way to Acre. . . . In the same year of 1869 when we claimed our forgotten frontiers with a riverboat company, the Americans linked the Pacific with a railroad from Missouri, audaciously wending through mountains and deserts.

Let us also parallel them in this episode of national life of a great Republic.

Let us accept the lesson of Bryce.[25] Magisterially tracing the pattern of Yankee expansion, the historian shows us that faced with the enormous distance from

> the East Coast, the people residing on the Pacific would have inevitably formed another nation if the engineering resources of the day had not permitted a permanent intimacy with the rest of the country. Our case is identical, or even more serious.
>
> The new boundaries of the upper Purús, upper Juruá, and Acre must also reflect the unrelenting action of the Government in a labor of incorporation that, as a practical matter, demands facilities, communication, and the alliance of ideas promptly transmitted through the vibrant enervation of telegraphs out to the remotest outposts. Without national integration as a firm and driving objective, Amazonia is merely an arena of natural selection. The spirit of von Humboldt thrilled to a dazzling vision of a setting where "sooner or later one must concentrate all the civilization of the globe." Amazonia otherwise could stand out from Brazil, naturally, irresistibly, a world detached from a nebula—through the centrifugal force of its own energies.

This last essay regarding the "civilizing molecule" of Brazilian settlement takes the region from its emptiness through a nation building "from below." The *bandeirantes*, with all their historical brio, did not effect the day-to-day transformation from inchoate wilderness to the comforting civilization and urban delights of Lábrea, and they lacked the will (as did the nation itself) to press forward into the great wildness. Instead it is the taming power of the unsung, maligned backwoodsmen, attached to the land, who transform amorphous terrain into national territory. It is the tappers, not the urban elite, who are the midwives of Amazonian Brazil, the greater part of the nation.

Two countries, two relations to landscape, both racially heterogeneous: one constructs its history only in plunder, roaming the landscape, builders only of ruins. Peru, says da Cunha, is a society without qualities or traditions, populations not yet a people, a despoiling culture, an incarnation of the convulsion and turmoil of earthquakes. In contrast to this "apprenticeship in calamity" rises the Brazilian "civilizing molecule": the domesticating "attachment of Man to the Land," the development of a new and transformative civilization in the remotest part of the planet: not the mythical "lost cities" but new ones with schools, newspapers, a cherished theater. Multiple races and nationalities, instead of a divisive history, meld into the national culture under the tutelage of the *sertanejos*. Brazilian *caboclos* shape immigrants into "our language, our ways, our destiny." Rather than the *bandeirantes* or imperial armies, it is the modest tappers who are both masters and protectors of this distant territory, claiming it for the fatherland in an entirely novel way, building an arcadia instead in a turbulent world, a world of progress against one of plunder.

These essays were da Cunha's great salvos into his future.

"I admit illusions . . ."

Da Cunha desperately needed employment. He dreamed that he might manage the construction of the Madeira-Mamoré railway, that lethal trench that rivaled the Panama Canal as a death maw with its endless swamps, peculiarly deadly malaria, and the usual scourges of typhoid, influenza, and construction accidents.[26] The border demarcation commissions too were a potential source of work. There were also hundreds of lesser sinecures that accrued to the new republic—but he had managed to irritate many of its patrons with an outburst or a snide column, or simply his air of tetchy self-righteousness.

Da Cunha had yearned for Acre with its "Republic of Poets." The previous year, as he wrote his friend Luis Cruls, astronomer and director of Rio's observatory "I've been nourishing a dream for the last few days, of traveling to Acre. But I don't quite see how to manage it."[27] He hoped that Cruls, who had directed the boundary team of the Javari in 1902 (with Pando, who would later become president of Bolivia), might be able to put in a good word for him among the powers that be. "As for me," he wrote to his friend Vicente Carvalho, "I admit illusions: I have thousands of promises, thousands of hopes even as I believe that tomorrow I will be thwarted and annoyed in the number of my choices. If this doesn't happen, I fall back on engineering, and the little scraps of my surveying."[28] On Coelho Neto's advice, he sought out his former Praia Vermelha schoolmate Lauro Müller, minister of public works for the state of São Paulo. He described the attempt to activate this bit of patronage as being like navigating the circles of Dante's Hell. As he mounted each landing, the lost souls of unemployed engineers howled, glared and gnashed their teeth. While he had a pleasant meeting with Müller, nothing was forthcoming from it.[29]

Questions of demarcations and of limits constituted the keystone of Rio Branco's international policy. Demarcation positions, though, were still structured by patronage and required the continuous hints, the asides, the words to the well connected. Da Cunha felt that José Veríssimo, among the founders of the Brazilian Academy of Letters, as an Amazonian, was most able to understand and appreciate his passions and insights about the region. Euclides's introspective letters to him are where the longing that took hold of his tropical writing finds its most earnest expression:

> For me, best to continue on to Mato Grosso, or to Acre or to the upper Juruá or to the two farthest banks of the Mahu River: these are admirable ways to enlarge

my life, to make it at least useful and perhaps even brilliant. I know I can do a lot. After the latest diplomatic efforts, these outposts are today like the state of Amazonas before Tavares Bastos, and if I haven't his admirable vision, I at least share the same passion to reveal the marvels of our country. If, by some chance, the organization of these (boundary) commissions is delayed, I could proceed by myself—with the idea of describing the physical aspects and natural resources of these areas. . . . I don't think it would be that difficult. . . . The simple name of Alexandre Rodrigues Ferreira (forgive me for my self-aggrandizing and audacious parallels) is an esteemed example, one that has existed for more than a hundred years. Besides this, if foreign countries can constantly send scientists to Brazil, how absurd would it be to assign the same task to a Brazilian?

Veríssimo: I'm ever more eager to move this forward: What better service could I offer to the country? Besides this, I don't want Europe, the boulevards, the glories of position: I yearn for the backlands, the rough track, and the hard and sad life of the pioneer.[30]

Veríssimo, and some of his other friends from the Institute of Geography and History and the Academy of Letters began efforts to move da Cunha into fertile artistic terrain and to help him out economically. Veríssimo had a venerable friendship with the chief of staff of the Baron of Rio Branco, Domício da Gama. Da Gama was an intimate of Rio Branco: he had been with him on various legations and had been his confidant in earlier demarcation intrigues and successes, like Missiones. This trusted colleague of Rio Branco would become ambassador to Peru (a highly strategic position of utmost interest to da Cunha and to Rio Branco during later negotiations on Peru's Amazon) and later to the United States. This chain of acquaintance proved providential: da Cunha was posted to the Purús boundary commission. While Veríssimo thought that the task was rather below da Cunha's talents, he and his companions all felt that that such a voyage would lead to another book, as Canudos had inspired *Os Sertões*.

Da Cunha and the Baron

The Baron and da Cunha were matched in intellectual affinities. Both were devotees of military history. The Baron had published books in his twenties on the Platine military forays and had visited and studied the battlegrounds of the Paraguayan, Argentinean, and Brazilian borders. He shared da Cunha's experience as a reporter: he had been a war correspondent in the final phases of the Paraguay War and had been a columnist for the *Jornal do Brasil*'s historical section and a contributor to the Parisian journal *L'Illustration*.

He had taught history at the Pedro II Institute, a position that da Cunha yearned for.[31]

Rio Branco's sensitivity to history and geography was one of the elements that had underpinned his extraordinary diplomatic successes, and this intellectual kinship united him to da Cunha's understanding of the subtleties of the border fight. With Rio Branco's support, da Cunha was ultimately to develop a western Amazonian regional history that would outline deeper dimensions of the dispute, as will be discussed further on.

Rio Branco was a cultivated bon vivant whose experience in the diplomatic corps had polished his taste for good food and conversation. He enjoyed the company of writers, pamphleteers, artists, and, of course, clever and beautiful women. While his glittering salons seemed directed at mere sociality, Rio Branco had learned much in Paris about using such events as means of developing and orienting public opinion. He was thrilled to have da Cunha in his circle, especially for his usefulness in the delicate matter of the particular demarcation expedition to which he was assigned, because of da Cunha's fiery geopolitics and style of nationalism. In addition to being one of the most esteemed authors of his day and a famous revolutionary, Euclides possessed the practical survey skills required for demarcation and the ethnographic, historical, and ideological skills for the task ahead. A person of his technical reliability and literary craft would be crucial for supplying the scientific survey and ideological framing for public attitudes toward the increasingly testy Peru-Brazil negotiations and the Bolivia-Peru adjudications in Argentina. Da Cunha's newspaper writings were already setting the terms of the debate.

Domício da Gama had taken Euclides up to the Villa Westfália for their first meeting at Rio Branco's country house in the elite resort of Petrópolis. He took da Cunha to the Baron's famously messy office at nine o'clock in the evening, leaving him seated "constrained and shy like a student in an exam." The Baron was pleased to find someone with whom he could share ideas on topics dear to his heart: frontiers, Brazilian diplomatic history, international relations. When da Gama came back an hour later, "Da Cunha seemed if anything more uncomfortable and intimidated, as though oppressed by the respect the Baron inspired in him, the great man, so simple and generous, but so unfamiliar." Da Gama came back again at eleven, "but the Baron kept him there until two in the morning. It was very much to his taste, this business of conversing until dawn, gossiping and smoking cigarettes made from cornhusks and rough tobacco." Both were insomniacs; later, according to da Gama,

Figure 11.1. Rio Branco and da Cunha (upper left), as well as the main international aides at Itamaratí.

they would extend long workdays at Westfália into nights of speculation: da Cunha in bed with the sheet up to his neck and the Baron sprawled out on a nearby chair, a lit cheroot to hand.[32]

The Baron apparently approved of the candidate: instead of a mere assistant, da Cunha was made head of the Brazilian delegation to the binational Commission for Reconnaissance of the Upper Purús. Although da Cunha would never overcome his shyness around the Baron, they had begun an abiding and powerful intellectual connection.

Da Cunha was delighted, as he wrote to Veríssimo: "To leave for the Purús is still my greatest, most beautiful, my boldest ideal. I leave without fears, and absolutely nothing, (unless it were a physical disaster that made me an invalid), nothing can deter me from this purpose."[33]

This progress on the professional front was not matched by improvements in his private life. He had hoped that Ana and his three sons could be installed in the lovely neighborhood of Laranjeiras, with Ana's now widowed mother in attendance. But things were not going smoothly at all. He wrote his father in desperation:

> I see now that it's impossible to proceed as I had planned. . . . Not because of me, but because of the unhealthy disharmony between Saninha [that is, Ana] and her siblings. I won't waste time in telling you of the deplorable incidents. It's enough to say that small differences exploded yesterday when I was away: one of Dona Tulia's sons (married, who is staying with us with his wife, in the most absolute indolence) spewed every impropriety at Saninha that came to his mind. On coming home, I was informed of the fact, and had to contain myself. I cannot continue here. I am truly distressed by the scenes in this family. My mother-in-law is profoundly unhappy with her children. I need to get away immediately from all this. But I have no door to knock on other than yours. I ask you to take in my family, and that you save me from a very unhappy situation to which I contributed nothing except my good faith and my belief in the human heart. If you deny me, which I absolutely know you will not, I will be obliged to turn down my commission, and as I cannot stay here nor can I return to São Paulo, I will be irredeemably lost.
>
> Believe me when I say that there is no other way out. I ask you help. Saninha has changed a lot, she's suffered a great deal and has come through bitter disappointments. She has for you only her highest esteem and is almost like a daughter to you, and except for some defects of character which have really diminished, she really is worthy of your regard for her honesty and heart . . .[34]

Despite his praise for her, the acrimony among Ana, her mother, and her siblings extended to da Cunha as well. It was probably with a certain amount of relief that he boarded the ship *Alagoas* on December 13, even though it might mean weeks of unremitting seasickness.

Da Cunha's travels were but the last stage in a complex set of disputes that had origins in the very foundations of Iberian polities in the New World. Many other justifications might be invoked in South American conflicts, but one sure element would be border disputes. The Baron had placed da Cunha at the head of an important boundary commission because that watershed involved the richest rubber forests on the planet and was the last, supremely necessary, demarcation for completing Brazil's territorial trajectory. The urgency of the situation was extreme, and someone of lofty sentiments like da Cunha was especially useful for putting a high-minded gloss on the pursuit of the lands of the "tantalizing latex" *Hevea brasilienses*.

PART 3

As Selvas

Into the Litigious Zones

12

"Impressions completely new to me"

First Encounters and Old Conflicts

The trip north was relatively uneventful, although it was punctuated with bleary seasickness. Da Cunha sent postcards at his various ports of call and was extremely relieved to finally find himself in Belém, the gorgeous tropical capital of the Amazon. He was delighted by its gracious, mango-lined avenues, its sumptuous mansions, its public gardens and plazas, and its residents of "European habits."[1] His own fame and publicity surrounding the boundary commission preceded him. He met with various dignitaries and soon found himself at the Natural History Museum of Pará, where he met with its director, Swiss naturalist Emilio Goeldi, who had played such an interesting role in negotiations over the boundaries of Amapá. Da Cunha dropped off a letter from Veríssimo, and perhaps he consulted with Goeldi on boundary strategies in the comfort of the beautiful buildings that can still be seen today.

What most moved him was his meeting with the great tropical botanist Jacques Huber, and his first real encounter with the Amazon as he sailed around the estuary in a touristic afternoon outing. He later described it this way in a much-celebrated presentation at his induction into the Brazilian Academy of Letters:[2]

> It's been two years now since I first entered the estuary in Pará at that almost indefinable point in the Amazon delta where the river becomes ocean. Against my expectations, I wasn't overwhelmed. What I'd imagined as monumental was flattened in perspective, a stretch of water with neither the picturesque effects of waves nor the mysteries of depth. The liquid surfaces, muddy and smooth, stretched without definition north and south between two strands of eroded land, equally undefined and without even the slightest undulation to relieve the view. In the midst of these indecisive shoals, like badly drawn islands or, better, caricatures of islands half flooded by tides and heavy with swamps, there was a sort of shipwreck of land. It was as though, half drowned, a castaway had crawled up and grasped convulsively at the twisted mangrove roots. Above these narrow strips of water and land was only the sky, empty and bending down to the perfect arc of the horizon, as in the midst of the Atlantic.
>
> I quelled my disappointment. With the obstinate intention of finding everything prodigious, of feeling the masculine lyricism of a Frederico Hartt or the "glorious" impressions of Henry Walter Bates, I retired to a corner and filled pages of my notebook with the most elaborate adjectives, stirring nouns, and glittering verbs that I could muster. But I ended by ripping up those pointless scribblings, ultimately inexpressive and empty. . . .

> Later, I stepped into the launch, and into Belém. I soon found myself at the Pará Museum, within whose walls all the marvels of Amazonia are condensed. There I met two men: Emilio Goeldi, who is the spiritual grandchild of von Humboldt, and Jacques Huber, less known, but a preeminent botanist. Here was no frozen-faced savant with depressing wisps of this hair as in those German portraits.[3] Huber is a subtle spirit served by an organism both robust and sleek: *vir quadratus*, as a naturalist ought to be, because the natural sciences today demand thinking titans whose muscles evolve along with their minds. The match between his sensitive, powerful intellect and the integrity of a robust physique is the consequence of his arduous forays into the wilderness. This scholar can complete an expedition of six hundred leagues—from Belém to the Ucayali—in less time than any one of us would take to travel to Gávea.[4]
>
> I spent two unforgettable hours with him, and returned to the ship with a monograph in which he analyzed the region that had seemed to me so barren and monotonous. I delighted in this volume the entire night: and at dawn the

next day—one of those "glorious days" about which Bates writes—I went up to the deck and there, with my eyes burning with insomnia, I saw the Amazon for the first time. At last there rose up in me the commotion of feeling that earlier had eluded me. How singular seemed that smooth and muddy surface. Now at last I was reminded of the superfluity of sky above and of water below of the last unfinished page of Genesis, still incomplete and marvelously writing itself. I then understood the innocent yearning of Cristobal da Cunha for the source of that great river to be Paradise. . . .

I considered again those undefined shoals, those islands or half-islands diluted by gentle waves—and I saw the gestation of a world. What I had first seen as a desperate crawl and pathetic grasping was in fact a leap of triumph. The plants were saving the land in a struggle through which shone a singular intelligence. Here was the phalanx of *anigas*—those giant aroids[5] with their stiff leaves, brilliant and sharp as though chiseled, united in palisades against the assault of the waters. There, set against the slow-swirling drifts of sediment, were the sieves and filters of *canaranas*—aquatic grasses—and *aturizal* thickets. Along the banks extended the twisted roots of the mangroves, in whose mesh the diluted mud transformed itself into a resistant soil, whence soon emerged the bush lianas and, draping over everything, tresses of *apuirana* vines and *juquiri* scrub. Thus in the end terra firma was carved out, forming itself slowly in the *buriti* palm swamps, where the palms open their enormous, whispering fronds to the blue sky. Foreshadowing the forest to come, destroying at a stroke all the monotony of that unending flatness, arise majestic *ceibas*—silk cotton trees—high and round and softly undulating in endless rolling landscapes, hills of green.

I understood then, under the resplendent and limpid sky, that land emerges from the womb of the waters. Thence from the rising sap springs the forest, conjured upward by the magnet of light. I continued my journey under the spell of a new enchantment, but with a dispiriting worry.

There was nothing "literary" about my new, truly artistic impression: it had not been inspired by expert stylists. The "poet"—Huber—who aroused it had neither rhyme nor meter. The eloquence and brilliance were frankly imparted by the extraordinary display surrounding him. I, son and desperate lover of the land, could not find in myself the choice words with which to describe it, but he could, using only language drawn from the austere lexicon of technical diction. I discovered then how difficult is a trivial thing in these times when the earth is crammed with books: *to write*.[6]

A few days later he was off to Manaus, a place that would become for him a kind of incubator of both greatness and despair. As he steamed up from Belém (an easy trip to take today) he observed this "unfolding world," a world,

Figure 12.1. Jacques Huber, da Cunha's "thinking Titan."

as he wrote Veríssimo, "as yet unready for man,"[7] a sharp contrast with the stasis and atavistic features he'd imagined for the Bahian outback. There, in the Sertão, the place and people were forgotten by history and resided in an arid, lost detour in the larger human trajectory. Here he was confronted with the vast apparatus of creation and emergent history. But his magnificent emotions were soon to be counterposed by the tedious details of government bureaucracies and thousands of daily frustrations as he moved to the staging point of his expedition.

"There will be a book . . ."

In Manaus, da Cunha moved to the lovely chalet, Vila Glirícia, of Alfredo Rangel, his friend from Praia Vermelha, in a suburb that abutted the forest. Rangel had been making his fortune as a surveyor for rubber estates, a profitable and necessary task. Although the larger international boundaries remained sketchy, the urgencies of profit required detailed property lines. Rangel was away, and da Cunha was reading, doing errands, and waiting.

Euclides hoped that his Amazon writings would become a companion volume to his *Os Sertões*.

> So here we are, still waiting for our departure day . . . it perhaps will still be further delayed, so sluggishly are our difficulties being redressed. It suggests an eloquent contrast: The great English explorer Chandless, when he arrived in Manaus to explore this exact same Purús, encountered the most efficient and efficacious help from the highest provincial authorities to the most humble *caboclos*. And this was when we were in the heat of the Christie Question![8] And Chandler was British! And merely a traveling member of the Geographical Society of London! We, Brazilians, invested with an official commission, confront endless obstacles. Times have truly changed a great deal, my illustrious friend . . .[9]

And to his friend Artur Lemos he wrote:

> Were I to write you now I could only sketch miniatures of Chaos, incoherent and tumultuous—an incredible mixture of endless flooded forests and vast resplendent skies: between these two extremes there are innumerable modalities and a new world completely unknown to me. . . . The news by telegram about my new book is true: I write like I smoke: it's an addiction. But I'll try and give the impression of a writer obsessed by the topic. And if really I do write the book I'm not going to give it a title that relates too much to the places where von Humboldt advanced his prophecies, or where Agassiz committed his greatest errors. I'll write a kind of "Lost Paradise," for example, or something else whose amplitude frees me from a precise definition of the facets of a land that, to be well understood, requires the unremitting efforts of a lifetime.[10]

Most of his letters at this time are rife with complaint. There were unimaginable delays, as there still are for anyone who has ever waited for official authorization for travel in the Amazon. Da Cunha had been wildly overoptimistic about the pace of officialdom. Yet these setbacks, while deeply annoying, served him in his larger Amazonian apprenticeship. He was ensconced in Rangel's chalet with a good library and a dear friend, Firmo Dutra. He was feverishly distressed and exhilarated by Manaus: "In the tumultuous mecca of the rubber barons, my life is unsettled and tiring. At the same time that I must attend to the countless demands of my post, I suffer from an assault of impressions completely new to me. The obligatory delay is not good, and

I don't know how long it will last. . . . We've already missed the best time to advance into the wilderness."[11] Da Cunha had not yet experienced the arrival of the Peruvians, who, he noted in mid-January, appeared with wrecked boats that would take two months to repair.[12] He grumbled: he'd imagined a rapid foray up the Amazon to his mysterious wilderness in the Purús. Three thousand miles from Rio, the vulgarity of Manaus was imposing: the rabid and devouring commerce, the tumult, all manners of business and dress from the loincloths of natives to the British formal garb complete with white shoes, a jumbled city of native huts, *malocas*, and opulent palaces in rampant disorder. He was wild to leave, and during this time there were regular letters to the Baron, full of polite inquiries: the departure date? the marching orders? These were still being hashed out.

Petty annoyances took their toll, but he was also distressed by the drastic drop in the river level. Shallower waters would make the upcoming voyage far more dangerous as sandbars, drowned trees, and treacherous rapids emerged from under the enormous cushion of water that swept to the Atlantic as the great channels drained. The wait would also hamper the all-important technical survey, which depended on chronometers, theodolites for ascertaining longitude, and other sensitive astronomical and survey instruments whose accuracy would inevitably suffer under the constant buffeting of rough travel. In addition, the expedition's duration would be increased by the difficulties inherent in a far more treacherous river channel, the endless and necessary portages. Their guiding map, that of Chandless, would require constant review: their coordinates would require verification as the commission labored upriver.

While waiting in Manaus, he wrote to his long time comrade, the literary critic Henrique Coelho Neto:

> . . . I've not told you about the days I've spent here waiting for my wilderness, my brutal and redemptive wilderness that I intend to enter with the British intrepidness of a Livingstone, and the Italian desperation of a Lara, in search of a new chapter in the messy novel of my life. I should already be mastering the headwaters of the magnificent river and be sprawled out, exhausted, on the first round foothills of the Andes. But what do you want? We're kept here by the entanglements of our indecipherable administration, and only on the 19th or the 20th of this month [March] will we receive the our marching orders. This Manaus, carved in long and ample avenues by the daring of the "Thinker,"[13] affects me like a small and narrow closet. I live without light, half done in, in a daze. I've told you nothing of the people, nothing of the lands. Later, then there will be a book: "The Lost Paradise," where I'll seek to avenge the glorious Hylea

for all those who've plundered her since the seventeenth century. Surely I was meant to be a Jeremias for such times—I lack only the tangled, tragic long white beard.[14]

Meantime, he was collecting sketch maps from the tappers and land surveyors on the Purús, comparing them with the Peruvian maps from the Institute of Geography in Lima sent to him by da Gama. He hoped that perhaps he could precede his Peruvian cohorts, who seemed reluctant to depart from the pleasures of Manaus. A solo start was impossible, however, within the framework of a joint boundary commission. He was also worried about his home life: Ana had not written him since his departure, two months past.

Given da Cunha's well-known newspaper articles discussing the Peruvian temperament, he perhaps should not have been too surprised at the diffident behavior of the members of the Peruvian Commission. He wrote again to Veríssimo:

I note that they are in no hurry at all, they persevere in an adorable placidity. Hot-blooded Spaniards, ebullient and raucous, ready for all the dances and the refined throngs of this chaotic Manaus; they are Quechuas—morbidly lazy—when it comes to departures. I have come to think that they are scarcely interested in our enterprise, and only barely, resentfully, endure it. And how they detest us! What's interesting is that I arrived at this conclusion thanks to my native *caboclo* slyness. Each one of these affable guys, when they encounter us, dissolves into amiable smiles, with endless compliments and sugared phrases, while they hide in their hearts a treacherous loathing. In the end I adapt to my own hostility, I dissemble, and shudder at the terrible irony of the hypocritical and dangerous cordiality in which we live. The future, perhaps, will confirm these conjectures; and without delay, I would, were I the government, try to secure the basins of the Javari, Juruá, and Purús, where one day fleets of boats and canoes will glide on the currents. Don't see in this patriotic apprehensions that I don't have. But there is a certain conclusion: there is no place in the world like that where Peru and Brazil adjoin in such majestically opulent outposts. The conflict—and let's hope that the palliatives of the current arbitration prevail over the insane, absolutely irreconcilable clashes of the *seringueiros* and *caucheiros*.[15]

He would later note in letters to the Baron[16] that the Peruvians were hardly convinced of the mission's prospects for success, and they felt quite certain that the expedition would have to turn back due to the "insurmountable obstacles" that would necessarily emerge making the exploration of the headwaters impossible. Indeed, it may have been part of the Peruvian strategy, as da Cunha suggests, to keep the physical exploration of headwaters in contest until the last possible moment so that the arbitration would have to fall back on earlier agreements, the imaginative Peruvian cartography and the Idelfonso Treaty. This was a potentially successful plan, one that Peru had employed to its advantage with Bolivia and Ecuador. The power of this earlier legal precedent could work in Peruvian favor. The Peruvian resistance to completing the project reverberated throughout the travels of the Joint Commission.

Da Cunha begged Veríssimo for leave to go visit his family. A day later, the marching orders would arrive.

The night before he left for the Purús he wrote to Rangel, in contemplation of what for him might be a very short future.

> "Alone . . . completely alone in the narrow study of your bucolic hideaway. Or better yet, alone but with some ancient shades: Father João de St José;[17] the valiant Ricardo Franco,[18] the meticulous Francisco Lacerda e Almeida[19]—and I don't know how many others: imagine if you can our silent and formidable orgy—alive with past triumphs and stupendous endeavors—forever ended. That old monk, of Voltairian wit with his impeccable phrasing, tells of the old exploits of an old Amazon; the implacable colonel—the greatest explorer of all the Livingstones and all the Stanleys—speaks of his four or five odysseys into the backlands, including his last travels from the "visible equator" to the meridians of Mantiquiera!
>
> They while away the hours in this wordless chatter. . . . Firmo is out; at this moment he is prosaically under the adorable and passionate gaze of his fiancée. Manoel is inside, with his elbows on the table, brow furrowed, doing his ABCs in a crumpled workbook . . . it's a great, mysterious silence, laced sporadically with the scattered murmurs of the foliage—and then becomes more solemn, an overwhelming stillness. Of course, if from time to time the fretful hacking from your rooster with the chronic bronchitis, "the nutcase" according to Firmo's diagnosis, were not to call me to an irritatingly earthly reality, I would see your bookcase mysteriously opening, and bursting out from it the tortured poet Rollinet with "L'éternelle dame en blanc, qui voit sans yeux et rit sans levres"* on his arm. Such is the august serenity that, in this dead hour, completely dominates your charming villa.
>
> Our departure is near. The orders came yesterday. My fleet? Two launches

* "The eternal woman in white, who sees without eyes, and laughs without lips . . ."

Figure 12.2. "A nossa flotilla"—our flotilla (in da Cunha's hand). The Brazilian fleet for the Purús: small boats for such a great quest.

> (one still problematic), a *batalão*, and six canoes, bob triumphantly at the end of the *igarapé** "Raimundo," and yesterday were baptized by a tempest. I never imagined that this sluggish river could so treacherously hide such violent waves. A sharp gale from the southwest produced the tumultuous swells of a sea—and it is a sea! A sea between embankments . . . Happily my boats bravely survived. . . . I have the belief, largely metaphysical, that our life is always guaranteed by an ideal, a great aspiration that is realizing itself. And I still have so much to write.

He ended this letter on this note: "A favor, a sacrosanct brotherly favor: please visit my four darlings—my *saudades*—in Larangeiras:[20] my wife and the three little ones. Visit them and give them my news . . ."[21] Even six weeks later, in May, well into his travels, he had heard nothing. By telegram he begged Domício to send him domestic reports, but no one would find the family he missed so at the home where he had left them. Da Cunha's correspondence and telegrams became increasingly desperate. When he eventually returned to Rio, he had to cable the trading house to locate her.

* An *igarapé* is a creek or small river.

Life at the Pension Monat

Da Cunha's family had moved to a boarding house, Pension Monat, and later to a beach house. What had transpired in Ana's life? It is difficult to say: perhaps she tired of languishing with just her children and servants for company. Her other social option, her family, was so rife with conflict and so obnoxious about Euclides that they provided little by way of pleasant company. Euclides's attempts to fob her off on his own relatives had also been unsuccessful.

Ana had traveled to São Paulo to put her children Solon and Euclides Jr. in an English boarding school, and there she hired João Ratto to be their tutor. She stayed in São Paulo for a month to settle the boys in their new school; during that time she stayed with Angelica and Lucinda Ratto, sisters of the tutor, young women in their early twenties. The company of these lively and musical young women apparently agreed with Ana. After returning to Rio, she and her youngest son, Manoel Afonso, moved to the Pension Monat, a kind of boarding house that let suites to families, where Lucinda and Angelica (whose nickname was Neném) also had moved.

Lucinda and Neném were aunts of the orphaned teenaged boys Dinorah and Dilermando de Assis, mililtary cadets who were connected to the da Cunha household through kinship with Euclides's friend Veríssimo and, on Ana's side, through their mother. The two de Assis youngsters visited their aunts bringing gifts, and they quickly struck up a friendship with Ana and her oldest son Solon, a few years younger than the cadets. Ana's interest in sports, music, and shooting was matched by that of the young Dilermando.

Neném, who seems to have been an accomplished intriguer, soon suggested to Ana that it would be an excellent idea to have Dilermando move into the Monat and urged her to suggest it to him. He could accompany them on their excursions, providing the masculine protections of domestic kinship necessary for freedom of movement for women of a certain class in Rio.[22] Since the Monat lacked the space necessary for this new addition, Ana leased a house on Rua Humayta and decamped there with Manoel Afonso, Neném, Dilermando, and a couple of servants.[23]

Dilhermando was almost everything that Euclides was not. He was a gaucho from Rio Grande do Sul, the temperate zone of "European" Brazil, rather than from "black" tropical Bahia. He was tall, blond, robust, athletic, sportive, and though not unintellectual, hardly the pedant that Euclides could be. Also in contrast to Euclides, he adored the military, made his career in it, and would eventually rise to the rank of general. De Assis had been suspended from Praia Vermelha because he had taken part in antivaccination riots in Rio, and it was in this hiatus that he ended up visiting the aunts.[24] (Some of the cadets had opposed this public health measure and had run amok, breaking streetlights in protest, and were punished by enrollment sanctions.)

Ana was twice the age of the seventeen-year-old Dilermando, but they soon embarked on a passionate affair. Euclides had one view of "bronzed titans" while Ana had quite another. And whatever Euclides's sexual inclinations might have been, his descriptions of women in his letters and publications are very few and, especially in *Os Sertões*, quite derogatory. There are some passing comments on the mistresses of various rubber barons in his Amazon work, and a laudatory depiction of the wife of Peruvian president Garramón in military drag, bespurred and leaping into battle, galvanizing the troops.[25] While Euclides da Cunha clearly loved his family, at least part of his nomadic profession might be understood as a means of keeping his distance from them, or at least from Ana.[26] Now, after fourteen years of uneasy marriage, Ana was clearly inflamed by her young lover—and his contrast with her sere, sick, cerebral and absent husband. She had stopped writing to Euclides. Only Bazaar America and her closest intimates knew where she was.

Da Cunha's leap into the wilderness meant traveling to a realm whose details were largely unknown. The latex economies were mostly appreciated through their results: the phenomenal wealth generated by the innovation economies of the Euro-American industrial revolution that relied on rubber: tires, machinery gaskets, water proofed footware and clothing. Rubber and caucho also made huge fortunes for its commercializers.

Most observers kept to the river, and the story of what was transpiring in the great forest was occluded by the Green Wall, even if Wall Street and London's "City" had remoter backwaters clearly in their sights. Da Cunha's expedition was about to enter into one of the most historically and socially complex parts of the tropics. His meditation on the river set the framework for his descriptions of myths, history and economies, and the violence that that would build on this turbulent substrate.

13

"Such is the river, such is its history"

"The great river, notwithstanding its overriding monotony, evokes all manner of marvel, and equally inspires the innocent chronicler, the romantic adventurer, or the informed scholar."[1] Certainly what da Cunha understood better than most of the scientists and scientific tourists of the age was how the impact of centuries of mythmaking, speculative maps, and preconceptions had formed a scrim through which people viewed "their" Amazon. Thus alpine geomorphologist and creationist Louis Agassiz saw massive glaciation and divine creation producing the physical structure and biotic diversity of Amazonia. But other researchers such as Frederick Hartt promoted accurate theories that seemed equally as fantastic at the time, such as the tectonic rise of the Andes changing an ocean trench into the vast "Rio-Mar," the River-Sea of the Amazon. The Amazon was not yet the venue of intense debates over its social and biotic history that it would later become: it was still mostly a place of simple compendium and naive observation, but people had begun to theorize about it.

Once scheduled steam travel began, the Amazon channel from Belém to Tefé was described by countless travelers, but the Purús, one of its lengthiest tributaries, remained a cipher, due in part to the aggrieved native groups at its mouth, and, as da Cunha would put it, there were "events that perhaps lacked a historian." The Juruá and Purús Rivers meandered through a vast plain largely unknown to formal survey and science even though it was generating immense riches, so it was in a vulgar sense "without history." There was quite a bit a fragmentary information about it, and it would ultimately become da Cunha's task to organize "news from nowhere" into a coherent story. Most travelers to Amazonia focused on "prodigious nature," the grandeur and exotic differences of Amazonia compared to their northern ecologies. Euclides was not so taken by the imaginaries and tropical wonders of Amazonia described by the throngs of European and American writers who so shaped how the place was viewed. After all, he was himself a man of the tropics; tropical landscapes weren't so alien to him. Da Cunha's artistic, scientific, and political purpose was to focus on the place for "creole" nation building.

In the short essay "General Observations" in *À margem da história*, da Cunha describes the dynamic geomorphology of the Purús and the Amazon more generally. This section was probably meant to function for his "Lost Paradise" in the way "The Land" framed the deep structure of *Os Sertões*, which begins with a detailed account of the geologic history of the Northeast. This supplied the metaphors and images for the people and society that emerged there. This Amazon essay was probably meant to be the geological "substrate" for the new book, especially in light of da Cunha's other descriptions of a nascent and new civilization emerging from the "womb of the waters." What is most striking about these few pages is that rather than summarizing the works of other geologists as he had at Canudos, in Amazonia, he took field measurements, incorporated the "shipping news" of channel dynamics, and integrated into this his reading and critiques of the few naturalists who had taken up Amazonian geology and geomorphology: Hartt, Derby, Wallace, Bates, von Martius, and Agassiz. He challenged the applicability of the "cycle of erosion" theory of Harvard's William Morris Davis (the dominant geomorphic paradigm of the time) to upper Amazonia.

The essay "General Observations" in *À margem* was a literary version and complete intellectual revision of a scientific paper, "Um rio abandonado (O Purús)," published in the *Revista do Instituto Histórico e Geográfico Brasileira*[2] in 1905, shortly after da Cunha returned from Amazonia, and republished in *À margem* with a slightly different title, "Rios em abandono."[3] In the first version, da Cunha framed his discussion of Amazonian landforms in terms of William Morris Davis "geomorphic cycle" or the "cycle of erosion." This model, based on Davis's analysis of the Alleghenies in Pennsylvania, was informed by both Lyellian and Darwinian ideas of evolution, where erosion processes determined the stage of development. Landscapes went through youth, maturity, and old age, beginning as huge flat plains were subjected to erosion, then shaped into mountains, moving toward a mature form of river system and ending up as old depositional flat peneplains with hyper-meandering rivers, a landscape, on the face of it, rather like that of the Purús. Eventually they would be uplifted and the cycle would begin again in an unending repetition of "structure, process, and stage."

The simplicity and elegance of this model enchanted generations of scholars. Given da Cunha's taste for historical and stage theories and his training in geology at Praia Vermelha, and given that Davis's theory was the dominant and widely taught paradigm of landscape development at the time, it is not surprising that shortly after his arrival back in Manaus he thoroughly embraced it. In his early 1905 article he argued that the "Purús was one of the best examples of it [Davis's model]."

However, Anthony Orme, a historian of geomorphology, notes that these organismal metaphors were but one strand in the understanding of landscape, and as compelling as Davis's imagery and metaphor were, Davis was very weak on the actual mechanics that generated the landforms.[4] It was civil engineers (members of da Cunha's profession) who were providing quantitative data on flow, flux, and landscape change. Although their understanding of the physical mechanisms would later prevail, in the early twentieth century this quantitative line of research was still a distant rival within geomorphology. Other paradigms were also lurking in the wings, including Alfred Wegener's idea of plate tectonics (a theory Orville Derby and Frederico Hartt favored), and American geologist G. Karl Gilbert's development of the idea of isostasy, or the idea that the earth's surface rises as parts of it are denuded. The idea of crustal loading and release, though empirically well elaborated, lay outside Davis's analysis or interest.[5] Davis apparently had a domineering personality, as well as a dominant theory, so many other approaches were relegated to the sidelines.

While Davis is cited in the first sentence of da Cunha's 1905 essay and completely infuses it, his name is astonishingly absent from da Cunha's later "General Observations" except in an oblique reference to the development of estuaries. One might conjecture that da Cunha had changed his mind upon his return to Rio, after conferring with his geologist friends, musing on his data, and rethinking his ideas. He was very familiar with the classics of geology, the discipline he most excelled in as a student; it was of special intellectual and personal importance since his friend Orville Derby had promoted da Cunha for the Institute of Geography and History. Da Cunha had worked with Derby in São Paulo and stayed current in the literature.[6]

In "General Observations" da Cunha is responding to several currents in the history of ideas within geology and geomorphology: the catastrophists, who viewed transformations on the earth's surface as the outcome of a cataclysmic or another singular event, and the deists, who made Christian scripture, especially the Genesis flood, the definitive framing device for understanding timing and types of events in the evolution of landscapes. Da Cunha was also profoundly shaped by Charles Lyell, whose *Principles of Geology* was the foundational text for modern geology and also influenced Darwin's understanding of natural history.[7] According to Lyell's uniformity principle, past events and present landscapes were formed by similar processes unfolding over extensive periods of time, and causes that transformed the earth's surface in the past remain active in the present. His vision of landscape was rooted in continuities of incremental change rather than divine design or simple disaster.

Da Cunha's analysis would refer mostly to the almost continental-scale research of geologists Derby and Hartt, who embraced the ideas of plate tectonics. They explained much of the Amazonian geology in terms of the Andean orogeny that blocked off the great channel between the Guyana and Brazilian shield, creating an immense lake that occupied most of the South American continent for a significant part of its geological history. They also noted the continent's geological affinities with Africa. Da Cunha also invoked the work of Frederich Katzer, the German paleontologist (and Pará museum researcher) who commented on the relation and similarities of Amazonian geologies with those of Africa, causing a sensation in European circles.[8] Da Cunha's view was syncretic: while embracing Lyell's basic position about the continuity of processes in shaping landscapes, he followed the arguments developed by Hartt and Derby that saw the Purús and the Juruá generated from the rise of the Andes, their structure and geology having virtually nothing to do with Davis's paradigm. The hyper-meandering of these rivers would have been evidence of great age in Davis's model, but that of Derby and Hartt had it as very young.

Insurgent Rivers

Da Cunha developed an original reading of the riparian dynamics of the upper Amazon, one that would be corroborated by scientists a hundred years later with greater technical precision and be reported in distinguished scientific journals.[9] What he would describe as the "unfinished page of Genesis" was the insight that the hyperdiversity of upper Amazonian watersheds was created from the complex interactions of his "insurgent" rivers and the vegetation of that "nascent earth."[10] His insights about the channel dynamism and the youthfulness of the landscape have been reinforced by modern tropical geomorphologists working with techniques from remote sensing and ecological field surveys.[11]

Da Cunha's supplements to the commission reports would provide the river soundings from the expedition, but he would argue that they were ephemeral and ultimately useless. The landscape was too dynamic: "one flood can undo the work of a hydrographer." "Places where Chandless sailed are today covered by trees, and we rowed over areas he described as forests."[12] This change occurred in a mere fifty years. It was from these observations that he so lyrically developed "General Observations" into the first section, "Land without History," of *À margem da história*. He was ironic with his titles, because he would in fact give this land its "history."

Da Cunha saw Amazonia as a young land, one that was forming itself through repeated dialectics of creation and destruction, erosion and deposition; in a sense he was embracing the idea of "everyday" forms of catastrophe that set the stage for less dramatic processes of landscape evolution. In many ways his understanding of nature fits better with that of Stephen Jay Gould's model of punctuated equilibrium than with models of his contemporaries.[13] Thus, while he describes the landscape as "fumbling for equilibrium" (and here he invokes Gilbert)[14] and as a gradual world buffeted by periodic cataclysm, his evocation of the river captures a dramatic dynamic: "It is always disordered, . . . erratic, destroying, constructing—devastating in one hour what it built over decades." Da Cunha perceived the young landscape as evolving under his eyes, protean and unformed. As he saw landscapes dissolve and evolve, so too ecosystems, societies, and humankind would change in gradual adaptations within a world periodically unraveled by processes of natural catastrophe and creation. This was a metaphor, for him, of tropical civilization, and certainly an insight into modernism.

Da Cunha makes the argument that the meandering landscape was an emergent terrain, an inchoate "nature," a world still unfolding, a natural history "still writing itself." The primitive register is reinforced with grotesque reptilian fauna, animals "suggesting rungs on an evolutionary ladder." It is a place still seeking its form and metaphorically evoking that nature, still undefined, advancing, retreating, trying one channel, then evading it, evolving—an image of the Brazilian experiment and remaking that infuses da Cunha's Amazonian nationalism.

As channel processes maintain floodplain forests in earlier successional stages, larger-scale fluvial dynamics such as channel avulsion—where the river completely abandons its bed—modify vast intrabasin areas, and with these shifts introduce floodplain dynamics into new areas. "Land without History" reflects the constant reworking of the landscape by immense forces: "Geological history writes itself every day before the enthralled eyes of those who know how to read it"—this "nascent earth" is the outcome of the most common physical forces," says da Cunha in his salute to Lyell.

River dynamics went largely unnoticed by the scores of Amazon travelers, most of whom were intent on registering species new to science and paid scant attention to anything else. Da Cunha felt that those scholars who were most identified with "Amazonia" the Anglophone researchers under the direct or indirect aegis of Harvard, Kew Garden, the Smithsonian, and the Royal Geographical Society—had not grasped its real "nature," because mostly their expeditions closely followed the main steamship routes of Amazonia, descended at the same wooding stations and regular stops, hired children and a few amenable adults to go and collect plants, insects, butterflies, fish, and mammals while they took agreeable tramps in the woods and organized the packing of their samples, berating the locals for drinking the cane alcohol that was brought along to preserve specimens.[15] By looking at close range for novelty and exotica, they missed the bigger picture.

The central theme of da Cunha's "General Observations" is the invisibility of the history, its dynamism, and the fragmented nature of knowledge of the Amazon. This would apply as much to the river as to the civilization unfolding there: "Such is the land, such is its history: always insurgent, always incomplete."

Here is da Cunha's view of Amazonian geomorphology.

General Observations: Land Without History

> Rather than admiration or enthusiasm, the feeling that overtakes one at the lush Tajapuru River's confluence with the Amazon is foremost one of disappointment. The huge mass of water is of course impressive, inspiring that awe to which Wallace refers. All of us who have traced out an ideal Amazon since our youth, thanks to the lyrical pages of I don't know how many travelers who from von Humboldt to today contemplate the prodigious *Selva* with an almost religious amazement—must confront a vulgar psychological pattern: as we face the real Amazon it seems inferior to what our imaginations had fashioned for so long. Moreover, when this stretch of land is unleashed from its artistic images and the compendiums of stirring impressions, it is, taken on the whole, a good deal inferior to other places without number in our country. In this respect, all of Amazonia is not worth the coastline that stretches from Cabo Frio to the Cape of Munduba.

It is a vast panorama, but one that has been beaten to horizontal contours on which the shoreline is barely lifted. The appearance of the rest is that of an immense broken-down scarp modeled from the arenite hills of Monte Alegre and the Granitic Guyana Shield. It is as though the place lacks vertical lines. The excess of the landscape is such that in a few hours the observer gives in to the fatigue of the unnatural monotony and feels that his gaze is inexplicably foreshortened in that world of endless horizons, as empty and indefinite as those of oceans.

The overwhelming impression that I had, and perhaps this corresponds to a fundamental truth, is this: man there is still an impertinent intruder. Neither awaited nor desired, man arrived when nature was still arranging its most vast and lavish salon, encountering there an extravagant disorder. Even the rivers have not yet formed their channels, they seem to be fumbling for some kind of equilibrium, descending and diverging in unstable meanders, contorting in draws and oxbows where isthmuses continually break apart and rebuild themselves in the futile formation of islands and lakes for a mere six months. These rivers even create new topographic forms in which the defining aspects of land and water are confounded. The waters extend themselves into lawless bayous, without one knowing whether this is a river basin or a sea dissected with straits.

One flood can undo the efforts of a hydrographer.

The flora vaunts the same imperfect grandeur. At the silent middays (because the nights are fantastically noisy) those who advance into the forest find their vistas truncated by the green-black of the foliage, and from one minute to the next. Faced with arborescent fronds that rival the height of the palms, and trees with straight trunks and completely impoverished of flowers, they have the anxious sensation of retreating back to time's remote epochs, as if one were adjourning to the dark recesses of one of those carboniferous forests revealed in the retrospective optic of geologists.

Completing this picture in this ancient register is the remarkable and monstrous fauna ruled by corpulent amphibians, which further enhance this Paleozoic impression. For those who follow the long river courses, it is not rare to find imperfect forms of animals persisting like abstract types or simple rungs on the evolutionary ladder. The *cigana* roosting on the flexible branches of the tropical *oirana* willow* still has on its wings, adapted only for short flight, the armature of reptiles.

Face to face, nature here is prodigious but incomplete. It is a stupendous construction that lacks all internal design. What is important to understand is this: Amazonia is perhaps the youngest land in the world, an idea consonant with the

* *Salix humboldtiana*, the common Amazonian willow. *Cigana* is the hoatzin *Opisthocomus hoazin*.

well-known inductions of Wallace and Frederic Hartt. It was born from the last tectonic convulsions that lifted the Andes and has hardly ended its evolutionary process with its Quaternary *varzeas*, the floodplains that are constantly forming themselves and are the dominant form of that unstable topography.

Amazonia has everything but lacks everything, because it wants that fettering of haphazard phenomena to a rigorous framework from which the truths of art and science can so vividly emerge—and which underlies the great unconscious logic of things.

And thus this irony: those outposts most observed by the savants are the least known. Reading them, one verifies that not one left the principal channel of the great valley, and it was there that each achieved, dazzlingly, his corner of specialty. From von Humboldt to Goeldi, from the dawn of the last century to our days, all the elect anxiously and assiduously deliberated there. Wallace, Mawe, Edwards, d'Orbigny, von Martius, Bates, Agassiz, to cite only those who come first to mind, soon reduce themselves to mere writers of ingenious monographs. The very ample scientific literature on the Amazon well reflects its physiognomy: it is surprising, precious, disconnected. Those who rush to misread her will at the end of these efforts only linger at the threshold of this marvelous world. . . .[16]

This phrase by Frederick Hartt reveals how even the most robust spirits quail when faced with the Amazon's enormity. Hartt studied the geology of Amazonia, when he discovered his precise scientific formulas so shaken, so structured by dreams, that he had to quickly draw in his sails of fantasy:

"I am not a poet. I speak only the prose of my science!" He then continued to write, devoting himself further to his rigorous deduction. But two pages on he could not face the new excitements and reinitiated his enchantment. . . .

It is the great river, notwithstanding its overriding monotony, which evokes all manner of marvel, and which equally inspires the innocent chronicler, the romantic adventurer, or the informed scholar. The Amazons of Orellana, the Curriqueres, the Giants, of Guillaume Lisle, and the Lake of El Dorado of Walter Raleigh all created out of our past such a dazzling and almost mythological cycle that these emanations still inhabit the most imaginative hypotheses of science. There is a hypertrophy of the imagination that cannot adapt to the incongruity of the land, and this unbalances the most staid mentality. In the realm of objective endeavors are the visions of von Humboldt and a series of conjectures where all sorts of concepts are advanced or refuted, ranging from the dynamics of earthquakes of Russel Wallace to the formidable biblical forces of the antediluvian ice ages of Agassiz. It seems in Amazonia that the magnitude of the questions implies a certain languid discourse: inductions favor flights of fancy. Truths are hidden in hyperbole. In its inflated idealizations, tangible elements of Amazonia's surpris-

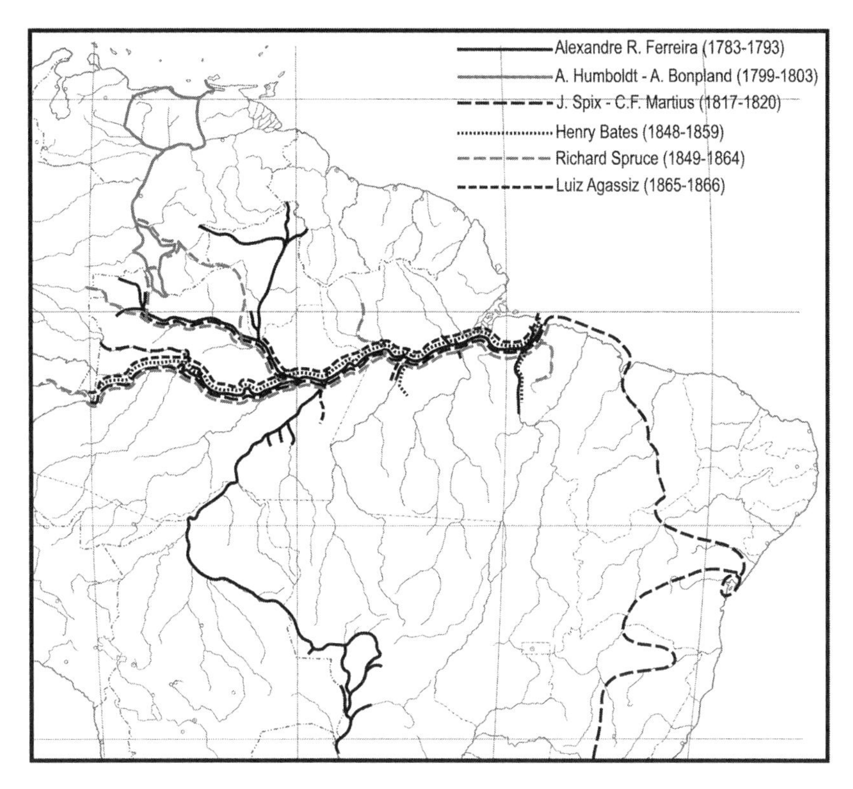

Map 8. Map of tropical travelers.

ing reality are portrayed such that the most unfettered dreamer can easily find himself in the company of the most exalted scholars.

One could go, for example, with F. Katzer[17] to find, classify, measure the ancient petrefate and gratolites in a lengthy pilgrimage ideal for encountering the remotest points of distant epochs—for a long time people debated about the classification of the massifs, enmeshing themselves in the web of Greek nomenclature, and surprisingly, the maxims of science unfolded into a kind of fantasy: the cold analyses ended as marvels. The condensed vistas of microscopes unleashed discoveries of a past that stretched back millennia through the stupendous contours of a dead geography. One's eyes focused on the indistinct horizons of the middle Devonian period, when an extinct ocean drowned all of Mato Grosso and Bolivia and thus covered most of Tropical America, whose tides lapped the ancient scarps of Goiás and the distant shoreline of the Brasilo-eliopic continent, which filled the Atlantic and stretched to Africa.

Follow as well the scientists of the Morgan commission,[18] where the historical geography even in its most austere sketches never loses its immensity, and the development of the two margins of the enormous tertiary canal that for a long time separated the Brazilian shield from that of the Guyanas. With the slow rising of the Andes in the west, one end was gradually closed off, transforming this trench into a gulf, an estuary, a river.

In the end, even considering the current state of Amazonian geomorphology, there are still other forces, excesses that disturb the dispassionate serenity of scientific observation. Even a glance at the simplest cases reveals a flagrant shift of ordinary processes in the evolution of topographic forms.

Everywhere else the land is a block where external forces mold the landscape, where the great rivers stand out as the principal factors that remodel and smooth out the natural roughness of the terrain. They compensate for the degradation of the slopes with the raising up of the valleys; eroding mountains and building up plains, they intertwine destructive and reconstructive actions in such a way that bit by bit the landscape is transformed, reflecting the effects of a prodigious sculptor.

Thus did the Huang Ho increase China by adding its delta, which is now a new province. Even more representative, the Mississippi amazes the naturalist with its vast expanse of enormous earthworks, which soon arrives at the boundaries that define the Gulf Stream. In their muddy waters flow dissolved continents. They change countries, they reconstitute and create territories. There is such a unified logic to these constant forces allied to these great natural energies, that to grasp them implies that sometimes one must follow every call of human endeavor: from the pages of Herodotus to Maspero,[19] one contemplates the genesis of a civilization paired to that of a delta, and the parallelism is so exact that they justify the exaggerations of those like Metchnikoff who see in the great rivers the preeminent cause of the rise of nations. . . .

What happens in the Amazonia is the opposite: What is privileged there is exclusively its destructive function. The enormous watercourse is destroying the land.

The mass of diluted lands is not reconstituted. The largest river has no delta. The island of Marajó constructs itself by a selective flora, a vegetation affected by its tidal environment. The river's unstable mud deposits are but the mirage of territory. If one denuded Marajó island, all that would remain would be the flat surfaces of swampy channels, drowning with the rising waters, or irregularly protruding fragments of sandstone. In light of the rigorous deductions of Henry Walter Bates that corroborated earlier conjectures by von Martius, what lies under the cloak of the forest is a ruin, the dismantled leftovers of a continent that in the remote past stretched and united the coast of Belém with that of Macapá. This is what one must reclaim, at least hypothetically, to explain the similar characteristics of the terrestrial fauna of the north of Brazil and of the Guyanas, today separated by the river.[20]

The Amazon, however, could reconstruct it in a short time with the three million cubic feet of sediments that it carries to the mouth every 24 hours. But it dissipates them. The Amazon's turbid current, thickening in the last thrust of its six-thousand-mile trip, with its disintegrating banks that daily collapse into the waters, decants itself completely into the Atlantic. And the residues of the demolished islands, among them that of Caviana, which was an ancient barrier that was breached in our own time, is imperceptibly dissolving, disappearing under the permanent assault of those powerful currents. Thus it gradually diverts itself more and more from the main outlet of the great artery, accentuating the shift to the north, continuously abandoning the backwaters that slowed its progress to the east and over which it used to flow in the past. It leaves still, in the areas recently revealed of the Marajoara swamps, the tangible proof of the lateral dislocation of the riverbed, which has given inexpert geologists the illusion of a landmass that is forging or uplifting itself.[21]

In reality, it reconstructs itself far from our shores. The river that invites all our patriotic lyricism is the least Brazilian of rivers. It is a strange adversary, one given to the inexorable routine of undermining its own country. Herbert Smith,[22] deluded by the powerful mass of the river's muddy water, which the traveler sees in the ocean long before viewing Brazil, imagined thus its momentous task: the construction of a continent. He explained that as it deposited its sediments in the tranquil depths of the Atlantic, new lands would emerge from the waves, and at the end of millennial efforts, the open gulf would be filled and would span Cape Orange to the Gurupi arch, in this way considerably expanding the lands of the State of Pará.

"The King is building his monument!" exclaimed the enraptured naturalist, adapting his clipped British syllables to a euphoric fantasy capable of startling even the most capricious Latin soul. He overlooked, however, that this singular hydrographic system doesn't stop after passing the Cabo Norte but continues, a river without banks, into the sea itself, in search of the equatorial current into which it pours its land-generating slurry. Its substance disperses in that immense underwater river stretching out to the Gulf Stream, concentrating and ascending with the flow, extending to the farthest zones. On leaving the Guyana coast whose lagoons, starting with those of Amapá, descend in level plains to the vast sea, it moves until it reaches the North American littoral of Georgia, of North Carolina.

In those places the Brazilian is a stranger, a foreigner. Yet he is treading upon Brazilian soil. He sinks into astonishment at the fiction of asserting such ownership. The irony of a country without land counterposes itself to another irony, more rudely physical, a "land" without land. It is the marvelous consequence of a kind of telluric disintegration. The land abandons man. It goes off in search of other latitudes. The Amazon river, translocating itself in an almost invisible voyage, constructs its true delta in distant longitudes and relentlessly diminishes, in an uninterrupted process of erosion, the vast areas through which it meanders.

One cannot point to enduring or fixed landforms. Sometimes, in the curvature of its channels, the backwaters flow slow enough that the sediments accumulate with the seeds they carry. Then the creative faculties of the river come astonishingly into view. The recently formed shoals whose surfaces emerge describe indecisive curves but soon delimit themselves, enlarging and lifting, rejecting the waters. The island that this generates grows under one's very eyes, studded with spits that lengthen and twist like the tentacles of an prodigious beast, unbound for the unfolding evolutionary battle, so vibrant and dramatic; containing all the convulsive movement of a momentous noiseless campaign in the shuffling of stalks, the twisting stems and branches, which weave, enlace, and confound themselves. The Aroids consolidate the inconsistent ooze with the webby fibers of their extensive rhizomes, which are in turn supplanted and expelled to the water's edge by the mangroves in violent and tumultuous graspings. The tall Javari palms, lording over the uplands, in turn overrun the mangroves, exiling them to the swampy margins.

But islands are formed to be destroyed, or at least incessantly dislocated. The same currents that generate them collapse them into the rising tides, restoring them in the ebb. The isles dissolve bit by bit, carried away by the river below like enormous, mastless pontoons, long of prow and high of deck, navigating night and day at imperceptible speed. Finally, depleted, they end.

The island of Urucurituba lasted ten years, thanks to a very large surface. It vanished in one flood.

The same is true of the riverbanks. The shores of the Amazon's unmeasured channel can hardly be specified. They are banks that evade the river. Normally they remain above the waters. Far beyond, the vast plains with their the scattered lakes of the *Terra Firma* lessen the violence of the currents and buffer the floods. There on a vast scale sometimes unfolds what appears as the immense construction of earth. The river, swollen in its great floods, punishes the banks and unburdens itself on the undefended flats. It uproots entire forests, heaping trunks and branches in the numerous depressions of the *varzea*, and in the backwaters of the flooded plains it decants the waters choked with detritus into a completely generalized thatch. As the waters drain, one notes that the land has grown and gets higher and higher with each flood, building up terraces, silting up swamps and flooded forests, sketching the outline of a rolling upland about to be invaded by a triumphant flora, until in one assault this entire lateral delta is undone. In one night (the 29th of July, 1866) the left margin of the river collapsed in a continuous line for 50 leagues.

This is an ancient and continuous process and still persists into the dim rays of our own time. The banks, ancient bluffs of Peru, where the legendary Amazons appeared to Orellana's expedition, are today reduced to degraded shoals visible only at extremely low water levels.

The erratic turmoil of the river is revealed in its unending curves, hopelessly entangled, reminding one in its uncertain itinerary of a lost traveler, who at fading horizons returns to all its old paths, or propels itself onto sudden shortcuts. The river thus surged through the suffocating narrows of Obidos[23] in complete abandonment of its ancient bed which one can still discern in the enormous tidal plain of Vila Franca. Or in other points it streams from unexpected side troughs into its own tributaries, thus becoming, illogically, a tributary of its own tributaries. It is always disordered, always insurgent, erratic, destroying, constructing—devastating in one hour what it built over decades—with the dread, the torment, and the exasperation of a monstrous, dissatisfied artist retouching, redoing, and perpetually restarting an obscure painting.

A metaphysician might imagine the singular carelessness of a nature that, after constructing the infinite modalities of the natural history, begins rather late to hastily complete her task, correcting, in forgotten retreats, the errors . . .

Such is the river, such is its history: always insurgent, always incomplete.

The dynamic geology is not deduced: you can see it, and history writes itself every day before the enthralled eyes of those who know how to read it. From this come surprises. Everywhere we so much fetishize the equilibrium of natural forms that now we appeal to a tumultuous hypothesis of cataclysms, so that these can explain sudden changes; in Amazonia, the most extraordinary and visible changes are the outcome of the simple play of the most common physical forces. It is a young earth, an infant earth, a nascent earth, and a land that is growing . . . every six months, each flood passes like a wet sponge over a poorly drawn painting, it erases or transforms the most salient and obvious landmarks, as though in the framework of its vast plains moved the restless brush of an capricious, prodigious artist. In Amazonia, in fact, this cruel contradiction prevails: in a bounteous land, in the midst of the cheerful plenitude of its diverse life, unfolds a moribund society.

Da Cunha's geomorphology was the substrate on which he imagined the dialectic between the great river and its civilizations. In the second half of "General Observations" he embarks on the formal forays that were meant to turn the Hylea into a Luso-tropical civilization. Here he begins to layer on some of the history for these lands "without one." He writes within the framework of an environmental determinism that seems at odds with his later positions, a fact that suggests that this was written earlier than most of his Purús materials such as *Clima caluniado*, which largely rejects climatic explanations for human character. However, as in *Os Sertões*, he may have been setting up a rhetorical structure of received ideas only to refute them later.

"At the Threshold of a Marvelous World"[24]

> The Amazonian wilds have always had a gift for impressing distant civilization. Since the earliest days of the Colony, the most imposing expeditions and solemn pastoral visits wandered by preference on its most unknown terrains. Into the wilderness went the most venerable Bishops, the most elegant Generals, the most lucid scientists. From this soil they sought to raise the most exotic spices and to elevate native cultures to the highest of destinies. The distant metropole swooned at the marvels of a land that more than compensated for the loss of glorious India.
>
> Vain efforts. The demarcation teams, the evangelical missions, with their fleets of hundreds of canoes, their astronomy commissions with their fabulous equipment,[25] their prelates and warriors arrived intermittently at the those remote havens and quickly established on the river levees the sumptuous tents of a traveling civilization. They directed cultivation, disciplined the locals, and thus shaped a land.
>
> They pressed on to other frontiers, or simply turned back, while the native longhouses, the *malocas*, collapsed into dust, returning to their primordial savagery.
>
> Already by the end of the 18th century, Alexandre Rodrigues Ferreira, carrying out his great philosophical voyage on the main channel of the Amazon, walked among ruins. The town of Barcelos, an old boundary and fortress town, riveted him with the impression of typically Amazonian progress—the imposing Palace of Demarcation—ample, monumental, and commanding—and covered with lianas. It was symbolic: everything fluctuates, everything is ephemeral, everything is paradoxical in the strange outposts where even the cities, like men, are nomads, perpetually moving from one place to the next, just as the land shuns them in the roiling current, or in its collapsing embankments.
>
> From one century to another in unbearable similarity are the same daring, yet abortive efforts. The impressions of the most lucid observers are perpetually

misled by the spectacle of a pitiable present counterposed to the illusions of a glorious past.

Tenriero Aranha, who in 1852 assumed the governorship of the recently founded Province of Amazonas, provided a historical review where he resumed the extraordinary progress now lost, referring to vanished industries:

> of cotton, indigo, the cultivation of manioc and coffee enough to satisfy local needs and with goods left over for export. There were factories of indigo, ropeworks of *piaçava* palm, industries of thread, cloth and hammocks of cotton, fiber, or feathers, the brick and tile makers, the civil and naval construction with able artisans elaborating temples, palaces, and imposing vessels.

If one goes back exactly one century to search for this enchanting period, one learns with great discouragement of a report done in 1752 by the Ensign Governor, General Captain Mendonça Furtado: "The Captaincy was reduced to total ruin." Appearances clash with realities, exposing the same failures. Or rather, they complement and confirm the same decadence in an impressive way.

In 1762, the Bishop of Grão Pará, the extraordinary Father João de São José, a cleric with a Voltairian sense of irony and the writing style of the pamphleteer Antonio Vieira, concluded after a review of men and things, that "the root of the vices of this land is sloth," and summed up the regional personality characteristics in this discouraging way: "lustful, drunken, and thieving."

A hundred years pass. One tries to see if things have improved. One opens the austere pages of Alfred Russel Wallace and sees that in some aspects he seems to have copied directly from that witty Benedictine, because the undisciplined society that passed before that astonished scholar was one of "drinking, gambling, and lying" in the same distressing heedlessness.

This impious indifference to superior qualities, the systematic renunciation of scruples, and the lighthearted and casual view of error extends down from the finest houses to the hovels of the rubber tappers. Browse through our old chroniclers, especially the imaginative Father João Daniel, and review the moral and physical obstacles that weaken their characters. And read Tenreiro Aranha, José Verissimo, and dozens of others. In these books are strewn all the dramas of depravity in human history.

Then there is the incoercible physical fatalism. Nature, dominant and brutal in the full expression of its energies, is an adversary of man. In the perpetual steam bath (as Bates puts it) one completely understands the allure of a complacent and free life, a life without risks, but without the delicate vibrations of the spirit spurred by the dynamism of ideas, and devoid of the superior tensions of a will whose acts transcend merely egotistical impulses. An Italian doctor, a superb talent—Dr. Luigi Buscalione[26]—who recently traveled in Amazonia, described two main phases of the impact of climate on the forest dweller. The first effect is a form of superexcitation of the psychic and sensual functions, followed then by the weakening of all faculties, beginning with the noblest ones. But in this appeal to the classic theory of climatic influence, he forgets, like so many others, the secondary, random, but appreciable effect of the very inconstancy of the physical base on which society is constructed.

The capriciousness of the river infects man himself. In Amazonia in general this happens: the wandering observer who has been ranging throughout the basin in search of its various marvels has at the end of hundreds of miles the impression of traveling in a loop, where the same beaches, levees, and islands present themselves, the same flooded forests, and bayous stretch out as far as the eye can see in empty horizons.

When encountered by the nomadic man, nature seems stable, but the immobile observer positioned on the river margins would be astounded by unexpected transformations: the scene, invariable in space, would change through time. To the eyes of the sedentary person who planned to submit nature to the stability of cultivation, it appears stunningly insurgent and changeable, surprising him, sometimes assaulting him, frightening him, and almost always chasing him away.

Adaptation worked then toward nomadism. Therein, in great part, lies the complete paralysis of the people who wander here in an enervated and sterile agitation.

Whether one likes it or not, for the Amazonia of today we should entirely restore, as definition of its collective psychology, the searing aphorism that Barleus[27] invented for the excesses of the colonial epoch: "Below the Equator there is no sin." The Amazonians know this. At the entrance to Manaus rises the beautiful island of Marapata. Lovely though it may be, it is a sinister augury. Most singular of "leper colonies," it is a colony of lost souls. Here, they say, new arrivals to Amazonia deposit their consciences.

Observe the reach of this popular fantasy. Another island at the mouth of the Purús lost its old geographic name and is now known as the Island of Conscience. The same with the island at the falls of the Juruá River. There's nothing lighthearted about this attention to lost conscience. Those who pass through these portals, on to the diabolical paradise of the rubber forests, surrender their best qualities and doom themselves, laughing at the deep irony of their situation.

And there, in the exuberant forests of *Hevea* and *Castilla*, awaits the most monstrous organization of labor that was ever imagined by unbridled greed. There the *seringueiro*, and we do not refer here to the rich Patrão, the *seringalista*, but rather to the worker tethered to the rubber-collecting *estradas*—the pathways where the tapper incarnates a tremendous irony: it is that he works to enslave himself.

"A land with a great deal of history . . ."

Amazonian may have appeared to some "without history," but it was at the center of powerful global historical and economic forces. Da Cunha and Alexandre Buenaño (the Peruvian chief of party) were not confronting an emptiness but a theater of war. The expedition would be traveling through settlements with national administrative posts and contentious customs houses—a place where gracious mansions overlooked crammed trading posts like Seringal Macapá, with its many retainers, its ladies riding sidesaddle on elegant white mules, and forests that were webs of trails and portages.

While letters from Manaus made reference to his "encounter with the wilderness," he well knew, because after all a Peruvian-Brazilian war was simmering, that the place was hardly "vacant." And indeed, for both Buenaño and da Cunha the entire object of this expedition was to see the Purús not as empty but rather as full, crammed with the unsung agents who carved out a living in those forests, the material expression of personal and national destinies. The Purús they confronted was hardly the "primeval landscape" of the immense biological park that now claims these terrains, but rather was tied to intense commercial forces, financial empires, and profound processes of economic and technical change that stretched over the globe. The realm of rubber underpinned this most dynamic economy and shaped its sociologies and ecologies in countless ways. Da Cunha and Buenaño were off to the exuberant forest, its landscapes shaped by "monstrous organization of labor" and a social ecology as he first imagined it, shaped by unbridled greed.

14

In the Realm of Rubber

Deep History of the Rubber Rivers

Da Cunha's commission traveled through the heart of the world's richest rubber forests at the height of the Amazon boom. It is very difficult for us to imagine what transpired in those forests over the course of a century, since the modern description of the Upper Purús is of untrammeled wildness. What is clear is the fates of millions—in tropical forests *and* industrial metropoles—were inextricably woven into the tropical biologies of latex trees, trees whose history figured in complex ways in native practices, ecologies, Amazonian geopolitics, and the making of the modern world.

Rubber is the one of the greatest gifts that Amazonia has given us. Tropical latexes made the industrial revolution possible. It was forged from steel, cheap energy, and elastic gums from tropical forests. Amazonia's manioc expanded food availability in Africa, and its sweet potatoes came to feed burgeoning Asian populations. The rise of tire-based transport, cabled electricity and telecommunications, tubing, and the vast array of gaskets that link moving parts of metal together in modern machinery would be largely unimaginable without this amazing stretchy, waterproof substance. Rubber is now present in our lives so integrally, so ubiquitously, that it's almost invisible. It has so many unique characteristics—elasticity, plasticity, memory, and resilience—that it was truly a novel material for the Europeans who encountered it at the end of the eighteenth century. Rubber is a polyisoprene, an extraordinary emulsion of complex hyperfolded hydrocarbons. It is the way that these are composed that gives these gums their unusual properties. By the early nineteenth century, it already figured in industrializing processes and mass goods.[1]

The great ethnobotanist Richard Schultes, whose passion for psychotropic plants was matched only by his interest in latexes, pointed out that only seven thousand plant species contain these polyisoprenes, the "caoutchoucs" that provide a usable rubber, and of these, only one, *Hevea brasiliense*, generates 98 percent of the natural latexes in modern commerce. This rubber is now mostly produced on Asian plantations.

Amazon Latexes

The term *rubber* in English embraces a wide variety of different latexes that came from several different plant genera and had different distributions, ecologies, elastic properties, and regimes of production. The latex industry dominated the Amazonian export sector for most of a century and changed its geographic focus, forms of economic integration, and the ways labor was organized and latexes were harvested. The dramatic terror slavery of the Upper Amazon and the relentless debt peonage that characterized parts of the industry at the turn of the twentieth century, when prices went stratospheric, do not actually represent all the social relations that prevailed in the rubber economy over time throughout the basin, or even necessarily in given watersheds. The Juruá and Purús, for example, had enslaved Indians tapping their *caucho*, as well as debt peons and yeomen tappers in the *Hevea* zones.

Although *rubber* (like *caucho* in Spanish) is today a generic and inclusive term, there were many terms for varied latexes in commerce, and the languages of the day discriminated a great deal among them. There were four main genera that occurred in the Amazon trade—*Hevea*, *Castilla*, *Manilkara*, and *Sapium*—and each included different species with varying qualities of latexes and many names, which were further differentiated by how the gum was handled and the form in which it was sold. Two species dominated the trade, *Hevea* and *Castilla*. The best-known and most valuable species was *He-*

vea brasilenses, or "Pará rubber" or "Acre fina" of commerce (in Spanish known as *jebé* or *sheringa*), largely found on the southwestern tributaries of the Amazon. An array of "lesser" *Hevea* species, such as *Hevea guyanense*, distributed mostly to the north of the Amazon along the flooded forests, and *Hevea benthamiana*, found along the Andean foothills of Peru and Colombia and well into the Ucayali, also entered the market. These latter varieties were known as "jebé fraca" or "Pará fraca"—"weak" *Heveas* that were pressed into commerce in the Upper Amazon, and possibly mixed in with Pará rubber when prices exploded.[2] "Pará fina" or "Acre fina" was the top-quality processed rubber; "Entre fina" was lesser quality, perhaps contaminated with other latexes; "Sernambi" was unprocessed scrapings, known in English as "scrappy." There were even geographic "brandings" such as "Putumayo rabos," although each estate literally branded its rubber as the product was shipped. *Hevea*s were tapped on a weekly schedule except during the wet season, when access to the trees on the floodplain was difficult and the latex would be diluted by rain and of poor quality. During the rains, tappers might migrate to cities or might engage in *caucho* extraction on the uplands.[3] Tapping *Hevea* involved a relatively sedentary population with access to several tapping trails.

Caucho rubber came from "caucho negro" (*Castilla ulei*) or "blanco" (*C. elastica*) and was also widely marketed and fetched respectable prices in the Amazon trade, although it was considered of lesser value. *Caucho* came to market in several ways, with designations referring to the tree source (caucho negro, caucho blanco) or form—slabs (planchas), balls (bolas), or scrapings (sernamby de caucho), and the aforementioned "rabos"—tails—or large sausages of rubber. During the boom, Amazon *caucho* was typically harvested by cutting down the tree and draining its milky sap into small excavations under the fallen trunk. It came to roughly a tenth to a fifth of the total latex output (depending on the year) that went through Belém, and a third to half of the Peruvian trade that left Loreto. There were regions—especially in the upper northwestern Amazon—where it dominated. The Putumayo, much of the Madre de Dios, the Caquetá, the Upper Ucayali, and the headwater areas of the Purús and Juruá Rivers were its main sources.[4] Jacques Huber, the master of latex botany, later noted *caucho*'s wider distribution in the Lower Amazon and the island zone of the estuary. These areas entered into *caucho* production in the 1910s to 1920s. *Castilla* entered Amazonian commerce late in the day but was the basis for the rubber industry in Central America—and indeed had been for millennia. Tropical latexes were always described as products of wild nature. This is debatable for *Hevea*, as we will see further on, but *Castilla* was a domesticated tree.

The Secret Life of *Caucho*

The French mathematician, explorer, and savant Charles de la Condamine (1701–74) described the marvels of rubber to the French Academy of Sciences in 1745 and in doing so publicly introduced northern Europe to these elastic gums.[5] He based his description on that of the French botanist Fusée Aublet[6] and had fact capitalized on native knowledge from the upper Amazon and the researches of François Fresneau, captain and engineer of the French army at Cayenne.[7] While it was news to the French Academy of Sciences, knowledge of the properties and uses of various latex trees was of great antiquity (and ubiquity) and was recognized by Iberian natural historians by the early sixteenth century. Lore advanced by de la Condamine had natives merely messing around with rubber balls and toys from the gums of "wild trees";[8] the reality was different. Columbus himself returned to Seville with one of these rubber balls (an important ritual object), an occurrence duly noted by indigenous defender Bartolomé de las Casas (1484–1566), who himself remarked extensively on the rubber ball game in Hispaniola.[9] Spanish writers noted rubber uses as early 1530,[10] but this information was kept under the veil of secrecy mandated by the Iberian crown and Vatican.

The word for rubber in Nahuatl is *olli*, and the Nahuas called the ancients from the southern Gulf Coast the Olmec, meaning "people from the place of rubber." The etiology is related to the term for movement or earthquake (*ollin*) because of rubber's elastic qualities. The regular use of waterproof gum for shoes, cloaks, hats, containers and cloths, and shrouds were recorded for pre-Columbian times and reported by Latin America's earliest observers. Fresh latex was drunk with chocolate as a medicinal, used as an incense, and wrapped around implements like axes and stone knives to cushion their hafts. It was also used for illumination, as it still is today in parts of Amazonia.[11]

There is a great deal of evidence that rubber harvested from *Castilla elastica* was not a casual product used only for amusement in pre-Columbian times. Meso-American people were already processing *caucho* by the second millennium BC. There are pre-Columbian images of the tree with tapping marks, and the plant was a heliophile that sprouted after clearing, an advantage in shifting cultivation cycles. It was a product with practical and ritual uses and well-developed native technologies.

The great technological breakthrough for rubber in the emergent industrial world was the 1839 discovery by Charles Goodyear of the process of vulcanization—the addition of sulfur to heated latex, which stabilized this substance while maintaining its flexibility; otherwise it tended to melt or become extremely tacky (and rank) in the heat, or became brittle with cold. This was a key technical advance to extending uses of latex. A different means of stabilizing latex had been achieved more than two thousand years earlier in native societies. MIT archaeologists and material scientists Dorothy Hostler, Sandra Burkett, and Michael Tarkanian sampled rubber balls from ancient Olmec sites in the Mexican state of Veracruz, where rubber products have been dated at 1300 BC. After documenting the practices of modern *huleros*, or *caucho* collectors, who squeeze the juice of the morning glory vine *Ipomea alba* into buckets of latex, these scientists analyzed the resulting rubber, which was far more elastic than the nonprocessed material. The strength and number of the polymer interchain reactions were induced by organic compounds in the vine and in essence produced a "cold" vulcanization, one used for thousands of years.[12] The capacity to change the viscosity and properties of rubber for differing uses has also been documented by these researchers. The juice of the morning glory vine causes cross-linking of polymer molecules, making the rubber elastic and removing compounds that turn latex brittle.[13] This made *caucho* latex useful for products we all use today—rubber balls, cushioning, waterproofing, rubber soles, etc.—although the system of meaning in which they resided was very different.

The great naturalist Oviedo commented on rubber in 1534 and observed the prevalence of ball games that are thought to have been invented by the Arawaks, who still played their ancient game until the late 1960s in the Xingu basin.[14] Oviedo described an ancestral form of soccer played with two teams on a central plaza with a stone ring set vertically in a stone wall at each end. "The Indians do not play with the open hand or even the fist: they received the ball on the shoulder, elbow, head, knee, and hips most often, and return it with considerable grace and agility."[15] The passion for ritual ball games among the great New World civilizations was evident in the omnipresence of ball fields in virtually every major pre-Columbian settlement: they extended from Arizona to the Argentine Chaco, a feature mimicked in almost every town, small and large, in Latin America today.[16] Even Teddy Roosevelt marveled at the game played with a rubber ball and no hands in his brief travels on a minor tributary of the Madeira with Cândido Rondon. Rubber balls made from *caucho* have been found as offerings in springs and *cenotes*[17] from the US Gulf Coast to Chichen Itza on the Yucatán Peninsula. Hostler and Tarkanian report that up to sixteen thousand balls were produced in specialized craft towns for regional trade and tribute.[18]

The thick cloudlike smoke produced by burning *caucho* was used as incense (think perhaps of burning tires), evocative of cumulus, was offered especially to the rain gods.[19] Kevin Terraciano, ethnohistorian and translator of Mixtec pre-Columbian texts, reports that native priests would go to the top of a mountain with an image of Dzahui, the rain god, place it on the ground, and then bounce a rubber ball around it, burning the ball and smearing resin over the icon before sacrificing the young child they had brought along by tearing out his heart and offering it to the divinity.[20] In other cases rubber effigies of many types were burned as offerings.[21] *Caucho* rubber was used for ritual offerings on paper or slathered on bodies, and rubber effigies formed part of the ritual matrix of the "elastic economy." This was in addition to its more mundane uses for waterproofed implements and amusements such as clown shoes for court jesters.[22] Modern Mexican and Mesoamerican rubber-soled sandals, today made with tire strips, hark back to a pre-Columbian *huarache* and were described in the diaries of Spanish explorers and missionaries. Linguistic evidence also attests to their existence: the Aztecs used a compound word that clearly blends the words for "rubber" and "sandals."[23]

Caucho rubber was far more widespread than is usually recognized, with significant roles in Olmec, Mayan, and Aztec ritual and daily life.[24] One indication of the importance of this latex was its value in imperial tribute. Abbé Francisco Javier Clavijero, writing in Mexico 1762, recorded that twenty-two cities supplied a tribute of some 16,000 loads (*cargas*) of rubber each, for an estimated annual tax of some 15,700 tons of this product,[25] a level of usage that was equaled in Amazonian exports only in 1891,when about half the world's production came from the Amazon.[26] The archaeology of *caucho* shows long-term management and domestication. Only inferential evidence exists for *Hevea* domestication, but the surprising richness of *Hevea* forests in the upper Amazon, site of some of the most complex and densely settled areas of Amazonian civilizations, suggests that they might be feral forests: that is, former production areas eventually turning wild with high densities of economic species. Amazonian natives describe myriad uses for *Hevea*: for ball toys, waterproof cloth, illumination, and shoes, as an animal attractant, and as a beverage—the latex was drunk, just like *caucho* and the gum of another widespread tree, *Manilkara*.[27] The sap of this last tree today figures in the innards of golf balls, but it enjoyed a vigorous market as a cable covering in the early telegraph days.

Deep History of *Hevea*

Richard Schultes, one of the most important collectors and taxonomists of Amazonian latex plants, portrayed *Hevea* as a tree only recently domesticated through the offices of Kew Gardens and South Asian planters, the outcome of Wickham's transfer of wild *Hevea* seeds to that institution in the 1870s.

But *Hevea* has a deeper and more complex history of domestication, one that didn't begin with Kew's botanists or Ceylonese plantations.

Domestication, as it is usually understood in the temperate zone, emphasizes changes in edible parts of plants (increased size, oiliness, or sweetness, reduction of toxins, heteroploidy)[28] as well as an inability to persist unaided by human intervention. Usually studies of domestication evaluate gradations from more primitive to more domesticated genotypes, providing hallmarks for documenting the differences between tame and wild species. Geneticists for the most part have concentrated on annual crops and generally viewed the large array of tropical trees that are widely used (and largely with multiple uses) as wild trees with "gathered products," part of the world of "non-timber forest products," or NTFP, in current resource development argot. [29] Tropical fruit trees were thus seen as more or less "self-domesticating,"[30] more like weeds or random processes than outcomes of local cultural and agroecologies. Like early explorers who viewed tropical abundance as uniquely part of nature's bounty, this view of useful tropical trees, often found in groves with high densities of individual species and frequently associated with other useful trees (a classic mix involves rubber, Brazil nuts, cacao, and suites of beneficial palms), as an outcome of biotic history—as wild, accidental, "natural" rather than as results of life shared with human communities. Recent research, such as that of botanist Charles Clements, has focused on tropical fruit domestication through a gradient from a wild state to incipient domestication to semi-domestication, and finally to domesticated species.[31] In Amazonian ethnobotany, the "binary" mindset of the wild and tame has increasingly given way to the ideas of domesticated *landscapes* where human activities influence and "enable" plant communities on various scales, where other organisms (like animals) participate in planting and distribution, and where selection may involve an array of attributes including persistence *without* intensive intervention.[32] Plant distributions, especially of oligarchic species like Brazil nut (*Bertholethia excelsa*), Buriti palms (*Mauritius flexuosa*), babassu (*Attalea speciosa*), piquiá (*Caryocar americanum*), and açai (*Euterpe oleraceae*), among many others, express the conditions of ecological possibility and human selection, as well as conscious and unconscious human shaping of environments.[33] These plants are described by indigenous populations as highly useful, often with multiple uses, culturally significant, integrated into ritual as well as economic life, and as markers of human landscapes even though our science often doesn't assess them as necessarily "domesticated."[34]

Many scholars view the upper Amazon from the Baurès area of Bolivia and a swath north to Ecuador as a possible "Vavilov" site, a center of domesticated diversity and wild ancestors. Surprisingly, many of the economic plants that were the tribute of the Central America empires—cacao and rubber as well as that subsidy to central American subsistence, peach palm, an important carbohydrate and fat resource—are from this region. Other widely planted crops like manioc, runner beans, peanuts, some chiles, tomatoes, coca, *ayahuasca*, guava, and *ingá* (custard bean) appear to have also been domesticated there. The region seems to have been especially important for tree crops,[35] and latex trees were probably also part of the suite of domesticates.

Ethnographic work with many groups suggests that Amazonian cosmovisions involve human participation in a "society of nature" where the meanings of the wild and the tame are modulated within a landscape[36]—a range of plants that produce latex, fuels, medicines, poisons, fibers, oils, animal attractants, and other noncomestible products that may be seen by outsiders and agronomists as "wild" even as they were probably subject to a form of indigenous "socialization" into landscapes, histories, and gardens.

One famous, and now mostly forgotten, botanist of domestication, Edgar Anderson, and his colleague at the Missouri Botanical Garden (and later of the USDA) R. J. Seibert, viewed *Hevea* as domesticated. They, like Schultes, worked on rubber as a consequence of the vulnerabilities of US supply when Asian sources were cut off during World War II. Anderson and Seibert were quite convinced of the domestication or semi-domestication of many of the species of *Hevea* in light of their exceptional ability to hybridize. Seibert noted many "hybrid swarms" in areas where human habitation was of long standing, such as the area around Iquitos.[37]

The Secret Life of *Seringa*

The importance of *Hevea* latex in industrial processes may have overshadowed the original logic of its domestication or semi-domestication. Seibert and Anderson, based on an extensive analysis of herbarium exemplars, regional travels, and ethnobotanical experience, argued that *Heveas* were initially grown for their nuts, which had to be processed in ways similar to those for manioc order to remove their cyanic acids. The French botanist who first described *Hevea* in Guyana, Aublet (whom we met in an earlier chapter), pointed out the Galibis ate them with enjoyment,[38] and Schultes reported the use of nuts in ceremonial meals along with smoked boar or tapir meat, and noted indeed that the seeds of almost all *Heveas* in the Northwest Amazon were eaten.[39] The renowned botanist Richard Spruce described *Hevea* deliciously: "The genus seems abundant throughout the Amazon but not all species yield caouctchouc of good quality. . . . The seeds are excellent bait for fish. Macaws eat them greedily, but to man and quadruped they are poisonous in a fresh state. The Indians on Uaupes render them eatable in this way: after being boiled for twenty four hours, the liquor is strained off, and the mass that remains has something of the consistency of rice long boiled. Eaten along with fish, it is exceedingly savory."[40]

Spruce also described the use of the oil extracted from the seeds for lighting, and another processing technique (one also used for manioc) of placing a basket with the boiled pulp and seed in the river for three days and then eating it "as is." and reported its "resembling cream cheese in appearance and taste." This parallel between processing techniques for manioc and *Hevea* is quite suggestive, since these ways of removing cyanogenic acids are not obvious or simple, and without them the food is very toxic.

Seibert suggested that the seed size and oil of *Hevea paucifilia* (which is an upland species largely found on the Rio Negro) actually made it the preferred domesticate when food was the point. As populations moved their shifting cultivation plots or gardens and trekked along trade routes, the tree, grown for its nuts, cross-pollinated with many wild species along clearings, creating "hybrid swarms" through genetic introgression of domesticated or semi domesticated *Hevea* with wild types. In addition, Seibert pointed out the common association of *Heveas* with Brazil nut, cacao, and peach palm. Beyond the genetics, Anderson and Seibert argued that *Hevea* was domesticated or semidomesticated on the grounds of what are now taken as more or less regular practices in indigenous landscape management for useful species: selection of seeds for large size, planting in agricultural fallows, manipulation of fallows, movement of plant materials over large areas, planting in forest gaps, manipulation of forest sites for oligarchic species, planting (and testing) of plants in dooryard gardens, and natural or manipulated hybridization.[41] A species that was useful to eat, for lighting oil, for fish bait, for waterproof latex, and as bird lures for macaws (prized as pets and for feathers) does not seem that strange as a candidate for domestication, although perhaps *socialization* is the better term. The elaborate knowledge of detoxification of seeds and the incorporation of *Hevea* into ritual life also suggest considerable historical and cultural intimacy with the plant.

Schultes, whose experience was in the Northwest Amazon, largely in Colombia (which was mostly *caucho* territory), was having none of this: he argued that indeed there were planted trees, as described by Seibert and Anderson, but that these were not planted by Indians but *caboclos* (Amazonian backwoodsmen), who used *Hevea* seed *only when famine threatened* (italics in the original). Schultes spoke of the "innate aversion" to destroying any rubber tree, even though they were not tapped on the Rio Negro where he studied this question. Schultes' historical ecology of Amazonia focused on the wild

Figure 14.1. Geoglyph, Acre.

and the mystically primitive. In his view, natives in the Colombian Amazon never collected from home sites because they preferred collecting on pleasant treks to wild groves.[42] The pleasure might indeed be in the camping trip but also in cultural memory and visiting ancestral sites. Trekking to wild groves through forests is a widespread native practice and usually involves going to former villages, former ceremonial sites, and areas "planted by the ancestors," reviewing history by using landscapes as a mnemonic device, and monitoring territories.[43] These treks also involve the movement of a great deal of plant material.[44] Schultes certainly agrees that "in areas where man has caused great upheaval in the natural vegetation, we have undoubted proof of crossing. . . . Specimens collected in such localities exhibit extremes of variation and all possible intergrades . . . but *clearing has happened only in recent times* [italics added] and can have had no effect on generic evolution." In this Schultes's understanding of Amazonian cultural history was, unfortunately, wrong. Schultes was radical in many ways, and had he lived longer he might have shifted his views about the nature of domestication and *Hevea*. In fact, the Upper Amazon, especially the Purús with its extensive geoglyphs, suggests that the region had had an extensive pre-Columbian clearing history.[45]

Socialized Nature and the New Economies

The earliest chronicles of the Amazon reported dense human landscapes separated into polities supported by land managed in complex ecological and engineering ways: weirs for fish ponds, bunds, dikes, causeways, roads, and turtle corrals for confinement of this delectable Amazon meat.[46] As Pedro Ursuá and Lope de Aguirre passed by the Purús in 1560, they fretted about the roads leading back to the interior and the numerous settlements.[47] The Purús watershed embraced major pre-Columbian occupation along with those of the Omagua at the confluence of the Rio Negro and the Amazon, the Baurès in the Bolivian llanos, and the engineering cultures of the Upper Xingu. The omnipresence of geoglyphs throughout the Purús watersheds wherever forests have been cleared requires a rethinking of the "nature" of this tropical landscape, its "wildness," and its historical "marginality."[48]

The frequency of geoglyph earthworks in the Purús watershed (at this writing more than four hundred have been found), their possible connection to the huge anthropogenic landscapes of eastern Bolivia, and the early descriptions of the Purús area as rich in populations and resources could mean that the "wild forests" of rubber, cacao, Brazil nuts, andiroba, and copaiba (trees which produce a medicinal and illumination oil) noted by many travelers and da Cunha in his reconnaissance report, were in fact "feral forests," landscape relics of large earlier native settlements, reflecting a long history of anthropogenic forests.[49] Cacao was so dense that da Cunha urged development of the cacao industry there. Historically, the rebellious Mura Indians were reviled because they made collection from the vast groves exceedingly difficult. Extensive areas of bamboo stands—an important marker for human impacts—have also been recorded.[50] Rather than a periphery, a backwater, or a pristine refuge as it is thought of today, the region was a cosmopolitan crossroads, as da Cunha suggests, connecting Andean and western Amazonian cultures in the complex domesticated landscapes of the pre-Columbian world.[51] Modern linguistic data support this view.[52]

The rubber period was perhaps a reengagement with elements of earlier production ecologies and their integration into modern commodity circuits. It was not an accident that in the early nineteenth century colonial administrators had closed the Purús because of the extraordinary quantity of commodities—the *drogas do sertão*—that had been flowing untaxed down it.[53]

"A Most Useful Product": Industrial Innovation and Amazon Latex

In Europe, a place without anything much resembling waterproofing other than oily wool, rubber soon found an essential niche. By the 1750s Portugal was outfitting its military with army boots and knapsacks treated with latex in Belém, and indeed, European militaries were eager markets for rubber products.

Figure 14.2. Production of rubber shoes on lasts for export. Note the tapping of the tree on the right, and racks of shoes in the background.

Syringes had existed among Amazonian natives (hence the Portuguese name for the tree, *seringa* or syringe), and by 1768, Pierre Macquer had learned how to make rubber tubes and catheters by molding the latex around tubes. Latex-coated silk taffeta balloons lofted the Montgolfier brothers over the Île de France in 1783, fueling the craze for this early airborne conveyance. By 1811 Champion was waterproofing materials for the French army. In the early 1800s Belém was exporting rubber shoes to New England, and this trade, localized on tributaries near Belém—the Mojú, the Acará, and the Bujaru—reached an astonishing 450,000 pairs by 1839,[54] when the US population was only 12 million. In William Henry Edwards's volume that so influenced Henry Walter Bates,[55] he described this industry in this way: "The man of the house returned from the forest about noon, bringing nearly two gallons of milk that he had been collecting since about daybreak. . . . In making the shoes, two girls were the artistes in the little thatched hut. . . . the shoe last was dipped in rubber milk and immediately held over the smoke . . . [where it] dried at once. It was then redipped and the process repeated until the shoe was of sufficient thickness. . . . The shoe is now cut from the last and is ready for sale, bringing in about 12 cents."[56]

These young "artistes" were probably working for T. C. Wales, the first major rubber purveyor in the Unites States, who started his enterprise in Boston in 1823, when he began a "putting out system," sending shoe lasts to Brazil and buying the rubber shoes made on them in return. Indeed, rubber footwear and waterproofing dominated the industry to the mid-1800s.[57] Charles Macintosh found that naphtha added to rubber could make the substance more stable (it was brittle in cold and could get very gummy and stinky when hot). By using rubber as a waterproof coating for cloth, Macintosh created raincoats that took on the name of their inventor.[58] Charles Goodyear's (re) discovery of vulcanization, the process that stabilized latex so that it remained flexible when cold and did not degrade when heated, vastly increased the industrial possibilities of the gum. The first pneumatic tire was made in 1845, though its adoption was limited as the technology itself needed more work. Still, Macintosh and Goodyear were eager participants in London's Great Exhibit of "Works of Industry of all Nations," displaying a kind of rubbery "Cabinet of Wonders" with their coats, boots, curtains, flooring, and tires. In 1852 John Dunlop developed a solid tire for carriage wheels, and by 1869, rubber tires were being fitted onto bicycles. At the Wolverhampton British Bicycle Championships, J. Moore won virtually all the heats, "showing his heels" as the sporting lingo put it, only to fail in the final as his back tire came off. These earlier rubber commodities focused mostly on apparel and waterproofing, for which there was great demand. But the rubber industry really took off as cities began to suburbanize, communication networks expanded, and medical and domestic uses developed, creating a great demand for industrial machinery, gaskets, electric and telecommunications cables, and tires for bicycles and cars.

In 1895 the first car tire was used in the Paris to Bordeaux car race with Michelin's product, to very mixed reviews (more than twenty-two tires were used to cover a distance of 720 miles). The industrial demand for rubber for transport was not limited to tires: by 1900 soft gum pads held in place by horses' metal shoes protected their feet from bumps and injuries on cobblestone and rock streets (a technology still in use today), and there was talk of rubber street pavers. Rubber for belting, as insulation for electric wires, in gaskets, and increasingly for medical and scientific applications grew dramatically with technical change, innovation, and the perfection of technologies.[59] Bicycles, which numbered five million in Europe alone by 1900, were a critical element in mass transportation as cities ballooned in size.

At mid-nineteenth century, Amazon exports had hovered around 2,670 metric tons; by the early twentieth century, they averaged over 34,000 tons. Nonnative Amazon populations swelled from a low estimate of about 250,000 at the mid-1800s to well over a million by the turn of the century—this in a period of relentless epidemics and murderous labor regimes in which countless people died. The real population numbers were higher.[60]

The Amazon rubber industry grew for more than a century, from about 1820 to 1920, with a flurry of high prices and high production at the turn of the century until the *Hevea* rubber price collapse in 1912, when Asian products hit the market, although production continued at high levels (but much lower prices) until the 1920s.[61] Technical innovations associated with the First World latex economy were as vibrant as those of any other sector of the industrial revolution throughout the nineteenth and early twentieth centuries, and their impact was unbelievably profound. The Amazonian part of this equation, however, was based on the most primal, indeed prehistoric, elements of human industry: a person, a knife, and smoky fire for curing the latex.

What seems clear is that latex trees of many kinds were integrated into pre-Columbian production and social systems in a variety of ways and for many uses, These systems were dramatically (and often brutally) recast when the industrial innovations of the Northern Hemisphere created an almost insatiable demand for this interesting resource. This demand depended on technical changes in transport and many ways of deploying labor that would create new social ecologies, sometimes building on old ways of mobilizing labor, sometimes inventing entirely new ones. Da Cunha, as well as other authors, noted that the region was largely divided into two major production ecologies: those of *caucho* and those of *Hevea*, and indeed he took this division as his central theme. It was the vulnerability of trees to disease after they were cut for latex that, da Cunha argued, determined the settlement pattern or lack of it during the height of the rubber boom. Tapping *Hevea*s was more or less sustainable, depending on techniques, and required a settled population of tappers,[62] but Amazonian *caucho* was a nomadic enterprise.

This framework of two economic trees (*Hevea* and *Castilla*), two latexes, two extraction techniques, and the labor regimes that were most associated with them became the central axis in da Cunha's Amazon writing. It was a nicely designed experiment in history and civilization, since the environment was the same (and thus the tropes of vulgar environmental determinism could be avoided), a mestizoized population was engaged in the extraction in both cases, and both economies were engaged in a highly globalized system.

Figure 14.3. Curing latex.

It was the social ecologies—or, one might say today, the political ecologies—that became da Cunha's subject.[63]

Latexes and Labor

The rubber economy of the late nineteenth and early twentieth centuries with its brutalized Indians and miserable debt peons incarnated a severe disjuncture in globalized economies between the mass demand for seemingly innocuous products—bicycle tires or waterproof garden boots—and a tropical resource produced under the most oppressive of coercions. There were regions, such as the Putumayo, that were unquestionably horrific, but to conflate all production systems over a huge area and over more than a century with this "devil's snare" is a mistake. The latex industry was widespread and did not require any single mode of extraction. Rather, local conditions structured the relations of production throughout Amazonia's "rubber century": distance to cities, river geographies, social histories, and forms of coercion and violence varied a great deal.

Systems of latex extraction changed over space and time: they were "softer" early in the export period and were less violent closer to the main urban areas and in the eastern Amazon; they became increasingly repressive later in the boom and farther up into the headwater and tributary forests, especially where access to a river could be controlled. As noted, there were also, as a general rule, differences between dominant labor regimes depending on whether *caucho* or *Hevea* was exploited. The extraction of rubber over this long period took on many modalities that are most usefully understood through the characteristics of their labor regimes.[64] These included "affinal forms," with collection carried out by family or clan, and tribute systems such as those associated with tribal societies. There was slavery, including chattel slavery, "terror slavery," and debt peonage. There were also some forms of modern labor organization: tappers could be paid wages, there were some contract systems, and there were even cooperative systems within the *quilombos*.[65] Rubber extraction could be part of a portfolio of small-scale collecting and farming based on family labor, a pattern that still exists widely through the Amazon, or rubber could be the single product in a vast empire of brutal labor extortion. Small-scale extraction carried out by families was a "no cost" or at least "no outlay" system and thus was competitive with the coercive rubber economies of debt peons and native slavery, systems that had to include the costs of provisioning workers, transporting over distance, and repression.

Historically, extensive extraction systems in peripheries were maintained by "nonwaged" forms of labor deployment, such as chattel slavery, terror slavery, debt peonage, and indenturement, or, alternatively, the "affinal economies of affection"—family, clan, and community duties—or obligations such as tributes or labor corvées. In these forms of extraction, direct labor costs—wages—did not really exist. For the various slaveries, however, there were the costs of *conquista* or capture, or the monies paid to a trader. In affinal economies, there were the "currencies" of reciprocity. For the debt peons and tributes, labor would be paid later with products.[66] Many analysts argue that profits in the rubber sector were realized less through efficiencies—it was, after all, men, knives, and smoke—than through processes of unequal exchange and commerce, localized trade monopolies produced by private steam vessels, "captive" rivers, and manipulation of markets, information, and speculation on the latexes.[67]

The export of latexes encompassed more than a century, a period of profound change in global economies and social relations in Amazonia. Chiefs or headmen could mobilize kin, clans, and even indigenous slaves in the gathering of latex. Based on community labor, these tribute economies offered latex in exchange for goods that were redistributed. Rubber was built into older patterns of commodity exchange between Indians and traders within mercantile circuits that had prevailed for hundreds of years. Bates, traveling in the 1850s in Amazonia, recorded the collection of "India rubber"—in his case, *Hevea*—in almost every native settlement where he spent any time; of all groups, the Mundurukú were famous for their integration into the commerce as autonomous traders.[68]

In the upper Ucayali and Purús, native peoples were part of a complex social dynamic where some tribal leaders themselves became *caucheiros*; others allied with rubber barons against their traditional enemies and supplied the slave trade. Some worked out reasonable systems with suppliers, while others fled as far as possible from the entire economy, which, given how widespread the industry was, was quite difficult to do. Most famous among native "rubber barons" was Chief Venâncio (see figure 17.2, a photo taken during da Cunha's expedition), who had organized *caucho* extraction on the Ucayali and Purús in concert with Carlos Scharff. Scharff, as noted earlier, was instrumental in fanning the flames of the Purús conflict and met da Cunha in his travels. Indeed Scharff is the model for the Caucho King in da Cunha's essay "Os caucheiros" in chapter 21. Scharff was later killed by "his" Indians—perhaps an occupational hazard in terror slavery. The Fitzcarraldos were famous for their Piro militia, also chillingly described by da Cunha.[69] The Arana enterprise was known for its ruthless *muchachos*, callous teenage natives used to hunt down and discipline recalcitrant indigenous workers and rival clans. The ethnographies of *caucho* production were varied but, more often than not, horrible.[70]

Women and children increasingly were sold in sex, servant, and slave markets as yet another type of extractive commodity. Bates noted a brisk trade in native children and asserted that the offspring of sixteen different tribes could be found in the commerce in Tefé. He himself was given a little girl, of whom he was very fond. He pointed to widespread illness, including malaria and influenza, among these young chattel and how many of them died in transport and in service because of mistreatment. His young girl also eventually died but at least was given the consolation of a full Christian burial.[71]

Raids for native slaves for latex industries were common but most characteristic of caucho. In 1861 José Antonio Ordoñez found sixty shackled Huitoto Indians from the upper Guaviare in Tefé, carried away by slavers for the rubber trade. His concern was not that they were unfree but that Brazilians had carried off "his" Indians, owned by him through the institution of *encomienda*.[72] Alcot Lange, who was waiting out the rainy season in Remate de Males (now known as Benjamin Constant) on the Javari, noted fifty Indians about to be sent up this tributary by the notorious Casa Arana, which operated on the upper reaches of this river. Treated brutally and wracked by

Figure 14.4. Caucho extraction: note the puddles of latex below the tree.

malaria, all but twelve died in the space of a few days.[73] Roger Casement fumed that one of Arana's administrators failed to meet him because he was off on a *correiria*, an Indian slave hunt.[74]

French explorer Henri Coudreau, whom we met in the Amapá scramble, described the system of Paulo da Silva Leite, a rubber baron based on the upper Tapajos who required tribal tribute in latex from the Apiaká Indians, which he transacted with regional traders.[75] Within the Spanish part of the basin, de facto *encomienda* and *mita* laws were still in effect. Corvées—labor tributes—in these areas were increasingly directed into latex gathering. The Suárez brothers, who dominated the Bolivian rubber trade and whose empire gradually spread to more than five million acres with some ten thousand laborers, began their operations through tribute exchange from the natives on the upper reaches of the Madeira-Mamoré River, but later used virtually every imaginable form of labor deployment to move the forest latex to market, ranging from slavery to wage labor.[76]

Where Caucho Was King

The first widespread exploitation of *caucho* for nineteenth-century international markets occurred in the 1870s in what is now known as Panama (then a province of Colombia), and the latex was known as "Panama rubber."[77] Amazonian *caucho* exploitation took off in the early 1890s. In the Upper Amazon, *caucho* was often found in large *manchals* or groves, suggestive perhaps of a deeper historical ecology. We know *caucho* could be sustainably harvested

PAMARY GIRL.—PURUS RIVER.

Figure 14.5. Pamary girl: a coveted extractive product for domestic use.

because of the large Meso-American economies that used this latex, and because of images from Mayan codices that portray tapping of *caucho* after a tree's bark is scored. But the modern extraction method was to simply destroy the tree and bleed the latex into shallow excavations under the trunk.

Full extraction of latex from a *Castilla* would yield some 15–20 kilos of salable latex per tree,[78] and in a moderately dense *manchal*, *caucheiros* could produce in a month or so what would take a *seringeiro* a year. It was a one-time extraction and involved complete deforestation of the groves. Joaquin Rocha described it this way in 1905: "There, when the forest is virgin, people cut down the *Castillas* in such quantity that the land is completely cleared and as open as though it were cut down for agriculture. . . . I have seen myself that there are many extensions of open areas in the midst of the forest with an abundance of seedlings of gum trees, and these are growing, coming from the seeds that germinated in the warmth of the sun on the land after the great clearing made by the first exploiters."[79] The problem, at least at first, wasn't a scarcity of trees but a scarcity of labor.

A Tightening Noose: Indigenous Workers in the Rubber Economies

The way that the fortunes of natives unfolded as the global rubber economy expanded can perhaps best be understood if we look at the history of increasing indigenous subjection on the Putumayo, the "river that God forgot" and the site of the Casa Arana's "British Congo," so named because the atrocities there rivaled those of the Belgian Congo. The river was initially opened to steam travel in the 1850s by Rafael Reyes, a trader in quinine. This commodity was relatively simply and benignly exchanged for metal goods with Huitotos, Andoches, and Boras, who also supplied fuel wood, food, and meat for the traveling trade.

Later, mulatto *caucheiro* Crisóstomo Hernández, a Colombian fugitive from justice (he was a murderer), married into the Huitoto; he loved them and lived among them all his life. Hernández used tribal interactions and native tribute collection to insert the Huitoto more firmly and reliably into the commodity exchanges of metal goods for latexes. He noted both the extensive groves of *caucho* and the populous world of the tribes, and indeed the presence of a lot of labor was exactly the allure of these watersheds.[80] Eugène Robuchon, who had been sent by the Aranas to reconnoiter on the Putumayo, estimated native populations in excess of fifty thousand.[81] While Hernández himself had a different relation to natives and extraction, he paved the way for the first vicious wave of *caucheiro* kings like Benjamin Larraniaga, a brute later poisoned by his workers, and Gregorio Calderón, who carried out the *conquistas*—a weighted term involving conquest (by violence) and seduction (by trade goods), and of course sexual conquest—that forced the tribes into commodity circuits, and developed the early institutional commercial structures of obligatory peonage that would become the platforms for the Casa Arana's later "rationalized" system of quotas and maximum labor control via terror slavery.[82] The War of a Thousand Days, one of the civil wars that have come to define Colombia, distracted national armies away from this interesting economic frontier and the territorial appropriation by Peruvian enterprise, and soon the Arana militia and unrefusable deals dispossessed the remaining Colombian *caucho* bosses and made the vast watershed of the Putumayo drainage de facto Peruvian terrain. This "everyday imperialism" was of intense interest to Peruvian Amazonian officialdom, who provided military backup (as Casement noted to his disgust), and was seen as a model worth replicating elsewhere, like on the Purús and Juruá.

The Putumayo of the Casa Arana was denounced as the "British Congo" because British finance floated the enterprise, which involved massive torture, sexual abuse, wanton killing, and literally working Indians to death. Julio Arana, however, imagined, as did his unfortunately named brother, Lizardo, that the system they devised was rational, industrial, even "Fordist" in its inspiration and vertical integration, even if labor control occurred through unmitigated violence and decimation of the machinery (trees) and the native labor to exploit them. Estimates of indigenous mortality on the Putumayo place it at over thirty thousand.[83] Maximum product was produced at the lowest possible cost, overseen by managers who were rewarded on commission. The material was moved on company boats—the Casa Arana ran twenty-two steamboats up the Putumayo, on which it held the steam travel monopoly—and sold through its clearinghouses in Manaus and Iquitos, with finances raised on international capital markets via the London seat of Arana's Peruvian Amazon Company. Julio Arana was among the wealthiest of the latex barons until a combination of Asian product and scandal unraveled his empire. He blamed geopolitical intrigue with Colombia as the underlying reason for his persecution.[84]

The sensationally grisly nature of Putumayo rubber extraction became known internationally largely through the revelations of Roger Casement, Irish nationalist, gay libertine, fervent anticolonialist, and indefatigable crusader for social justice, who had played a pivotal role in revealing (and ultimately ruining) King Leopold's Congo enterprises. Traveling with Casement on that trip was Joseph Conrad, who would ultimately use his African experience to write the anti-imperial novel *The Heart of Darkness*. Newspaper denunciations like those of Walter Hardenburg (*Putumayo, the Devil's Paradise*) and many brave Peruvian journalists appealed to Casement, whose inquests and travel on the Putumayo brought the horrors of this enterprise—the Amazonian Congo—to the glare of international attention. He was well placed in the diplomatic corps and was able to make his denunciations at the highest levels in Britain while mobilizing national outrage. There were movements to boycott rubber. Casement's strategy, in the end, involved what we today call "commodity chain analysis" and triggered a major international human rights crusade on behalf of the brutalized Huitoto, Bora, and Andoche natives of the Putumayo.[85] Casement has emerged recently as an anticolonial hero with new biographies, compendiums of his work, and profiles. His attempt to foment anticolonial movements in Ireland, however, resulted in his hanging as a traitor, in part due to his actions in the Irish rebellion, but also because of a miasma of scandal that surrounded his "black diaries," writings that chronicled details of his Amazonian travels (stops at wooding stations, how much things cost) as well as numerous homosexual exploits.[86]

It was not only people that were doing the dying. The relentless clearing of *caucho* also produced a major deforestation pulse during the heyday of Amazon latex. Ecological constraints and the failure of plantations in the nineteenth century helped foster the shift to other latexes like *caucho* and *balata* (as well as the "weak *Heveas*"), but the most efficient (and lucrative) alternative was to home in on the pure stands of *caucho*.

The *caucho* economy of the upper Amazon generated an important but largely unrecognized deforestation pulse, a period of "selective logging" unmatched until the late twentieth century. More than six million *caucho* trees were cut, and millions of other trees were damaged in the felling process. Recently, while investigating the effects of modern selective logging in Amazonia (research carried out as part of the Large-Scale Biosphere-Atmosphere Experiment in Amazonia [LBA], a NASA international research initiative), researchers reviewed several "subcanopy" impacts of this kind of extraction.[87] The selective cutting of valuable species may leave the forest canopy more or less intact but strongly affects forest functions through changes in carbon uptake, seed stocks, regeneration, ecological community structure, and resilience. Perhaps the most important finding is that that between ten and thirty adjacent trees are taken down for each "targeted" tree. LBA scientists concentrated on "harvest" rates of about one tree per hectare, roughly equivalent to modern *Castilla* densities, although these low modern *Castilla* distributions seem somewhat at odds with the descriptions of *manchals* of nineteenth-century observers like Rocha, quoted earlier. Harvesting in the more intensive *caucho* zones produced a "stealth" deforestation rate of between 36 and 180 million trees.[88] The "take all" approach resulted in local extinctions—and these helped drive Peruvian *caucheiros* into the lands of Brazilian tappers.[89]

If we take modern data as the norm, even though this may lead us to underestimate the impact, the selective clearing of *Castilla* during the height of the rubber economy may have been "invisible" at the time but produced such deforestation that today the satellite signature of the upper Amazon is successional forests rather than the primal *selva* these places are imagined to be. *Caucho* has become a species of floodplains rather than of the uplands, its former haunt.[90] Da Cunha was correct in describing the *caucho* system as "killing trees and men."

Where Rubber Reigned: Peonage and the Northeastern Diaspora

Attempts to intensify *Hevea* rubber production in Amazonia have consistently met with failure because of the vulnerability of *Hevea* trees in the Amazon to *Macrosystis ulei*, a fungus that decimates the young leaves. Traditional and native peoples have concentrated useful plants wherever possible, a phenomenon documented throughout the Amazon basin. Even today tappers in Acre and in the eastern Amazon along the Tapajos improve the density of *Hevea* trees by planting seedlings within the forests.[91] There had been many larger-scale attempts at making denser *Hevea* stands, largely unsuccessful, especially in the later part of the nineteenth century.[92] Attempts to adapt the tree to industrial production through plantations in Amazonia—the most efficient way tree crops can be organized to serve dynamic capitalist markets—were always thwarted in Amazonia by ecological constraints, even when, as in Fordlândia, Henry Ford's enclave for industrial modernization of rubber production, capital was no object.[93] While slavery was the general mode in *caucho* systems, debt peonage has characterized much labor mobilization in the Brazilian rubber economy in upper Amazon tributaries. The ending of slavery everywhere in the nineteenth century required new modalities of labor coercion. One of the most important of these was debt peonage which was also associated with vast diasporas all over the globe in colonial plantations and extractive economies. Very much under-researched as a response to the transition to waged capitalism, peonage was associated with, for example, Indian migration into the Caribbean and to South Africa and Madagascar, Japanese and Italian migration to southern Brazil and the United States, and, of course, *sertanejo* resettlement in bondage in the rubber forests.[94]

Debt peonage involved advancing goods and travel costs in exchange for repayment with rubber in a system locally known as *aviamento*. Monopoly of labor and of exchange dominated *Hevea* areas above river rapids and in the more distant headwaters, as it did for *caucho*. This system will be described in much more detail by da Cunha himself in later chapters.[95] There was plenty of blood mixed in with the milk of all the latexes, but slavery and peonage, the most lethal and abhorrent means, were far from being the only ways that latexes were produced.

Free Forms of Labor and the Rubber Economy

In contrast to the "blood rubber" and peonage narrative that has come to dominate the imagining of the Amazon basin's rubber economy, many parts of the Amazon, especially in lower reaches of the basin, had conditions that bore little relation to those of the more distant tributaries. Extraction there relied to a greater degree on family and kinship. These traditional forms of labor mobilization were relatively benign in the earlier phases of the rubber boom, when the rubber economy was still largely confined to the Lower Amazon. This is certainly the sense one has of it from early observers like von Spix and von Martius (1812), Edwards (1842), and Mme. Agassiz (1866), who noted as she took outings in small villages that "the men were up river collecting" but did not seem to hear any expressions of alarm from women about this, which certainly would have been the case as the century wore on. Agassiz was more concerned about military conscription.[96]

Next, the Cabanagem rebellion (1834–40) had undermined elite control in the lower Amazon basin, quite dramatically altering local power relations over forest workers, agriculturalists, and slaves, and produced highly autonomous communities of many types, including *quilombos*.[97] These groups were not "prisoners of the landscape" in the way that indebted northeastern migrants in the upper watershed might have been, nor were they relocated native populations. The locals had kin and affines in the area, and they were inhabitants of, rather than exiles to, Amazonia. Producers were part of families that relied on a portfolio of activities for their livelihoods, not just rubber, but participated in the varied activities of commercial extraction, as kindly Mr.Edwards noted.[98] Further, while the *aviamento* system—advancing goods in exchange for rubber—certainly operated, the system was more complicated and complex in Lower Amazonia, where proximity to many small-scale traders implied competition among them rather than the domination by a single *seringalista* with monopoly over a "closed" watershed. Henry Pearson, editor of *India Rubber World*, whose interest in all aspects of the industry was naturally quite acute, notes the number of *aviadores* and the relative ease with which one could become one in the Lower Amazon. The patron-client relation was often affinal and infused with a great deal of social meaning and mutual obligation. Historian Barbara Weinstein's work presages elements of contemporary research on Amazonian middlemen, who are often themselves tappers, so that the relations may have been more nuanced, a kind of "moral economy."[99] This autonomy was assured in the rubber systems closer to Belém through the ability to provide one's own subsistence, to market other extractive goods like palm fruits and dried fish, and to have more choice about to whom one could sell, so that one could, if necessary, avoid oppressive traders. Further, to move up and down the river with the tides did not require expensive steamboats, so "freelance marketing" was possible. The immense and complex mangrove swamps and islands also hid and protected communities and provided escape routes that made "spatial coercion" very difficult. This appears to have been typical in the well-traveled part of the lower basin and "zona das Ilhas," the estuary basin including the islands of Marajó and an enormous number of small islets, which actually rivaled the Acre territory in the volume of its latex products at the tail end of the rubber period.[100] Pearson reported that in the "island district," the area near Belém, including Marajó, in the Amazon estuary, as many as eighty "brands" or sources came in on one steamboat to an array of commercial houses, far different from some of the river monopolies of the upper Amazon.

The upper Amazon itself did not have one simple story. On the "sister" river of the Purús, the Juruá, historian Cristina Wolff provided evidence of a much more varied social landscape near cities. Based on the data of the 1904 census focused on the town of Cruzeiro do Sul, Wolff noted that 27 percent of the population was female, a striking statistic if one is accustomed to the masculine version of the lonely tapper engaged in his rounds in a relentless circuit of dearth and oppression.[101] In addition, Wolff's review of legal cases pressed by women (usually over false promises of marriage) revealed that almost half of the men involved defined themselves as agriculturalists, not *seringueiros* (although one was an artist). Photographs from the region in the early 1900s show a great deal of clearing and livestock.[102] This suggests more elaborate economies in these regions than we have come to expect. While there is no question that there were debt peons living in the outback languishing with beriberi due to a diet of canned foods and manioc flour, the situation was far more complex. Da Cunha also described several different types of extractive settlements including those of peasant yeomen, as we will see later, and a great deal of personal agency as well as enslavement. Historian Barbara Weinstein's research, like Stephen Nugent's, reveals the importance of autonomy to tappers, a point they themselves raise today.[103] Historian Greg Grandin remarks in his incomparable study of Henry Ford's failed plantation attempt at Fordlândia that labor control in the lower Amazon was quite difficult even in the 1920s and 1930s, not so long after the collapse of the rubber industry. Workers would work for wages for a time, but when irritated by time clocks (these were demolished during labor uprisings) or miffed at the soy products and brown rice served to them in the canteen in lieu of turtle, fish, and farinha, the men simply headed back to their homes to plant and fish.[104]

Quilombos and Rubber

The number of *quilombos*—fugitive slave communities—in Amazonia was substantial, as recent research has shown.[105] Many slaves and detribalized natives fled into *quilombos* during the Cabanagem to avoid the extensive violence and

Figure 14.6. Wickham's baby: a tree from the original seeds in Sri Lankan plantations.

to escape their chains. The rubber economy provided, as had gold elsewhere, a lucrative means of exchange to support autonomous and renegade cultures.[106] In eastern Amazonia, Amapá, Pará, and Maranhão, *quilombos* were engaged in complex systems of commodity exchange in which rubber, like gold, was one of the most important items. From *mocambos* on Marajó Island to the upper reaches of the Trombetas, through to Bolivia, slavery's fugitives tapped the "black gold" but used the small-scale traders, the *regatões*, for their transactions. In a world of rampant malaria, black and mulatto populations had a genetic advantage in malarial upriver settings where disease mortality was very high and rubber forests were quite rich.[107] Prior to emancipation, in some cases it may have been effective for *quilombos* to simply contract themselves as a community to one rubber baron to forestall chattel enslavement and the dispersal of their families, as happened in some communities near Mazagão in Amapá. These strategies were described for a Bolivian *quilombo* and some cases in the upper Trombetas.[108] Arnous de Rivière, traveling in Bolivia, noted that some *quilombo* villages were integrated into the rubber trade, based on contract exchanges as well as barter.[109] He thought that free American blacks might use *quilombos* as a model for new American black colonization in Amazonia.

This was the complexity and range of the rubber economy. Da Cunha would describe it as a vast octopus, sucking everything into its insatiable maw—not a bad metaphor for rubber at that time, and quite apt for the economies of the Purús.

It would have been hard for da Cunha to imagine the "litigious zone" he was marking out with his survey team as it is described today. The Upper Purús is now a national park the size of Costa Rica and is portrayed in the usual language of the distant and pristine. The complex social landscape encountered by da Cunha and the commission on their way to the headwaters—native slaves, debt peons, refuge communities, and family-based extractors—has been erased from the modern imaginary.

There were many things that would transform the realm of rubber that da Cunha traveled through at the very apex of its glory. He saw in the forest societies the foundations for a "Civilization in the tropics." The economy of this realm was built on the ecological substrates and ruins of earlier civilizations.

The Amazon rubber world was on the cusp of collapse. As Haraldo Torres, economist of extractive economies, has pointed out, when the prices of extractive commodities rise there are several processes that produce stagnation or undermine wild extraction systems: overexploitation, which reduces the output of extractive product (this was clearly happening with *caucho*); the social limits to coercion, which certainly were becoming evident with various denunciations of terror slavery and debt peonage; domestication and plantation development away from the commodity's center of origin and its coevolved pests; and finally, the search for synthetic replacements for the natural product. Rubber was subject to all of these.[110]

The Purús and its history had found its bard: da Cunha was making his way with the reconnaissance team from the mouth to the headwaters of the contested realms of rubber.

15
Argonauts of the Amazon

The Marching Orders

The orders arrived at the end of March, and in April the joint boundary commission embarked in crafts that were modest for a such a formidable journey: the Peruvian launch *Cahuapanas*; the Brazilian *Manoel Urbano* and *Cunha Gomez*, named after the Brazilian explorers of the Purús and the Juruá; and light canoes—*ubás*—for the inevitable time when channels would not permit even the three-foot draw of their vessels. The Peruvians had also five smaller craft among them: *batelões* and *ubás*—which became key elements of later intracommission conflict.

Both groups included about twenty-two of their own recruits: the Peruvians relied on those from their Navy based at Iquitos, while the Brazilian ranks were composed of *praça*s (enlisted men) from the Thirty-Sixth Battalion from Manaus, the human engines of the trip. The expedition included a specialized crew (mechanics, photographers, cooks, the doctor, including Euclides da Cunha's cousin Arnaldo da Cunha) in addition to the two principals, Euclides da Cunha and Navy Capitan Alexandre Buenaño. They traveled with theodolites, chronometers, compasses, barometers, and thermometers to measure altitude, latitude, longitude, relief, and river depths and to make basic weather observations.

Their main tasks were outlined in the simple language of their marching orders:

1. The parties charged with the exploration of the Purús River will leave from Manaus, verifying the course of that river, carrying out a simple hydrological reconnaissance up to the trading post at Cataí, whose geographic coordinates they will determine, as well as any other interesting points of their journey.
2. They will explore areas above Cataí in their entire extension, and the *varadouros* that must exist to the Ucayali.
3. An expedited survey of the Purús will be carried out to determine the approximate coordinates of the mouths of all the principal affluents, especially those called the Curanjá, Curiujá, and Manoel Urbano.
4. The Joint Commission will, insofar as possible, correct and complete the map developed by William Chandless and will verify the correspondence between the geographic nomenclature used in his chart with names currently in use. On your return you will determine the coordinates of the confluences of the Purús.
5. The commission will present a map documenting this work and a descriptive memorandum of the zones through which you traveled.[1]

Because Chandless had left this question unanswered in his explorations, the commission was also meant to prove or disprove the existence of Lake Rogualgoalo as the source of the Purús, Ucayali, and Madre de Dios.

On April 5, 1905, the Joint Commission for the Reconnaissance of the upper Purús assembled with its men, matériel, and scientific equipment and "set out in search of its destiny." They departed from Manaus, whose latitude and longitude had been precisely determined, and all their measurements were tethered to this fixed point, the axis mundi of the rubber world. They embarked from this physical landmark to follow a marker of another kind, the map of William Chandless, the British explorer, which they would verify both as guide and as model. There was also the terra incognita—the part of the Purús untraveled by Chandless, still among the most remote of the terrestrial mysteries. Finally, like so many earlier explorers from Sir Walter Raleigh on, they would be engaging with mythical terrain, in their case the question of Lake Rogualgoalo. They would traverse a geography from the most precise and scientifically known to the utterly conjectural and imaginary.

With these directives, the reconnaissance team set sail in a virtual brotherhood with the flotillas of late eighteenth- and nineteenth-century travelers, ranging from a handful of the most august scientific explorers (many of them aristocrats) under the patronage of crowns and princes, to the fact-checkers of empire (tropical surveyors and administrative travelers), to a diaspora composed of runaway slaves, outcasts, and captives from the ruins of native societies, the exiles of catastrophes of northeastern El Niños and migrants from a larger world in upheaval. There were businessmen reviewing portfolios and a rogues' array of "tropical tramps" and adventurers from all over who sought their fortune in the Amazon's rail and port construction crews or as carpenters, seamen, tappers, or skilled laborers. Travelers' tales reported a surprising cosmopolitanism, with Europeans of every nationality, outcasts of the Ottoman Empire, footloose Americans, Chinese indentured "coolies," Caribbean creoles, and indigenous peoples—Andeans and Amazonians in bewildering diversity. The Upper Amazon at the time was probably as multicultural as any place on the planet. There were also tourists and adventurers, living out Victorian fantasies of masculinity in a spiritual and physical face-off against wild nature.

The Amazon was not empty: ecclesiastics, natives, the residue of early booms, and traces of earlier events were also to be found. Native intellectuals were the sources for much of the natural history and landscape knowledge that would keep the explorers alive, and in some cases make their names.

These voluntary and involuntary migrations materialized in the Upper Amazon forests, all part of a massive transformation that changed that *terra nullius* into a vital part of a globalized economy, consolidated nations, and produced a geopolitical flashpoint. It was these travelers who, for da Cunha, truly claimed Amazonia, and whom he thought knew better than anyone its fecund, treacherous possibilities. It is useful to think of this "march to the West" as an unfolding of tiers of travelers: a handful of elite internationals, fairly anonymous (at least today) but essential sets of surveyors and administrators, then tourists, and finally the huge labor diaspora.

Elite Endeavors: Sagas of the Sages

By far the best known of Amazonian travelers in the modern day were the northern European and American naturalists whose descriptions inform the modern canons of tropical science exploration and nature writing. These travelers were a mere handful among the million-plus who were sucked into the latex vortex of South America. Da Cunha read their science but was dubious about their "explorations." These prominent travelers were the best positioned with patronage and connections to the great institutions of the day; they dominated the understanding of the region though as they stood, as da Cunha put it, at "only the threshold of the great Hylea."

Da Cunha was compelled by social history, and the Peruvian-Brazil negotiations required a focus on explorations by South American nationals and local populations to construct territorial claims, so his attention, unlike that of most writers of the Amazon exploration, was focused less on the marvels of the Amazonian biota and more on the more or less "invisibles" toiling on distant tributaries. His approach supported the political purposes of his travels and provided a counterbalance to the Euro-American framing of the Amazon as a place of wild nature and barely civilized locals. After all, da Cunha was entering a place producing more than half the world's latex, linked to global financial and commodity circuits—a place that was the heart of a highly conflictive geopolitics that engaged several hemispheric and European actors. Da Cunha's writings are awash in the names of Latin American and Iberian Amazonian colonial explorers, names that survive in the modern day mostly as toponyms like Maldonado, Urbano, Orton, Heath, Chandless. His attention (and access) to the vast gray literature produced by ecclesiastics, surveyors, administrators marks da Cunha as among the most thorough of Amazonian scholars of any time.

Anglophone and European scientists and collectors were fundamental to the writing of Amazonia as an archive of biological history rather than a place of human endeavor, and to configuring the great forests as an enduring "Natural" world counterposed to "Cultural" Europe.[2] Human life and its social formations in the tropics were, in the intellectual idiom of the time, outcomes of the hierarchies of nature as expressed in class. When local people appear in these works, it is as gracious hosts, earnest bearers, slaves, helpers, or rowers and watermen of varying degrees of reliability and quirkiness—useful exotica, but overall, in need of better management. This was the rural sociology of many Amazonian explorers who kept to the main channel, but younger "off-route" travelers—like Lange, Hardenberg, Woodroffe—who were caught in the snares of the upriver economies saw the intersection of international economic forces with local power as structuring the lives of countless thousands, and "natural" social hierarchies as simple ideological masking of economic brutalities.

Enlightenment Explorers and Nineteenth-Century Naturalists

The eighteenth-century European voyagers had forged a tradition of adventure, science, and measurement that increasingly defined "Greater Amazonia"* as a terrain of mythic endeavor, scientific exploration, and global ambitions. While the imperial narrative has largely dropped out of the popular perception of these travels (perhaps due to Rio Branco's success at deflecting incursions), it was certainly not far from the interests of the many observers of Amazonia from the Enlightenment to the twentieth century. Earlier expeditions set forth with surprisingly august protection and were manned by Europe's aristocratic intelligentsia. De la Condamine traveled with the blessing of the French crown. The Spanish royals issued a *laissez passer* to von Humboldt in the hopes that his mining experience might lead to new sources of wealth, especially if he was roaming around Raleigh's El Dorado in the Guyanas. The Brazilian explorer Alexandre Rodrigues Ferreira was contracted by the Portuguese crown. The young aristocrats von Spix and von Martius, who form a bridge between the Enlightenment travelers and the nineteenth-century naturalist-collectors, traveled in Brazil as emissaries of the king of Bavaria; the Agassiz family voyaged under the protection of Emperor Pedro II (although the funds came from financial tycoon J. P. Morgan). By midcentury, scientific institutions had replaced the coffers of kings, reflecting new forms of states, new global institutions, and new imperial statecraft.

Reports of the voyages of de la Condamine and, most famously, Alexander von Humboldt initially defined the tropical scientific travel genres and the early technical expeditions of tropical empire. These travels were seen as Enlightenment enterprises whose central concern was theoretical as well as practical knowledge derived from scientific principles. These highly public itineraries were sponsored respectively by the crowns of France, and Spain, which were hardly indifferent to the possibilities of Amazonia and the information collected by these scientific icons with excellent mapping skills. These aristocratic travelers galvanized Europe with their South American natural history and philosophical voyages and imbued European tropical travel with scientific luster and manly daring.[3]

These travels lacked the explicit tinge of imperial exploration that clung to the later deeds of Stanley, Livingston, Rhodes, and Brazza in Africa in the nineteenth century, but the studies of von Humboldt and de la Condamine figured in the arsenals of the Guyana Scramble. Both would identify signifi-

* Greater Amazonia includes the Orinoco.

cant regional resources that would shape tropical imperial history: quinine, latexes, petroleum ,and gold.

Science, Espionage, and the Deep History of Scrambles

The template for modern scientific expeditions in Amazonia first appeared in the late eighteenth century with the reports of de la Condamine (1735–44), who was measuring the length of a degree of meridian at the equator with astronomical techniques but took the opportunity to note curiosities, build pyramids,[4] outline potential resources, and discuss them in detail with the French crown. De la Condamine made a special point of visiting the French colony in the Guyanas, a place embroiled in regular boundary conflict with Brazil, and in fact it was there that he was introduced to *Hevea* latex (he'd encountered *Castilla* latex or *caucho* earlier in Ecuador).

Von Humboldt's expedition (1799–1804) remained on the perimeter of Amazonia, although he traveled well inside the Orinoco watershed and the Guyanas, areas of significant interest to Spain, his sponsor. He was concerned to transcend mere description and to "track the great and constant laws of nature." Measuring the Earth and the "congress of all the forms of knowledge" would provide the scientific framework from which "nature itself was to emerge as a dynamic equilibrium of forces."[5] It was, in his view, the task of natural philosophy to develop a history and theory of nature, one unleashed from the religious dogmas of the time, one that was scientific but also lyrical and expressive: "nature herself is sublimely eloquent."[6] The integration of aesthetics and natural science was a way of overcoming the limitations of mere description or pure sensation and united the insights and understandings of nature in Enlightenment science and emerging romantic sensibilities. As one of the most influential scientists and literary stylists of the South American tropics, he had an impact on the method of reportage on virtually all nineteenth-century expeditions that was incalculable, with even young William James yearning to replicate his manly style of rational science with romantic rapture. The unity between scientists assessing the natural world and imperial survey more or less bifurcated later in the mid-nineteenth century into divisions between the "purer" natural sciences of biogeography and taxonomy and the applied trades of colonial survey, though the lines between them remained famously blurry.[7]

Another eighteenth-century traveler, whose exploits received far less attention though his circuit was much more extensive and thorough, was the Portuguese- Brazilian Alexandre Rodrigues Ferreira, whose nine-year expedition (1783–92) in remote parts of the basin produced extraordinary botanical, zoological, and ethnographic results, not to mention exquisite paintings and drawings by José Joaquim Freire. Rodrigues Ferreira traveled regions seething with boundary conflict, including the Oyapoque River, the Rio Negro, the Pantanal, Mato Grosso, and Paraguay. Although Rodrigues Ferreira is known as "the Brazilian von Humboldt," unlike that icon, he had no theoretical or truly "philosophical" or Enlightenment agenda for his travels and elaborated none. His job was imperial exploration and resource assessment for Maria I of Portugal.[8] His scientific notes sat in Lisbon; then, during the Napoleonic Wars, they were taken to Paris as war booty for their usefulness as "intelligence" and were studied carefully (and even plagiarized) by French naturalist Geoffroy Saint-Hilaire.

Wallace, Bates, and Spruce, explorers of modest means, traveled with the blessings of the Royal Gardens at Kew, while the Britain's Royal Geographical Society sponsored Chandless, Fawcett, and the brothers Schomburgk, among many others, and ran a training school, a kind of "Hogwarts" for explorers.[9] The gloss of scientific exploration was a more prominent feature of Amazonia travelogue writers compared to the main African "headliners" like Mungo Park, Stanley, and Livingstone, who were overt agents of imperial enterprise and charter companies. Scientists like Spruce, Bates, and Wallace were not formally staking claims, although they were providing information (and germplasm) to a larger imperial apparatus even if they were not placing survey marks.[10] Exploration, science, and imperialism were profoundly melded in these nineteenth-century tropical travels, a point that is hardly profound at this point in colonial scholarship but is often overlooked in Amazonia, because the republics and colonies under review by imperial powers maintained their territorial integrity and sovereignty[11] or else lost them to Brazil. Some explorers like von Humboldt and von Martius generated concepts that fueled the colonial development imaginaries, like the von Humboldt tropics "flooded by civilization" and von Martius's ideas of specialized racial roles in Brazilian development,[12] notions echoed by other travelers. Unlike the African adventures of Park, Stanley, and Brazza,[13] Euro-American expeditions into Amazonia were not the vanguards of royal colonial projects but rather forged new relations between imperial institutions, bureaucracies in emerging circuits of capital, commodities, and knowledge.

"Great and constant laws"

Tropical scientists like von Humboldt, Bates, Wallace, and Spruce made lasting contributions to modern science and in their rewriting the "nature" of nature moved natural philosophy beyond the earlier encyclopedic traditions of description and collection and into the realms of inquiry that generated "great and constant laws" and provided evidence for new theories. It was a different form of science—based in biogeography, ecology, and evolution. Von Humboldt embodied approaches that were positivist, both in ideas about the directionality of history and in upholding "objective" methods. This "scientization"—rigorous measurement and the search for systems of laws—was a key step in the professionalization of the natural sciences but also interacted with totalizing historical theories. Simple collecting, after all, was more or less an amateur's game, even if it was how Bates, Wallace, Spruce, and Clements Markham, men of humble beginnings, paid for their tropical explorations. Even those from modest backgrounds like Bates and Markham were launched into aristocratic realms and could be knighted and become heads of the most important colonial institutions: Bates became the director of the Royal Geographical Society. Markham headed the India Company, Britain's most important charter, and later the Haklyut Society, devoted to the publication of global travels by British subjects and translation of relevant works on overseas nature and culture as a means to enhance the tropical colonial archive and inform its practices.

If von Humboldt was the progenitor of the modern scientific field naturalist, Joseph Hooker (1814–1911), the director of Kew Gardens (a very close friend and well-placed ally of Charles Darwin), can be seen an agent through which scientific institutions (especially the botanical garden) shifted from being depositories for specimens and attractive destinations for weekend outings to active sites for the creation of new knowledge and "imperial natures." Hooker, who himself carried out collecting expeditions in the Himalayas, the American West, Morocco, and Antarctica, had a powerful vision of a "botanical empire."[14] He subsidized many tropical collections and was especially interested in a "philosophical botany" that would answer questions about the interaction of organisms and their relation to place (and how these might be replicated elsewhere on the planet, an issue of interest to Imperial England) rather than focusing on the simple craft of collection. In this sense, as historian of science Jim Enderby has argued, the collectors and natural historians supported by Kew and kindred institutions moved from being "passive providers of specimens or inert recipients of metropolitan knowledge . . . to becoming active participants in the making of new scientific knowledge."[15]

The naturalists' enormous scientific efforts contributed to the genre of nature writing that had shifted from extolling God's greatness to more secular scientific description as well as the Humboldtian and romantic adoration of nature itself. The "prospect" and the aesthetic of the individual experience of nature nourished the uplifting possibilities of spiritual, scientific, and artistic impressions. These could inspire poetry and, at the very least, stir noble feelings. In writings about New World nature, these approaches were infused with exoticized landscapes and mysterious perils that seemed to many of these writers to cry out for the generous guiding hand of Northern colonialism.

These works also helped nourished Victorian ideas of masculinity in the Anglophone world. If Africa had its big game hunts, Amazonia was a worthy proving grounds for the mettle of young men with its giant forests, many insect discomforts, monstrous alligators, piranhas, native savages, and tiny *candiru* fish that could swim up the urine stream into our intrepid traveler's penis. While these adventures lacked the carcasses and trophies of African exploits, they were reliably unpleasant and produced ample evidence of manly toughness. Both Africa and Amazonia had cannibals, serving for some travelers as the most "other" as well as the "top predator." This model of masculinity has had a durable life on best-seller lists, from William Lewis Herndon and Lardner Gibbon, Alcot Lange, Charles Barrington Brown and William Lidstone, Up de Graff, and Theodore Roosevelt to Peter Fleming and David Grann's biography of Percy Fawcett's *The Lost City of Z*. Scientists and adventurers so dominate our view of Amazonia that it has effectively obscured another set of literate and relatively well-informed travelers, people who worked deep in the interior.

Fact Checkers of Empire

The European romantic movements, the emerging aesthetic of the tropical picturesque, and scientific collection were the flip side of a far more mundane phalanx of official national surveyors lugging theodolites and chronometers, making notes, charting their courses. Unlike the naturalists, who tied their studies of the unknown to an exalted science and a speculative tropical future (the scientific genre almost always seemed to require commentary on transforming the exuberant nature into productive enterprise), the surveyors were meant to bind their landscapes into a grid, to transform terra incognita into recognizable terrain depicted in the simple dimensions of formal mapping for the claiming of territory. The purpose of Buenaño and da Cunha's efforts was, after all, to culminate in a map.

The explorations were used to determine which valuable products might be commandeered and who might help or hinder in this enterprise. Knowledge that might contribute to colonial efforts and resource extraction was the mundane but crucial information on soils, crops, disease, medicinals, geologies, navigability, peoples, economic botany. The "interests" of other colonial

Figure 15.1. A fact checker: Alexandre Buenaño, of the Peru-Brazil Joint Commission.

powers were also of objects of intense curiosity and were included in the "charges" given to these explorers by their patrons and their nations. These travels could also enlighten relevant governments about aptness of places for colonization itself. The number and scale of these exercises are often quite surprising. Peru's commissioned explorations of the river routes, Vias Fluviales, sent many mapping and resource survey expeditions into the tributaries of the Ucayali and Madre de Dios. The officialdom of Amazonian states was concerned to sponsor travels to keep an eye on what was escaping their provincial taxes and to provide more general information about the region.

This "lesser strata" of Amazonian explorers (da Cunha was one of them and revered their work) remains largely underappreciated, in part because they wrote in Spanish and Portuguese and their purposes were not to resolve some abstract scientific question, to collect exotica, or to have riveting adventures. Indeed, the fact checkers hated travel drama and their often lethal outcomes. Mostly they hoped to avoid the close shaves so loved by armchair explorers, since they would have so many of them. Their task was to solve some seemingly mundane but crucial physiographical problems on which entire tropical imperial enterprises might hinge. These involved straightforwardly empirical topics: who the natives were, how many were there, what languages were spoken, what were the soils like, what and how things were grown, the real names of the rivers, whether a reported mountain range actually existed. Such questions seem of little import today, precisely because they have been so definitively answered via the empiricism of these "fact checkers" of empire. Because these travelers were often addressing specific national rather than "natural" problems, they are the largely unknown titans in the pantheon of Amazonian exploration. They included Lardner Gibbon, William Chandless, Richard Schomburgk, Louis Cruls, Ricardo Franco, Silva Coutinho, Fausto Maldonado, John Randolph Tucker, Henri Coudreau, and Frank Church, all of whose efforts infused the "Amazon Scramble." These ranks of observers set sail under the auspices of regional colonialism, as in the case of da Cunha, or carried out similar travels cloaked in a more international register, such as the efforts of Chandless, Herndon, Coudreau, and Schomburgk. They used their maps strategically and clearly understood their roles as chessmen in "great games" unfolding in the region.

Another significant and largely ignored source of survey, natural history, and information on the natives was the ecclesiastics, whose reports on the lay of the land were as careful and meticulous as those of military explorers. They too were staking out terrain and territory. Da Cunha called Padre Samuel Fritz "Amazonia's first geographer" for the quality of the maps he generated on the Upper Amazon. While the information of chroniclers who were traveling with conquering expeditions are invaluable early sources, these were what we would call today "rapid appraisals." Far more detailed work was carried out by an army of Jesuits, Dominicans, and Franciscan brothers who traveled into the deep interior, stayed put for decades, wrote detailed reports to the Vatican about their lives and discoveries, and mapped the great domains from the Paraguay, Chiquitania, and the Moxos to the Maynas and the Orinoco missions of the continental interior. Antonio Vieira (1608–97), João Daniel (1722–76), and José Gumilla (1686–1750) are among the more prominent of the priests who wrote large tomes and traveled widely during the "Jesuit century."[16] This literature includes ethnographies, maps, and resource studies prepared as part of the administrative structures of the religious world, since for both the Spanish and the Portuguese, extending the faith was by definition expanding the empire.[17] The ecclesiastical works were used widely by the other tiers of travelers and traders, who frequently sought shelter in the missions as they traveled through the "uncharted Amazon," usually with a map showing the location of the next mission.[18] Eminent scientists like von Humboldt and Herndon regularly followed the mission routes, since these were places that were reliably secure, and the fathers knew native languages, were able medics, and could provision and provide labor and guides. They were often intermediaries at the nexus of the traveling worlds of labor, local administration, and international ambition.

Tropical Tourists

There were phalanxes of what we might today call adventure tourists, part of nineteenth-travel fashion, who toured in edifying searches for masculine adventures that would fuel their intellectual and artistic enterprises and round out their personal experience. While some of the young European intelligentsia rushed to Damascus and Tangiers, others set sail for Brazil or Peru, trading the opulence of the Orient for the exotica of the tropics. Gustave Flaubert and George Gordon, Lord Byron have become almost clichés associated with this sort of travel in the Middle East, where they would gaze on Egyptian or Ottoman ruins and acquire the precious artifacts, cultural commentary and revealing anecdotes that became the raw materials for "Orientalism." Von Humboldt and Bonpland ruminated on whether it would be better to accompany Napoleon's Nile expedition, which traveled with a battalion of scholars who would document every aspect of the country from its plants to its pyramids. Finally they opted for the offer of the Spanish crown and went off to the New World.[19]

The young aristocrats von Spix and von Martius, under the aegis of the king of Bavaria, set forth (1817–20) to gaze on nature, experience hard travel, and make collections and measurements of all types as they mimicked the style if not the substance on von Humboldt's travels.* Barrington Brown and Lidstone, agents reviewing rivers and wood availability for the Amazon Steamship and Navigation Company, had suitable breeding and natural history training for such enterprises, and one can imagine the two pondering

* These two young noblemen were able collectors, amassing more than sixty-five hundred botanical specimens.

whether, as for von Humboldt before them, it would be more amusing to stake out the Nile or the Amazon. Others like Robert Avé Lallemant, or plantation manager Daniel Kidder steamed up to Tefé and back down the great channel to Belém in the great steamship circuits, making routine observations on the flora and fauna. Even the US environmental icon John Muir, frail and aged, was not immune to the allurements of Amazonian tourism, producing what were among the more banal observations on Amazoniana and the weakest prose ever generated by this otherwise lyrical and insightful writer.[20] Even Muir knew his material was pedestrian; he constantly complained to his editor that it would be difficult to pull much out of his trip. He liked mountains and didn't really know what to make of the sweltering landscape.

Travelers drifted through the main channel, visited enterprises, and wrote of their voyages and the marvels they observed, repeating the social nostrums of the day about the tropics and its people and making some commentary on the cuisine. This type of literature seems to have its own endless DNA and reproduces itself with mostly predictable vignettes of unusual fruits and animals, moving and picturesque moments, the charms of the native women or lack thereof, the close shaves, and, as often as not, how good the shooting was.[21]

Native Intellectuals

In Amazonia an array of homegrown scientists, surveyors, and chroniclers as well as armies of local inhabitants supplied the local knowledge of rivercraft and the access, logistics, and information on which the successful outcomes of expeditions rested. All European travelers depended on an "invisible" set of native intellectuals who made it possible for them to travel in these zones: they arranged for the labor that would haul these explorers and their gear; they were the scouts and the *mateiros*—knowledgeable woodsmen—who would do the hunting, navigate the treacherous waters, explain the landscape, and carry out most of the collecting. In the case of Alfred Russel Wallace, Luis, his black cook, was one of his main botanical and ethnoscientific informants. Everywhere Wallace stopped he pressed locals into the service of collecting for him.[22] These local guides knew the natives (and often were at least part indigenous themselves), the native languages, the entrepôts, and how to keep an often fairly hapless bunch of collectors alive, so travelers could dry their plant samples (often identified by these same companions), fill bottles of alcohol with their animal specimens, and dream of the sale of their collections and lucrative speaking tours. These local guides and interlocutors, often possessed of great knowledge of natural history, were only obliquely

Figure 15.2. Apatú, Henri Coudreau's Amazon guide.

acknowledged or not at all, even as their skills provided the logistic and intellectual foundations that underpinned European "discovery." Graham Burnett has explained the quotidian reliance on such natives by Richard Schomburgk in his expedition, his research, and his colonial recommendations, all enabled by vast indigenous traditions. Von Humboldt describes in detail the roles of bearers and native informers in his Amazon efforts;[23] Bates relied on his collector Vincenti.[24] Coudreau depended on the *quilombola* Apatú as his guide through the interior between the Amazon and the Orinoco. Wade Davis, in his biography of Richard Schultes, describes movingly how important these local informants became as Schultes's friends, companions, and intellectual muses.[25]

Collecting was one thing, but often the most arduous and treacherous work in these travels was navigating the rivers and tributaries. This was generally carried out by Afro-Brazilian boatmen in the eastern and Caribbean Amazon and tributaries of the Madeira, perhaps a consequence of West African resistance to both *Vivax* and *Falciprium* forms of malaria in these famously fever-ridden rivers. Often members of *quilombos*, they had a long history of occupation in remote interiors and headwaters and as interlocutors and traders on the major rivers that made their skills and detailed knowledge of the rapids and the cataracts incomparably valuable.[26]

One of the Brazilian guides for several distinguished expeditions in the Upper Amazon was Manoel Urbano, a *cafuz*—a product of black and Indian parents who traveled with Silva Coutinho, Chandless, Agassiz, and James in the 1850s. Chandless, the exception that proved the rule, was to state that his entire exploration merely reprised Urbano's remarkable earlier travels on the many rivers of the "land of the Amazons." Even Agassiz, who was habitually dismissive of blacks and mestizos, was to credit him with a "rare intelligence."

Figure 15.3. Other travelers of the diaspora in the gold fields of the Guyanas.

Travelers of the Diasporas

Indeed, Urbano was an example of another form of traveler, the outcome of the huge diaspora (and holocaust) that convulsed the region. The surge of population into Amazonia was a death knell to many native groups, and the use of terror slavery and relocation of native populations as labor was equally devastating.[27] There were also the runaways thrown into the maelstrom of Amazonian revolt by the Cabanagem and the conditions of slavery. And as we have seen, there was a vast flood of refugees from the human and climatic harshness of the Northeast. There were other less-known diasporas—for example, those fueled by ethnic strife and political instability in the Levant with the unraveling of the Ottoman Empire, which sent Syrian, Lebanese, and Jewish traders into the ambit of the rubber world.[28] The large number of wage workers who flocked from the United States, every part of Europe,

Figure 15.4. Manoel Urbano, genius loci and guide to Labre, Coutinho, Chandless, Agassiz, and James.

the Mediterranean, and the Caribbean as well as China and Japan to work on the Madeira-Mamoré railroad spoke of global political instability.[29] These unknown millions largely escaped the notice of more elite travelers except as comic relief or local color, although da Cunha would frame them as his protagonists in the machinery of imperialism and as national heroes for his "Lost Paradise."

Analysis of the role of scientific exploration has largely focused on how these excursions shaped scientific efforts in remote landscapes and transformed the imperial countries in terms of ideologies (ideas about race and environment), scientific paradigms (the rise of new classification systems, the theory of evolution), and theories of human history.[30] Because New World scientists had seminal roles in the construction of modern knowledge systems through their taxonomic, evolutionary, and ultimately ecological contributions, literatures that were keenly focused on the imperial ventures and the continental-scale diasporas within the global economy as well as active processes of nation creation perhaps merit more attention than they had received thus far. These themes were exactly da Cunha's subject.

From the most aristocratic or scientifically sophisticated traveler to the most reviled native slave, the Amazon seethed with people moving on its rivers and through its forests, though often noted by outsiders as mere wisps of smoke rising from the forest, glimpsed from the deck of a steamship. The various journeys can be seen as a kind of fugue of economic and imperial practices: They involved new economies and systems of resource use. There were new labor regimes. New bureaucracies came into being, systems of proconsuls and militaries rose from the tropical ooze—all of this mediated and midwifed by steam travel and new technologies in the metropole. Underpinning the surges of population into the vast outback was the massive outflow of coveted latexes, products that put the region at war and had drawn our particular group of travelers, the imperial fact checkers Euclides da Cunha and Alexandre Buenaño, into the fray.

16

In Hostile Territory, Part 1

Official Report of the Joint Boundary Commission

The Report of the Binational Boundary Commission

The description of the journey is best left to Euclides da Cunha himself, the main author. What follows is the "official story" redacted as an inoffensive diplomatic document jointly written by Buenaño and da Cunha in Manaus, although, given da Cunha's authorship, it was not without its ironies. A far more scabrous tale unfolds in his personal communication with Baron Rio Branco, the unofficial story, which follows in chapter 17 in his "Purús confidential" report.

> Carrying out our exact orders, the survey commission met in Manaus, verified each other's titles, coordinated our chronometers, and extended our stay until the 5th of April, when the Joint Commission for the Reconnaissance of the Upper Purús set out in search of its destiny. This necessary delay was occasioned by the exceedingly belated arrival of our instructions: these were received only a few days before our departure. The time spent in Manaus was dispiriting, calling into question whether we

Figure 16.1. Officers of the Brazilian expedition. Da Cunha is third from the left at the back.

could even carry out the voyage with which we were entrusted within the time limitations. The setback vastly increased the obstacles: on so long a river when the channel drains, conditions for steam navigation decline even as the distances that we would have to travel by canoe would increase. In spite of this, we took advantage of the time to prepare the vessels in the best possible way—and both commissions, wedded to an exact and speedy accomplishment of our task, were ready at the same time to proceed, together, with what we could achieve of the preliminaries of our instructions.

We left at the most inopportune time, exactly when regular navigation to the Upper Purús would cease because, as is well known, steam travel is at the mercy of the flooding of the river from November to March and the draining outflows that follow from April to November. In spite of this, the ascent to the confluence of the Rio Acre was uneventful, even though excessively slow.

The entire commission reunited at the confluence of the Acre and Purús Rivers at 7:00 a.m. on the 9th of April. The two *comissários*—head officers of the Peruvian and Brazilian missions—agreed to the general procedures that they would adopt to begin their tasks, including part of the memo drafted at this time.

It was necessary to navigate night and day, only carrying out the hydrographic measures in the day; sections that were traveled at night, even though not registered, would be included in the reconnaissance upon the return. This was a means, essentially, of making up for lost time by taking advantage of the remaining floodwaters. In any case, we were still in the best-known regions of the Purús, and we needed to do what we could to attain the most distant points of the headwaters, the essential object of our mission. These general principles would shape our future studies and would be, as in fact they ultimately were, constantly modified by circumstances and our greater experience.

Thus, from that point on, the Peruvian Commission, whose single steam vessel made travel much easier, began uninterrupted reconnaissance and was for a short stretch accompanied by the Brazilian Commission, as it had contracted the streamer *Tracuá* to tow the *batelão*—cargo canoe—*Manoel Urbano* on the 13th of April in the village of Boa Vista do Bacuri.

Up to this point the voyage was extremely slow. It improved later when the two Brazilian launches sailed tied together in order to establish greater uniformity in the speed of travel and to better verify the ongoing research to which we have already referred. Unfortunately, the steamship that was towing the *batelão* rammed a submerged log and was in danger of sinking, which contributed to more delays in our travel, such that only on the 5th of May, exactly one month after our departure from Manaus, could we continue to the mouth of the Acre for the headwaters.

There we took advantage of a stop for three days, where we carried out the coordination of the chronometers, as well as the initial observations of the regimes and physical character of the rivers. As these efforts required a protracted stay, we agreed (because the instructions required only a quick survey of the Lower Purús) to begin the coordinated observations and other details only after the Acre River. There was also the basic reality that the area we were traveling through had already been amply studied, and there were well-determined points farther along that would permit the calibration of our chronometers with greater accuracy, even subject as they were not just to ordinary causes of variation but to other accidental effects and special conditions that we would experience on the voyage.

We coordinated our equipment, and in light of the rapid fall in the water level and the lower volume of the Purús once it had lost its main tributary (the Acre), it was established that we would travel only in the day, given the perils of moving upriver at night due to the submerged logs that began to appear in ever greater number—the river was awash with them.

At the same time we agreed on a code of signals so that the two parts of the commission could communicate easily and as circumstances dictated. And so the passage proceeded without notable incident except for the required stops for buying firewood and the various, ever greater precautions that were needed to evade perilous collisions with the logs and debris that increasingly congested our channel. To better coordinate our navigation and the survey, which depended in part on a certain regularity in our advance, we tied the two launches *Cahuapanas* (the Peruvian boat) and *Number 4* together; these were accompanied by the launch *Cunha Gomes*, along with the *batelão* that was being towed.

On the 11th of May we reached the mouth of the Iaco, where we encountered the *Neptune*, the last steamer to make it down the river, fleeing the severe depletion of the river waters. Navigation soon became extremely irregular, requiring constant soundings and stops, not just because of the logs, which were excessive and prolific everywhere after we left Novo Destino, but also because the shoals of hardened laterite called *torrões* (clay shoals) and *salões* (deep pools) were continually making our passage more uncertain. At the villages of Teruan and Catiana, the *Cunha Gomes* was beached on these shoals.

We had anticipated the end of steam navigation at the mouth of the Chandless, after which we knew we would not be able to take advantage of our fleet because even the minimal draught of the vessels was too great for the shallowness of the waters. But exactly on the day on which we hoped to arrive at the confluence, when we found ourselves in the curve of São Braz, a disastrous accident happened and changed the entire course of the expedition.

The passage at São Braz, unlike the others we had already crossed, offered the alternatives of becoming either shipwrecked or stranded—we could seek the convex side of the breach, which had the powerful current shooting over shallow sands, or we could navigate on a chute of the concave part of the banks, where the advantage of greater depth was completely annulled by the profusion of logs whose threatening trunks could be avoided only with great difficulties. In weighing the dangers, we naturally preferred the latter case. This was how the *Cahuapuanas* and *Number 4* arrived safe and sound through the dangerous section, although not without running aground on the sharp curve of the meander, from which they escaped only after much maneuvering. Luckily these were still trying to get free of the shoals, when the *Cunha Gomes*, traveling slightly behind and towing the heavy *batelão*, approached the chute, where navigation was extremely precarious due to the churning violence of the current and the submerged logs that crammed the river. In spite of this, the *Cunha Gomes* got through this channel without incident. But then as it traveled through the oxbow (and in spite of a preliminary sounding), it ran lightly aground on some sandbars.

The vessel backed up in order to dislodge itself, which it did easily. But as the

Figure 16.2. Da Cunha wrote: "The *Manoel Urbano* just before sinking. Transferring the cargo."

current was very intense, the launch should have advanced quickly right after backing up, thus overcoming the current so that that its heavy towload would not smash into the underwater logs that were still close by. This did not happen, because the engine failed at exactly the moment when it should have fired most powerfully. Instead the launch, with the *batelão* in tow, careened vertiginously over the masses of piled logs on the left margin, which slashed our *batelão* in a swath of 3×7 m. The launch was dragged by the current and soon carried a short way to collide with an enormous log of *Cumarunarana*, where finally, wrenched apart and wedged in the sand, it began to take water and sank in an irretrievable shipwreck.

Greatly helped by the crew of the Peruvian boat *Cauapanas*, who backed up our efforts with maximum dedication. After the heroic labors appropriate to such occasions, we managed to save little more than half the rations, and at least there was no personal disaster to lament.

The Joint Commission was thus completely immobilized one day before reaching the confluence of the Chandless, not just from the disaster but because the launch *Cahuapanas* for its part also could not get afloat. The commissions had to reorganize and at the same time develop new methods that would enable the completion of our assignment.

The Brazilian Commission was thus sharply reduced, as its members were now much too numerous for the resources that had so abruptly declined by half. I arranged with Sr. Buenaño, the head of the Peruvian Commission, for us to use our little launch to the best advantage to transport our supplies, and proceeded forward towing the rest of the food after dividing it among those who would stay. The adjutant, the quartermaster, and the secretary and the crew of the *Cunha Gomes* remained in São Braz. The measures we agreed on were ready: the Brazilian Commission would include da Cunha, an aide, the doctor, a sergeant, and eleven soldiers and laborers and would continue the voyage to the Chandless. On the 23rd, we met up with the Peruvian mission, diminished only by the crew of the *Cahuapanas*.

When we next reunited on the 25th of May at the mouth of the Chandless, the commissions agreed to measures that the situation demanded: among these were communication with our respective governments, providing them with a real picture of the emerging conditions and, given the unforeseen character of these problems, information that might justify the development of new instructions. It was an indispensable step. The news about the state of the river ahead was extremely disheartening. The Purús was at low water and largely blocked to navigation. A few miles ahead, provisional barracks were set up by the stranded Peruvian-Brazilian administrative commission. Three steamships, the *Santos Dumont*, the *Phénix*, and the *Cassiana*, had run aground not far from us, trapped in the sands. Every day, canoes and *montarías* descended the river headed for Manaus, their crews and passengers united and unvarying in their assessments of the unfavorable level of the waters. All this justified an urgent communication that was entrusted to a subaltern of the Brazilian force, who on the 26th of May left for Manaus, also charged with acquiring more supplies.

In spite of disaster, our communications were merely preventative; we were not contemplating stopping or going back, but rather thinking about how to proceed although now using canoes and small *batelãos* for our flotilla. We did not delude ourselves about the difficulties awaiting us, but when we examined the map, we noted that we had already advanced a great deal. We were about 1,500 miles from the mouth, or about three-quarters of the way up the Purús. There remained still the southwest route just a bit more than 2 degrees longitude and less than 2 in latitude: a traveling distance of some 450 miles. But our new means of travel, imposed on us by the recent events and dictated by the stage of the river, made our apparent proximity to our objective completely illusory. We would, after all, have to achieve our goal against the current.

In fact, traveling at a velocity of about five miles a day (and this would be no small feat given the tasks we had to carry out—tasks that would increase to the same degree as we advanced into the unknown—we concluded that only with ninety days of forced travel could we arrive at the headwaters. And thus we set out on this extensive journey, leaving the confluence of the Chandless at noon, at a pace that totally counterposed in its slowness the ever expanding enormity of our route: May 30, 1.7 miles; May 31, 4.4 miles. June 1, 5 miles . . .

This sluggish pace was mostly related to the procedures that we had adopted for taking hydrographic data, where the orientations we took with the compass were linked to the indirect distances we were taking with the Lugeol lens, which forced us to stop at all the inflexions and meanders of the river. If we continued with this system we would extinguish our supplies well before our objective. We modified it, substituting the indirect measures of the lens for those we obtained by measuring the velocity of the canoes, and by repeated base measurements taken along the beaches most apt for these operations. Thanks to this change, the speed of our ascent grew, accelerating even more as we reached the headwaters.

On the second day of June, at one in the afternoon, we arrived at the outpost of Refugio, where the Peruvian administrative commission of the Neutralized Zone was camped under the command of Colonel Manuel Bedoya; they were stranded there because their launch, the *Phénix*, had run aground. The next day after a rapid advance through the trading posts of Triunfo, Velho, Porto Mamoria, and Cassiana we arrived in the evening at Novo Lugar, where for the same reasons the administrative section of the Brazilian Commission (directed by Borges Leitão) was also marooned: the *Santos Dumont*, coming from Manaus, had also foundered.[1]

A major halt was in Novo Lugar, where the Brazilian Commission only decamped on the 7th of June (this delay was necessary because of transfer of thirty trunks of material from the *Phénix*). The Peruvian Commission advanced two days ahead, traveling slowly, and waited for us en route. At Novo Lugar there was a raging epidemic of beriberi, a circumstance aggravated by illness of the doctor of the Peruvian administrative commission, who died a few days later. The situation was such that the Brazilian head of commission, heeding a request from Commander Borges Leitão, agreed to let the Brazilian doctor stay there.

Our travel settled into the pitiless regime that we imposed on ourselves in order to complete our mission: our days invariably began at dawn, with short stops for meals, and only ended at sundown. We generally camped next to each other on the same beach, Brazilian and Peruvians, and thus stimulated by reciprocal example that never degenerated into discord, this had as its only consequence an exceptional travel speed, which we had not foreseen. In fact, at the end of a few days we broke camp in the first light in order to read the compass and advanced on until night. Meanwhile, adept with barge poles, the crews of the canoes defied the river that every day demanded ever greater care and greater efforts, as the growing dangers of submerged logs and extensive sandbars often required portage of the canoes. We owed the speed of our travels to the unvarying mutual inspiration of our crews in spite of the periodic stopovers that the nature of our work imposed.

The commission reunited again on the 9th beyond the hamlet of Funil and then proceeded on to Sobral, where they arrived on the 11th after having passed the settlements of Cruzeiro and, on the 8th, Hossanah, an abandoned Peruvian trading post, and on the 9th an improperly named "Furo de Juruá," a stream whereby one passes to the *varadouros* that link the Jurupari, an affluent of the Tarauacá.

After Sobral, the last Brazilian trading post on the Upper Purús, navigation further deteriorated, with ever more collisions with logs and constant running aground on sandbanks or shoals. On the 13th we arrived at Muronal, the first Peruvian trading post of the Upper Purús.

Happily, no serious illness had appeared in our encampments:[2] our crews were strengthened by the arduous regimen to which they were subjected, and by the palpable improvements in the climate in spite of sudden swings in temperature, where days of searing heat would be followed by freezing and humid nights, which at times made taking the measurements excruciating, in spite of the clear, serene skies. On the 14th of June we had to decamp at three o'clock, breaking with our preestablished program. The morning broke cold and was followed by a torrential downpour that awoke both camps and, with tremendous blasts, wrenched apart our shelters. Counter to what one might have expected, instead of rising temperatures, they plummeted throughout the day. Starting at 24 degrees Celsius at 9:00 a.m., the temperature plunged, so that the next morning the temperature was 13 degrees, absolutely anomalous for those latitudes.[3]

On the 16th of June we passed some abandoned Peruvian hamlets (União and Fortaleza) and arrived on the 17th at a *tambo* of Peruvian *caucheiros*, Sta. Rosa, on the confluence of what the Chandless chart calls the Curinãa. We continued on the same day. Between the settlements at Sta. Rosa and Cataí the region is apparently deserted—only *caucheiros* work there, prisoners of the forest, and only some abandoned farm and garden plots reveals old villages or trading posts. We passed through the region in a bit more than four days, reuniting on the 22nd of June in Cataí, the seat of the Peruvian-Brazilian customs administration. After this stop we arrived on the 25th at the hamlet of San Juan, inhabited by Piro Indians and Peruvians from Loreto who extract *caucho*. In this entire area the shoals and the emerging logs became inconsequential, no longer causing alarms or tribulations as they had at first.

By the 25th, the Brazilian Commission was reduced to just nine people including the chief and the auxiliary engineer, because five prisoners had to be sent back to Cataí, soldiers who had revealed themselves little inclined to obey the

Figure 16.3. The Peruvian Commission on a hard haul on the Cujar.

orders given them. In spite of this diminished personnel, the speed of our ascent did not decline appreciably, which continued on until we arrived at the Curanja River, on the 28th in the afternoon. Curanja is the Curumãa of Chandless.

We delayed five days at this obligatory stop, where for the first time since the shipwreck we again compared chronometers of the two parts of the commission. There we stayed until the 3rd of July, principally to carry out the indispensable observations necessary for the calibration of our chronometers, taking advantage of the site and its definite coordinates. There we confirmed with completely trustworthy information the predictions made in Manaus about the unfortunate timing of our departure and the disquieting obstacles ahead of us. It was too late, however, to return, and united in the same thought, we resolved to continue our upriver trek, which we had began on the 6th of July.

In contrast to what we expected, the natural difficulties did not particularly increase, becoming in fact even less obvious even after the loss of the flow of the large tributary of the Curanja to the Purús. Our journey continued on with its primitive momentum, as you can verify by a simple review of the distance between the stopovers we made: 10 July, we passed Sta. Cruz, 11, Cocama, 13, Independencia, 14, Shamboyaca, 15, Tingol Leales, 16, Kaki, 17, Ordem, and finally on the 18th, Forquilha do Purús, where the hamlet of Alerta had been erected, the most prosperous post on the whole southern part of the river. Then, even though suffering through a series of difficulties on the ascent (especially serious for the Brazilians, whose supplies were meager in the extreme, having not yet encountered a place to resupply), the commission resolved to complete the upriver travel, continuing the next day up to the headwaters via the Cujar River.

Understand the difficulties we had to conquer in this advance on one of the last forks of the great river, exactly at the moment of its lowest water, and in addition, due the geology of the landscape, there was but one channel, with interminable and successive rapids that churned through it. You can judge for yourselves the effort expended up to July 30th, when at night we once again joined the rest of the commission at the confluence of the Cavaljani, the last of the dichotomous branchings, so characteristic of the Purús.

We were finally at the point of the great river where we advanced to places that had never been scientifically explored. In fact Wm. Chandless, with his prodigious tenacity, only came up to here but did not proceed in the direction that we were about to embark on. He advanced on the northernmost branch and up just a few miles, while we would proceed in a path that inverted directly to the south. This circumstance contributed in no small way to our renewed excitement. It was really but a long stretch of the Purús, certainly well known by all the local *caucheiros*, but it hadn't appeared yet in the geographic sciences, and we noted the first and perhaps only mistake of the illustrious Chandless: he traced the Cavaljani with a completely false direction, from the east to the west.

This little tributary was, however, almost totally lacking in water and needed other means for the ascent. The Peruvians took advantage of the small *ubás* commandeered from the mail delivery from Iquitos, whom we encountered at our last encampment, and this permitted them to cede one of their old *ubás* to the Brazilian Commission, since these were much more appropriate to the river conditions than the heavy canoes of *itaúba* wood in which they had been traveling. But even with these modifications, the journey was made only with extraordinary exertion. Except for very few sections, in the backwashes that were scattered through the river course, the boats mostly had to be hauled over the sandbars with an arduous dragging until we reached the confluence of the Pucani (August 3), the southernmost branch to the Purús. It was sometimes necessary to cut staves of a special plant called *cetico** and place it over the riverbed, and then shoving the boats upon them we traded poles and oars for levers, slowly, painfully rolling our boats gradually upstream through long sections of dry channel. In this

* This is a species of *Cecropia*, an early successional plant in the New World in general and on the banks of river terraces in Amazonia.

way, the short itinerary of 20 km required three and a half days, which is about three miles a day.

Arriving on the 3rd of August at the confluence of the Pucani, which clearly is the most southern of all the sources of the Purús, we did not delay our exploration of the *varadouro*, which we did easily on the 3rd, the Peruvians on the 4th, and we returned at once, with a speed imposed by the lash of hunger, to Forquilha, where, reunited again on the 10th of August, the two parts of the commission agreed about the execution of the last part of their expedition, the ascent of the Curiúja River.

The outflow of this river was in its most intense phase, and it was clear that no boats other than the *ubás* would be able to navigate its diminished waters. The same thing happened as occurred before with our trek up the Cujar River: all concurred that the ascent was impossible, but unlike with the previous river, we soon saw that with the rapid drop in water levels no victorious ascent would be forthcoming, And in any case the food supply of the Brazilians was frankly completely depleted: in that locality one could only find a bit a of manioc, and on this one could only subsist for a short time, as it was inadequate as the only foodstuff.

In spite of this, with the last bit of our remaining strength, the Joint Commission left for the last little section of the river that remained to be discovered on the 14th of August. The Brazilian Commission, faced with the serious problem of feeding its starving personnel, had rations for but five days at the most, so they set forth with the intent of a swift advance and equally rapid return. This was the only solution to the painful and irredeemable conjuncture at which they found themselves.

We could only press forward if open navigation on the Curiúja were feasible, and it absolutely was impossible in that period. The low-water river periodically divided itself between extensive swamps, dissected at times by the sandbars that abounded and either dominated the riverbed or opened into narrow chutes that pressed against the bluffs and river terraces; these brought into sharp relief two deplorable realities: the fading of the last energies of the exhausted explorers, and the enforced slowness of travel, which should have been rapid in order to safeguard their lives.

From the very first hours of the first day of that journey it became clear that the necessary speed would be impossible, and so the Brazilian Commission returned, since it was materially impossible to continue a journey that in the best hypothesis would need at least at ten days, that is, twice the time their supplies would last. Given that the Peruvian Commission had left most of its supplies at Curanjá, it had only those necessary to arrive at the *varadouros*, and based on the

Figure 16.4. Peruvians with a tapir.

assumption of losing part of these rations in the obviously dangerous navigation, circumstances did not permit offering to share these with their Brazilian colleagues.

Thus, with everything so impossible, the Brazilian Commission began its return and worked on its report, and following the instructions that tasks carried out separately would lack official weight, turned in the direction of Manaus, always with observations and surveys that would support the efforts that had been made on the way up.

Happily, the part that was left unstudied was neither large nor of great importance, as it referred to the *varadouros* of the Curiujá, recently opened by the *caucheiro* Carlos Scharff. This section was of no practical consequence, for beyond the difficulties of navigation of that river, there were other impediments: the *varadouros* traversed extremely steep lands with so many obstacles that the best one can say about it was that it was abandoned.

We think it is necessary to explain what a *varadouro* is. This is the name that is applied to pathways that are quickly opened and pass from one river to the next in the shortest possible arc, and often greatly reduce distances by linking segments of the same river. The *varadouro* should offer the advantage of being relatively flat, at least in the areas we traveled, so that it allows the latex extractor to transport boats and various types of cargo. This occurred with the Cujar. The traveler who passes through it moves from the waters of the Ucayali to the Purús, and vice versa, and continues traveling in the same boat that he dragged through the isthmus. However, only with great difficulties could this occur on the Curiúja, and so it completely lost its importance. It was abandoned in favor of the *varadouro* of the Cujar, somewhat farther to the south, now the preferred crossing. Happily, that small section that still remained to be explored is subject to the most secure and detailed information, and had no importance in the larger culmination of our work. The commission returned definitively to Manaus, where they arrived the last days of October. There they dedicated themselves to the labors of the assignment, adorning their observations with the results that we will now outline more concisely.

This amiable ending of joint and happy triumph and manly solidarity was hardly the story da Cunha would recount in a less public document. This is the subject of the next chapter.

17

In Hostile Territory, Part 2

Ex-party Report from da Cunha to Baron Rio Branco

Senhor Ministro:
So as not to miss the first secure mail that offers itself, here is the quick report that I have the honor to send to Your Excellency. It naturally has all the imperfections of a work hastily realized, but has no other end than to give Your Excellency an outline of our efforts. I will have the opportunity to complete and amplify the study in all its points later. Given my hurry, I also ask that you forgive me for not always having subordinated myself to the norms of official redaction.

The map that travels with this missive will diverge little from the definitive one. Unfortunately, as much as we tried, we could not extend it to Lábrea without undermining the overall quality of the work. It is at a scale of 1:100,000 and very useful for addressing your main concerns. The enclosed photos are just samples that we took in great numbers and that we will send to Your Excellency at an opportune time. Also, in our next communication I'll send the detailed statistics that I produced on the

Figure 17.1. Da Cunha leaping from his canoe onto the beach, wearing headgear to protect him from gnats.

Purús between Barcelona and Sobral, which like other clarifications we have left out for lack of time.

I very respectfully greet your Excellency in the name of all our expedition companions, and with greatest consideration.

Euclides da Cunha

Field Report*

Senhor Ministro:

On my return [to Rio] from the Upper Purús, whose reconnaissance I carried out, I will expound on the key facts that occurred during our travels, but at the moment I have to attend to many tasks, so I will limit myself to those topics that most merit attention and which will be detailed or clarified later. During the trip I had the honor of sending Your Excellency many communications. These were, in general, incomplete because of the haste with which I wrote them and the absence of the guarantee of a seal on the letters, which had to pass through many hands prior to arriving in Manaus.

I will remind you of information that I have already presented to you in those missives, and I begin this exposition with our departure from the confluence of

* This section includes three da Cunha fragments that pertained to this journey but were not published as part of the report. I have added an endnote for each to indicate the insertion and provide a full citation.

the Chandless—up to that point I know you were completely informed of our situation, thanks to the official letters I sent to you with a secure messenger.

We began our departure on the 30th of May at midday, and in spite of the great difficulties that were easily foreseen, and the delays, we were excited. My commission, as I communicated at the time, was reduced by circumstances to just fourteen men (the head of the mission, a helper, the doctor, one sergeant, six soldiers, and four workers), while that of Peru, which was complete, included among its officers twenty-one men, not counting the laborers. This disparity seemed to me without importance. The relations between the two commissions seemed very cordial: we had traveled out in total harmony, and thanks to this circumstance, we expected a successful journey, no matter how daunting it might first have seemed.

Only on one point was there frank disagreement, the crossing of the *varadouros*. This culmination of all our efforts was, even in Manaus, deemed impossible by the Peruvians. Better informed than I by the patricians of the Ucayali and Upper Purús, Sr. Buenaño had for several months expounded on the serious obstacles that would require that we turn back. It was an impressive inventory ranging from the perils of the rapids and cataracts and the impassable routes to the fierceness of the indomitable Campa Indians. These future impediments took on such menacing contours that on the eve of our departure he launched the idea of a memorandum signed beforehand by the two head officers. By explaining the countless hazards and obstacles, this memo could explicate and attenuate the implications of the inevitable retreat. I confess that in the beginning I was frankly in favor of this idea, but I rejected it later—I rejected it precisely when we came together to work out the memorandum. I explained my attitude, declaring to the head Peruvian officer (*comisário*) and the other colleagues who were present that even though well intentioned, our memo could inspire negative and prejudicial comments whose motives and repercussions we could not foresee. I do not regret what I did. Unfortunately, the incident seems to have displeased that officer (Buenaño), who made constant references to it, making clear his disappointment in veiled commentary. And always insisting on the impossibility of reaching and crossing the *varadouros*, so that as we left the confluence of the Chandless, this was the single source of discouragement.

But we advanced, engaging in our hydrographic work with the Lugeon lens for monitoring distance and the compass for direction. This process implied obligatory stops that we could carry out for only a brief time—at the end of three days we had to abandon it because it implied a pace—three or four miles a day—completely at odds with the length of our trip. The accuracy of the instruments themselves could not justify such a delay. By mutual agreement, which was how we always proceeded, we changed the procedure—instead of using the distances provided by the lens, we measured instead the speed of the canoes, calibrating the rate by direct measurement of the beaches that we moved along. As Your Excellency will see, this method produced admirable results, far superior to the rough character of the technique.

We thus continued along in relative harmony, to which I contributed more than my colleague (Sr. Buenaño) because I made the constant sacrifice of listening to his endless complaints and bitter grievances that alluded to our military successes in this zone in the period 1903–4. I tolerated them not only out of respect for those for whom he mourned but also so as not to disrupt or destroy so much effort through an argument whose consequences could lead to a frank rupture. To avoid this at all costs was for me a major consideration, and this stance, where I felt that even the slightest reference could not be disputed, indicates very clearly that I carried this posture to its maximum limits, to the point where I really had consider my own self-respect.

In any case, we arrived at Novo Lugar—where the administrative commission directed by Captain Borges Leitão was provisionally based due to the rapid decline in river flow—without the most minor incident altering our relations or our work. During the day this consisted of uninterrupted reconnaissance, and by night observation of the skies to ascertain latitude and longitude. Our chronometers were damaged by the shipwreck, so we limited ourselves to the first activities, waiting until we arrived at Curanja, where, based on the coordinates developed by Chandless, I could make a judgment about their general state and daily movements. This was the only alternative under the circumstances—our rudimentary system of canoe transport meant we could just barely carry our provisions, so I could not bring our heavy astronomical theodolite, which under favorable conditions would have generated absolute longitude. Instead, we carried the sextant, a nautical instrument somewhat strange for engineering, but which I only used on the one hand because approximate coordinates satisfied the terms of our orders, and because we recognized—the Peruvian chief and I—from the beginning that for astronomical determinations the accuracy of the data furnished by Chandless was of inestimable value given the ongoing imprecise calculation of locations that is the nature of all practical astronomy. Besides this, more valuable than the data from Curanja, we had the coordinates for Forquilha do Purús, where I could accurately rectify our findings.

We arrived at Novo Lugar with no troubling incidents and continued on with friendly diligence until a fact, an apparently insignificant fact, revealed that my colleague had, in tandem to his scientific concerns, others at odds with his mission. There, while being presented to a neighboring landowner by Borges Leitão, Sr. Buenaño refused to shake his hand, to the consternation of all, given the circumstances. He later explained himself, excusing it in this way: he knew that the man had taken part in the skirmishes that had occurred there between the Portuguese and the Peruvians. Such a lack of discretion in a man of such exquisite manners soon revealed to me the depth of the passions that drove him and warned me to greater caution in order to preserve our future relations. In Novo Lugar, I waited for "x"* days for the upriver transport of some thirty cases of supplies dispatched from Manaus, which had been finally found in a warehouse arranged by the commercial house after the *Phénix*, which had carried them, ran aground.

The Peruvian commission preceded us by two days, and Sr. Buenaño promised to travel slowly so I could catch up with him. Thus, as we left Novo Lugar, we began our ascent once again, leaving Dr. Tomas Canduta, our doctor, at the request of Captain Leitão. On day___, I continued upriver alone to the rubber estate of Funil, where I arrived on the morning of ___. The Peruvians were very far ahead of us, so I disembarked for just a quick look at the place but was detained by a totally unexpected discovery.

Your Excellency knows that site celebrates our successes, which so inflamed this region: it was also there that many Peruvians had been shot. What is confirmed is that the corpses were not buried: they remained exposed until they were completely decomposed, and the bones of the victims remained scattered. Sr. Buenaño had docked with his troops, piously collected the remains of his compatriots, and left them a memorial. This action was most noble and would merit our most heartfelt approval, had not Sr. Buenaño profaned his act with the unfortunate impress of his obdurate hatred. There remained this improvident epitaph carved into a sheet of zinc:

F. La Fuente
F. Ruiz
D. Ocampa
P. Retagui
M. Montalban
Peruvians shot and burned by Brazilian Brigands

Considering these words—which I noted but promptly discounted—I immediately worried about the deplorable effects they might have in the midst of a population whose memory was still alive to the events of those days. I thought of removing the improvised gravestone, but as soon as they were aware of

* These were redacted in the original. Other dates and distances shown as blanks and dashes also were redacted.

this, the Peruvian *comissário* would dispatch a messenger.[1] And thus the Funil *seringal* would mark the end of our mission. There could be a severe rupture which I feared in view of its possible outcome, because upon the first outburst I would have at my side not only my traveling companions but also the numerous neighboring compatriots, which would give me incalculable superiority of force. But I understood that this was above all a disadvantage: we were still in areas populated only by Brazilians, and we had the might, but however much loyalty we might invoke in that emergency, there would be no lack of those who would perceive in this fact a treason, an assault that could compromise my country. In addition, I feared the disruption of the negotiations that I understood were on the table and of whose status I was ignorant. I understood also that one should not annul so much effort and expense by giving too much value to what might simply have been an error, a moment of bad judgment, infinitely below the honor of our country. . . .

Weighing all these, I decided to leave things as they were but to resolve them upon my return. I turned to the tasks at hand, to go forward, to advance as far as possible, and to surmount the myriad obstacles of the mission entrusted to me. But I continued under the specter of that epitaph. I caught up with the Peruvian Mission two days later—and when Sr. Buenaño quickly sought me out, alluding to what had transpired in Funil, I quickly countered that for me the subject was too painful and disruptive to the task we were trying to achieve, and I asked that he change the subject. He acquiesced, but I sensed that my attitude further aggravated his dislike of me, which he had had from the beginning, even though veiled with an affability that I always found highly surprising.

Thus we continued up to Cataí and Curanja. The trip became dreadful. Having but one helper meant that, however dedicated he was, he could not free me of my various tasks, which ranged from astronomical observations to the most detailed review of camp accounts. I had to manage our provisions, which due to the scarcity of food had to be rationed. My labors were further interrupted by the need to maintain order in the ranks, which was a sharp contrast to the correct foreign troops camped at our side. Arriving in Cataí, where I was racked and bedridden with fever, I reviewed the gravity of our situation. The remoteness of our current camp, the obvious scarcity of our supplies, the labors that we had carried out and those that lay before us, the depressing prospect of that advance into the wilderness whose objective now seemed almost obscure had visibly disheartened about half of my party. Noting all this, I dared not delay there. I continued on, ill, trading my bed for the prow of the canoe, where I meant to carry out the reconnaissance as we traveled upriver. But even as another attack of fever, more violent this time, prostrated me, at least our ascent was not slowed. However, on

Figure 17.2. The commercial house of Carlos Scharff on the Curanjá.

the morning of ___, a few days before we arrived at Curanja, disaster struck: the mutiny of five soldiers, whom I had to send back as prisoners, to Cataí.

In this way we arrived at Curanja, traveling in lands now populated only by foreigners and facing an immense and hostile region with just nine men—who arrived flayed, with mangled, bleeding feet, corroded by the sands. This was because we had had to haul the canoes innumerable times over successive sandbars due to the extreme declines of the river level. Besides this, I soon understood that even for our reduced number (we were nine in total), our supplies were very short—they would not last for two months. As if this were not enough, our boats were becoming ever more inadequate for the rapidly draining waters. They were the two heavy canoes of *itaúba* wood in which we had been traveling from the mouth of the Chandless. The tree trunks that had clogged the river ever since Novo Destino seemed ever more numerous, choking the straits, which barely permitted the narrowest canoes to navigate through them. During these trials, our chronometers, already battered from one shipwreck, continued to be further pummeled as we jolted against logs just slightly below the surface . . . and from Curanja onward, besides these, they would be subjected to blows incomparably more damaging from boulders which were increasing in size the more we advanced. The river volume of the Curanja, roughly equal to that of the Purús, foretold a doubling of efforts. The main river (the Purús), already so depleted, would be almost impassable when it lost that tributary. . . .

Your Excellency, you can ascertain from this sparse information the seriousness of our situation. In spite of this I did not think of turning back. But to continue, it was imperative that we use our time to best advantage. Any delay would imply two immense inconveniences that loomed larger each day: the lack of food and the drying river. So I resolved to start the next phase of our expedition even though the indispensable calibration of our chronometers really demanded a calmer and longer stay. Fortunately, there still remained ahead an intermediate point—the confluence of the Cujar with Curiúja, where we could do this. During the several days of delay in Curanja, I carried out observations that lacked the precision of those who dispose of the time to wait for clear skies and favorable conditions. As a result, our standard and that of the Peruvians were off by 18 seconds, and that to a large degree was a result of the vicissitudes though which our equipment had passed.

At Curanja, where we were very well received, we were more disheartened by news about the surrounding region we were preparing to enter. It was concluded that it was impenetrable, accessible only with the light balsa *ubás* of the *caucho* hunters, manned by domesticated Amahuaca Indians. Sandbars, logs, and boulders clogged the river, impassable waterfalls reappeared, while in the riverbed itself, vast floating meadows would thwart the boldest foray. The riverbanks were entirely impossible, and on top of these tribulations there was the hostility of the treacherous Amahuacas and fearless Campa Indians. Locals cited the recent homicide of an employee of the Casa Arana in the Cujar *varadouro*, and in addition to this verifiable case, others without number were invoked, further demoralizing us and convincing us that we would never arrive at the end.

But still we continued from Curanja to the headwaters. I confess to Your Excellency that we went forward intent on encountering the obvious, tangible impossibilities that would completely justify a return that I supposed would be inevitable. The Peruvian *comissário* , resolute in this attitude since we left Manaus, maintained his usual position: we wouldn't be able to cross the *varadouros*. But in contrast to what I thought, the navigation neither improved nor worsened. The Purús did not appear to have lost a tributary after the port of Curanja. It varied little in its width, maintained the same depth, and, inexplicably, increased its velocity. Our travel to Forquilha, where we arrived on ___, was achieved in ___days, and this was among the most rapid rates of ascent that had ever been done. The palpable change in climate—incomparably superior to the region we had just left—was no minor contribution to the difference. Even the plagues of mosquitoes, gnats, and chiggers that had so tortured us downstream were gone. The general regime had a tonic effect—there were no sudden changes in temperature nor the heavy humidity that we'd had to endure—so we soon felt invigorated and inspired for the trek ahead of us. Unhappily, in this section another incident threatened the cordiality that I and the Peruvian chief had struggled at all costs to maintain.

As you see in the attached drawing, Peruvian outposts dotted the upper reaches of the Curanja. Some, like Chamboyaco and Cocama, had the air of little villages. I found these places of great interest, and ever since the first one, Sta. Rosa (at the confluence of the Curinaá and the Chandless), I never lost an opportunity to go out and chat with the unusual people whom I came upon. Even on this point Sr. Buenaño and I differed. He had never gone into a Brazilian outpost, even at the last of these, Sobral, where the owner, at my request, went to the Peruvian camp to offer Sr. Buenaño the use of his house.

Lacking such antipathies, I was inspired on these visits by an eagerness to know this rough and interesting society. Thus, as I docked at Sta. Cruz, an outpost of the *caucheiros* (which I show on the map), I had my first disappointment and noted an animosity entirely discordant with the way I had been treated at other posts. As Your Excellency knows, the Peruvians have an almost mechanical gentility: they surround recent arrivals with smiles, salutations, flattery, and offerings as though reciting an old lesson by heart. They have, if you will, an automatic courtesy such that only the imperative of a powerful sentiment would make them neglect this reflexive habit of pleasing. On this occasion, what I had begun to notice in Sta. Cruz, and saw later disgracefully confirmed, was their incandescent hatred of Brazilians, which could not be masked even by their skilled gallantry.

I noted it in everything: the coolness with which we were greeted, the sullen responses to our questions, and even the astronomical prices placed on the most insignificant items for us. On the other hand, I do the Peruvian *comissário* the justice of admitting that he acted unintentionally and didn't know how to mollify this treatment. Having arrived and disembarked before me, they did not show the same attitude as at Sobral. Hardly had I been presented to the master of the estate than Buenaño withdrew under the pretext of having ordered the site to be photographed. I determined then not to go into any Peruvian villages without a preliminary invitation, to avoid further contretemps. Beyond this, I vowed to exhibit an intentional lapse in courtesy. In the long passage through the Lower and Middle Purús, Sr. Buenaño never saluted our flag, even in front of the barracks. I decided never to salute the Peruvian flags profusely displayed now at all points at the passage of the two commissions.

We continued on to Sta. Cruz, and following my earlier resolution, we traveled in a way that maintained the distance between ourselves and the Peruvian Commission, avoiding the embarrassment of arriving at a hamlet where they had to stop and I had to leave. But this came to pass at the settlements of] "Independencia" and "Cocama." Here at Cocama, the Peruvian officer, after a speedy trip, caught up with me in the evening. Avoiding the hamlet, I proceeded ahead again, only to be joined by the Peruvian Commission on the evening of ___ and our arrival at the Campa settlement of Cinco Reales. Both commissions were camped on the beach, and nothing revealed the slightest disharmony. At dusk I was visited by Sr. Zavala y Zavala, who, in the name of Sr. Buenaño, informed me that we were entering regions populated by savages—*infieles*—and made the suggestion, which I quickly accepted, that both camps organize sentinels. The next morning, we could not follow the Peruvians so early because our engineer, Arnaldo da Cunha,* had fallen ill. To not nurture false speculations I ordered this fact to be communicated to the Peruvian *comissário* (he was still on land)—and hoped that this courtesy would certainly indicate the my best intentions. But it produced the opposite effect. Sr. Sá,† who was irritated by the speed of our advance a few days before, now seemed to view in this delay some insubordination or weakness on my part—and wanted to put it to the test in one way or another.

There is no other way to explain the case, which I outline to Your Excellency, so that you can review the sources of our deplorable discord, which I deeply lament and which bothers my conscience, and with which I did not concur.

When I arrived at Cinco Reales, many of the Peruvian Commission were already there. Sr. Sá had with him a tame Campa, and thanks to this interpreter we could amiably deal with Venâcio, the chief who dominated those parts, radiating his influence and imperatives over the other native chiefs of the region. The village was very pretty with its vast banana grove unfolding and embracing a hill that towered on the right. With numerous Peruvian vessels at the port, the population crowded on the beach around the unexpected visitors in a most animated way, and it would have seemed odd if I, sticking to my earlier position, did not visit but continued upriver, thus seeming to devalue them and giving occasion for prejudicial interpretations. This I wanted to avoid.

I ordered the canoes to stop for a quick formal visit. I disembarked and soon presented myself to the head of the Peruvians, who even had the courtesy to introduce me to the leader, the "curaca," Venâcio. When, after a few minutes, I returned to my canoes in order to continue the journey, I was surprised by a vehement outburst from Sr. Buenaño, who inquired in a strident voice about my attitude in recent days, demanding that I explain myself and wanting to

* Euclides da Cunha's cousin.

† This is the shortened name for Nicolas Zavalos e Zavelos, the second lieutenant and subchief of the Peruvian commission.

know why I didn't stop in Cocama, in such a way that everyone there asked him whether we were fighting.

Overcome by surprise, I approached my interrogator and denied him the right to address me in that manner, as we were of equal rank. I said my words in a voice louder than his, avowing that the disparity in the size of our forces and the circumstances of traveling among foreigners would only give me more vigor in my energetic and swift response to whatever word or act clashed with the seriousness of the charge I had undertaken, or with the natural nobility of the Brazilian character. I could not proceed in any other way. I stood before my reduced troops, and any symptom of weakness would have completely demoralized them, extinguished my moral force, and in the end made impossible the greater efforts and sacrifices that were indispensable for our ascent. Our dialogue continued in the same sharp, quarreling tone for some time, and ended with explanations from both sides that seemed to assuage the earlier resentments. The two commissions continued forward together, leaving both camps absorbed in thought. At dusk, as we had become separated from each other and were late, Sr. Buenaño had the delicacy to place a lamp at the end of the beach where they were camped to show us the way. He returned to his old affability, and when we arrived at the confluence of the Curiúja, it was certainly at his suggestion that Carlos Scharff, the owner of the outpost "Alerta," sent one of his employees out to the beach to invite me to stop over there. After many repeated and insistent requests to stay in his house, I finally acceded. In accepting this kind reception I principally obeyed the proposition of avoiding any elements of discord. It was useless to belabor a refusal that might later figure in hostilities. . . .

This is what transpired in the village of Curiujá on the third of July. The pillars of the community offered a banquet to me and the Peruvian chief of mission. I accepted with pleasure; I was under the illusion of a sympathy that would vanish shortly. I went to the commercial house of Carlos Scharff, where the festivities were to be held, and which was managed by his bookkeeper, the German A. Scharff. I was soon surprised by a profusion of Peruvian flags in contrast to a complete absence of our own, in spite of the fact that it would have been easy for the host to acquire one at our camp.

Noting this fact, I thought to remove myself and waited for the first opportunity to do it without fuss or scandal, when I observed that the greenery decorating the walls in the room where the festivities were to be held included those of paxiuba palm, which had leaves whose inside surfaces were of an extremely intense yellow that contrasted with the green of the foliage. It was a solution.

Figure 17.3. Dinner amongst the *caucheiros*, including the local king, Carlos Scharff. Da Cunha is fourth from the left.

Really, a bit later, after being seated at the table I suddenly made a speech without waiting for an opportune moment during the toasts, and in a rapid tribute I thanked them for the invitation, and this for two essential reasons:

First, as an American, I felt happy at the display of cordiality between men who were of almost brother races, and perhaps destined for intimate alliances in the future in response to the growing imperialism of the great nations.

Second, as a Brazilian, how profoundly moved I felt in light of the "inabsolute." I clarified what I meant by then telling them what an extraordinary nobility of feeling had impelled them when instead of seeking out a Brazilian flag from the mercenary heart of some commercial factory, they had instead searched in the soul of the forest, taking from it precisely the tree that most symbolized the superior ideas of rectitude and stature, and I ended with these words: "Because, Peruvian gentlemen, my country is as righteous and as great as the palms." . . . I couldn't express . . . the effect of these words, nor the embarrassment with which the Peruvian chief and others complimented me, declaring that I had "understood very well their thinking. . . ."[1]

We stayed there for a few days, an interval that I used to carefully calibrate the chronometers. It was the first time we'd really done it, because, in fact, the adjustments in Curanja had no value whatever, due to the abnormal conditions in which they were carried out. I brought this to the attention of the head of the Peruvian Commission: he had witnessed the unfavorable conditions in which I worked, where the days were in general quite cloudy and I had innumerable other tasks to attend to. The entire situation was aggravated by the unfortunate circumstance of my assistant, Arnaldo de Cunha, having become gravely ill there. Sr. Sá kept at me about the memo in spite of its total lack of importance, and this fact, taken together with others, demonstrated an attitude that was hardly compatible with the solidarity of effort we needed to maintain.

This hostility, which had been veiled until then except for the platonic indications at Cocama, unmasked itself finally, entirely, when we tried to agree about what to do after this stop. We were in Forquilha where the enumeration of the enormous, countless difficulties they had indicated earlier began again. Wherever we wanted to advance, whether to the north via the Curiúja or by the south and the Cujar, the crossings were impossible. It was this they had affirmed in Manaus, it was this the *comissário* monotonously repeated as we traveled, it was this that was confirmed in Cataí, and that new information at Curanja reinforced, and finally, it was this that they declared in light of the current news of practically all the inhabitants of Forquilha. We could not advance. The Cujar, which would lead to the so-called official *varadouro*, the preferred route those who communicated with Iquitos, awaited us with sandbars, submerged logs, and seventy-five cascades, some of more than two meters in height. If we succeeded, we would then arrive at the Cavaljani, where the obstacles would increase along with the difficulties of the falls.. Afterward the arduous passage to the Pucani, and then to arrive finally at the *varadouro*. On the Curiúja, the same impediments . . . After this . . . the savages, the *infieles*. Two hours before we had arrived at that point [Forquilha], we had seen on the left bank of the river in a clearing the corpse of an Amahuaca woman that had been simply flung there. As we learned later, the Indian had been betrayed by the other barbarians who moved close by, according to many, a permanent and mute threat.

It was natural that we would deliberate on how to proceed given such special circumstances—and the conclusion that was most logical and irresistible was to return.

Considering the state of my outfit and principally our lack of supplies (which couldn't be renewed at Forquilha because there, on this occasion, you couldn't even find a basket of manioc meal), I wouldn't contest this position. I'd calmly return. We had already expended tremendous effort to arrive after months of punishing travel in our canoes.

But I was surprised by the new attitude of Sr. Buenaño. In fact, exactly at the moment when they were about to realize their old prophecies, Sr. Sá changed. The transformation was inexplicable—when the difficulties and hazards were vague, inconsistent, and really unacceptable in Manaus, en route, in Cataí, and even in Curanja, Sr. Sá decreed: we would not be able to transit the *varadouros*. And when those perils and predicaments were so eloquently, so impressively confirmed, in the pronouncements of those who struggled with them almost daily, and by our own observation, since from a single window we could contemplate the continual fall in the water levels of the two rivers and the impressive and disturbing evidence of a murder—Sr. Sá reverted to a surprising optimism and professed in a firm and categorical way: indeed yes, *one* could cross the *varadouros*. But he was not generous. His affirmation was not just limited egoistically to the singular; he completed his statement with a negative equally as imperious: that *I* could not cross the *varadouro*. The situation was clear. The Joint Commission would split apart in an inequality of energies. The Peruvian Commission, strong, able, self-abnegating, was ready to continue but was unable to do so because the Brazilian commissary, ruined by shipwreck, with a reduced staff, working primitively, without supplies, could absolutely not proceed further. This Sá loudly affirmed.

I understood that Sr. Buenaño completely grasped our real situation. It was most grave. Without being able to resupply in Curanja, where they swore there was nothing to be had, having lost hope of replenishment there (Forquilha), where they also declared to us they had nothing, I confess, Your Excellency, that I doubted that we had enough provisions for the return. In addition, I was suffering from severe neuropathy that still persists, which disheartened me: my affliction made it impossible for me to carry out the decisive action that the moment required.

Our colleague, however, had posed the question in a way that was utterly unacceptable. The return would require that we sign a new memorandum in which the conditions of the two commissions would be revealed and it would be explicitly stated that the Joint Commission could not complete its obligations entirely because of Brazilian failings. Thus he had his vengeance on the abortive document of Manaus. . . . And, he argued, they had the proper craft for the ascent—the *ubás*—they had provisions for a long time, their group was three times the size of mine: they could pass through the *varadouros*, they were ready to pass, while we, ill-equipped, we could not. He was still captivated by that memo of Manaus where he had predicted the retreat and believed that ____ miles distant from that city, almost without resources and buried in the wilderness, I would be forced to compose—alone—that which I refused to do in company. He was fooling himself. I refused the proposal. I declared to him that I was still in pursuit of the "possible impossibilities," whose existence I did not doubt, but which I had yet to see face to face. I prepared myself for departure from the Cujar on the 24th of July, upriver in search of the *varadouros* of Ucayali.

I do not exaggerate when I say that we went at half-rations. We commanded an immense region, entirely uninhabited, and the provisions we had—to last at the maximum for twenty-five days—was reduced basically to jerky and manioc meal, which we finished at the end of twelve days, a bit of sugar, which lasted only three days, half a barrel of rice, and the leftovers of some crackers purchased in Curanja. I give you this list intentionally. It is revealing. You can judge from it, if not our good faith in carrying out our mission, at least the bravery of an advance that was more than anything else a vigorous response to an impertinent challenge.

We left on the 24th of July and quickly reached the limits of knowledge we had about the region. At the lower reaches of the river, prior to the first rapids, the Cujar stretched out in a straight channel, sometimes flooding its riverbed in a way disproportionate to its treacherous waters. It was dammed by continuous sandbars, which spanned the river from one bank to the other without the narrowest breach so that we might avoid the arduous work of hauling the canoes across them. This task was so punishing that I often had to jump out and assist. A new hindrance, apparently underrated by the locals, appeared in the form of the characteristic vegetation of the river margins, draped in "buchitas" [*Calliandra trinerva*], an admirably elegant member of the Leguminosae family whose branches spread out horizontally, resting on the water, and choked large tracts of the sloughs of best access. In this way, even before wading through the rapids, we repeated the interminable battle that began at the mouth of the Chandless and was aggravated by the inappropriateness of our heavy canoes, very different from the sprightly *ubás*, the only vessels that could really navigate that river. Unfortunately, at Forquilha, where I had tried to buy some at any price—and where there was a great number of them—my efforts proved useless. We arrived at the first rapids and saw at once, in addition to the large series of their own obstacles inhibiting passage, another problem that would force further delay: the portage of our instruments on the banks of the river. These were already badly damaged by our earlier careening navigation. From there forward, in an inflexible cadence, where we proceeded degree by degree, we confronted barrier after barrier, which we not infrequently conquered by shoving, slowly dragging the canoes over the rocks when they didn't require the use of cables and pulleys to haul them along the banks against the fury of the waters.

After Forquilha, the nature of the geology changed dramatically. While there were no traces of primitive formations, everything indicated that we trod on layers much older than the lower basins and characterized by intense metamorphic action. Boulders jutted out everywhere, almost covering the riverbed. Rocks that were evidently sedimentary were of two main forms, either finely granulated or coarse conglomerates that in their exceptional hardness suggested quartzes and granites. The combination or separation of these formations created the distinct character of the cascades, which either plummeted dramatically in one drop, now in successive steps, now into mighty rapids, or descended into chutes bristling with outcrops, passing through huge dismantled boulders. We changed our tactics in order to master them. We didn't analyze the geology in detail. I don't want to abuse the patience of Your Excellency by relating the monotony of the ascent of seventy-three falls, forty-six large ones and twenty-seven small, which we indicate on the map.

Then there was the final stop. When we arrived at the headwaters on the 28th of July, we encountered the Peruvian Commission, which had already passed through the obstacles and was camped out upriver.

I immediately recognized the difficulties: the river, now dammed up by outcrops of the conglomerate that I already referred to, would be blocked during the flood season by a major cataract. But now, in full ebb, the entire river plunged through a fissure on the right, with a violence that increased with the narrowness enclosing it. But to the left all the way to the bank was the riverbed, mostly dry, pounded and plowed up by immense blocks and riddled with boulders. I knew then that we would have to drag our canoes through this rock bed after unloading them. The task appeared greater than our strength. The Peruvian *comissário* made it clear that he shared this conviction, in declaring to me that they continue camping nearby, just two beaches up.

As usual, they did not offer even minimal help. And I, as usual, did not ask for it. I told them, ironically (and to mask my own despair), that we would climb the dry falls or dismantle them,* and at the end of several days' labor we arrived at the Peruvian encampment. Having come this far, it was clear that no natural obstacle would make us turn back. I then noticed that paradoxically the system of regular cascades was quite beneficial for upriver travel in the dry season. They created dams at regular intervals, without which everything would have sunk into impassable flat swamps. Finally, with our boatmen exhausted from dragging the canoes through the muddy stretches left by the drained waters, we heard with delight the noise of the falls upriver, and this fired them to redouble their efforts, later compensated by a time of open, free navigation.

Thus we advanced to the confluence of Cavaljani, where we arrived some days after departing Forquilha. We were finally at the headwaters of the Purús. The

* This Brazilian engineering joke is often repeated when an obstacle must be confronted: one either overcomes it or blows it up. It also involves a pun, *montar* and *dismontar*.

river, as one can note on the map, still revealed its interesting dichotomy, so well expressed in the forks at the Acre and those of the Curanja and Curiúja. It divided itself into two almost equal branches, one to the south, the Cavaljani, the other to the north, which still carries the name Purús. It was on that branch that Chandless proceeded, and the river dried up a few miles past the fork. So we went via the Cujar, now shallow and epitomizing all the unfavorable conditions of the bifurcating river.

. . . We were but nine, me, a dedicated aide, Dr. Arnaldo do Pimento da Cunha, a sergeant, a soldier, and five representatives of all colors, united at random in Manaus. We arrived at this far point completely defeated. Our commission would soon disband, bound as we were only by circumstances—we had been shipwrecked on the way, and what was saved from the catastrophe would barely feed that reduced and foolhardy group. By luck, . . . we arrived at that remote position where we had traveled for days on quarter-rations—leftovers of jerky and manioc meal, which was both our salvation and our despair, without other additions to extend our unnatural diet.

The greater our misadventures, the greater our impediments, the greater the rate that our other resources diminished. The river, ever shallower, almost stagnant in the sandy straight stretches and foaming in endless rapids, required ever increased labors and real sacrifices.

We could no longer navigate: the two heavy canoes of *itaúba* went in a dragging pulse, as though we traveled on land, and numerous times the oars or barge poles transformed themselves into levers, making traversing these shallows even more difficult. When night fell, these men who had slaved all day immersed in water camped out sullenly, without a drink of coffee or *cachaça* to mitigate that brutal regimen. They could barely set up camp. The next dawn, limping and staggering because the river sands had sliced their festering feet to tatters, despairingly, they resumed the battle to ascend the river that seemed as if it would never end—so extensive, so monotonous, always the same in the uniformity of its banks, so that we had the illusion we were making a circular voyage. We set up camp, and after ten hours of torment in a cruel and interminable penance, we seemed to have returned to the same beach we had left.

In contrast to this misadventure, the Peruvian Commission, which accompanied us, was complete, well supplied, and robust. They had not suffered the distress of a shipwreck. They were twenty-three intrepid men directed by a captain of exceptional valor.

Thus, every night on those far-off beaches there was this contrast: on one side, the miniature camp mute and drowning in gloom; on the other, only fifty

Figure 17.4. Brazilian camp—a meager setup.

Figure 17.5. Peruvian camp.

meters off, a boisterous and well-lit camp where the songs of the satisfied *cholos* of Loreto resounded.

The separation of the two was complete, the relations almost nil. The Castilian diffidence, the inheritance of our elegant neighbors, was surprising in contrast to the other more heroic, wretched group, engulfed in self-pity and retreating into its penury, meticulous about completing its business without owing the minimum or most justifiable help to the foreigners with whom it was associated.

On arrival that evening at the headwaters of the Cavaljani, I thought the enterprise was lost. The way words were spoken, the irreparable despair, the sarcastic side comments let me know that the next day there was only one movement, the vertiginous return, rolling to the straits and waterfalls that had been so taxing to overcome, ending our efforts in retreat.

The next morning I sought them out, bent on the impossible task of convincing them to make yet one more sacrifice. They were huddled in a circle around a dying campfire and received me without rising, with the impunity of their misfortune. Two of them trembled with fever. I spoke to them: Honor. Obligation. The Fatherland and other magnificent words, resounding stentoriously, monotonously, uselessly.

They remained impassive.

I silenced myself in an exasperated sadness. And as if to increase my misery, I noted exactly there, crossing on the right, were the Peruvians, who were readying themselves for departure. We broke camp and took the gear back to the canoes. Soon the oars and the *tanganas*—long canoes raised by the rowers—briskly harpooned the air.

And crossing past the groups, the Peruvian sergeant, with a grave and solemn step, cut perpendicularly across the beach in the path of the leading canoe as though he were in a public plaza in front of a formation. Taking his right arm, he unrolled the Peruvian flag, which should have been raised at the moment of embarkation, but which my arrival had hastened.

A stiff southeasterner blew. The lovely cloth of red and white quickly extended itself, ruffling in the wind.

And it occurred to me to point out the contrast to my beaten-down comrades. But on my return, I didn't recognize them. All were standing at attention. The simple image of a foreign standard triumphantly raised like a challenge galvanized them. In one rush, without any orders, they were already prepared for departure. In seconds our flag, which had been rolled up on the ground, was now vertical in its turn in one of the canoes, proclaiming itself in front of our eyes.

The divine promises of hope! . . .[2]

Figure 17.6. The Peruvian Commission unfurling its flag at a *varadouro*.

It is certain that we would not have been able to move forward to the *varadouros* with our motley fleet, which had arrived at that point only through unrelenting toil, if an unforeseen circumstance had not favored us. Shortly after the arrival of the Peruvian Commission, which had two hours' advantage on us, the mail boats appeared from Iquitos—four small *ubás*, lightly modeled, which could be dragged through the *varadouros*. Sr. Pedro Buenaño resolved to requisition them, judging that his own *ubás* were inadequate for the shallow waters of the Cavaljani. The mail would wait at that point, and they gave up their boats. The Peruvian *comissário* sought me out in my tent to apprise me personally of the happy news of how much their conditions, already incalculably superior to my own, had improved, and before dilating upon them asked me with elegant gallantry, puffed with the satisfaction of the moment, for my instructions. I answered that our governments had given us our orders and that we must obey them. I could only answer: go forward. At that moment, without a lapse in his most elegant manners he repeated the painful refrain that he had adopted since Forquilha: they would pass, they were prepared to pass, and I, alas, would not.

I countered with my usual "possible impossibilities" and asked him, in the presence of Arnaldo da Cunha, whom I had order to assist in the deliberations, for the two *ubás* that he judged so deficient, so that we could examine them. Sr. Sá in a later report said that he offered them to us. The truth, however, is that he somewhat resisted my request, fearing that the boats that didn't belong to him might get damaged in the crossing, et cetera. But he relinquished them to us, he could not do otherwise, as he had deemed them to be totally inadequate.

Indulge me, Your Excellency, to insist on these details, which are most expressive.

Now in possession of the two *ubás* and the portage of the minimal material that was strictly necessary, and after having triumphed over the justifiable reluctance of my outfit, which was disheartened by the prospect of new and greater travails with ever more paltry resources, on July 31 at 8:00 in the morning we sought to enter the Cavaljani. We then passed into the most bizarre part of the entire voyage.

The two canoes, on meeting the confluence, again ran aground in the waters of the Cujar and for forty-eight minutes sat immobile, buried in sandbars, completely unyielding to our desperate efforts. This all transpired about four meters from the Peruvian Commission, whose field camp had not yet been dismantled. Their people and officers mutely contemplated us from two paces away, and from there did not offer to extend a hand, not one hand, to help us.

Finally a breach opened in the sand, and we slogged by foot into the Cavaljani. An hour later the Peruvian *ubás* passed us—they had also run aground, but much less seriously. And still we were at the confluence. Three hours later we had gone but twenty meters. It was eleven, and the Peruvian Commission disappeared ahead of us.

I thought that we should proceed on foot, with each person carrying what supplies they could, but to walk the rest of the way—even the suggestion was futile. We decided then to further reduce what we carried and put all the personnel in one lightly loaded *ubá*, with the rest of the equipment and supplies entrusted to the sergeant, who would stay there. Thus we resolved the problem. At 3:00 p.m. we caught up with the Peruvian Commission and camped on the same beach. At the end of ___ days, we had finally arrived at the slope that led to the *varadouro*.

We arrived at 12:50, got out of the boats, and entered the narrow channel of the Pucani. This interval was critical. We could not stop, as the provisions for nine men in an absolutely uninhabited area was reduced to four cans of condensed milk, three kilos of dried meat, and two cans of chocolate. And we were in total wilderness.

In passing by the tent of the Peruvian chief, I told him of our resolution to advance ahead to the *varadouro* because "there was no way that I could delay"—I was sickened by the idea of explaining to him the painful straits in which we now found ourselves, but we could tolerate this scarcity better than we could stand the charity that they never deigned to offer.

We walked through the torturous Pacani, three meters broad and in general quite shallow, crossing the deep pools into which we were intermittently plunged, and passing by shortcuts through the forest that flanked the riverbanks. With no guide other than the empty cans of supplies and gunpowder that we periodically encountered, we did not transit a similar passage that rose on the left margin of the river, so that by luck at 3:15, as we arrived at the last pool, we suddenly came upon, straight ahead of us, the head of the *varadouro*, positioned above a steep scarp.

There were four paxiuba palm huts where travelers gathered and cached their merchandise. All around in all directions, empty cans of all sorts, bottles, rags of old clothes, and broken tools revealed that what passed for relatively heavy traffic passed through this necessary stop. The *varadouro*, one meter wide, opened in front to the south. It rose up a bluff, much steeper on our side, but soon sloped gently down in three small plateaus toward the Ucayali. We were at its culminating point. . . .

. . . The sun descended to the Urubamba, and in one glance our dazzled eyes took in the three largest valleys on earth in a marvelous expanse of horizon, bathed in the luminosity of an incomparable afternoon. What I mainly noted as I gazed at the three quadrants stretching indivisibly was that embracing them entirely to the south, north, and east was the striking vision of our country, which I had never imagined so vast.[3]

Da Cunha and his emaciated crew returned more or less triumphantly to Manaus, with extraordinary amounts of work ahead of them. Most of the intellectual effort would be taken over by da Cunha. Firmo Dutra reported that da Cunha began to compose his "Paraiso Perdido" while the experience was still fresh in his mind and the authority of the travel itself was still powerfully animating. Da Cunha had arrived from another world—just barely—one that Manaus refracted, but that was entirely different from every other place in Brazil. As with his *Os Sertões*, it was the land that would frame the narrative, and it was the "austere lexicon of science" that would provide the adjectives.

PART 4

Cartographer at Court

18

Return of the Native

October 19, 1905, telegram to Baron Rio Branco:

> I communicate to you our arrival in the city [Manaus], with the commission completed in all its essential points. I continued on to the valley of the Ucayali, crossing the *varadouro* of the Curiúja until we completely depleted the supplies and the water, but with the most certain information about the small area we did not cover. We mapped the river up to the headwaters twice, on the way up and on the return, rectifying many elements of the existing maps. We determined the principal coordinates. Except for small discomforts, we all arrived well back in the city.[1]

They had indeed made it back, though Euclides da Cunha was drastically sick and hallucinating his "woman in white" through the fevers of malaria and malnutrition. He was back at work after a few days, but as he wrote to Veríssimo, "I sense that those long days of anguish, of misery and triumph, that I lived through at the headwaters of the Purús have endangered my life. Misery and triumph—only face to face can I recount my obscure and tragic battle with the wilderness—and anyway, one cannot discuss these things when one's head aches from logarithms."[2] His days were now taken up with detailing the final maps and the reports that he and Buenaño had to develop, sending down the album of more than one hundred photos (now unfortunately mostly lost) and lamenting the lack of photos of the Brazilian hero of the Peruvian War, Antonio Ferreira de Araújo.[3] The coordination of the elements of the maps was relatively harmonious, especially given the delicacy of the charge, the animosities, and what was at stake. "The topographical surveys of the two commissions in various sections are of a surprising concordance,"[4] he wrote to Rio Branco. In spite of the contretemps on the river, formality and courtesy returned as the Brazilian and Peruvians labored quickly in order to return to the pleasures of Christmas and to be quit of each other.

Euclides would find he was not going home to exactly the domestic situation he might have wished. His was hardly a return to a faithful Penelope, fighting away suitors as she ached for her husband's homecoming. His wife's teenage lover, Dilermando de Assis, described da Cunha's arrival in this way:

> In January of 1906, Saninha [Ana's nickname] was surprised one morning in her love nest[5] by a visit from an employee of the "Bazaar America." The proprietor, Bastista de Fonseca, was the agent for her husband and supplied Ana with the necessary funds for her support. The clerk brought a telegram that had been sent care of the firm. It said, "I'm in the bay on board the *Tennyson*. Send someone to fetch me." This was the end of the Idyll. That day, for the first time, the cadet and the writer would meet each other. Thinking that in this way he might dispel suspicions, Dilermando, Solon (da Cunha's oldest son), and some servants went to the quay to collect Euclides.[6]

Ana and Her Cadet

After da Cunha departed on his expedition, Ana had left the house leased for her and, as I noted, had decamped, first to the Pension Monat and then to another house where she lived with her youngest children, her lover, his aunts, and some servants. Her silence for the duration of his expedition, her removal to a new address without bothering to inform him, and her isolation from friends and family suggest that she was ready to end their marriage. In an extensive airing of historical dirty laundry, her family has argued that Euclides left his household with little money while he was on expedition and that Ana moved to the Pension Monat out of fiscal desperation.[7] The da Cunha side points to the accounts of the trading house Bazaar America as proof of a generous allowance[8] and paints her as a cynical and frustrated schemer. Perhaps it was some of both: she loved Dilermando and she needed money.

Dilermando described their passion this way:

> Living together soon led to closeness, and the lack of experience or good judgment permitted the most intimate proximity. A life no longer cloistered opened up new horizons, the reading we did together awakened fantasies, adolescence sparkled with its own enchantments . . . our isolation eased the play of the empire of the senses. I lacked the protective and well-considered advice that could warn me away from this godless path that would soon convert into a passion . . . and then, there were so many other circumstances, material and moral, too many to number, for good or for ill, and all conspired to awaken new, delirious sentiments. And thus hidden away, in this unbounded intoxication, my crime was consummated. Because it is only this way I can view it, the transgression of the Law: for having loved at seventeen a married woman whose husband I didn't know, and who was absent in the most distant outposts, who was not even being remembered by so much as an lifeless photograph. It was fate, it had to be like that, and thus did it transpire that fall of 1905.[9]

This amorous prose certainly trumps Euclides's banal billet-doux upon falling in love with Ana: "That evening I entered with the vision of the Republic, and left with yours." Ana had now given up contact with most of her and Euclides' family members and fled to the delirium of first love and profound sexual awakening for both her and her teenage lover. With the Ratto sisters as confidantes, and with kinship links to Dilermando glossing over the reality of the domestic arrangement, she reveled in the pleasures of her romance.

The telegram from the *Tennyson* ended the serene phase of their affair. Dilermando initially stayed at the house so as not to arouse suspicion, but within hours of Euclides's arrival the domestic explosion was well on its way. Dilermando states that a letter from Ana had informed Euclides of her infidelity but not with whom, nor of its carnal nature. Euclides became aware of their romance but perhaps not of the fact that Ana was three months pregnant.[10] Along with anonymous revelations and widespread gossip, there was a story given by Ana herself in 1909 at a police deposition: she stated that she wrote a letter to her husband where she affirmed that she was unworthy of him, had "betrayed him spiritually in his absence, not knowing whether this infidelity came from being finally free of his poor treatment of her, or whether it was simply the outcome of lack of affection from Euclides; she thought they should prolong their separation by either divorce or commis-

Figure 18.1. Dilermando de Assis.

sions."[11] After receiving this letter, her husband had asked whether she had profaned her body, to which she responded that she had profaned him only in spirit.[12] Using several subterfuges to hide her pregnancy, she continued living with her husband and, as she reports, performing her wifely duties.[13] Amid the continuing acrimony between husband and wife, Dilermando soon moved out of the house.

Within two weeks of his arrival in Rio, da Cunha was writing his old housemate in Manaus, Firmo Dutra, that he might soon be on his way to Venezuela or Guiana to continue border surveying. He was immersed in scandal. Even his father, who resided on a remote coffee farm, was not distant enough to avoid commentary about Euclides and Ana. Moralizing gossip, as Euclides's friend novelist Joaquim Machado de Assis noted with pitiless acuity in his works, was one of the central pastimes of all strata of Brazilian society. In the small worlds of the Rio Branco court and the extended da Cunha and de Assis families, probably everyone speculated on the triangle and had an opinion, and most probably reveled in Euclides's distress, whether from the whinnying position of the moral high horse, the luxurious stance of *schadenfreude*, or the realm of frank ridicule.

The household had been behaving strangely for a while, enough so that even Manoel da Cunha noted it: "You told me nothing, but I understood that there was a lack of trust, but I didn't want to impose on anyone, so I left, quickly and vexed. Not only for this but also because of the strange way you are treating your wife and children, especially Solon, whom I most esteem. I thought . . . that my example and advice might have changed the your way of living, but I encountered the same tantrums, the same disorder of the past . . ."[14]

Euclides wrote this to his father:

> As to the other, very delicate point in your letter: there are no, absolutely no, grounds for what you believe. How everyone exaggerates my intentions. . . . I am not falling, thank God, into those repugnant and ridiculous jealousies that are completely unjustifiable; I wouldn't even be here, writing this letter, if I doubted for a moment the honesty and harmony that surrounds me.
>
> It is just an old issue. Considering the complete moral collapse of this country, where the smallest acts are subjected to the falsest interpretations, where I oppose myself to things that everyone else allows—frivolities permitted by everyone—but which "everyone" then usually turns into the most grotesque thoughts. In sum, the discord in my home derives only from my audacity in hammering on that old saw "It's not enough that we be upstanding; it is also necessary that we *appear* upstanding."[15]

Ana's pregnancy would rupture the figment of appearances.

The events of that time left little by way of documentation, and the principals who have opined on the death of the baby have axes to grind with each other. The Ana de Assis faction viewed da Cunha as a temperamental, vindictive brute who maintained appearances but tormented her psychologically in numerous diabolical ways. Dilermando, who would later produce books on the subject, viewed da Cunha as a complacent husband but also given to verbally abusing and humiliating his wife.[16] The da Cunha relatives have, insofar as possible, sought to polish his domestic image in addition to his reputation as one of Brazil's leading intellectuals.[17]

Keeping up appearances must have been straining for everyone. In this general exercise in falsity, Dilermando visited the da Cunha household regularly each Saturday until March 1906, when he left to continue his training in Porto Alegre, far south of Rio. A sense of how odd these visits must have been for all involved is suggested by a now lost letter that the distraught seventeen-year-old cadet sent to Euclides, who in addition to his other glories was now an aide to the Baron of Rio Branco. Euclides had probably insulted Dilermando in some way, an act not unimaginable in the circumstances, especially given Euclides's famously short temper at home and the circumstances. Euclides wrote back:

> My response is simple: there is an enormous, absolute error in what you think. The question is very much other and you are completely extraneous to it. See what happens when one makes deductions from isolated facts and words? Anyway, in spite of being exasperated by innumerable irritations, I don't think I treated you badly. At your age, one is never a low person. Do not believe that I did you such an injustice. My house continues to be open to those who are worthy and good. I could not close it to you. When you know the reason for my annoyance you can evaluate the injustice that you did to yourself and to me. See you Saturday. Study, and continue to be the same youth of noble sentiments . . . etc. etc.
>
> Euclides da Cunha.[18]

Dilermando de Assis was certainly not an unbiased observer of the events but was perhaps not lacking in insight about them, and he was in a position to know many details. He described the ambience of the da Cunha household thus:

> The husband apparently lacked the acuity necessary for understanding that the incompatibility was not faked but very real and profound. Admitting, perhaps, the validity of the conjugal accusations she made against him, Euclides told her that he gave little importance to what she might think, given that her body had not been profaned. She continued to share his bed. She avoided encounters with her young cadet, and tried by all ways and means to hide her pregnancy. This, however, soon became impossible: the unconfessed truth emerged, brutal and

> accusing. The husband, now certain of her infidelity, hurled the cruelest insults at her.[19]

Dilermando, who was transferring to the military college in Rio Grande do Sul, took his leave of the couple in a formal letter. Euclides opened it and read it out loud in front of the household as well as some guests who were there by happenstance. Ana listened, humiliated by the shock. Her husband exclaimed, glaring at her anguished expression, "Look at her face! And tell me if it isn't that of someone who has lost the being they most love."[20]

The months that followed were terrible, filled with threats and fear.[21] Dreadful scenes were a regular feature of their domestic life. Nevertheless, Dilermando and Ana maintained a passionate correspondence after his departure through the classic ruse of the post-office box (she was "Olinda Ribeiro") and through various domestic ploys (e.g., Dilermando visited when Euclides was visiting the Baron at his summer house). When Dilermando moved back to Rio, they were able to continue the obviously compelling sensual part of their romance.[22]

Da Cunha was living a life full of deceptions. Young Mauro was born on July 11, 1906. The entire household "fell ill" with the birth: Euclides purported to be wracked with fever and unable to leave the house, according to the notes he sent to da Gama and the Baron. Ana was in a postpartum as well as romantic depression, and the domestic atmosphere was utterly bitter. Little Mauro, who was born at a normal weight and was recognized by Euclides as his son and registered as such on the birth certificate, was dead within the week. The formal cause of death was "inanition"—starving to death.[23] Usually this is due to some inability to nurse, whether through weakness or disease or a lack of milk from the mother.

The child was certainly not premature, although in his correspondence Euclides averred that it was so. In statements under oath to the judge Manoel Costa de Ribeiro, Ana stated that

> during the pregnancy she sought the assistance of Dr. Erico Coelho [a friend and doctor to the da Cunha family], to whom she related the following: that while her husband was away, a friend had become impassioned with her, and forced her to have relations with him, and she thus became pregnant. She sought some medicines to abort, or some other advice. The doctor could not attend her because of the late state of the pregnancy, but advised that she should go home, have the baby, and send da Cunha to speak with him after the birth, since he would point out to da Cunha the possibilities of having a baby of six months' gestation. If he could convince da Cunha of this, she would be saved, otherwise, she would see what destiny had in store for her.[24]

Ana also visited a midwife, Emilia Movan, who proposed to her that just about at term Ana should suggest that the da Cunhas change households, and she could thus explain the labor as premature induced by the fatigues of relocation.[25] This was in fact the route Ana followed. The day that they moved, she waited for her husband to leave for the office so when the child was born she might find some way to hide it or "to make it disappear." But it happened that Euclides stayed at home that day, probably not by accident. In the end, when she gave birth only her husband was in attendance. Seeing the flaxen-haired child born alive, Euclides, according to Ana, exclaimed that the baby was "the monstrous offspring of that monster who betrayed me"[26] and left for another part of the house.

With the help of the cook, Ana began caring for the infant. About midnight, an enraged Euclides barged into the room, posing the central question in this way: "You slut, you disgrace, daughter of that *tarimbeiro** who has the shame of having put you on this earth, swear on your father's ashes whether this child is or is not mine."[27]

On her knees, with the child at her breast, Ana confessed the child was Dilermando's. Euclides, according to Ana's testimony, then stated "that he wouldn't kill her because he didn't want to defile himself with the stains of her blood, but that he would have to kill her lover. She would," he said, "have to live a life of suffering," and then he forbade her to nurse her new baby and averred that the birth would not be noted; indeed it was hidden until the death of the child some eight days later.[28]

The position maintained by the Ana de Assis faction was that Euclides took the nursing child, locked Ana in her bedroom, simply let the child die of starvation, away from the milk and care of his mother, and later buried the tiny corpse in the garden.[29] This macabre account is contested by the da Cunha faction, who point to a grave with a headstone for little Mauro and suggest that there was some genetic problem; some speculate that the medicines Ana obtained from popular curers may have in fact damaged the fetus.[30] Euclides did formally recognize the child as his own, even though, as Ana reports, he treated her insultingly and cruelly. Dilermando, uncharacteristically, had little to say on this point.

The baby was merely the first victim of this deadly triangle.

In his correspondence da Cunha maintained his public persona and was cheery about his fatherhood, dissembling about the state of the relationship. For example, he wrote apologetically to his friend Coelho, "I got your card and

* *Tarimbeiro* was the term used for military men who ended up betraying the emperor. In this set of insults it refers to the general treachery of her nature.

am convalescing from a painful health crisis. Also, my wife has been gravely ill due to a premature birth, and I cannot tell you the anxious days I have passed. Even greater torture, though, was to harmonize the dry necessities of my profession with all my afflictions; during the last fifteen days I lived between my maps, the medicines, and the infinite attentions required by the delicate situation of my life's companion."[31]

The domestic and political scenes were equally discouraging. Although Ana had sworn to give up Dilermando and to become a different woman, the relations in the da Cunha marriage were impossible.[32] Euclides was looking for a way back to the Amazon, whose trials and deprivations seemed a delightful alternative to the demands of the court and the ambience of his household. He reported to Dutra: "I don't know if the news has arrived there yet, but I have been nominated to supervise the construction of the Madeira-Mamoré railroad. Really, things are coming to that, and if a serious obstacle that I encounter—the opposition of my father—can be derailed, I'll be there soon, shod once again in my seven-league boots."[33]

Da Cunha's family had to be placed under the responsibility of someone, and as he had before the Purús commission, he turned to his father: "The commission is the most serious and brilliant of all that an engineer can aspire to today. . . . According to the Minister, I would be the fiscal supervisor—whose autonomy would be absolute—and who would represent the government along with that great continental railroad that will transform South America. I have accepted. I can't resist this attraction. It would be one more sacrifice, but it would be another daring launch into the future. I know that you would not reprove this, my last audacious act. Besides, I would complete my still partial observations on Amazonia." As to the family, he continued: "Saninha obstinately declares that she does not want to be left alone again. I would have to take her and the youngest to Belém. Solon and Euclides would stay at the boarding school where they are now, under your care."[34]

It is not clear that Ana *was* adamant about traveling with him to Belém; actually, given the nature of their domestic affairs, it is quite difficult to imagine that she would not have preferred staying in the south as he went north once again, hopefully to disappear once and for all in the maw of Amazonian afflictions. In Belém she would be beyond the reach of Dilermando, the external niceties could be maintained, and Euclides could monitor her behavior more easily. The incandescent letters of Dilermando and his descriptions of the time suggest that their passion was fueled by the difficulties of their circumstances.[35]

The financial rewards would have been significant, and da Cunha was enjoying the prospect of being the engineer who would be in charge of the project and thus continue his own epic, but unsurprisingly, his correspondence is always tinged with depression: "Soon I'll be back engaging once again my tragic duel with the wilderness. . . . I don't know whether I'll triumph once again. It doesn't matter."[36]

The Madeira-Mamoré railroad, like the Panama Canal, was a huge maw that ate men by the thousands. Accidents, malaria, yellow fever, cholera, and drowning killed indiscriminately: Barbadians, Italian migrants, American adventurers, Indians, tappers—anyone foolish enough to join the labor crews was prey. The swampy terrain resisted the efforts of all to make the first connection lines from the Bolivian Amazon to the Madeira ports below the falls of the Devil's Cauldron—the falls of St. Antonio.[37] It was said that each railroad tie had cost a life. Bolivia had given up huge amounts of territory in order to get rail access to the Madeira, where its rubber could then be taken by river to international ports rather than being packed over the Andes on llamas. This railroad to the port below the rapids was the key to Bolivia's imagined Amazon future. With the usual ironies of history, it was completed just as the great Amazon rubber economy started its collapse.[38] It is unlikely da Cunha would have survived, as most who went to work on the line did not.[39]

In the end da Cunha refused the position, in part because his father was opposed to his departure, perhaps due to the predictable marriage problems and the continuing Dilermando scandal. A removal to the north might have made the whole sorry affair appear even worse. But also, as da Cunha noted in his letter to his Amazon friend Firmo Dutra, he hoped for another assignment that he thought would be more useful to him: establishing the demarcation border of Venezuela and Brazil—this, he confidently asserted, he would lose only if Rio Branco was forced from the government. He had been angling for this position and discussed it with Oliveira Lima. "I am eager to do it, in fact extremely so. . . . A stay in those mountains where the illusion of El Dorado once emerged would certainly compensate for those long and sad times I passed in the infinite monotony of the Purús."[40] The Baron had other plans for him, however.

In the letter to Firmo, da Cunha also observed:

> The Government of Amazonas is interested in having a more reliable map of the Purús, and in knowing how to go in to Peru through its most unimpeded entrances [the *varadouros*]. Buenaño was right in his irritation that increased as we advanced, defying even hunger: in a *casus belli* with Peru (which is not unimaginable) how would we advance up to there, disoriented by the network of *igarapés* in that great river? . . .
>
> I've started! Finally! To outline my *Paraíso Perdido*![41]

His earlier newspaper essays were being published in Portugal as *Contrastes e confrontos*, which he described as the "bastard offspring of my spirit, more neglected, but even so perhaps more worthy of our love." It had been a horrible year. On New Year's Eve, alone, he wrote Escobar from Rio: "I've so much to tell you, but it's quarter to midnight and I don't want to hear the final twelve strikes of this 1906 where I worked so extraordinarily hard and saw promoted in front of me so many, the cheery and indolent, that throng this Brazil! Maybe it's better. I want to feel in my depths the ache of the deepest disappointments; and if these are cruel enough, perhaps I'll achieve something in 1908, or, 2000. 2000 equals posterity: that, in the end, is the only certain prize and the one worthy of true fighters."[42]

He was about to embark on another set of battles pertaining to the Peruvian commission, the lineaments of which would produce the first real diplomatic history of the Upper Amazon. But his domestic situation would become unimaginably worse.

Ana yearned for Dilermando de Assis, and after his departure they corresponded. According to her deposition,[43] Euclides treated her badly, and she begged Dilermando to come back to Rio. She wanted to have another child with him.[44] He was able to return to Rio, and they met in various places in the city (including at the tomb of the unfortunate Mauro). She arranged for him to visit her at home when Euclides would be at Petrópolis working with the Baron. By February 1907, they had resumed their affair. The fecund Ana was soon pregnant again, and perhaps once more trying to abort.

Euclides wrote this to his brother-in-law in June of that year: "Saninha, in early pregnancy, has been very unwell after a threatened miscarriage. She passes days in bed on the advice of several doctors, and cannot even leave home to visit my mother-in-law. Travel now would be disastrous for her. Besides our own doctor, others have prescribed maximum caution for her."[45]

On November 16 young Luis would be born, blond and blue-eyed, "an ear of corn amidst the coffee." Luis, as an adult, would later take the surname de Assis in recognition of his true paternity. From that time forward, Ana stated, "Euclides was regularly insulting to her and to the young child ('that shit from the big sergeant'), and made Ana hide the child when they had visitors."[46] Given this depressing domestic scenario, it is probably not surprising that he flung himself into the labors for Baron Rio Branco, escaping to the archives to reconstruct a past and imagine a new future for a new Amazon.

19

Maps, Texts, and History

Background to Euclidean Cartography

Euclides had to generate the intellectual apparatus of the Brazilian boundary negotiations using the science, poetics, and politics of history and landscapes. His approach required the development of several distinct types of framing documents including maps, supplements to the field survey, and a series of essays on comparative cultural and diplomatic history. The first two types of documents—the charts and their commentary—were the technical elements distilled from the Amazonian experience, interpretive reflections, and additions to the fieldwork. The Baron had also suggested that da Cunha write a series of essays for publication in the *Jornal do Comércio*—the *Wall Street Journal* of Brazil—that would be partisan and would analytically and ideologically shape the larger debate. These were produced between June and September 1906, an enormous amount of research and writing to generate so quickly. The essays were part of a public relations blitz to recast the understanding of the Amazon boundary question and argue the case overtly while Brazilian negotiators wrangled over it privately in the chambers of the plenipotentates. Da Cunha also had his own agenda: a yearning for what we would today call "equitable development." He would have to hew a precarious balance between Brazilian triumphalism and his concerns about social justice.

What was at stake was, as da Cunha put it, "the largest territory that had ever been contested between two nations, some 720,000 km^2 in one of the very least known parts of the planet," where conflicting colonial documents and maps inflamed the clamor over rights to Amazonian terrain and rubber forests. Not only was it the largest area in contest, but also Brazil's last. Once finalized, whatever the configuration, Brazil's border would be completed, and the outlines of the nation defined. This was a culminating element of Rio Branco's statecraft.

The official Peruvian position, as I have noted, was that Peru's lands extended to the historical lines established by the Treaty of San Idelfonso (1777), and thus Bolivia had had no legal right to transfer to Brazil what Bolivia claimed as its terrains. This conflict between Peru and Bolivia was put to arbitration in 1902 with the Argentinean president as mediator, but then languished.

The effort from da Cunha's side focused on making the Purús definitively Brazilian, so idioms of nationalism and imperial justification, as well as his own hopes for a more socially just Amazonia, saturate these writings. Da Cunha produced several maps that were key parts of this exercise, along with the formal cartographic depiction of the Purús that had been prepared jointly with the Peruvians. The first was a macro vision of territorial arrangements and settlement. The next was an inverted map of the Upper Purús that charted the headwaters and the *varadouros*—interfluvial shortcuts to the Ucayali basin. This was a classic military "lying map," but da Cunha had much to say about the ideological and political significance of the *varadouros* in the way they knit together the landscape. The third was a sketched map of the novel property regime, the rubber trail or *estrada*, the basic production unit of the rubber estate. The hand-drawn map filled in empty spaces away from the banks of the river and onto the uplands under the forest canopy. With its micro-scale description of demarcation, it revealed how a new property institution was unfolding, and it disclosed, through da Cunha's commentary, the deplorable labor relations within, as the converging circuits of the *estradas* inscribed latex's entry into global commodities exchanges. The maps of the *estradas* can also be read in a more Foucauldian manner, as means through which labor was disciplined through the spatial organization of production.

Da Cunha used the map as a diagram of and a metaphor for the oppressive peonage regime that was quite common on the Purús.

These maps are illuminating as cultural texts because the explicit "Euclidean" commentary that goes with them permits us to understand and deconstruct their politics and implications. The three maps can be seen as complements of each other at different scales, as having legal sway, and as summarizing the mechanisms of territorial incorporation in international, national, and institutional terms. These maps were deployed to show to Peru the Brazilian nature of institutions and settlements. They also showed the Brazilianness of Amazonia to Brazil.

From the Lands of the Amazons . . .

The 1705 map of Royal French cartographer Guillaume Lisle placed the "litigious zone" squarely in the "Pays de Amazons." This land of warrior women was undergoing rigorous mapping exercises at the end of the nineteenth century, as the Upper Amazon nations partitioned the mission claims and rubber lands among themselves. The fact that the formal map for the negotiation was jointly produced by a Peruvian-Brazilian commission gave special luster and legitimacy to what was meant as a "nonpartisan" document. On December 16, 1905, da Cunha and Buenaño signed their approval in Manaus. Each country provided its own contextual material to the collective map, structuring the way that the politics of the map would be read. While the chart had the authority of a scientific document—that is, the measurements of where "things" in the landscape were were largely correct and produced through the accepted scientific procedures of the day—it was an exercise that was permeated with subtexts. The Peruvian document contained no additional information about the map other than its coordinates but published as the "framing" materials an extensive dossier of the bureaucratic exchanges as well as hostile letters traded by the chiefs of party of the Brazilian and Peruvian commissions as they traveled the upper tributaries. In contrast, the Brazilian report included "Complementary Notes of the Brazilian Commission," which involved the history of mapping, geographical information, and settlement of the Purús, which I present in later chapters. Given the magnitude of this territorial contest and how overwhelmingly important the discursive elements were, it is perhaps not surprising that Rio Branco deployed the most powerful writer he could, one who happened to be a romantically lyrical nationalist.

The Peruvians would insist that legal procedure and the precedent of the San Idelfonso Treaty placed the Purús well inside their national boundaries. They viewed the conflict as really about respecting venerable territorial boundaries that had carried over from colonial times.[1] The Peruvian framing documents, despite recording some sharp exchanges, included a dossier of bland bureaucratic practices such as "On such day we will coordinate the chronometers," "We agree to take measurements only during the day." The salience given to this correspondence indicated that the value of the map for them was largely cadastral and, given the particular moment's unpleasantness, procedural and military.

The Peruvian government was reiterating a strategy deployed in an earlier conflict with Ecuador, where Peru contested Ecuador's "false sale" to a British charter company of terrains Peru had taken to be its own on the basis of colonial treaties. Peru argued that the hierarchical authority of the Viceroyalty of Lima retained a kind of "federal" power over lower administrative units—*audiencias*; thus Lima's historic claims should prevail over governing bodies of mainly regional influence. This Peru-Ecuador conflict had been decided in the Peruvians' favor, invoking Idelfonso demarcations.[2] Bolivia's case was similar: it had been part of the *audiencia* of Charcas. It was a dangerous precedent for the Upper Amazon: if the logic of Peru-Ecuador prevailed, the earlier Idelfonso line would remain the boundary, and Brazil would have to retreat from these lands or fight for them.

The Brazilians, on the other hand, believed they were surveying lands they had acquired legitimately, because Idelfonso was a contract between the extinct empires of Spain and Portugal, and more recent boundary agreements with the republics of Peru and Bolivia had been signed during Brazil's empire, which also no longer existed. New boundary regimes had to be put into place, presumably informed by treaties signed by the various modern republics.

The key element was *uti possedetis*, but other components had to be assembled as well.

The idioms of Brazilian nationalism would have to be presented in terms of ethnic, settlement, language, administrative, spatial, and symbolic continuities. Political philosopher Ernest Rénan's ideas were pivotal for da Cunha's essays on nation building: a fusion of races, peoples, a legacy of memories, a moral consciousness that gains strength through sacrifices in the common work of civilization. Rénan's further point about resolving boundary difficulties by consulting the populations was a "modest empirical solution" that da Cunha would use in shaping his arguments.[3]

For the imperial claims directed against Peru, the narrative logic would revolve around the colonial axis of Civilization versus Barbary, embellished with the usual tropes: improvement versus nomadism, political cultures (modern republicanism versus revanchist caudillism), social cohesion versus spatial dispersion, and the other weaponry of the imperial ideological arsenal, namely comparative patterns of discovery, historical cultures, nature, morality, and social evolution, and finally social ecology. The map, the travel reports (official and otherwise), and the additional commentary would become the sinews of a spatial history and narrative of the "Brazilianness" (as opposed to Peruvianness) of the Amazon. Historical cartography, settlement history, folklore, and distinctive activities of daily life in the region would infuse the place with traditions, institutions, and peoples that reflected Brazilian pasts and futures. The geopolitics of these forests would have effects on rubber supply, future diplomatic conditions, and economic possibilities in the rich Upper Amazon. Railroads and ports were being built, and the throngs of upriver travelers had reported many other kinds of resources and potentials.

Patriotic Epistemes and Tropical Statecraft

Da Cunha's task was to create a "patriotic episteme" formed of Creole and native sources to fashion a nationalist narrative for the settlement of the Upper Amazon.[4] His job was to recast the political economy and ecologies as part of an authentically and historically progressive Brazilian "March of History," but in an Amazonian register that coupled it to and echoed the rest of the nation. Da Cunha's emphasis on Brazilian sources and actors was a way of valorizing local knowledge and histories and highlighting the lineages of administrative action in Amazonia, as well as expressing his own populism. Rather than constructing an exotic backdrop for European explorers, he locates the "natural" actors of Amazonia *within* it, shaping its history in knowledgeable, familiar, and practiced ways. He was not interested in exotification, whether in the modes of Brazilian nativism or the essentialisms that infested writings about Amazonia, but rather sought to portray its inhabitants as naturals to it, outcomes of its history and localities, modest bearers and creators of a distinct, but Brazilian, tropical civilization, and one intellectually on a par with the highest levels of colonial science.

The international questions were of import, but creating an analysis that would shape *national* public opinion about the Brazilianness of the region was also a serious task. The political and cultural distances between northern and southern Brazil were enormous. From the optic of the industrializing South, the Amazon, with its dark populations, Indian tribes, and its refugees from climatic disaster who extracted tree gums through precapitalist labor regimes with Stone Age techniques involving knives and smoke, was embarrassing and hardly fit with the sleek Europeanized version of themselves that southern Brazilians had fashioned. Southern Brazil's European immigrants (meant to whiten and "improve" the mestizo population), its emerging industrial economy, and its stylish urbanism mimicked continental fin-de-siècle tastes and modernism. The Amazon, on the other hand, seemed atavistic: instead of advancing out of slavery into wage labor and industrialization as had southern Brazil, it appeared to have regressed from agriculture to extractivism and had reinvented slavery in the primitive modalities of debt peonage, dirt farming, and Indian terror. Amazonia seemed, actually, to be evidence of all the hypotheses of racial and climatic degeneration. The development discourse and political identity that worked for the South involved a rejection of most of the social and natural attributes of the North. If the ideas of modern southern Brazilianness were taken as the norm, Amazonia would seem less of a Brazilian place, since it lacked similarities with the rest of the country. This cultural disjuncture would make western Amazonia more liable to appropriation, and this had not been helped by lax attitudes toward the region among the foreign ministers who preceded Baron Rio Branco.

Da Cunha had to blur the obvious differences between Amazonia and the rest of Brazil by invoking its national kinship shaped through common mythologies: the *bandeirantes* who bound the North to the South through their epic travels; the ideas of an emerging Civilization in the Tropics—an aspiration that had especially animated Brazilians since Pedro I's "Tropical Versailles"; the heroic travels of the Northeast's *sertanejos*, weaving the Atlantic Coast to the Amazonian hinterlands, all within a Brazilian version of manifest destiny. With Os *Sertões* da Cunha had, after all, given the blankness of the backlands a historical and cultural animation as part of a national saga within the peculiarities of a powerful and distinctive Brazilian nature. His task was now to do this for the Amazon.

This was all allied to the deep logic of *uti possedetis*.

Certainly *Os Sertões*'s "Land, Man, and History" served as a template in the production of the background documents. These began in the mapping of mythical realms that gradually are transformed into a national epic not by warrior kings but by a massive, inchoate diaspora. Here the drama of the voyage, the heroic combat against the unknown (whether of scientific mysteries or personal destinies), would substitute for the spectacle of battle at Canudos, shaping the dynamic landscape as metaphor and historical theater. Instead of suffering bloody defeat as at Canudos, the *sertanejos*, his "bronzed titans," would emerge as history's heroes, the protagonists of da Cunha's next, but never finished, "avenging book," the "Lost Paradise."

The supplementary materials were published in June 1906. The completion of this part of the dossier was imperative for Rio Branco, given the controversies swirling around him. His appointment seemed precarious, and there was some talk that he would not survive past the term of President Alves. The British Guyana adjudication had not gone at all well (Brazil had not prevailed), and he was perceived to have bungled the Upper Amazon arrangements.[5]

Da Cunha was neither a naive reader nor a naive maker of maps, and he understood the context of the times. He fretted that the Baron would be replaced by someone else, a prospect he deemed catastrophic.

Euclides wrote this to his friend Escobar in June 1906:

> Even though my [field] commission is over, the Minister has not dispensed with me and has charged me with organizing some maps. So I live now entangled with the old drawings of ancient cartographers, the most disloyal and dishonest characters who have ever appeared in Geography—in the midst of those scoundrels who sketch in rivers and raise mountain ranges at random in order to complement the overall aesthetic of the design, I am passing through cheerless and exhausting days. . . .
>
> Happily I continue to regard the Minister with whom I have worked—the only great man in this place—with same admiration and sympathy. . . . No one can substitute for him. . . . I know about half of the dimensions of the questions that occupy us in the far north, but even with this partial notion it is enough for me to assert that the replacement of Rio Branco would be a calamity. There is such reshuffling of intentions among our neighbors. The historic vices merge with so many cartographic doubts: these then multiply the deceits that abound in our accords, conventions, and treaties from St. Idelfonso down to our days. To decipher such intrigue requires a long-term understanding, an understanding that is difficult to acquire. I don't know who could attain it from one day to the next, nor how a simple nomination decree could outfit just anyone with the necessary qualifications. I know that the litigations under way are extremely serious and able to generate the greatest and most painful surprises for us.
>
> Imagine just this case: one-fifth of the most opulent Amazon, due to some misguided direction by an inept statesman or the caprices of an arbiter, could pass from our hands to the Peruvians'. . . .
>
> The rabble of sycophants is making these campaigns unsuitable for the sincere and worthy. The noblest acts are subjected to the most deplorable interpretations. . . . The atmosphere here is poisoned. . . .
>
> Peace then, this rude pen of a *caboclo*! Or better yet. It's better to go and outline the first pages of my *Paraíso Perdido*, my second book of vengeance. If I can do it as I imagine it, it has to be (and forgive the vanity), it has to be an enigmatic being, one meant for posterity, and truly incomprehensible to these low kinds of men.[6]

"Our own geography remains an unwritten text": Critical Cartography and the Maps of da Cunha

Maps are often viewed as simple transpositions of three-dimensional spatial elements into a two-dimensional schema that conveys something about landscape. Whether decorative or communicative, maps have been seen as technical or artistic renderings through mechanisms of survey and simple mathematics. But maps are representations, they emerge out of a social matrix to express "interests," and they are produced with intentions. They are not unbiased spatial images but ways of ordering specific kinds of information. Maps, like histories, scientific reports, and travelogues, are texts.

With the interpretive turn to "critical cartography," maps have come to be treated as visual elements that shape and manage the understanding of places, components in the imaginative and technocratic procedures mobilized in claiming, imbuing, valorizing, and surveilling space. Maps are mostly "states' way of seeing," And this was especially the case in the nineteenth century. In this view, maps are not value-free scientific products but reflect the covert ideologies and histories inscribed in landscapes. This use of cartography has opened up significant and valuable methods for understanding the dynamics of frontiers and imperial trajectories.[7] Colonial and boundary maps especially express subtle (or not so subtle) politics, propaganda, and aspirations.[8] Like the essays that accompanied the reconnaissance reports and the narratives of the da Cunha survey itself, the maps should be understood as commissioned, authored works that lent themselves to levels of interpretation even as they also interpreted the place. They were simultaneously expressions of knowledge (and particular forms of knowledge) and projections of politics, as well as backdrops in imperial imaginings of a "theater of history."

The historian of cartography J. B. Harley has pointed out in a general analysis of "deep iconography" that while there are obvious conventions in the "text" of a map—its grammar—and the symbols on paper clearly *represent* something in the landscape, there are also meanings that go beyond the simple surface and these idioms to articulate political and spatial power.[9] As Harley puts it, "whether produced under the banner of cartographic science or just overt propaganda, [the map] cannot escape the involvement in the way that power was deployed."[10] Within these general outlines there are several modalities ranging from simple propaganda and the efforts of survey and internal administration to the more subtle, more quotidian creation of place, what Paul Carter has called "spatial histories," in the transformations from the "invisible" or *ignota* to imperial terrain.[11] This approach has been singularly useful in focusing on the diversity of forms of the colonial enterprise and in revealing how historic myths and imperial desires interacted in the semiotics inherent in maps.

The content of maps is influenced by their communicative role—that is, who is supposed to be reading them, which patrons request their production, why, and who the cartographers were. They did a great deal of the work involved in the ideological and visual shaping of place in da Cunha's day, not only because maps were obvious symbols of territory and empire, providing the backdrop for imperial and national gestures of conquest and management, but also because they instantiated in their austere lines past debacles, achievements, and future ambitions. While only images on paper, they often became law on the landscape. The boundary disputes and adjudication processes related to Amazonia are singularly interesting because they explicitly reveal the ideological interests of the maps and provide the texts for particular readings of them.[12]

Imperial Mapping

The questions posed by maps went well beyond whether they were "good" or "bad," "accurate" or "false"; the meaning lies behind their virtue or precision or lack of thereof. As rhetorical devices, maps took authority from being "scientific" rather than simply decoration or propaganda, but the technical means of producing the map, whether through celestial orientation, traverse surveys, or copying, did not leach political content and domination from the acts of representation. Maps ratified claims, whether or not locals had any idea about these assertions of dominion, as with the partition of the New World by the Tordesillas Line or the maps used by European states for the Scramble for Africa, or, for that matter, for the terrains of North America. Geographical gerrymandering embodied accidents of discovery, ceremonial and commercial assertions of sovereignty, and imagined territorial futures.

Shoreline areas and views from the rivers would almost by definition be best known, since ship pilots were the practical mapmakers of the day. Field cartographers were mostly "aquatic creatures," so it is not surprising that interior spaces of continents were ciphers marked by mythic toponyms that acted as placeholders for arenas of future conquest. Such toponyms might indicate indigenous populations (Land of the Chunchos) or resources (El Dorado). It was symbolically and politically less provocative in imperial matters to situate oneself next to the "Land of the Amazons" than in "possible Spanish territories." Forbidding snakes surrounding the maps (or, alternatively, crowned heads of Europe implicitly claiming "the Map") and drawings of naked savages enjoying human barbecues on inland spaces became cartographic clichés among the map copyists. The threats implied by the maps were the first lines of defense against incursions when the areas were vast, knowledge minimal, and militaries sparse. But maps were also used as seductions: Esmeraldas, El Dorado, the Amazons themselves were useful ways to lure adventurers and imperial freelancers to test the reality against the chimeras of fables and rumor.

At the level of the state, the great South American interior was a vast obscurity where myth, chronicle, and theory were easily conflated. As da Cunha put it, "what we know of the *sertões* is little more than its rebarbative etiology, *desertus*. . . . Like medieval cartographers idealizing portentous Africa, we could easily inscribe on large swathes of our maps our own searing ignorance and dread: *Hic Leones*. . . . Our own geography remains an unwritten text."[13]

"Mistakes," too, had their logic, as the inventively mapped Tordesillas Line relentlessly moved to the west for hundreds of years on Brazil's maps of its interior, effectively incorporating millions of hectares into Brazilian national territory.[14] A lying map was a dangerous yet useful thing, one that could lead travelers to doom, or mislead them, or transfer huge terrains from one country to the next. Early maps were persistently and creatively copied. After the Luso-Hispanic empires embargoed information about the continent, maps were increasingly creations of rumor and aesthetic tradition. The misdrawn lines of one map could become territorial assertions as they were transcribed onto another. Lying maps were also parts of security strategies. The map of the *varadouros*, the interfluvial pathways of the Upper Amazon produced by da Cunha in March 1906, was a semi-lying map, meant to confound field military copyists as tensions heated up. Its production marked da Cunha not only as a highly skilled cartographer but also as one thoroughly schooled in cartography's complex uses.

It was in the "deep" cartography of the Upper Amazon and through the question of errors that da Cunha would both assert authority for his own maps (and the Chandless Chart produced with local sage Manoel Urbano) and claim competing geographies and cartographies to largely be fictions, lacking the authority of "having been there."[15] That maps could be texts of multiple and layered meanings, and could be produced and read with rhetorical and political eyes, was clear to the both the author and the cartographer in da Cunha, who as an artist and a scientist understood the powers both of fiction and of what was taken as fact.

Wars Made Maps (and Maps Made Wars)

Cartography was known as the "science of princes" due to the direct relation of systems of territorial knowledge to questions of state power. The usefulness of maps in defense and warfare has a deep history, and this political use of them is obvious and well known. Mapmakers marched with armies. The impetus for the Purús mapping exercises, after all, had been the Acrean War and the Peruvian skirmishes. The claims of old maps underpinned this conflict as much as modern resources. If negotiations failed, then reconnaissance maps would become inputs to military campaigns.

Maps visually carved out national space and made purported nationals of those who resided within the outlines of the cartographic frame. Brazil had cartographically "incorporated" everyone to the east of the Paraguay River and south of the Amazon as Brazilian by 1750 via the map lines of the Treaty of Madrid. Regardless of how the inhabitants might have classified themselves (as native nations, for example), they were now technically Brazilian. Brazil was inventive in its spatial logics, relying on juridical rationales when convenient, on military skirmishes, and on settlement histories read through the lens of manifest destiny, even when their true meaning was flight *from* Brazilian slavery, as was the case of Amapá. Maps showing settlement were an established weapon in the larger Brazilian expansionist trajectory, as we have seen, even when the land under claim was "inhabited" by half-wild cattle, runaways, hastily transferred colonies, downgraded missions, and the like.[16] The outcome of the Brazil–British Guyana controversy in favor of the latter revolved around aggressive cartography of Richard Schomburgk's surveys and Brazil's delay in transferring Indian tribes to the region to function as *muralhas do sertão* (bulwarks of the backlands) and "Brazilianize" the landscape.[17]

The usefulness of physical settlement for territorial appropriation was in part why charter companies, colonist schemes, and large landed enterprises were viewed with such alarm during the Amazon scrambles. Charter companies and colonization schemes might be able to establish definitive territorial sovereignty within the area they claimed as a consequence of the terms of the territorial contracts. Such colonies often engaged in "statecraft" processes like establishing their own currencies, stamps, and flags. Da Cunha was happy to invoke US parallels to justify Brazil's own territorial ambitions, but having large American populations in situ during a period of North American imperial enthusiasm was far more problematic. In Africa, Belgian charter companies had completely superseded the various Kongo nations within the boundaries of Belgian charter rights and had made the Congo a crown colony; the Belgians were currently active in the Madeira watershed as well. Amazonia had to be settled by Brazilians, or would-be Brazilians, or it risked territorial losses from within. This anxiety was clearly expressed by da Cunha, who saw in the terrains of the scramble a zone of appropriation and, disconcertingly, the birthplace of a potentially new (non-Brazilian) nation if his country didn't prevail.

Maps Make Subjects

Maps could make a "people," and people could make a "nation," but states (and their maps) also made subjects. James Scott's *Seeing like a State* has emphasized the degree to which states sought "legibility" of places and populations by creating more uniform tenurial regimes, more easily identifiable landscapes, asserting dominion over places through enumeration of populations in censuses and land ownership and land use surveys. By normalizing and rationalizing such systems, states ratified and legitimized certain forms of knowledge and spatial organization. Knowing who was where, and how many, and what the resources might be enhanced the potential for control. Taxes on products, land, and population and other forms of revenue might be extracted once the state knew what (and who) was there. The Upper Amazon had considerable revenues that were moving through the watersheds and *varadouros* untaxed. And in the great outback, where economic and political "nature" abhorred a vacuum, absence of a state meant the presence of someone else: the freebooters and entrepreneurs of empire, neighboring polities or "pseudo-states" such as charter companies.

Neither the Peruvian nor the Brazilian state had much administrative or military capacity in these regions,[18] and in fact it was only the smoldering conflicts that had spurred these polities out of their torpor to chart an area that had been previously mapped in the 1860s by Chandless as an emissary from the British Royal Geographical Society. The *seringalistas* and *caucho* barons ran the show and had their own militias in any case. Peru and Brazil could find in the maps useful elements for larger state strategy, since the Purús map served as an inventory of land use from which more exact information on production might be derived. In its naming of places and establishments it functioned as a cadastral survey and census. The map showed dense populations and a plethora of Brazilian toponyms, with its mapped and unmapped *varadouros* weaving the whole region together.

Euclidean Geometries: Reading the Maps of da Cunha

The marching orders to the commission were clear about what had to be "on the map." These were not simple cartographic questions, although on the surface they appeared innocuous. There were some geographic questions regarding relatively stable landscape elements (distances from Manaus, for example), but the concerns over other apparently unproblematic components carried within them far deeper subtexts implicit in corrections to the 1865 Chandless map, mapping of the *varadouros*, and "observations of interest."

The Chandless map, generated with the assistance of Manoel Urbano, was, as noted, produced for the Royal Geographical Society as part the effort to understand the "Madre de Dios question" and can be seen as part of the broad push at the time to map the tropics and document its "scientific" and resource natures. The Chandless map, accurately and scientifically produced, was a useful compendium of people, places, and commodities.

The Peru-Brazil Joint Commission's marching orders required it to review and revise the Chandless map where necessary, and it was the template for the Buenaño/da Cunha chart. Chandless's elegant map delineates native territories, trading posts, and historical landmarks (missions, battlegrounds, etc.). There is the smattering of settlements mostly named for wetlands, lakes, river features, Indian villages, the mouths of tributaries, local history (where the Purús Indians had been slaughtered), Manoel Urbano's *feitoria*—small workshop and processing factory—and the ruins of the missions. It illustrates a number of anthropogenic landscape elements such as the channel and *furos* (aquatic shortcuts) and documents numerous *varadouros*, including those that led over the next watershed, the Juruá. It names the rivers as black water or white and indicates or hints at all the connections and relations within the upland and the riverbanks, mostly through the naming of native places.

The Buenaño/da Cunha map with its "corrections" is reproduced here as map 9. This map sits suspended in its grid with no other referents than latitude and longitude. The area away from the channel is mere blankness. It documents the place-names of all the estates—the fuzzy transverse lines that give this reduced map the aspect of a caterpillar. The areas where the density of names declines are those inhabited by *caucheiros*. By the time the expedition traveled, both sides had regional administrative posts as means to reduce tensions during the adjudication, and these are indicated on the map. Other particularly important sites (the ruins of missions, administrative seats, the oldest establishments) were "geolocated" through astronomically determined

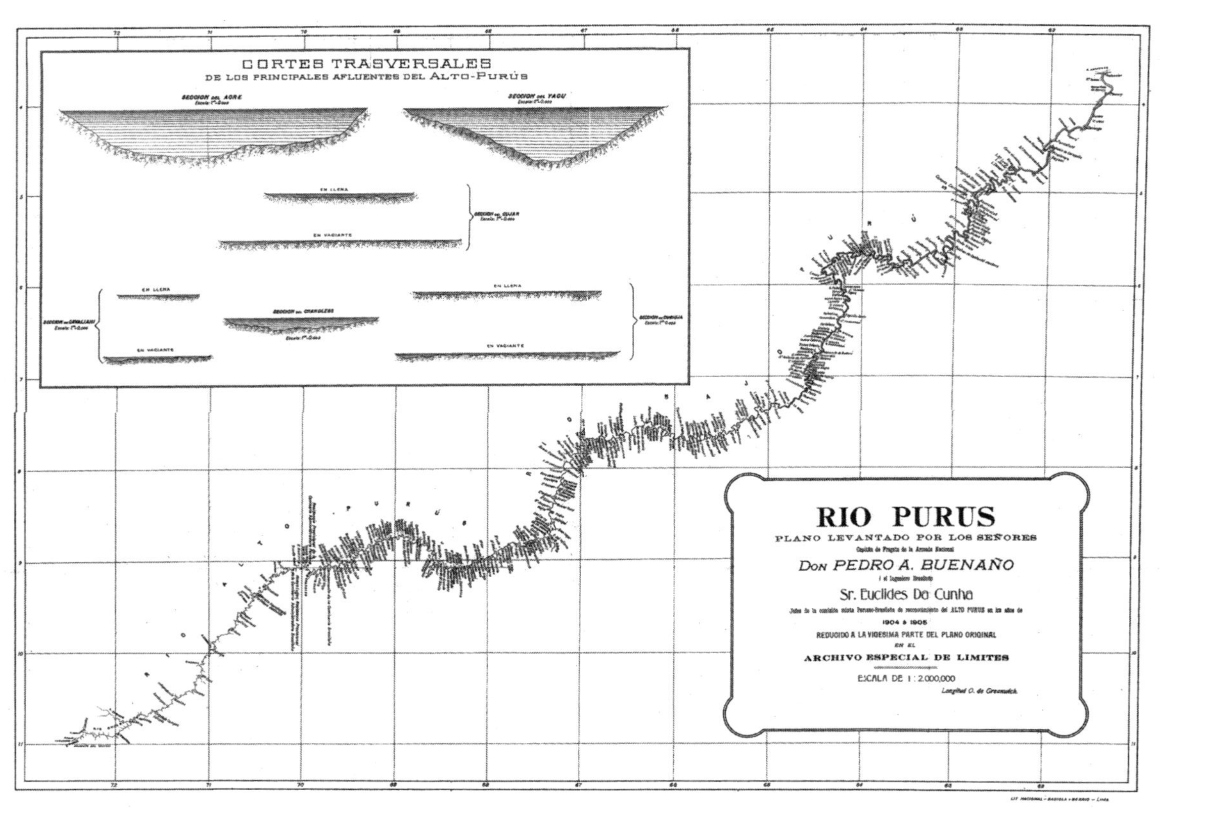

Map 9. Buenaño/da Cunha map.

points of latitude and longitude that accompanied the travel report but were not on the map itself.

The Buenaño/da Cunha map was an effective Brazilian imperial document, one whose joint production gave it the imprimatur of legitimacy. The Peruvians, as explained earlier, were interested in undermining this cartographic effort but were also in many ways indifferent to the meanings of this particular act of scientific mapping. They believed that the juridical antiquity of their claims would not be superseded by the charting of river courses deep in the heart of what they took as "their" terrain. They saw the exercise as mainly carrying out political requirements of the Treaty of Petrópolis and as a way to gain useful military and economic information on a region about which they were poorly informed.

Every possible idiom of imperial right would be invoked, but this map and its discursive accompaniment revealed a powerful story of unremitting Brazilian settlement that stretched up the river for more than two thousand kilometers and more than four hundred settlement sites and trading posts. With each place-name the fuzzy, sinuous map line spoke of geopolitical presence, especially compared to the occasional village far upriver in the Peruvian landscape. Da Cunha's other maps, those of the *varadouros* and of the *seringal*, would, as we will see, imaginatively expand that fuzzy line into an immensity of settlement and connection, reaching, by implication, back into the terra firma, woven together by the *varadouros* "that spread everywhere." The simple visuals of the map also expressed naked power: Peruvians were vastly outnumbered, they controlled little of the watercourse, and should Brazil choose to do so, it could (and did) close the river, obligating Peruvians to transport their entire production over to the Ucayali waterways, or even over the Andes.

For Brazilians, the documentation of the settlement of the landscape by Portuguese speakers would have several important effects. First, it expressed an "imperialism" without state imperialism, the everyday forms of territorial occupation that were implicit in the names, and bore witness to a conquest not by rampaging armies but by the footsteps of legions of humble northeasterners. Next, the names themselves were deeply expressive of an expanding spatial history, emblems of nation creation produced in the ordinary practices of daily life, the planting of orchards and naming. These were not sites of imperial baptism but rather the modest christenings of nascent societies.

Fifty years after the Chandless map, the Buenaño/da Cunha chart was crammed with names expressing everything from geographic clichés (for example, Boa Vista—nice view) to gastronomic names, surely of interest to earlier extractors, and possibly residues of native practices: Peach Palm, Cacao Tree, Brazil Nut Stand, Pirão (Pestle), Umari (a fruit), Tambaqui (a large catfish)—a delicious sampling of Amazonian domesticated and wild edibles. There are agronomic descriptions: Large Meadows, Red Soil, High Lands, Good Gardens, Black Wood. The channel is predictably full of the names of saints, invoking them as protectors and the place itself as offerings, as well as nostalgically evoking towns and hamlets elsewhere in Brazil. There are geographic transpositions: Europe, Barcelona, Palestine, New Olinda, Belém. Da Cunha noted in his histories a shift in existential conditions ranging from "Suffering" to "Bom Lugar"—good place. It was a world where Brazilians were applying names, imagining goals, inhabiting the country. The voids of the Chandless map had been turned into active places of the production of daily life and global commodities, crammed with Brazilian settlement. But what had been the "fullness" of the Chandless chart, anchored to its Indians, was now emptied.

In the supplemental documents (see chapter 20) da Cunha would repeat the story of improvement, of attachment to place, of taming the wild, of an adapted new race, the best promise of a tropical civilization putting down roots and creating a nation and an economy from a swamp. The semiotics of the fuzzy line with its relentless names would invoke a history, a culture, and a destiny. This forward arc of modern nationhood would be compared with the sparse and nomadic trajectory of the Upper Purús, whose inhabitants and habits had made it wilder and poorer than when they found it through the subjection of natives and depletion of resources. Visually the map revealed a populous world morphing into a mostly empty one and was meant to be read through the uplifting narrative of an emergent Brazilian civilization countering a sparse, moribund Peruvian barbarism destined to extinction with the last *caucho* tree.

Varadouros

The reconnaissance of the Purús was under specific orders to explore the *varadouros*—the land shortcuts—to the Ucayali. These routes were of military interest to both camps, due to the Peruvian garrisons based in Iquitos as well as the economic usefulness of overland access to the deep-water port in Iquitos and the connections among the great rubber rivers of the Upper Amazon. Further, there seemed to be more administrative penetration in the Upper Purus by Peruvians than Brazil might have liked. One might recall from da Cunha's informal report of the voyage to the Baron in his that the Peruvian commission was able to commandeer *ubás* "from mail service from Iquitos." The route of this tropical "pony express" with its sodden letters was not down the main channel of the Amazon, some 1,000 km, and then 2,500 km up the

Maps 10a and 10b. Details of Chandless map, 1867, and da Cunha map, 1906. These were charted in the same place about fifty years apart. The da Cunha map reveals a much greater density of habitation.

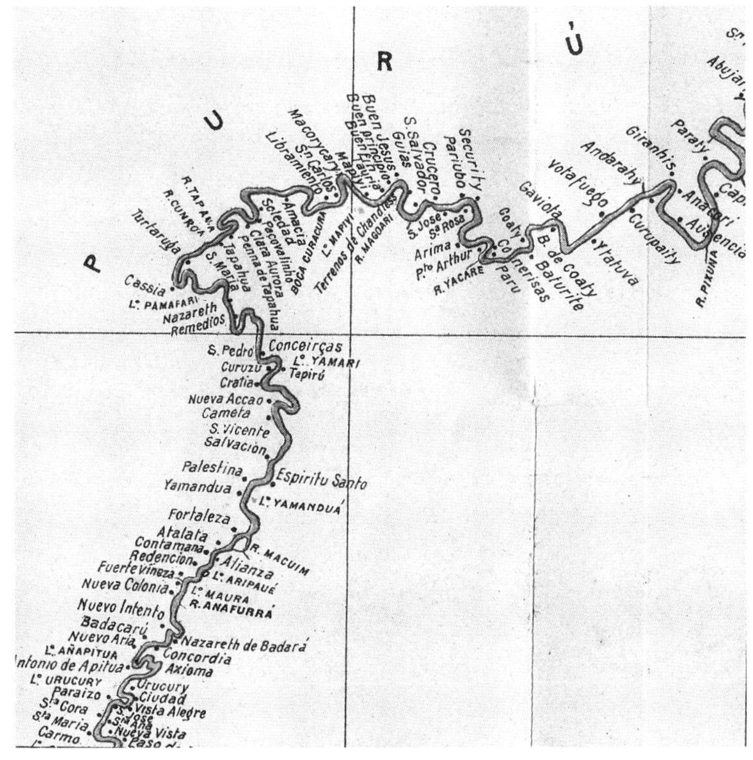

Purús to deliver mail at the headwaters. The post relied instead on the system of *varadouros* to go from the Upper Ucayali to the Juruá or the Purús in a fraction of the distance and time. The implications of this were clear. An ability to shift quickly from one watershed to the next in that landscape of endless meanders would provide a clear economic, administrative, and military advantage. This strategic geography was made more obvious as tensions heated up after da Cunha's return to Rio, when the State of Amazonas requested a more detailed set of maps, one with all the *varadouros* on it, not just a few obvious ones, that led to the Ucayali.[19] These passages themselves were elements in the arsenal of claiming territories; they had economic importance as substantial infrastructure works, as the image of a *varadouro* on the Madre de Dios reveals (fig. 19.1).

In this light, the meaning and history of the *varadouros*—connections between watersheds and meanders—provided by da Cunha in the travel report (see chapter 20) are significant as rhetorical and political mapping strategies. Da Cunha would (incorrectly but imaginatively) credit their invention to the Paulista *bandeirantes*.[20] He portrays *varadouros* as a *Brazilian* landscape artifact, an element as indicative of cultural continuities as language. Through this wide extension of a "Brazilian" activity he provides a framework of expansion and unification into the uplands beyond river channel. He described it in this way:[21]

> Between one water course and the next, the forest's expanses are as isolating as though they were mountains . . . they separate and divide. . . . One sees then this inversion: man, instead of mastering this land, has become enslaved by the river. The populations never expanded outward but merely extended themselves. They stretched out along infinite lines and returned always by the same route they started off on. Thus they immobilized themselves by the appearance of an illusory progress, as in the retreats and advances of the adventurer who sets out, goes to the end of the earth, but explores and returns on the same trail, monotonously renewing the same itineraries. . . . In their short but very active history, the region's new staging points, except for a few minor variants, continue to press steadily along those routes already opened to the southwest: three or four risks, three or four river voyages, and the uncharted oceans of the forest.
>
> This discouraging social consequence, created by what had seemed so auspicious a natural feature, namely the great arterial rivers, corrects itself by a series of transverses connecting its valleys. The idea was not original. It began with our Paulistas.[22] It is the path, the shortcut, that goes from one fluvial slope to the next. At first tortuous and short, suffocating, down in the forest's thickness, the *varadouro* reflected the indecisive steps of an emerging, vacillating society

Figure 19.1. Varadouro Souza Vargas at the headwaters of the Rio Mishagua, 1906. The track was a system used for moving small craft and material in overland portages.

that abandoned the comforting lap of the rivers and chose to walk to itself. . . . Today the narrow one-meter-wide trails of this society, trails hacked out with machetes, stretch everywhere. They divert themselves in countless directions and crossroads, linking the various affluents of the headwaters: the Acre River to the Purús, thence to the Juruá River, and then to the Ucayali. They trace the contemporary history of the new territory in a way entirely counter to primitive, impotent, fatal submission to the natural lines of communication. . . . Taking to the trails, man in fact is not submissive. He is an insurgent against affectionate and treacherous nature, which enriches and kills him.[23]

This "invented tradition" was used ingeniously by da Cunha. First, it allowed him to reinforce his idea of an emergent Amazonian civilization evolving its own autochthonous cultures and destiny as "a people no longer enslaved by the river." By linking *varadouros* to the Brazilian epic of seventeenth- and eighteenth-century mixed-race *bandeirantes*, he invokes and extends both the racial and the primordial territorial history of Brazil's legendary mestizo heroes and mythologizes the contemporary tappers who reenact this singularly "Brazilian" story. The connections among the Upper Amazon rivers thus echoed the ways that the *bandeirantes* had made their circuits of slaving and discovery through eastern and southern Amazon terrains and brought them into the ambit of southern coastal Brazil. A fretwork of *varadouros*, a historical and geographical continuity created by the modern *bandeirantes*, would entwine the upland forests with the Brazilian polity. The maps thus outline the tracings of a type of landscape "innovation" that by its ubiquity would help claim the interfluves for Brazil. In its mythical content and spatial integration, this connection would reprise and, more importantly, *complete* the tasks of the "lusiadic" travels of the *bandeirantes*. The *varadouros* would, da Cunha argued, be akin to the building of railways to the American West or the Siberian East. The heroic past would be connected to a modern future through the actions of an activist state employing modern technologies that could build on ancestral pathways.[24] In fact *varadouros* were often parts of an extensive indigenous system of connections between settlements and movement through headwater areas, a latticework of communications.[25]

Silences and Lies

Maps are also interesting for what they do *not* reveal—cartographers point out that their silences can be as evocative as their traces. In the Buenaño/da Cunhas map, the notation of *varadouros* is surprisingly infrequent, even though, of course, they were "everywhere." But more interesting, perhaps, is the map that da Cunha sent down with his private letter to Rio Branco and recopied and signed in its final form in March 1906. This map shows a few of the main *varadouros* that lead from the headwaters of the Purús into the Ucayali and the Madre de Dios, but its most notable feature is that the elements are inverted. Longitudinal lines run from left to right. The Urubamba/Ucayali and Madre de Dios/Manú meander through what in most conventional mappings would be Brazilian territory, while the Purús twists through "Peruvian" terrains. The Manú River appears to be the main tributary of the Amazon. The "observation" notes that the map is inverted "following conventions of the Northern Hemisphere." Basically the map is upside down and reads as

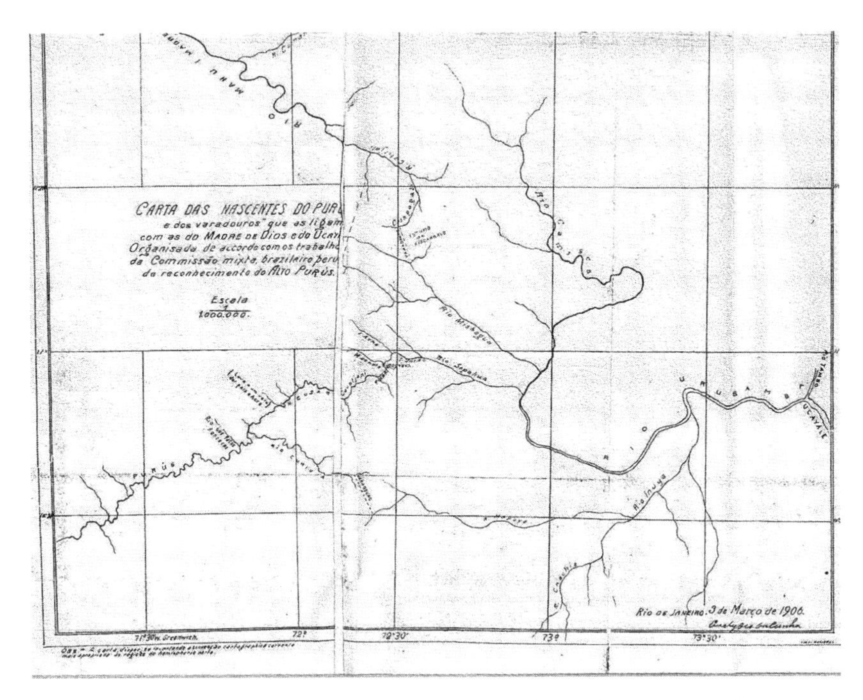

Map 11. "Map of the headwaters of the Purús and the *varadouros* that link them to the Madre de Dios and the Ucayali. Prepared in accordance with the work of Joint Reconnaissance Commission of the Upper Purús." The dotted lines represent *varadouros*. These were incomplete mappings, since the Chandless map shows many more. The text at the bottom left states: "This map inverts itself according to cartographic conventions that are appropriate to the Northern Hemisphere."

though the observer were placed above the equator and facing south—but even so read so, it remains intentionally inaccurate. Since the other maps produced by da Cunha follow standard conventions, this map may have been produced in this way to foil potential military field copyists.

New Lands (and Forms of Tenure) in the Tropics

The third map produced by da Cunha outlined a new property regime that defined the spatial structure and labor system of the rubber world. This little sketch map seems innocuous enough, but within it, once again, da Cunha articulated an emergent cultural/institutional novelty, one that put formalized human *place* onto what might to outsiders seem to be inchoate nature. He describes a technique and form of land tenure unknown away from Amazonia, invented by the locals and evolved from ecological peculiarities of the forest. This survey related to the distributions of trees rather than the grids

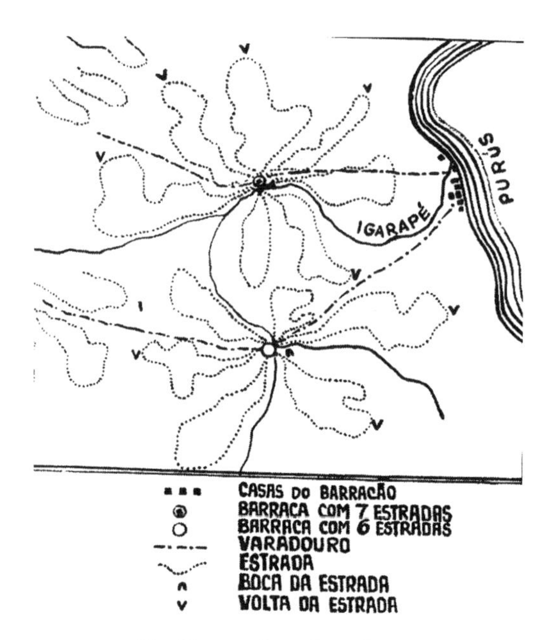

Map 12. Estradas and *varadouros.*

of land tenure and reflected forest realities unknown to European and agrarian survey. The sketch map outlined property boundaries not as land *area* but as *arboreal economic space*. Only decades later, in the 1970s and 1980s, would these usufruct tenurial forms convert to different property regimes amid bloody contention. The tenure model described by da Cunha would eventually become the foundation for conservation units known as extractive reserves.[26]

Da Cunha's map, presented in the fashionable journal *Kosmos*, in January 1906 shortly after his return, is reproduced here as map 12. Each *barracão* (rubber post) implies six or more *estradas*—tapping trails—and their extensive areas behind the river port. This was the everyday form of occupation, stretching out, as da Cunha saw it, like an octopus, incorporating the landscape away from the river in its economic tentacles. This was the ambiguous subtext to the triumphal Brazilian tenure that extended from one river basin to the next.

By the time Euclides steamed up the Purús, there was already a huge industry of surveyors. Alfredo Rangel, at whose charming book-crammed villa he had awaited his departure, had used his military survey training to make a fortune measuring out the vast estates to secure claims of uncharted lands for the *seringalistas* of the rubber forests. Plácido de Castro, the caudillo hero of the Acrean War, had been a surveyor as well and probably owed his success in guerrilla battles to his familiarity with the rubber estates of the region—he had mapped out many of them and actually explained *seringal* survey methods to da Cunha.

The large geopolitical map of the Peru-Brazil Commission provided the macro context of national territoriality, while the new mapping systems of the *seringais* showed the tenurial system that spun out behind each of the named posts of the Purús survey map and the populations implied behind each toponym on the river. Like the system of *varadouros*, this was a totalizing and integrating imagination of landscape settlement. It projected the territorial sovereignty of the micro, a commercial dominion following the *varadouro* lines, deep into the forests of the river interfluves. The ethnography of this process is meant to illustrate the emerging practices, part of an evolving and novel institutional repertoire. The central point of this map was to fill in empty spaces in the maps, the extensive backwoods, with interior boundaries and holdings. The little paths and partial *estradas* suggest an infinity of conduits stretching from watershed to watershed, which in fact they did. Da Cunha was perfectly aware of the ambiguous political subtext of his own maps. Here is how he described making a map of an *estrada*, thousands of which circled through the rubber forests.

> The opening of a rubber estate on the Purús is a task inaccessible to the wisest agricultural surveyor, so capricious, variable, and malevolent is the geometry required for the division of different lots. In fact, the value of land is relegated to such insignificance compared with the unparalleled value of the tree that it has engendered a novel agricultural measurement—the *estrada*, the rubber route or trail—which alone summarizes the most varied aspects of the new society haphazardly perched on the banks of the great rivers.
>
> The unit of measurement isn't the meter, it is the rubber tree itself, and in general 100 trees in unequal intervals constitute an *estrada*. It is understood by all that disparities in form and dimension of that singular pattern are the only embellishments to the nature of their work.
>
> There is no need to trace out another. Lost in the exuberant and rich forests with intentions only of exploiting the coveted *Hevea*, the tapper soon understands that his efforts would flail uselessly in the tangled jungles were he not able to orient and bind them to secure trails, normalizing his labors, giving rhythm and structure to his efforts and pattern to toil so apparently disordered and crude.
>
> The *estrada* is, beyond this, indispensable so that the numerous comrades, clients, or indentured slaves, all fated to work in solitude, do not become confused

or disoriented, deluded by detours in the forest. The rubber trails resolve the question. But tracing them out is itself the first challenge faced by those who want to try to open a rubber estate.

The first task is to quickly set up the rough structure of a barracks, always overlooking the main river, on an embankment of solid ground, and after a preliminary reconnaissance of the estate that surrounds it the owner seeks out a seasoned *mateiro*—backwoodsman—to whom he entrusts the task of dividing and evaluating the holding.

The *mateiro* plunges, without even a compass, into the forested labyrinth with confidence formed of a rare and surprising topographical instinct. He explores, in all senses of the word, the stretch of forest to be exploited; he notes the slopes, absorbs the complex physiography, ranging from the flooded forests (the *igapós*) to the terra firma forests, above the floods, and decides where he will trace out future *varadouros*; he rigorously reviews the *estradas*. And in the same exploration, without the encumbrances of transcribing complicated field notes, he chooses all the places on the creeks where the little shacks of the workers should be placed.

Once this general examination is through, he calls on his two indispensable assistants—the *toqueiro*, or marker, and the *piquiero*, or trailblazer—and they quickly raise a temporary lean-to, a *papiri*, with long fronds of the Jarina palms, and then put their shoulders to the task.

The process is invariable—the *mateiro* proceeds ahead and signals the first rubber tree, which he homes in on as soon as he leaves the hut. This will be the trailhead of the *estrada*. There the blazer, woodsman, and marker gather together. The *mateiro* continues on alone until he finds the second tree, ordinarily not so distant—about 50 meters. With a special cry, he then informs the *toqueiro*, who moves toward him and the new tree, while the blazer accompanies him step by step with his machete and clears a rough path that prefigures the final rubber trail. The marker helps for a time by helping to clear the path, as long as another shout from the woodsman doesn't yet summon them to a third tree. And so it goes to the most distant part—the *volta da estrada* or return point of the trail. From there, using the same method, they return through various detours, going from rubber tree to rubber tree, closing the irregular curve, which ends at the point of departure.

They normally finish this work in three days and the *estrada* is roughly opened. Departing once again from the same place and adhering to this same system, they open yet another *estrada*, centralizing all the paths on the same trailhead, near the hovel that will eventually center the *barracão* or trading post. The *mateiro* then seeks out the next area, intelligently chosen, and proceeds with the same operation until the entire holding has been crisscrossed with paths and the estate subdivided. There, tethered by the tracks to the main trading post raised on the edge of the river, one sees the shacks and rubber trails surrounding it, contorted like the tentacles of an immense octopus.[27]

This is the image, monstrous and expressive, of the tormented society that thrashes in those forests. The bold Cearense arrives there with an unbridled optimism for fortune. After a brief apprenticeship that corresponds to the circuits of the rubber trees, he passes from "wild" to "tame": what this means is to pass from the illusions that first dazzled him to the total apathy of one defeated by an inexorable reality. He builds a hut of *paxuiba* palm at the inlet of a picturesque creek, or perhaps in a clearing hacked from the menacing forest. Far from the baronial trading post where the *seringalista*—estate owner—resides in permanent opulence and complete parasitism, the tapper has the premonition that he will never free himself from the *estradas* that embrace him, on which he will trudge the rest of his life, coming and going in the endless, numbing, monstrous cycle of his exhausting and sterile trek.

This fearsome octopus, like its pelagic miniature, is insatiable, served by many constraining rings or *voltas*. It lets go only when the *seringueiro*, bereft of all illusions, has had all hopes stripped away one by one—when one day, in one of the tentacles, the repulsive, malaria-ridden corpse is discovered, completely forsaken.

Genealogies of the *Estrada*: Discipline and Punish in the *Seringal*

"This description," says da Cunha of his maps of rubber estates, "is the monstrous image of a tormented society." Maps articulate many different, subtle iconographies of power: the power "over" the map exerted by patrons and politics, the power expressed "with" the map—the administrative and military dimensions, including the extensions of "juridical powers" and symbolic icons and naming. The third arena of power in maps revolves around what J. B. Harley calls the power "internal" to the knowledge systems that go into maps, their relation to other scientific practices of production of maps—what is included and what isn't, the hierarchies of importance, and so on—and how these express control over the way the world is imagined through cartographic "conditions of possibility."[28] The social configuration described by da Cunha, his understanding of how maps work as social metaphors and descriptions of the means of control, is relatively clear because he kindly provides us with texts for reading them, first as a spatial technology and next as a kind of social death.

Michel Foucault's discussion in *Discipline and Punish* remains one of the benchmarks for understanding the social and spatial mechanisms of Jeremy Bentham's prison design, the Panopticon, and remains among the most powerful metaphors and analogues for explaining the nature of power and control through surveillance. The character and ubiquity of surveillance involve an extension of "carceral cultures" into the realm of everyday production and social management, a continuum including the microphysics of power in modern institutions and bureaucracies. Indeed, Bentham himself saw the utility of his architecture for hospitals, workhouses, factories, and schools as well as for penitentiaries.

While modern social theorists emphasized discipline exerted by the built environment and the controlling "gaze," practitioners of the punitive in the late eighteenth and especially the nineteenth centuries were concerned with quite a different form for control of deviant populations: the tropical penal colony. Studies of colonialism and slavery generally have not been particularly interested in what seemed overall to be small-scale and minor variations on much grander forms of imperial integration, but the idea of transporting convicts to distant tropical places as forced labor was an appealing idea if slavery or free colonist settlement were not possible for whatever reason. The penal colony itself resonated with colonial powers because they could address the problems of deviance in the metropole and the labor problems of the tropics with one grim institution, as we have seen in the case of French Guiana.

While convicts were the intended population for Europe's tropical penal colonies, it is useful to remember that not every a prisoner was a criminal. The structural changes implied in emerging capitalist relations and enclosure in rural areas and the rise of aggressive urban industrialization created a class of migrants and the dispossessed, who thronged cities and were caught up in and survived in the economic marginalia now called the "informal economy" and petty criminality. Revolutionary movements of the nineteenth century regularly produced political "undesirables" who could be conveniently domiciled thousands of miles away in tropical prisons. The rise of steamboats and the end of galleys generated a large convict population that also had to be dealt with. Stashed at the ends of the world, penal colonies sanitized the metropole of the social "detritus" inherent in economic crises and transitions. In terms of "carceral cultures," it was not the Panopticon but the Void—of exile into invisibility that defined tropical gulags. Social excision, isolation, and nature itself defined the axis of control.

As noted earlier, the backland populations of the Northeast, the *sertanejos*, were facing a complex of socioeconomic disasters at the fin de siècle. The decadal pulse of El Niños unmoored the populations from their subsistence plots and petty commercial plantings and ravaged their meager assets. The restructuring of the period liberated these ex-slaves into realms of uncertainty made more uncertain by the economic contraction of the northeast sugar, and cotton economies. The denizens of appalling refugee camps had little recourse besides petty crime, migration, and bondage: doing one's time, paying one's debt. The transfer of "undesirables" from their concentration camp–like *currais* to labor-starved *seringais* was an ideal solution for socio-political problems that states and local societies were unable or unwilling to address. If Kafka's colonial apparatus wrote the punishment endlessly on the body of the accused, the tapper inscribed his sentence with his body on the landscape in its endless circuits. Here is how da Cunha explained the meanings of his map:[29]

> Those exuberant outposts of rubber and Brazil nuts harbor the most villainous organization of labor ever designed by avaricious conceit. The *estradas* are themselves a diagram of rubber society, characterized by one of its most depressing attributes: obligatory dispersion. Man there is solitary. Even in Acre, where the greater density of rubber trees permits the opening of 16 rubber trails in a square league, this entire enormous area is easily exploited by just eight people. From these come the limitless latifundist holdings, where in spite of the permanence of active exploitation one notes the great desolation of the wilderness. One estate embraces 300 *estradas*, which corresponds to roughly 20 square leagues. All this anonymous province carries at the maximum a force of 150 laborers.
>
> From the first slash of his tapping knife, the *seringueiro* is trapped within an unendurable cycle: the exhausting struggles to undo a debt that constantly expands, always equaling the energy of his toils. In the dreary round of his existence he is completely alone. In this respect, the exploitation of rubber is worse than that of *caucho*: it enforces exile. Dostoyevsky's darkest narratives of Siberia could scarcely capture the tapper's torment: a man confined to the same trail, tethered to the same trees, for his entire life, setting off each day from the same point along the prison of his dark and narrow passage. For him, the task of Sisyphus is to move not a stone but his own body along the cramped arcs of the endless circuit of this wall-less dungeon. It takes but an hour to learn the task that will consume him the rest of his life. In fact, the *seringueiro* . . . reveals this anomaly: he is a man who works to enslave himself. . . .
>
> Flee? He doesn't dream of it. What restrains him is crossing an unmarked distance. Seek another station? Among patrons there is an agreement not to accept the workers of others unless their debts are paid off—and in Acre recently there were many meetings for formalizing this alliance, creating heavy fines for recalcitrant patrons. . . .
>
> This circumstance, this unleashing of activities in a highly dispersed routine, matched to the other anomalies that we have discussed, contributes substantively to the stagnation of a civilization that thrashes about as it drowns in the

> thick woods, sterile, without destiny, without traditions, and without hope, in an illusory advance that always returns monotonously to its point of departure, like the depressing *estradas* of rubber trees.

It was not by accident that da Cunha's first public presentation of his Amazonian explorations contained a denunciation, even as it glorified Brazilian ingenuity. This paradox and ambivalence is key for understanding his work. But he was never purely an ideologue, and it is his deep empathy that pervades one of his most moving fragments of Amazonian writing, "Judas Asvero," and captures best the anguish of this internal exile:

> On the Saturday preceding Easter, Hallelujah Saturday, the *seringueiros* of the Upper Purús liberate themselves from their days of sorrow. On this day the tappers hallow all evils and, following their basic view of life, embrace the sanctification of the worst lapses. Above, God, infused with delight at the arrival of his resurrected Son, now finally freed from human treachery, smiles complacently at the fierce happiness unleashed below.
>
> They have no solemn masses, nor luxurious processions, no footwashing, no touching ceremonies of abnegation. During Holy Week the *seringueiros* continue in the torturing sameness of their immutable existence, fashioned of identical burdensome days of penury, of half-starvation, which for them is a kind of interminable crucifixion—an endless Good Friday extending infinitely, relentlessly though the entire year.
>
> Some tappers vaguely remembered that in their distant native villages during that funereal period of Holy Week people stopped all activities, deserting the streets, paralyzing business—the flames of tapers flickering in the agonized processions, voices faded into prayers and meditation, as a great and monstrous silence fell over the cities, towns, and villages, even well into the backlands, where an anguished populace identified with the prodigious suffering of God. Enraptured, people assumed that these seven exceptional days of sorrow were transitory everywhere, and so established the greater happiness of the other, more numerous days. But in the forest, days of mourning are human beings' entire existence, monotonous, painful, obscure, and anonymous—oppressive rounds of bitter and unalterable paths, without beginning and without end, inscribed in the closed circuit of the rubber trails. The shadow of a singularly pessimistic view of life falls on their simple souls and darkens the most dazzling illusions of faith: No Redeemer will save them, He has forgotten them for all time. Or maybe he just didn't see them, remote as they find themselves on those solitary rivers, rivers whose very water is the first to flee those sad, unfrequented, and forgotten recesses.

But they neither rebel nor blaspheme. The rough *seringueiro*, in contrast to the artistic Italian, doesn't abuse his God's abundance with excessive remonstrance. The backwoods position is stronger, purer. They resign themselves to misfortune. They don't grumble, they don't pray, although sometimes anxious appeals will ascend to the sky carrying the frank bitterness of resentment. But the *seringueiro* doesn't complain. He has a tangible practical notion, that of fatality—a notion without excuses or metaphysical dilutions, a massive and inexorable notion. And he submits to this burden without the subterfuge of kneeling and cowardly pleading. It would, in any case, be a useless exercise.

In him dominates the rough criterion of a conviction perhaps too objective or ingenuous, but one that is irreducible—one whose reality shadows him at every instant: it is that distance itself excommunicates him from other men. Even God prefers not to sully his view by casting his vision into those bayous and backwaters, so it's not worth the effort to become a penitent (which is a cautious way to be a rebel) and demand a place in the undefined hierarchy of blessedness. In any case, there are other competitors for those places who are happier, better connected, and more numerous. And making the whole enterprise more efficient, they are more visible in the chapels, churches, and cathedrals of the rich cities, where they can display the pomp of a suffering embellished in black or retreat into an extravagance of tears, vaunting their sorrow.

There in the forest the tapper continues, impassive and mute, stoically, in the face of his great misfortune. But beyond this, he alone is allowed to punish himself for the cursed ambition that led him to those places where he delivered himself, shackled, a slave, to brazen and treacherous traffickers. His sin, which is its own punishment, transmutes his life into an eternal penance. What remains to him is to unmask and wrest his spirit from the obscurity of the forest, revealing it nakedly in its most terrible form to a distant humanity.

To do this, the Church has given him a sinister emissary: Judas—and only one day of celebration, Hallelujah Saturday, the Saturday affixed to the most sacred outrages, the most confessable disorders, the mystical turbulence of those embracing the apotheosis of vengeance.

But the usual straw effigies of this outcast on this day are banal and nowhere up to the complex and grave mission of the tappers. The effigy Judas arrives beaten down by centuries of rejection, so trampled, so tattered and abused, that his infinite misery is transformed into something trite, cheaply monopolizing the universal odium, further diminished by the insipid general loathing. They emphasize in streaks of charcoal the most vivid and cruelest lines on his grimy face, a torment so tragic and in so many ways close to reality that the eternally damned one appears to resuscitate at the same time as his divine victim. This practice defies a more spontaneous repulsion and understandable vengeance—and satiates the resentful souls of believers with a perfect image of Judas's misery and terrible agonies.

The *seringueiro* embraces this effigy as his own statue, assisted by his children, who delight loudly in giggles and run in search of a few straws and bits of repulsive outfits made of old borrowed clothing, enchanted by this acrobatic task, which breaks for them in one blow the sad monotony of a quiet and invariable existence. They make Judas as they always have, an old shirt, a pair of pants coarsely sewn and full of patches, with outstretched arms and jointless legs akimbo, impaled upright in the center of the patio. On top is a ball of rubber ungraciously representing the head. This is the vulgar mannequin produced everywhere, which satisfies most people. But this is not enough for the tapper. For him it is from this rubber block that he will bring forth his sculpture, the tapper's chef d'oeuvre, a disturbing creation of his primitive genius, informed by his ceaseless adversity. Where others perhaps distinguish the admirable traces of a highly subtle irony, for the *seringueiro* the carving is a concrete expression of a painful reality.

He thus goes round the deformed visage, emphasizing and decorating the nose, deepening the eye sockets, sculpting the forehead, accentuating the cheekbones, modeling the jaw in a slow and careful massage—he paints his eyebrows and opens slowly, with slow, patient fingers, the eyes—usually sad and filled with a mysterious gaze. He designs the mouth, shaded by a coarse mustache and downturned at the corners. They dress him later in shirt and pants of cotton, still serviceable, and provide him with old broken-down boots. The *seringueiro* steps back a half-dozen steps, contemplating Judas for a few minutes. He studies him. All around the kids, now silent, wait expectantly, watching and marveling at the unveiling of the creation.

The tapper returns to his homunculus, retouches the eyelids, molds an expressive grin in the arch of the lips, shades in parts of the face, thus modeling it, repositions the head, folds the arms, adjusts the shirt . . .

A reconsideration, deliberate and slow revision with view toward total effect—the exact impression, the synthesis of all those lines, and the renovation of all that labor with the persistence and anguish of an unsatisfiable artist. Retouches, more delicate, more careful, more serious, the tenuous extension of shading, an almost imperceptible trace on the mouth, an insignificant twist of that neck decorated with tatters.

And the monster, bit by bit, in a most subtle transformation, gradually becomes a man. At least the illusion is startling.

Suddenly the sculptor makes gestures somehow more stirring than the anxious cry of Michelangelo:* he rips his hat off and fixes it to the head of Judas, and the kids all leap back in with a cry, seeing in the disjoined and sinister creation the figure of their own father.

It is a painful triumph. The *seringueiro* has sculpted the cursed one in his own image. He has taken revenge and finally punished himself for the damning ambition that brought him to this land, taking vengeance on himself for the insulting moral weakness that is all that remains of the rebellious impulse, ever more muted by the degraded life to which an infantile credulity has now yoked him, a slave to the swampy lands and the traffickers who duped him.

This, however, doesn't satisfy our *seringueiro*. The material form of his own desolation should not remain useless in the small clearing of his hovel, suffocated in the impenetrable obscurity that hides the tableau of his perpetually anonymous suffering, invisible even to the eyes of God. The river that passes by is his conduit to the rest of the world: his annihilation is at once exteriorized and expressed through this strange, mute herald so that that the whole world might contemplate his misfortunes, his unfair humiliation, his worthlessness.

Below, constructed in front of the shack the night before, one sees a raft made of four buoyant logs, roughly worked. It waits for the gruesome traveler. The *seringueiro* takes him hurriedly, dragging him down the muddy slopes of the riverbanks gullied by the floods.

In a short while the demonic figure is straightened, fastened to the stern of the raft.

The tapper makes the last preparations, he adjusts Judas's jacket one more time, arranges a sack of gravel and rocks on his shoulders, places a rusty knife or useless pistol in his belt, without bullets or with only used cartridges; then, making the most curious recommendations to the ghoulish effigy, the *seringueiro* pushes the fantastic craft into the current.

And the new effigy, the figurative Judas, advances vaguely to the middle of the river. Then the nearest neighbors gather inquisitively on the heights of the levees, make rude catcalls, and in this macabre farewell party, salute the outcast with repeated rifle volleys. The bullets hit the surface of the river, rippling it; they perforate the craft, lashing it and finally reaching the terrifying passenger. He rocks a moment on his floating pedestal, thrashing in the shots, indecisive for a few minutes as though gauging a route, until he rejoins the general flow of the current. And the graceless, horrifying burlesque, with his disjointed gestures of demon and traitor, challenging curses and scorn, continues his mournful voyage

* This refers to Michelangelo's ecstasy at his own creation: upon finishing one piece, he cried "Parla!" exhorting the statue to speak.

without destiny or end, to descend forever, unbalanced in the whirlpools, made dizzy by the endless rotations, at the mercy of the currents, bobbing forever on the great waters.

Judas never stops again. As far as he advances, the errant specter continues to scatter desolation and terror in his wake—birds fly up in fear and regroup themselves, mute, in frondy recesses. Heavy amphibians lumber uneasily into the depths, terrified by the shadow, which at the end of the afternoons and in the early mornings lengthens, extending itself mournfully on the surface of the river. Men run to their weapons and, in a fury born of fear, cross themselves and shoot at him pitilessly. Judas cannot pass even the poorest hut without receiving a hail of bullets and a stoning.

The bullets salute him all around, they thrash him, water streaked by rocks surrounds him in undulant circles—the raft rocks, and following that motion, Judas waves his arms and appears to thank in clumsy bows the bitter manifestations of shots, pungent sarcasms, cries, taunts, and especially the curses, which revive in the worn-out words of backlanders this curse, which has echoed for 20 centuries:

Get outta here you bum! Get!

And Judas doesn't stop. He's away by the time the waters rise—freed from his persecutors. He slips silently into a slough, straight and long, he turns at the gentle bend of a deserted beach. Suddenly, as he traces another turn, another hut, women and children are startled by him at the riverbank and race headlong up to the shack in sobs and wailing. And soon after, from above the rifle shots, the stones, the mocking, and the curses.

Two or three minutes of alarms and uproar until the wandering Jew passes out of the range of the rifles, ever descending . . .

And he keeps floating downriver, and finally he no longer continues in isolation. He allies himself with his comrades in misfortune, other frightening effigies on the same diminutive rafts delivered by random currents coming in on all sides from the tributaries, varied in their features and gestures: some are very rigid, tied to the posts that support them, some floppy, shifting with every swale like drunks, some arms raised and threatening, blaspheming; others humble, curved in postures of profound abjection. And sometimes, yet more wretched, those who dangle at the end of a thin and curved mast . . . hanged. Thus they pass in pairs, descending, vaguely descending the river.

Sometimes the river captures them in an immense circle, a whirlpool, its current turning as it travels in slow circuits along the margins, tracing the ample spiral of an imperceptible and treacherous eddy. The vagrant specters penetrate these vast enclosures of stagnant water, rocking, and get stranded there for a while. They amass and circle in slow and silent review—they mix together; they cross for the first time the implacable and false gaze of their make-believe eyes and confound themselves in a convulsive agitation of paralyzed movements and rigid postures. There is the illusion of stupendous tumult without sound, and of a strange council—frantic, locked together in secrets, in muffled inaudible voices.

Later, little by little, they disband and disperse. Following the current, which straightens out from the meanders, they go in single file, or one by one at random, processionally, to the river below, descending, descending, ever descending.

And so da Cunha interpreted his maps. But because there were other charts in play, and other interpretations of history, these too had to be engaged. He would do so in both articles and supplements, as he wrote and rewrote western Amazonia into history.

20

"Events that perhaps lacked a historian"

Reflections and Supplements to the Formal Report of the Joint Survey Commission

Maps, Lying Maps, and History

Da Cunha needed to appear evenhanded as he framed the supplements that accompanied the map and the formal survey. The travels had stimulated a great deal of research and reflection, and he substantiated his positions through careful study of the Itamaratí archives and maps. His discussion of the Madre de Dios, Ucayali, and Upper Purús perhaps had documentation from Vias Fluviales, studies of the waterways of the Upper Amazon carried out by the Peruvian government in the early 1900s. These were essentially consultant reports, detailed, quite rare even at the time, and probably sent to him by da Gama, who was then based in Lima. Da Cunha makes explicit reference to these studies and the explorers who carried them out.

The writing and reflections that constitute the Brazilian supplements focus on a comparison of Brazilian and Peruvian enterprises in the region. Da Cunha's strategy involved discrediting the rival national claims through the maps that were their foundational documents and then through settlement history and the extent to which Brazilians had shaped the Peruvian enterprise, even, as he states, going so far as discovering the uses of *caucho* and thus saving the Peruvian Amazon from a catastrophic decline after the end of the quinine boom. These technical and social idioms defined the Brazilian supplements to the Joint Commission's report.

Mythical and Real Geographies of the Purús

Da Cunha's cartographic critique began with a dramatic statement by the chronicler João Daniel: "Between the Javarí and the Madeira is no one." This rhetorical device was the ironic axis on which Euclides would organize his central arguments: the lack of knowledge, the realm of rumor, the lying cartography that prevailed until the explorations by William Chandless and Brazilian Manoel Urbano, and, as he wrote, how this land became populated. It was, he affirmed, a place of conjecture, a world of hidden travels. What was known was that the natives were tall and had gold, and the sailing was quite good—gossip one might hear anywhere in the basin, since it was certainly generic enough. The place *was* occupied: manioc, maize, cacao, peach palm, Brazil nuts, and turtles were the famous domesticates of those upper reaches. The narrow scope of explorers and the poor copies available of imaginative maps made most assertions about the place questionable. Father João Daniel, the author of *Tesouro descoberto no Rio Amazonas*, confined as he was to his inquisitional cell in Europe, never traveled there, and each inaccuracy about the Purús engendered further and wilder speculations. These errors contained the implicit assertion that the Purús had in fact never been truly "discovered." It was *terra ignota*, a world of surmise, a land that even in the nineteenth century had not undergone the formalities of claiming and was subject to no authority. This lack of "geography" made the Chandless/Urbano expedition and its maps crucially significant.

Da Cunha's review of the mapping of the Purús takes in the great names of exploration history in the Upper Amazon: Pedro Teixeira (1596); his chronicler, the Jesuit Cristovão d'Acuña; the cartographer Guillaume Lisle (1723); Portuguese naturalist Alexandre Rodrigues Ferreira (1779); the luminaries of Anglophone nineteenth-century exploration—US Navy man Lardner Gibbon (1852), naturalist Louis Agassiz (1867)—and Peruvian geographer Mariano Paz-Soldan (1823). All, he thought, had fallen short. None knew the river. In da Cunha's geographic history, he lays the laurels of "discovery" at the feet of a *cafuz*, the black-Indian mestizo Manoel Urbano, who in his explorations of the rapids, the *varadouros*, and the natives becomes a modern version of the *bandeirante*, the bridge between the mythical past and the modern day. It is Urbano's passages that mark out these realms as "Brazilian," traveling the connections between the Madeira to the south, the Juruá and the Tefé to the north; Urbano even traveled above the headwaters of Tefé and the Coarí in his great arc of discovery. Even today this would be a remarkable achievement of sheer endurance, very difficult to carry out.

Urbano would "discover" and possess these areas (and, as we see later, colonize them) through the tried-and-true technique of marriage into the tribes; he left a swath of children and affines from one end of the river to the other. The significance of his travels, their legitimation by state explorations, and their subsequent inscription into imperial science by Silva Coutinho for the Brazilian state,[1] Chandless for the Royal Geographical Society and Agassiz for the Harvard Museum became the mechanisms through which political authority over the Purús was symbolically rendered, precisely and scientifically described. We have no way of knowing whether in fact the travels of Manoel Urbano encompassed the geographic amplitude asserted by da Cunha, who was basing his report on an addendum to the Chandless report,[2] but Urbano was the guide for all three expeditions just as the regional latex economy began to take off.

The Coutinho expedition had a German botanist and illustrator in attendance, the "first entry of European science into those regions." They carried out simple hydrological and geological studies and listed the tribes, but the botanist, Wallis, unbelievably accident prone and unlucky, never produced the longed-for sketches. He apparently went mad and eventually drowned, or perhaps drowned himself. Coutinho's formal survey, though incomplete, placed the question of the Purús firmly within international geopolitics. It preceded Chandless's expedition by a year and was probably stimulated by the rumblings around the "Madre de Dios question."

It is quite likely that Chandless used the Urbano/Coutinho map as a template, and he used Urbano as a guide as his Bolivian rowers oared up almost three thousand kilometers of waterway. The Chandless/Urbano map had symbolic value well beyond its markings. It had the merits of the modern mapping exercises that defined colonial trajectories throughout the tropics in the nineteenth century, and it had the virtues of scientific production. But it also resonated with earlier forms of claiming that were decisively Portuguese. In the heyday of Portugal's seaborne empire, its rituals of territorial possession claimed dominion through navigation—using numeric orientations—and "fixing" the place on the earth by the position of the sun in the sky.[3] This ability to accurately locate, to provide means to navigate the realms of "new skies and stars," was a key Portuguese expression of territorial authority.[4]

Chandless may have been sent out by the Royal Society, but he oriented his points with a mixed-blood Brazilian guide to whom he gave full credit for the success and caliber of the mission. Urbano's explorations, as they are described by da Cunha, integrate the myths, semiotics, and practices of *bandeirantes*, the state, science, sexual conquest, and ceremony as forms of claiming political rights over the Purús. This effort, instantiated in the maps and, as da Cunha underscores, its difficulty, was later undone by lazy and casual copies that relegated the much superior field study to cartographic oblivion while regional cartographic fictions exploded (and by extension, the questionable nature of maps advanced to Argentinean arbitration by the Peruvians). Da Cunha speaks, naturally, with the authority of one who has traveled the river.

Here is his review of the mythical and practical historical geography of the Purús, which becomes, true to his erudition, both a history of speculations on place and nation building and documentation of the actual expeditions.

Real and Mythical Geographies: From the Mouth to the Headwaters

Like the great majority of tributaries on the right side of the Amazon River, the Purús appears completely foreign to our history. The phrase of Padre João Daniel in his imaginative *Tesouro descoberto* sums up the knowledge of the old chroniclers: "Between the Madeira and Javari, a distance greater than 200 leagues, is no one, no whites, nor *tapuyas*, nor Missions."

The emptiness is informed less by real conditions than by lamentable gaps in our knowledge. Our chroniclers, on the whole, blinkered by their narrow itineraries or particular objectives, lacked any vision that would have allowed them to embrace more expansive interconnections. We can cite numerous examples that reveal all the idiosyncratic aspects into which they divided and disjointed the annals of Amazonia. Whether the rigorous data of the astronomers or the ingenuous narratives of the missionaries, all were shackled to the regime that guided them. Even Alexandre Rodrigues Ferreira, the greatest polymath of the colonial era, in his *Philosophical Voyage* reined in his noble spirit with pointless minutiae and rarely lifted his eyes beyond his field instructions. And this is why, in general, explorers limited their voyages to the main channel, with variants on the Rio Negro or the Rio Branco, and stayed there. The great majority of narrators, distant from occurrences elsewhere, ignored them; and though perhaps of lesser magnitude, these might have contributed substantially to a more profound integration of processes and events that are still poorly defined, fragmentary, and dissonant.

Be that as it may, tracing an irregular line from the northern hills of Amazonia to the east and then tracking south at the head of the Napo, and descending from the Amazonas to Pará, one has sketched practically all the geographies of Amazonian scholarship. The south, excluding the Madeira (which was historically linked to Mato Grosso by the audacious sallies of the Paulista expansion), remains the wilderness: "a waste of waters," as William Hadfield wrote in 1877,[5] repeating the regrettable exaggerations of the old fantasies that gloss those landscapes with the luster of the arcane and mysterious.

The Purús, in particular, since the beginning, has been the victim of old chroniclers. It entered history for the first time with a singular and fantastic imprint. Really, all the facts suggest that this was the remarkable "River of the Giants" to which Father Cristovão d'Acuña refers:

> . . . A renowned River which the Indians call Cuchiguara. It is navigable even though some ports have rocks. . . . There are a lot of fish, many turtles, and an exceptional abundance of manioc, maize and everything necessary to facilitate exploration of it.

He refers later to its population and cites among others the Curú-Curú, corrupted obviously from the Purú-Purús, and of the Curiqueres:

> Giants 17 hands tall, very brave, who are naked: they carry great disks of gold in their ears and noses, and to arrive at their villages two months of continuous walking from the mouth of the Cuchiguara are required . . .

Thus was launched the mythical geography of the Purús, so it is no surprise that shortly thereafter, a mapmaker largely responsible for the continuing cartographic eccentricities, Guillame de Lisle, the first geographer of the Royal Academy of Sciences in Paris, summed up in 1703 then-current notions about Brazil and gave the great river a capricious depiction. His map shows the Purús under another name, "River of the Omepalens," stretching in a sharp line to the south to 18 degrees latitude, where it branches out into imaginary headwaters somewhere past La Paz. And from these beginnings, there is a short explanatory note about the new and singular beings:

> The Mutuanís, or Giants rich in gold, these inhabitants are two months from the mouth of the river . . .

Thus persisted, as we see, the fictions of the credulous chronicler of Captain Pedro Teixeira.* Among these errors, nothing more is revealed than the propen-

* At least as expressed in the writings of Cristovão d'Acuña, his chronicler.

sity for the marvelous, a feature particular to those times. Padre João Daniel, in the same volume from which we extracted out earlier phrase provides the Purús with so a true description, that he appears entirely innocent of hyperbole of everything in the 18th century:

> The Rio Purús is so big that it has thirty days of good navigation because it hasn't many cataracts . . .

These two elements, the nature and extent of the channel, suggest the existence of earlier explorations, forays to the area that perhaps lacked a historian.

This is also suggested in a letter from Antonio Pires de Pontes Leme, an astronomer for the royal Portuguese demarcations. Assessing it, one notes that the entrances of the tributaries of the Purús have the same configurations today, although one of them, the Paratani, is much longer and stands out as though it were another river, even noting that the main river follows from latitude 6°30′ a course very similar to its current channel, but loses its line further south until the cartographers gave it, at random, a deviation to the southwest, parallel to the Madeira. One observes on the same map a large Paraná-Mirim* (at latitude 5.40) mixing, by way of a tributary of the Capana, the waters of the Madeira and the Purús. These coordinates practically correspond with the mouth of the Paraná-Pixuna, and even though this communication between the rivers doesn't exist, this congruence of positions is yet another indication of more complete information than vague deductions of the Indians. More recently, Manuel Aires do Casal, in his *Corografia brasilica* (1817), though he concurs with the errors that persist to our time, does present, near the headwaters, doubts that place him in the vanguard of modern geographers:

> These rivers, the Tefé and the Purús, are not derived from the mountains of Peru where some say they begin: this is proved by the existence of a communication between the Ucayali with the Mamoré via the Rio Exaltacion and the Lake of Rogualgoalo: but whether they derive from these, as others would have it, or whether the origins are in the north must still be determined.

Lake Rogualgoalo was for a long time considered the inexhaustible source of numerous rivers whose headwaters wended their way through the Bolivian highlands between the 10th and 15th parallels. Thus it was that in 1852, Lieutenant Captain Amazonas, referring to the origins of the great river and considering incorrect "the idea that they form in the mountains of Cuzco via the communica-

* A *paraná* is a side channel.

tion of the Ucayali with the Mamoré, by means of the Exaltacion River," inclined toward those who judged the Purús to be an outlet of this already mentioned lake.

Professor James Orton,[6] in 1868, substituted this error for one even greater and the most surprising of them all: he presumed that the Purús was the legendary Amarú-Maiu, or River of Serpents of the Incas, and traced it from the Andes, where it nourished the romantic valley of Paucar-Tambo before spreading out in the flat plains of Amazonia.

Lardner Gibbon and Tadeo Haenke[7] considered the Purús an extension of the Madre de Dios, thus countering the inexplicable blunders of Mariano Paz-Soldan, who in 1862 in his *Geography of Peru* and its atlas presents the Madre de Dios and the Inambari as the tributaries of the Marañon.

With such conflicting assessments, one can understand why the Royal Geographical Society in London would commission one of its members, William Chandless, to resolve the controversy, or as one used to say, the "problem" of the Madre de Dios and of the Purús. But before this, Brazil would also commit itself to the systematic exploration of the River.

In fact, except for the fruitless travels of João Cametá (1847) up to the Ituxi, and of Serafim da Silva Salgado, who went slightly above the Iaco, the Purús was really opened in 1861 by Manoel Urbano de Encarnação, an inexhaustible source of heroic undertakings. Urbano, a fearless and wise *cafuz*—a mixed blood of native and black ancestry—matched his resolution and courage with intellectual vibrancy ("a great natural intelligence," in the words of Chandless), which contributed enormously to the power he had over all the river tribes, and so could open up the region for one of the most remarkable chapters in our geographic history. His contributions were extraordinary and merit more pages than this quick sketch.[8]

In accordance with his instructions from the provincial government of Amazonas, the first goal of his extensive travels was to verify what had long been conjectured as a channel between the Purús and the Madeira well above the giant rapids of the Madeira. Manoel Urbano traced almost the entire itineraries of many other, later explorations.

Leaving Manaus on January 27 of that year, he arrived after fifty-five days of slow canoe travel at the mouth of the Ituí, and then arrived thirty-two days later at the Aquirri (the Acre). He inched up this river for twenty days of arduous travel, stopping finally when the extreme decline in water levels annulled all the efforts of his Pamari Indian boatmen, who were, at the end, dragging the canoes. He returned at low water to the main river and for forty days followed on the vagaries of the current to slightly beyond Rixala (to the rapids of San Juan, as it is known by the Peruvians), arriving close to the mouth of the Curumãa (Curanjá), having covered about 2,800 kilometers of the Purús, a distance that had never been traveled before then.

An immediate outcome of this expedition was that the reality of no connection between the two rivers—the Purús and the Madeira—was definitively proved, and that the tributaries between the Acre and Curinãa (today called Sta. Rosa) became known. Besides this, they discovered a stream that led to a *varadouro* to the Juruá (by means of the Jurupavi and the Tarauacá), and as this transit was accomplished above the headwaters of the Tefé and Coari, these simple circumstances were enough to correct the understanding of the latter water courses, whose elements until then had been extremely overstated.

Manoel Urbano then directed his research along other lines, always in the search for faster connections between watersheds. He entered the Mucuím River, and in a trip of some twenty days, triumphing over successive rapids and falls, and securing the trust of the disdainful Pamanás Indians, he arrived at the left bank of the Madeira at the Teotónio rapids after following a *varadouro* of 10 leagues. Returning to the Purús, he followed the Ituxi and surveyed it until the part with the rapids beyond the inlet of the Punicici.

While carried out by an uneducated man, but one equipped with exceptional acumen, these travels provided the first reliable data vis-á-vis the Purús, its three main affluents, and the tribes that populated them. These immense itineraries between the various points and the general orientations of the different segments of the river soon surprised even William Chandless.[9] The general nature of the land and the number and character of the tribes were little altered later on.

It is natural that these efforts in many ways influenced the government that would resolve to continue these endeavors, so brilliantly begun. This is in fact what happened when on February 13 the governor of Amazonas, Dr. Carneiro da Cunha, gave engineer Silva Coutinho the order for a reconnaissance of the Purús and its most important tributaries. This was a complex mission. Beyond the hydrographic survey, the orders were to review the geological structure of the valley, the flora, the properties of the terrains most given to cultivation, the number and characteristics of the tribes, and the most efficient means to couple them to civilization, and to top it off, to attempt again the passage to the Juruá via the *varadouro* discovered by Manoel Urbano. Urbano accompanied Silva Coutinho as well as the German botanist Wallis, the first representative of European science to enter the Purús.

Coutinho summarized his findings in a detailed report of March 1, 1863, which, in addition to a general study of the river, its lakes, tributaries, islands, shoals, rock outcrops, the houses one encountered from the mouth up to Rixala, included ample information on the natives, production, nature of lands, etc. This reconnaissance, in which the hand of Manoel Urbano clearly manifests itself, includes many notable chapters, except, lamentably, those of Wallis, whose artistic contributions circumstances did not permit.[10]

Silva Coutinho, in addition to the interesting formal data he disclosed, substantially revealed the destiny of these areas. In spite of having ascended the river only as far as Huitanãa, as he turned back for lack of supplies, he outlined with eloquent simplicity the importance of these unknown outposts:

> The importance of the Purús is too great to abandon its reconnaissance. When Europe discusses with immense interest the "Question of the Madre de Dios" we should not, with our special concern in this issue, remain indifferent. The richest regions of Peru and Bolivia can only connect with the Amazonas by means of the Purús and the Juruá, rivers that have no significant falls and which offer easy travel along their entire course.

Thus came Chandless, one year later, specifically to resolve the "Madre de Dios Question"—one aspect of the old problem of linking the Amazon basin to that of the La Plata. Even though Chandless left without this crowning achievement, he carried out the most serious of all the explorations of the great river. For the first time, fixed astronomical coordinates of its principal points were carried out, and while various other tasks were not completed, that simple contribution suffices to place him among the premier scientists of Amazonia, if not among the top ranks of other observers who have studied our country. It is difficult to encounter anyone else so persistent, so conscientious, so lucid and so modest.

His extremely arduous voyage of eight months, assisted only by some Bolivians and Ipuriná Indians who rowed his canoes, was probably the calmest of all the great geographical expeditions. There were no major calamities, no emotional episodes or unexpected incidents like those that typically embellish expeditions into the "unknown." It is striking and interesting for the remarkable results generated by this expedition: the extremely rigorous but clear barometric readings and determination of coordinates. In this last aspect especially, it is the best model of geographic work in our country.

Compare him with anybody who was charged with the same task of exploring any one of the Amazonian rivers. Few other areas match the obstacles for an observer: the extreme humidity that clouds the skies at exactly the hours most apt for measurement, even when the weather is clear and stable; even the best days always begin in haze until about 8 o'clock, obscuring the position of the sun for determining the hours, and at the end, the sky is veiled in mists and vapors through which one can barely make out the stars. At the headwaters, the narrowness of the channel, choked by large trees, limits the choice of stars, closing the skies to a 45-degree angle, and basically undermines the best situations for these survey tasks; the logs that choke the riverbed, from the middle section up to the headwaters, result in ceaseless jolts and shocks that are very damaging to the chronometers, already impaired by portages around the worst rapids and cataracts. The capricious and sinuously twisting channel demands constant attention and an exhaustive reading of the watercourse, which changes from one minute to the next, and one must record all these innumerable observations in a field book. There are ever-increasing causes for error in the ultimate results—the barometric anomalies, still inexplicable today, not only make attitudinal readings dubious but diminish the importance of corrections to the calculations of altitude. And as if there were not already enough difficulties, the observer (regularly obliged to veil himself in mosquito netting) is often denied the indispensable serenity necessary for these endeavors because of the onslaughts of a succession of biting gnats in the day and the stings of mosquitoes at night, the welts of chiggers, and the torture of the "no-see-ums" that one has sometimes has to endure, stoically immobile, in order not to lose the precise moment of the passage of a star or contact with the sun.[11]

Chandless overcame these difficulties alone. He had no one to even read his chronometers. Balancing the inevitable errors that increasingly add up in calculating longitudes by means of distance in such disadvantageous conditions, he not only observed these by means of absolute longitudes at periodic and large distances but also compensated for this by means of double observations of the same points on the ascent and the descent.[12] In this way he rectified the hydrographic survey and coordinated the topographic and astronomic observations in a rigorous integration.

His maps later required few modifications and became the dominant model for the numerous geographer-copyists who used, adapted, and, notably, distorted them. Unfortunately, this remarkable expedition has not had the prominence it merits. Having studied in detail the entire Purús and Acre, Chandless, because of a slight shift in his route, could not decisively determine the headwaters of the first and could not resolve decisively the *divortium*—the division—between those headwaters and the upper branches of the Madre de Dios and Ucayali, although he deduced it. He did not present it as sure fact, since only direct observation would serve him:

> From the small size of the branches at the farthest points I reached and their rapid diminution, it's clear they cannot come from a very great distance . . . certainly not from the cordilleira. . . . The Madre de Dios is certainly not the source of the Purús . . .[13]

. . . Comparing the maps of the headwaters with our own, we can show that if the illustrious geographer, on reaching the last bifurcation of the South Fork (the Cujar), had proceeded to the right on the Cavaljani, the river with greater water volume that further extends the Cujar and opens into the southern quadrant, this route would have taken him in less than eight days simultaneously to the valleys of the Ucayali and the Madre de Dios, after traversing a small hill which is on our side of the Pucani River, the southernmost branch of the Purús connecting to the stream called the Machete, one of the uppermost branches of the Ucayali.

His findings would have been even more astonishing because at the same passage and on a single day he would have arrived at many extremely valuable conclusions:

1. He would have showed the independence of the Purús and the maximum length of its origins to the south, without arriving at the 11th parallel.

2. He would have seen that the source of the Madre de Dios and the Ucayali is on those banks that split away from the narrow Isthmus of Fitzcarraldo: their close proximity partly justifies the old errors that persisted over so many years.

3. Comparing these with the Purús, which is separated by a small hill and less than 2 km of *varadouro*, would have not only explained the countless times that the great Amazonian tributary was confused with the Madre de Dios but also revealed the stupendous geographic fact that from here, in an extremely reduced area, originate and radiate out three great fluvial arteries, well outside the zone of Andean uplift and at a relatively low altitude, perhaps 500 m above sea level.

In spite of this, today Chandless's exploration is still the most serious ever carried out on the Purús. Those who followed in no way modifed its general truths. We cite only the great explorations by land of Labre and Alexandre Haig (1870–1872) in order to mark out a road between the Port of Labré and that of Florida in the Beni; the merely descriptive travel accounts of Barrington Brown and Lidstone (1873), who arrived only at the cliffs of Huitanãa; and the joint Brazilian-Bolivian commission (1897) to place the marker lines of the Beni-Javari.

In sum, for many years the geography of the Purús was inscribed in the lines mapped out by Chandless in 1867. Then, what is even more remarkable, it regressed.We cannot explain through what perverse enterprise a brilliant case of cartography, the notable map of a famous explorer, copied in a thousand ways, produced and reproduced without number by mapmakers, ended up completely falsified. The geography of the Purús reverted to the times that preceded Manoel Urbano. To the degree that new maps—generated by those lacking the courage to face the great river, who then shuffled the lines, erased others, capriciously bent the main channel, dilated its length beyond the 12 degrees latitude or more, changed tributaries from one bank to another, altered names, closed off estuaries . . .

> Without exaggerating, we can say that the Purús, described so lucidly by Chandless, has bit by bit returned to the fabled Chusiguara, veiled in absurdities, incited by the marvelous fantasies from Cristovão d'Acuña to Guillaume Lisle, dreamed up by the chroniclers and cartographers who followed.
>
> After Chandless, the only reconnaissance that was carried out from the main route of the Purús to the headwaters was that which was the motive for the present study: the reconnaissance of the joint Brazil-Peru commission, whose survey and results are largely a copy of, and are mainly a complement to, the efforts of that explorer.

In the interests of balance and out of his respect for South American Amazonian travelers, da Cunha also analyzed the expeditions that came from the Peruvian side of the basin. Here the Andean face of Amazonian history is revealed through its surveyors, cartographers, and adventurers, not least of whom was the famous Carlos Fitzcarraldo (known to filmgoers as "Fitzcarraldo," after Werner Herzog's eponymous film), whose ambition to move a steamship from one watershed to another was portrayed as obsessively insane.[14] From the perspective of the time and place, this ambition seems less odd, especially in light of the extensive regional *varadouros*, over which a great many crafts and tons of rubber and other commodities were hauled. The *varadouro* of Fitzcarraldo (later called the Isthmus of Fitzcarrald) was less than two kilometers long. Even in the 1940s people were moving boats from tributary to tributary: Harvard ethnobotanist Richard Schultes did just that on the Apaporis River in the northern Amazon.[15]

Fitzcarraldo seems to have been done in less by the scale of his ambitions (since he was usually successful) than by the kind of bad luck anyone could have in the torrents of the giant tributaries. Peruvian expeditions were more frequently visited by disaster, since they had to descend Andean waterways through treacherous rapids and turbulent currents; Brazilian explorers, though they faced daunting odds as they sailed and rowed up the rivers, did not have exactly the same dangers. Da Cunha describes the Peruvian part of the journey.

To the Headwaters

> In the foregoing pages we reviewed the uncertainty about the origins of the Purús, matched only by the great delusion of geographers who often suggested that it was an extension of the Madre de Dios, and we noted that the close proximity of the headwaters of these rivers was the cause of this confusion.

In fact, the last branches of the Purús (the Cujar and Curiujá) and the easternmost tributaries of the Urubamba (Sepauá and Mishuá) and the northernmost sources of the Madre de Dios (Caspaljani and Caverjali)could be linked by a meridian segment of less than 20′. It is not surprising that the understandings of the origins of these rivers are interconnected and complementary. The explorations carried out on the Madre de Dios were in a short time complemented by those of the Purús.

Setting aside for the moment the notable expedition of the Inca Inpangui, who descended with ten thousand warriors to the fabled Maru-Maiu (today's Madre de Dios) from the Rio Tono to the provinces of the Moxos, one can date the first formal exploration of the Madre de Dios to 1860–61, exactly the same time that surveys began on the Purús.

At the same time that Manoel Urbano was embarking on his great tasks, Faustino Maldonado left Nauta, traversed the valley of Paucartambo, traveled the Tono to the falls of Pitama, which he crossed, moving toward the mouth of the Pinipini. Then, with help from only the Conibo Indians, he constructed a raft and sailed the currents until the confluence with the Beni, where via the Mamoré he arrived at the Madeira River, and there continued his descent. Unfortunately, the daring enterprise had a catastrophic conclusion in "Hell's Cauldron" when the brilliant pioneer was shipwrecked and perished with most of his crew. But the results he obtained were admirable, and it is difficult to understand why the Madre de Dios was confused with the Purús for so long, except for the influence of the distinguished Peruvian geographer who was the greatest exponent of this absurd proposition.*

Those shores were not subject to continuous attempts at exploration that prevailed on the Brazilian side. In twenty years there is little to cite: the unfortunate expedition of Colonel LaTorre, who succumbed in an attack by *chunchos*—wild Indians—not far from Cuzco (1873). In 1880–81, Dr. Edwin Heath[16] completed the efforts of Maldonado in arduous round-trip travel to the Reyes River at the confluence of the Beni–Madre de Dios. At last there was a definitive verdict on those two great rivers that had long inspired the inventiveness of cartographers.

Research continued. In 1890 a Peruvian *caucheiro*, Carlos Fitzcarraldo, overcoming extraordinary difficulties, discovered the *varadouro* from the Misauau (the last, easternmost branch of the Urubamba) to the Caspaljani (last of the northernmost tributaries of the Madre de Dios) and hauled the launch *Contanama*, in which he had ascended the first river, thanks to the robust Piro Indians, through it to the second. Thus he would transit from the waters of the Ucayali to those of the Madre de Dios—now known as the Isthmus of

* Da Cunha is referring here to Peruvian geographer Mariano Paz-Soldan.

Fitzcarrald—and so revealed the narrow strip of land that separated these two immense basins. Thus, by 1891, the origins and general directions of the rivers that flowed through those banks was well known. There remained only, to the north, the Purús.

One Peruvian version, very open to question, suggests that a gentleman from Loreto, one Leopoldo Collazos, was the discoverer of the passage between the Purús and the Ucayali. Leaving in about 1899 from an outpost on the Urubamba, the explorer, accompanied by thirty infidels, set forth on the Upper Sepauá and slipped through the last tributaries until the Machete creek, where in the last days of August he appeared, traversing a slight undulation in the terrain, on the Pucani and the Cavaljani, and at the headwaters of the Purús.

Others, perhaps more reliable, affirm that this honor falls alone to the distinguished brother of Fitzcarraldo, Dr. Delfim Fitzcarraldo, who established himself in 1892 on the Urubamba with a Brazilian associate, Lieutenant Colonel José Cardoso da Rosa. Whichever it was, in 1900 the great question was settled: the three rivers were completely independent, but their headwaters were so close together that *the passage from one to the other could be effected not just in the light* ubás *of the Indians but in the same launches of explorers.*

Da Cunha had shown the headwaters to be utterly permeable. The next section of the report pertains to the *varadouros*. Euclides describes the last parts of the passage to headwaters in more geological detail,* reprising material about the difficulties of the travel from the joint report and the materials of his confidential communications to Baron Rio Branco. He would later have a broader mediation and policy prescription for the *varadouros* (his insights I summarized in the previous chapter),[17] but the momentous experience of the discovery of the *varadouro* between the Purús and the Ucayali was transformative for him.

From here forward, the channel begins to inflect to the south and continues in two almost imperceptible curves for about a kilometer . . . then suddenly it enlarges into its last impound, an irregular circle about 30 meters in diameter, deep, excavated from the hard scree of the river terraces. The great river expands at exactly the moment when the explorer imagines it to be ever more strangled between the river cliffs, bifurcating and diminishing at the final sources of the headwaters. Imagine yourself ending up at this diminutive lake, and how

* This segment on the geology of the Upper Purús was meant for a specialist audience, so I have not included much of this description.

above arches the clear and open sky, this after emerging from the half shadows of the Pucani. One has at first the impression of arriving at the culminating point of one's travels. But this is not a foothill of the Andes, nor yet a knoll whose proportions one might exaggerate after advancing across 3200 kilometers of almost invariable flatness. Encountering this rise, one most notes the clearing for a trailhead, about a meter wide, sharply inclined, at a slope of 28 degrees.

It is the *varadouro*.

At its edge are four *tambos* of paxiuba palm, which is what the *papiris*, or palm huts of Amazonia, are called here, where people take shelter and store rubber and trade goods. All around are strewn empty tins of every kind of canned food, bits of tools, tatters, all scattered widely about, indicating the stopovers of a regular and constant trade. The *varadouro* begins in a southern direction on a steep hill, but in five minutes of energetic ascent one arrives at the highest point—the parting of the waters, the *divortum aquarum*, of two of the largest rivers on earth,

Unfortunately the forest canopy limits the view and doesn't help one, at that point, to take in the surrounding lands. One just notes that that unimpressive mound, with a relative height of maybe 50 meters, reigns over all the places below it: the Purús to the northeast, the Sepauá and the Urubamba to the west, and the farthest tributaries of the Madre de Dios rising to the southeast.

From there continuing always to the south to the Sepauá valley are the farthest branches of the Ucayali. The slippery clay soil with its superficial polish reveals the effects of constant dragging of *ubás* over the land. Except for the cutting of one or two trees, one does not see evidence of the minimum effort necessary for conserving and maintaining such a critical passage. At occasional points some logs are aligned for an imperfect rustic bridge, or to correct the unevenness of the land, and over six erosion gulches some randomly felled trees served as precarious bridges, requiring alert and careful crossings. The descent is much longer than the ascent on the Pucani side, and is completed by crossing three large flats separated by narrow and deep ravines. In half an hour one can travel the entire *varadouro* of some 1.5 kilometers. Rectifying the distances of the undulating lands against the horizon, one sees that the distance that separates these enormous watersheds is less than a kilometer. This escapes the scales of most maps. The Purús and the Ucayali are almost united at that point, a place from which they embrace a fifth of Amazonia, in an unmeasured expanse of the continent that completely takes in the basins of the Tefé, Coari, Juruá, Jutaí, and Javari. . . .

Considering our map, you can see that the same traveler, in the same boat, can today in a very short time pass from the waters of the Purús to those of the Ucayali through the Isthmus of Sepauá, and from this to the Madre de Dios via

> the Isthmus of FitzCarrald, thus completely resolving all the disagreements, the doubts, and even the greatest errors that for so long so confused the understanding of the origins of the three great rivers.[18]

Populating and Depopulating the Purús

The next supplement produced by da Cunha was his discussion of administrative history, titled "Populating the Purús," which is also about Amerindian depopulations. This section describes an "absent-minded imperialism," one that doesn't exactly oversee but reacts to the immense historical changes occurring in remote Amazon forests. This segment was a narrative of Brazilian triumphalism, mostly read through formal bureaucratic events outlined in native decline and the explosive entrance of rubber tappers into the watershed. The tale begins with the defeat of the nomadic Mura Indians and the displacement and loss of countless tribes, and ends with the landscape of desertion of the Peruvian *caucheiros*. As da Cunha tells it, it is the imperial arc of Civilization versus Barbary, settlement versus nomadism, that shapes the region from a native redoubt into an enriched and enriching frontier. "From the Mouth to the Headwaters" documents the rush of northeasterners, while "At the Headwaters" describes the Peruvian version of this saga. This analysis was hardly immune to da Cunha's derogatory views on Peruvians and were meant to sway negotiations as the writing increasingly took on the historical themes of *sertanejo* triumph and the emergence and energizing nationhood of a new Luso-tropical civilization.

Vanishing Indians

At the turn of the century, Brazil's natives were either swathed in a romantic naturalism that had them conveniently absent in daily life but mythically vibrant in an undefined national past, or subjected to various forms of slavery, hunted down, and massacred in the interior provinces of the country.[19] Da Cunha's measured, elegiac statements are unusual for the time. His abiding closeness to Cândido Rondon since their days together at Praia Vermelha affected his views of Indians.[20] Rondon, a Terena Indian, was an unrelenting positivist, Brazil's first indigenist and seminal designer of its native legislation, and almost its president.[21] Da Cunha's surprisingly knowledgeable discussion of aboriginal groups and migrations suggests a modulated, well-informed reading of native history, quite unusual for the time.

The Upper Amazon, the famous zone "between the Madeira and the Javari," was hardly empty of cultural history. Da Cunha was correct in saying that the banks had teemed with natives. The first European observers of the mouth of the Purús found it so densely inhabited that they were afraid to stop. They noted statues "like those of Cuzco" and many roads leading back to terra firma.[22] The region was, as da Cunha points out, the crossroads of numerous ethnicities from the Rîo Negro, the Madeira, the great interior plains of the Moxos, and, of course the Andes.

This complex of cultures formed of Pano, Arawak, and Tupí linguistic groups would figure throughout da Cunha's narratives: the Shipibo, Piro, Mashco, Conibo, Kaxinuá, Mayoruna, Campa, and Amahuaca included populations from the Ucayali that easily shifted into and through the Juruá-Purús watersheds. The Pano, the most numerous ethnicity on the Purús, claimed a kinship or vassalage to an "Inca" cultural hero and set their place of origin as near Cuzco. In ritual elements (animal sacrifice, blood offerings, and solar divination), medicinals (including coca), therapeutic materials, as well as manufactures of various kinds, and of course warfare, there were clear relations between Pano cultural groups and Andean societies.[23] The interaction between the Pano and Andean cultures was documented by one of the pioneers of Amazonian archaeology, Donald Lathrap.[24]

The Purús basin was also occupied by Arawak groups, the Ipuriná, Canamari, Piro, Campa, and Amahuaca, who brought with them sophisticated manioc agricultural traditions and the landscape modification techniques that are found throughout the Moxos (in Bolivia), the Upper Xingu, and the Purús.[25] The Arawak formed a kind of cordon sanitaire between the lowland and the Andean ethnicities.[26] They were also, apparently, capable allies. As Inca emperor Atahualpa awaited his fate, it was reported to Pizarro that thirty thousand Carib (Arawak) "who eat human flesh" had joined hundreds of thousands of warriors from Quito.[27]

While the terra firma was strewn with large-scale earthworks, the banks of the river itself were also densely populated. Along the great channel of the Amazon River, the mouth of the Purús had significant populations. Teixeira's chronicler, d'Acuña, counted more than four hundred villages and two very large cities. Roads swept back from the banks of the river to where, it was said, "Andean sheep," perhaps llamas, were known.[28] Any linguistic map viewed today will show what da Cunha remarks upon: an extraordinary diversity of linguistic and cultural units concentrated at the headwaters of the Upper Amazon, the outcome of relocation by missions, slavery, migration, and death.

Da Cunha begins his essay on the settlement of the Purús with the taming of the Mura. The Mura were a defiant and bellicose tribe who had controlled the entrances to both the Madeira and the Purús Rivers, having moved into the void created by the demise of the dense Omagua settlements that had previously occupied the main channel. The "isolation" of the Purús and Juruá in spite of the ease of travel up them may well have been linked to this tribe of able guerrilla fighters and water masters who dominated the region. The "official" version of the remarkable transformation and "domestication" of the Mura via iron implements obscured the decades of violent attack and effective guerrilla warfare that had impeded the entrance of whites into the region for generations.[29] But da Cunha is correct to begin with their dramatic capitulation: administrators and clerics of the region were stunned at the sudden surrender of this warlike tribe. But disease along with internecine and protracted warfare with the Mundurukú made the missions' and Brazilian garrisons' promise of protection against their arch enemies appealing.[30] The Mura were not destined to survive this alteration of circumstances, even as thousands thronged to missionary or military settlements. There they succumbed to the usual maladies, and a few short years later they lost their missionary interlocutors when the Pombaline reforms exiled ecclesiastics throughout the New World. Mission lands and their native charges were left to the depredations of theft, death, and slavery by local directorates.[31]

Da Cunha describes the more famous of the Purús tribes but also recounts their "displacement or absorption" in the unfolding of events more than half a century later in what was actually a far less benign process. In fourteen years, by da Cunha's calculation, native settlement had declined by two-thirds. But the real story is the economic dynamism that "perhaps lacked a historian." What is clear in da Cunha's narrative is that the Purús was always a rich river, whether thronged with natives or not. Freelancers were plundering its gifts to such extremes that the river had to be embargoed early in the nineteenth century. The Purús always had the attention of Brazilian administrators, whether Jesuit or imperial, and indeed was so important that it became the first Upper Amazon site for regular steam travel: this began in 1869 and opened the floodgates of spontaneous Brazilian colonization in quest of sarsaparilla, turtle meat, and oil for the expanding world of sea travel and the burgeoning rubber economy. The flood of immigrants "unyielding, undeflectable, and committed to stable dominion," articulated the Brazilian imperial discourse: attaching man to the land through useful toil, as described by da Cunha, in the exemplary urbanism of the town of Lábrea. But more to the point, four-fifths of the river was populated by Brazilians along its length "without the hiatus or blemish of one abandoned area linking all the *seringais*." This paean to a peasantry and a process mimics, at least in da Cunha's ideological framing, US yeoman expansion and manifest destiny. Though he acknowledged the problem of *latifundias* and elsewhere the horrors of the rubber estates, these, he felt, could ultimately be ameliorated by state actions and greater political integration with the rest of the nation.

The Brazilian context stands in contrast to the Peruvian case as da Cunha explains it, with the international negotiations firmly in mind. Unlike Brazilian engagement (as early as the eighteenth century), the Peruvians were recent arrivals on the Purús, refugees from the Ucayali, insinuating themselves into Brazilian lands with complete indifference to place. No national feeling animates them: they are unwilling to die for the *patria* and happily abandon their posts under military pressure. Their occupation is ephemeral, predicated on the abuse of natives, sexually immoral. They incarnate only the ambition of greed, so even their towns are provisional. Where settlement and agriculture occur, it is the result of native effort and ownership—Amahuaca women and Campa chiefs. In short, there is "not a glimmer of social cohesion." This contrast will become more thoroughly elaborated in da Cunha's imperial ethnography, provided in the next chapter.

Populating the Purús: From the Mouth to the Headwaters

Reading the "news of the voluntary reduction of the fierce Mura nation in peace and harmony" in the years 1784 to 1786, and reading the lengthy correspondence between Lieutenant Colonel João Batista Mardel and João Caldas on how to deal with the people "who inhabit the channel and the lakes from the Purús to the Juruá," reveals a venerable and persistent interest in the settlement of those regions. But an appraisal of the ancient documents would be tedious. It is enough to know that in 1787, after an extraordinary campaign in which no weapons were employed other than the preferred gifts of the natives, the aborigines of those areas were pacified, entirely captured by civilized people. The Purús especially, with its incomparable richness in precious spices, was thus opened to the disordered and primitive regime that still rules Amazonia. This is revealed in one act of the last governor of the Rio Negro Capitancy, Manoel Joaquim do Paco; it is eloquent in its extravagance. In 1818 he closed the river; he prohibited those seeking the *drogas do sertão*—backland botanicals—to travel it: "They set out with their eyes blinded by their lust for precious fruits. They want above all to make their fortunes with the profusion of their product." The governing junta of Pará soon revoked this curious resolution, which was after all quite indicative of the importance that the great river had already assumed by that time.

Unfortunately, the years of adventures into the Purús and its hinterlands left little documentation. We catch glimpses, though, in the fragile and disjointed reminiscences of its most aged inhabitants, which inspire little confidence. Only in 1854 do the first really certain data in reference to the Purús appear with the report of the president of Amazonas, Hercules Pena, in which there are references to the mission on the Purús (São Luiz de Gonzaga) entrusted to Father Pedro de Ciariana.

From that date forward, the occupation was continuous and palpable, so that by September 7, 1858, another president, Francisco Mendonça Furtado, could justify in his reports the need for establishing regular navigation to those outposts. In fact, the population increased, but as it was still precarious and wandering through unmarked territory, there was no way to remotely approximate its numbers. One conjectures that residents didn't decrease only by the circumstance of the creation in June 1857, near Guajaraba, of a health post for those victimized by the pernicious fevers that infest those shores.

Among those early occupants was a man whose own ethnic background prepared him for founding a new society in these newly discovered lands: Manoel Urbano de Encarnacão. We have already described his admirable role as a conqueror of the wilderness. But his activities as the founder of villages are more important. Even today, his scruples and tenacity, matched only by his integrity and natural generosity, are part of the folklore of the Purús.

He was the mediator between the new populations searching out the Purús and the wild tribes who occupied its banks. Remember that for centuries the Purús was perhaps the major route on which the most remote tribes from the most distant parts of the continent continually traveled. The Muras, nomadic and wild, who so alarmed the colonial governments, are not native—they descended from Bolivia via the Mamoré and are perhaps relatives of the Moxos, who were in turn besieged first by incursions of the Incas and then by other tribes from the south of our country who had been terrorized by the Paulistas. The Jamanadís, residing deep in the forest and avoiding river margins, still retain the memory of the slave raids that expelled them from the Río Negro. There are the Ipurinas, in whom Silva Coutinho glimpsed traces of the Ubaias of Paraguay; the aspects and clothing of the Canamaris, as described by Manoel Urbano, vividly recall the stiff, unsewn *cushmas*—woven robes—of the Campas who live at the headwaters.

These tribes swarmed both banks of the Purús. The Muras of the mouth of the Parana-Pixuna were found at Beruri, on the Lake of Hiapuã, at Campina, and in Arimã, where, after 1854, they were settled by Padre Ciariana. The territory of the Pamaris and the Jubaris, who are known under the general name Purús-Purús, extends from the mouth of the Jacaré River to Huitanãa. Gifted artisans of light canoes and incomparable oarsmen, they live exclusively from fishing turtles and the fruit-eating *pirara* catfish, from whence comes the singular malady that mottles their skins with white blotches.* The longhouses of the brave and

* This blotching is in fact not caused by eating catfish but by a depigmenting skin fungus.

OLD JAMAMADY INDIAN.—PURUS RIVER.

Figure 20.1. Jamanadí Indian.

vigorous Ipurinás stretch from the Pacia to the Iaco, where their huge circular houses sometimes contain more than 100 people, each commanded by a *tuxauá*—a headman. From there on up the river are the Canamaris and the Maneteneris, and the hidden forest tribes of the Pamanás and Jamanadís.

On the Purús today one no longer sees them as did Silva Coutinho, Chandless, or Manoel Urbano. The Ipurinás are among the most numerous, but without the characteristics of the past. The Purús-Purús, whom we saw, in no way reminded us of those intriguing savages completely unshackled from the banks, living in enormous floating longhouses on a permanent voyage, anchoring at random off sandbars and beaches.

Figure 20.2. Pamari houseboats on the river.

They have been dislodged by intensive immigration or were absorbed by it. In 1852 Silva Coutinho, advancing only as far as Huitanãa, passed fourteen villages from the mouth (today known as Redenção) up to Canotama, which Manoel Urbano had tamed with help of the Pamiri Indians. In 1866, the director of Indian affairs, Gabriel Guimarães, reported only five partially stable *directorias*: Alto Purús, Ituxi, Tapuarua, Arima, amd Hiapuã. In that same year, remember, the president of the province advanced his initiative for regular navigation. which finally was contracted to the Upper Purús by the Amazon River Boat Company, with its maiden voyage up the Purús in 1869.

The populace, initially mobile, began to settle and organize: a customs collecting point established in Canotama drew in revenues in 1867–68 of 692,647 pounds sterling, significant given that the income of the entire Purús fifteen years earlier was but 1,214,827 pounds.* And finally in March 1868, an outpost of the provincial police was established.

New pioneers appeared. Caetano Monteira and Bonaventura Santos advanced

* These values in US dollars of the time would be about $4,675,367 and $8.2 million respectively, significant sums at the time. The value of the 1867 US dollar in 2005 dollars is about $13.93.

upriver in the launch *Canavari* to the most remote points of the river, and the daring frontiersman Leonel Joaquim de Almeida was an admirable model for the tough Cearenses who would soon follow in his steps.[32] In fact, shortly thereafter, steam navigation was inaugurated, and a powerful wave of population washed up on the Purús in an uninterrupted advance that has yet to stop—unyielding, undeflectable, committing itself to the stable dominion of the lands through which it passes and is animated by a rhythm that impels it forward to the furthest headwaters.

This movement, begun in 1870, had a guide, Colonel Antonio Rodrigues Pereira Labre. Ably assisted by Manoel Urbano, who warmly received him in Canotama, the adventurous Maranhense soon proceeded up the Purús, passing Huitanãa, the terminus of the incipient navigation of that time, and went on to stake out the confluence of the Purús-Ituxi. At that point, on a bluff on the right side of the river, he cleared a bit of forest and raised in one day a hut covered with palm thatch. He founded a city. Lábrea grew quickly out of the wilderness, keeping his name and soon turning into the most privileged entrepôt from which further conquest would advance.

In 1873 Brown and Lidstone were not surprised as they advanced up the Purús to constantly see wisps of smoke filtering through the foliage of the riparian forest, revealing dwellings where latex was being cured, and in the settlements of Mabidir and Sepantini, more than 2,092 km from the mouth, they passed opulent estates, exporting 18,000–30,000 kg of rubber. In order not to be tedious, we will not review every population expansion into the region, one of the most dynamic of our country, if not of all Latin America.

Rigorous numbers will substitute for the most detailed description. As we mapped it, we chose the more remote points of that great river and, in a stretch of one-tenth of its enormous extension of 2,624 km, we noted it was exclusively colonized by Brazilians. Now, reviewing this table, we see that in the decade of 1873–83, the occupation extended up through to Triunfo Novo (2,212 km from the mouth) propelled by indefatigable explorers such as Antonio Bacellar, Casinio Pereira Caldas, and Antonio Leonel de Sacramento. In view of the development of the whole river, including the Acre, to whose mouth steam navigation arrived for the first time in 1878, a simple comparison with the exports of the last three years with the last decade of exports from the Madeira reveals that the Purús was already the richest of all rivers in the Amazon, whether in rubber, Brazil nuts, copaiba oil, dried fish, sarsaparilla, or *cumuru* oil.

This startling progress, except for a few insignificant interruptions, continued apace, at least in reference to rubber, where the Purús produced 1,950,000 kg in 1884, 1,648,000 in 1885, 1,967,000 in 1886, and 1,990,000 in 1887.

One can see also in a quick glance at the map annexed to this study [see chapter 19] that already at that time more than four hundred rubber estates graced the margins of the Purús (not counting the Ituxi, the Pauini, the Acre, and Inauina, the Iaco, etc.), and a city, Lábrea, was formed as a *comarca*—a formal administrative unit—in 1881, as well as the settlement of Canotama. Many of the rubber estates are in fact villages where one can discern solid constructions, certainly unlovely, but ample and comfortable, contrasting mightily with the primitive hovels of Paxiuba palms. . . .

In the next period, 1882–92, the speed of occupation did not slacken. Considering the last stretch of river, we note that even in this most remote area twelve new estates were established. The total exports of Purús in 1892 weighed in with 3,359,455 kg of rubber, more than double that of 1885, and Lábrea appeared with the largest import and export balance of all the *intendencias* of Amazonas, including Manaus. As the turmoil of earlier episodes died down, the newly established society in those new lands equilibrated, disciplined itself, and expanded the main economic activity. It embellished the restrictive routine of rubber production with small-scale cultivation of common foodstuffs, still only for local consumption. Around the large barracks they made the first clearings, reviving the landscape with the regular cultivation that binds the occupants to the land.

But the exportation of latexes in their various forms, which range from the finest products of *Hevea* to *caucho* to *sernambi*, continues to be the most secure measure of general progress. The Purús exports doubled in the decade from 1892 to 1902, and simple reference to the production of the last three years, 5,500,000 kg in 1900, 6,016,000 kg in 1901, and 6,750,000 kg in 1902, showed that this river alone accounted for more than a third of the production of the entire state of Amazonas.

The waves of colonists dominated almost all of the upper Purús. Except for tributaries like the Acre, where in that period the pace of occupation resulted in grave conflicts with Bolivia, which is not our purpose to document, and restricting ourselves to the main channel of the Purús, we see that from 1898 to 1900 yet another five *seringais* were established on its most distant points. Sobral, which sprouted in 1898, today marks the furthest sentinel of this enormous campaign against the wilderness. Who arrives there has traveled 3,889 km or 400 leagues, from the mouth of the Purus and has tangible proofs that four-fifths of the length of this majestic river is completely populated by Brazilians, without one hiatus or the blemish of one abandoned area, linking all the *seringais*, uniting them, defining a still rustic society, but one that is vigorous and triumphant.

What is realized there and continues to be realized is a vast natural selection. In this duel with the unknown, a simple craving for riches is not enough; rather, what is required is a will, a stoic fearlessness, and a privileged physiognomy.

Only the strong survive. There, exceeding numerically by virtue of better organic equilibrium due to a more rapid adaptation based on their sturdiness, their eagerness and indifference to danger, are the admirable northeasterners, mixed breeds, the *sertanejos*, who truly discovered Amazonia.

There is no doubt that in that emergent society, the vices and disorders inherent in great social dislocations reappear there as they did in the early days of the Transvaal, in the turmoil of the Far West or the mines of California. Poor distribution of land: on the one hand the *latifundias* stretch out on lands bounded only by rivers; they are limited economically to the hands of a few owners. The rough *seringueiro* is brutally exploited, living forgotten on a bit of land on which he wanders long years, and in his precarious situation he urgently needs social legislation that would guarantee him better returns for such immense effort. The solitude in which he is entombed at the most remote outposts, which is aggravated by the lack of communication, reduces him to a serf, at the capricious mercy of his masters. Justice there is naturally slow or nonexistent.

All these evils are overt and prove above all the mere fact of distance. They would disappear if the region were incorporated into the rest of the country, and for this reason [the nation] requires now, urgently, the development of navigation to the last inhabited point, and telegraph at least between Manaus and Boca do Acre. Taking these measures would be extremely rewarding given the current economic returns of the region and do not demand extraordinary outlays or efforts.

To the Headwaters

The Brazilian ascent of the Purús now reaches to Sobral. But as the annexed map indicates, above that hamlet are others: Sta. Rosa, Cataí, Curnaga, Sta. Cruz.

These are Peruvian posts.

These posts in no way signify the definitive domination and regular settlement of the land. The Report of the Joint Commission already described this, and nothing remains to be added to the limpid precision of its descriptions of the legendary ephemerality of the *caucheiros*—and lines of that report's have the added merit of being signed by the esteemed Peruvian *comissário*.

It would hardly be generous to revisit a topic where our preeminence is both comprehensive and reproaching. We note merely a few significant circumstances.

Peruvians began to occupy the area only after 1900, occupying but three sites beyond Sobral—Hossanah, Cruzeiro, and Oriente at the mouth of the Chandless—and insinuating themselves peacefully in lands already previously occupied by Brazilians for a long time, What permitted this was the innate generosity of those rough *sertanejos* who saw *caucheiros* less as strangers than as companions in

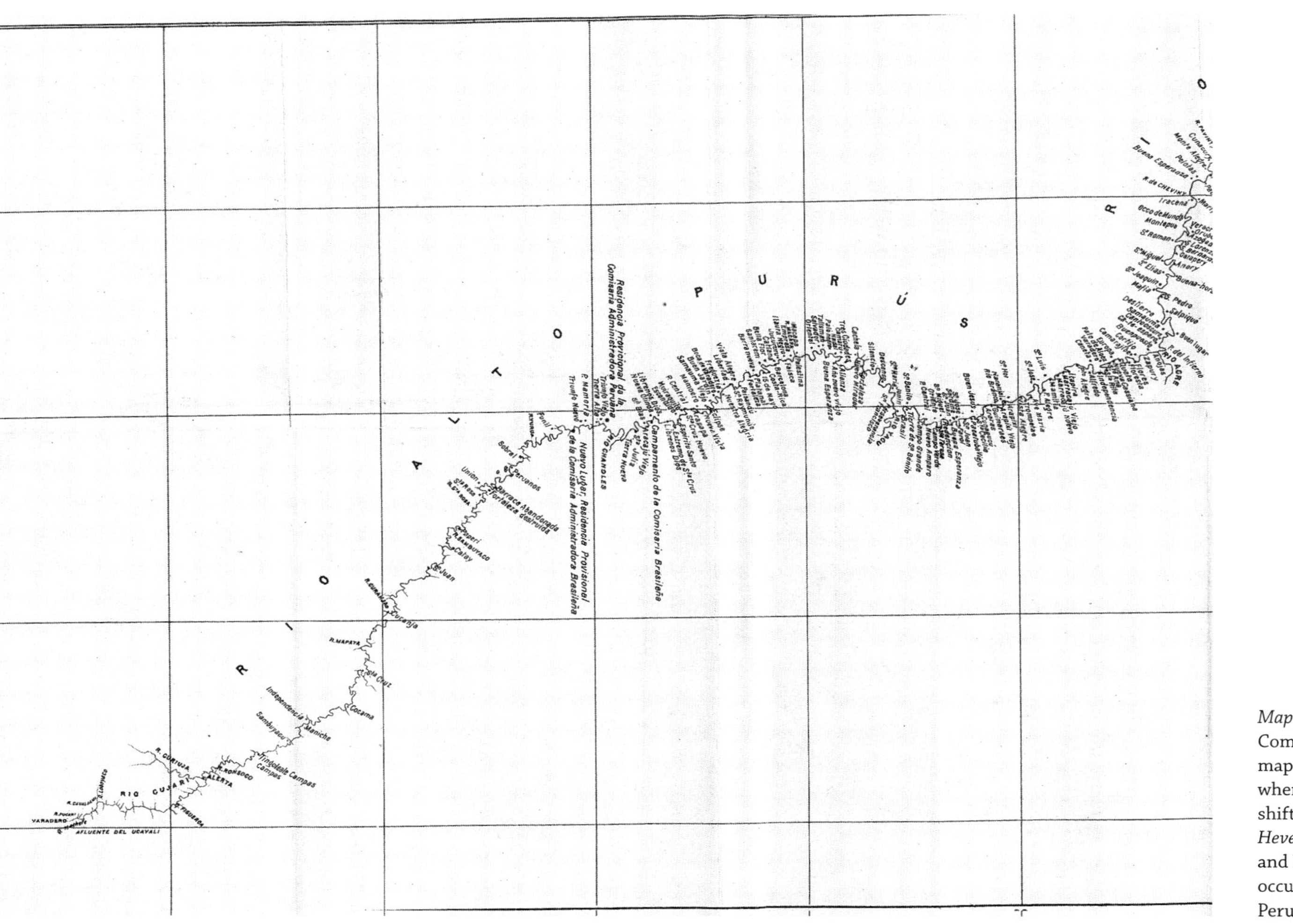

Map 13. Joint Commission map at the point where occupation shifts from *Hevea* to *caucho* and Brazilian occupation to Peruvian.

the same enterprise against the onslaughts of nature. But after two years (1903), the intention of politically authorizing the outcome of benevolent tolerance was tested with the attempt to establish a Peruvian *comísariá*—a national military outpost and customs house—at the mouth of that river, with all the official apparatus. It was then that the disparities of character that so distinguished the *seringueiros* from the *caucheiros* made for an inevitable conflict, which we restrain ourselves from describing. These are widely known, and their many episodes would only enflame the inherent serenity of these pages.

Only observe that the invaders, fleeing from battle, ceded all the territories they had been allowed to occupy and travel through, and retreated to Sta. Rosa at the mouth of the Corinaã, the northernmost extent of their settlement. Between Sobral and Sta. Rita, the lands of the neutralized zone, one can still make out the ruins of two abandoned *caucho* posts, União and Fortaleza, both abandoned by the *caucheiros*.

The abdication, in this case imposed by battle, would have occurred anyway, quickly and pacifically, as soon as they cut down the last of the neighboring *caucho* trees. The hamlets of the Peruvians, even the large ones, like Curanjá or Cocama, are mere camps.

In the entire area from Sta. Rosa to the farthest headwaters of the Purús, there is not one house made of tile. Hovels of palm thatch constructed in ten days proclaim the unstable presence of a nomadic society, one that despoils the land and then deserts it. One can compare it to the restlessness of the natives, now mostly crushed. In general, there are five Peruvians, mainly Loretanos, for one hundred Piros, Campas, Amahuacas, Conibos, Shipibos, Coronauás, and Jaminuauás whom one stumbles across in various types of activity and indolence, all conquered by the shotgun, all deluded by extravagant contracts, all now yoked to the most abject slavery.

The family doesn't exist—in most, if not all, hamlets one cannot point to a single legal couple—and immanent everywhere is a sense of breeziness, the casualness of a perpetual eve of departure that infuses those makeshift outposts where men plan to stay but one, two, or three years at most, to get rich and then never go back. They erect those *tambos* in new clearings, and noisily animate that corner of the forest for a while; but then, when exhausted and ruined, they vanish into the suffocating foliage of the lianas.

Curanjá had 1,000 inhabitants two years ago, today it has 150, and it will be abandoned shortly if the *caucheiros* are not successful at dislodging the fierce Coronauá, who are still the masters of the headwaters. Cataí, a hamlet opened up by a Brazilian, old João Joaquim de Almeida, at the frontier of Cassianã, will soon be a ruin if it isn't chosen as the seat of the joint frontier administration. In Shamboyaca, almost at the mouth of the Rio Manoel Urbano, the best

> agriculture—a vast manioc plantation covering a small hill—belongs to a Campa Indian, the "Curaca" Chief Antônio, who established it on a *caucho* post. (The Campas, thanks to their personal courage, maintain a primitive liberty unless deluded by the intricacies of the contracts they accept.)
>
> Cocama and Sta. Cruz today are very animated but will not last three years more: their life spans are coupled to the last trunks of the *castillas* that still abound in their environs. In Tingo Leales, an immense banana plantation and a permanent field of cotton belong still to a Campa, Chief Venâncio, who immigrated from the Ucayali.[33] Finally, at Alerta, where the main residence is reduced to a vast *tambo* of Paxiuba, there is no agriculture worthy of the name except for manioc and cane planted by the Amahuaca women. On that estate the land is of a rare exuberance, with soft hills stretching along the margins of the Purús and the Curiujá, offering a magnificent base for more prosperous and permanent activities. But this requires different passions that transcend a lust for easy wealth: those who arrive there are prepared only to punish themselves for three years for the right and the means to an opulent life in *other* climates. Nothing transcends that single-minded obsession, beyond which one sees not a glimmer of social cohesion.

This was the devastating conclusion of da Cunha's analysis of Peruvian settlement history on the Purús.

"Brasilieros": Peruvians and the Hidden History of Brazil

Da Cunha's other descriptions of Peruvian occupation, derogatory as they were, were perhaps too controversial to include in the final report and so ended up in essays upon his return in the *Jornal do Comércio* in 1907, and hastily reappeared in the chaotically compiled *À margem da história*. This essay, "Brasilieros," compared the effectiveness of settlement of Brazilians and Peruvians and drew a dispiriting conclusion: Peruvian enterprises had often followed on Brazilian efforts, and most had been pathetic failures. Da Cunha recounts the entrance of Peruvian administrative efforts in Amazonia. He reviews the quinine industry and the collapse of the largest Peruvian Amazon industry before *caucho*. From the remote forests of Peru and Amazonia, the drugs from "secret" forests, developed by natives and further elaborated by the Jesuits, were in frantic demand as imperial economies extended into malarial lands, and as colonial armies died in droves.[34]

The desperate need to control fevers was key in nineteenth-century medicine. Quinine was used to treat not just "intermittent agues" but almost any fever. Gustave Flaubert's adored sister Caroline, suffering from childbirth fever, was given massive doses of quinine, which of course did not save her. More than a million and half US soldiers in the Civil War were diagnosed with malaria, and as Philip Curtin has shown so eloquently for colonial troops in west Africa, mortality among temperate-zone troops who found themselves in the tropics was very high.[35] "The ague" affected military men and colonial governments everywhere and was decisive in many colonial wars, as historian J. R. McNeill has shown so conclusively. Clements Markham, one of the great scholars of the Andean zone, claimed the biopiracy coup that transferred cinchona bark to Kew. Unfortunately, his specimens had low levels of quinine in them and were ineffective. The seeds of the powerful antimalarial strain were actually collected by Charles Ledger with the help of a native friend. Kew Gardens, when presented with the seeds, refused to purchase them. The Dutch did, however, and proceeded to plant them in their colony in Java, thus dominating the trade in this most valuable medicine until the Second World War.[36]

Da Cunha's essay on Peruvian Amazonian occupation revisits his first essays on the culture of the region and what he sees as the central dichotomy between Peru's coastal culture and the potentialities of the Amazon. Even unhinged *caudillos* could assess the economic allurements of the lowlands. Da Cunha emphasizes the emergence of an immense state apparatus mobilized to support a new tropical economy to animate Peru's national identity, but the "Jesuits' bark"—quinine—on which the whole enterprise was predicated vanished overnight as a viable commodity, leaving in its wake the most abject abandonment; the region was left to stagnate with only "panama hats" and a few gold nuggets to define its commerce. This first attempt at a tropical Peruvian civilization, aborted in the geopolitics of plant theft, was saved, according to da Cunha, by Brazilians who helped develop the industry of *caucho* and steam travel. The rest of the essay documents robust *sertanejo* colonization in contrast to the pathetic subsidized German (considered the ideal colonist race) colony that ends up in rags and religious fanaticism, evidence for da Cunha's argument regarding Brazilian adaptation and "counter-colonialism," and an empirical rebuke to the advantages of white colonists. Even when Peruvians take over a functioning enterprise, in a few years the place is reduced to rubble, useful mainly for firewood burned by casual travelers. This theme of Peruvians as "builders of ruins" echoes through all da Cunha's writings on the Peruvian Amazon, culminating in his final essay, "Among the *Caucheiros*," in the next chapter.

Da Cunha notes that a hidden Brazilian history resides between the lines of Peruvian tropical expeditions "wandering out of our own historical annals," unifying the Brazilian spirit from the east and from the west, and so making even the "Peruvian Amazon" at its heart a Brazilian enterprise.

Brasilieros, or the Problem of the Oriente

Peru has two fundamentally different histories. There is the familiar one from books, the one that is theatrical, histrionic, the one readily condensed into the farcical romances of proclamations of freshly minted marshals. The other is obscure and fecund. It unfolds in the wilderness. It is more moving, more serious, more majestic. It extends to the Purús' most remote terrains the glorious traditions of the struggles for independence, and continues them without pause to our days. In all its varied aspects, this history could summarized in a single title, one adopted by the best publicists of that Republic: The Problem of the "Oriente"—the challenge of the eastern jungles. . . .

To Peruvians there is no need to resort to the elaborate arguments of a sociologist or the happy intuitions of a statistician: one needs only the material goad of the environment. Constrained to a strip of land set between the mountains and the sea, for three centuries Peruvians languished, deluded by the pomp of the conquistadors and the viceroys. Peru today is the main heir of the virtues and the vices equally notable in the Spanish nobility—and in decay since the seventeenth century. Peruvians finally understood, through the reflexive instinct of defense, the overarching necessity of abandoning the seclusion that had isolated them from the rest of the world.

Thus they began the Andean crossings.

It would be too tedious to recount the pilgrimage to the mountains, the successive assaults invested in the five tortuous roads winding precipitously through the curving mountains, surging up slopes thousands of meters high, and uniting the coastal ports of Mollendo and Paita[37] with the coveted outposts of the *montaña* at the extreme edge of Amazonia, extending through the valley and *pongos*—river chutes—of Manseriche to the foaming whirlpools of the Urubamba. After crossing the last easternmost range and arriving at the Ucayali basin, even the most oblivious pioneer would note that it rivaled that marvelous Manseriche valley and was a place able even on its own to invigorate exhausted Peruvian nationality. There is the incongruity that results from the physiography itself: the best parts of the country, among those countries that most that identify themselves as "Pacific" nations, really have but one true sea for linking long-distance trade and civilization; it is defined by the three long, unimpeded channels: the Purús, the Juruá, and the Ucayali.

No Andean engineering miracle can substitute for them. The train line to Oroya and other lines that match this daring track, curving up the jagged scarps, threading through tunnels, drowning in clouds, and traveling over suspension bridges perched above the innumerable chasms, cannot create a practical and secure system of transportation. The exceptional technical conditions were disastrous for industry, since they made it forever impossible to transport goods from the "Oriente" without excessive freight costs, even when the opening of the Panama Canal dispenses with the long voyage around the Horn.

Thus the passage to the Atlantic through the Amazon and its tributaries was the clear solution to the problem. And the new outposts, sprouting up in what is now the Department of Loreto, soon began the intensive labor of domination, which continues to this day.

They opened up the roads required by the rich fluvial zone; they planned, in spite of many setbacks, many military and agricultural colonies; they mobilized the revival of the apostolic missions, the admirable traditions of the Jesuits of the Maynas;* they developed a vast system of land regulation; they built a port at Iquitos; and to stimulate occupation, they abolished all taxes to induce domesticating man to inhabit that most feral land.

Meanwhile, geographical expeditions began in 1834 by P. Beltrán and Wm. Smith were followed by those of Castelnau, Maldonado, Raimondi, and J. Tucker, and, in our day, G. Stiglich, who tenaciously and inexorably spread out through all the compass quadrants in the complex task of making a type of rapid survey for a new nation.

The strident upland caudillos reviled these tranquil explorations. On the littoral, always rife with sedition and insurrection, the chronic incompetence of revolutionary governments institutionalized itself. There, distorting the noblest aspirations of the recent campaigns for liberation, the reckless brigands catapulted to power and gave themselves over to a pernicious militarism, which there, as everywhere, is the festering wound of sickly nations. But meanwhile, in the desolation of the *montaña*, with or against the currents of unknown rivers roiling through dizzying bends of the *mayunas*, canoes were bid farewell, and like arrows, they shot into the famously powerful currents of the *pongos*,† into the tumult of the foaming rapids. The geographers, the bureaucrats, the missionaries had outlined the contours of a revitalized state, one where the noblest traits of

* The Maynas were the region of the missionary system in the far Upper Amazon of Peru and Ecuador.
† *Pongos* are narrow channels and rock canyons of rivers as they descend the Andes into the Amazon.

the Peruvian race could be purified by the apprenticeship to danger and hardship and thus revive the weakened national character. They gave those scarcely defined geographical coordinates a vibrant extension into History. Because the "Problem of the Oriente," after all, includes in its numerous unknowns the destiny of all Peru.

Even the maniacal caudillos knew this. Distracted as they were by their own recklessness and constant vacillations, in the short intervals between two firing squads or one battle and the next they agreed to consider these desirable outposts, and many of them suddenly transformed themselves in light of the alluring and lucid revelations of the statistics.

One can cite numerous extensions of politics that restore as well as demolish, that accent the physical contrast between the recondite West with the inflexibility of the moral order of Peru, whose energies were malignantly dissipated in the emotional hysteria endemic to the formal world of ministerial proclamations, and the resplendent tropical Oriente, where reborn hopes were dawning.

Here is an example.

In 1841 the Republic was in complete havoc. D. Agustín Gamarra ruled. This tyrannical *zambo*—mixed blood—exhibited in his capricious acts all the imbalance of his mestizo temperament, goaded as he was by the fears and impatience of a transient prestige that he had won by luck in guerrilla assaults. His government, which inaugurated in Peru the new legal process—the coup d'état—that eventually unseated the virtuous José de la Mar,* was naturally extremely turbulent. His restoration was imposed by Chilean weaponry and Manuel Bulnes† on the ruins of the ephemeral Peru-Bolivian confederacy. Gamarra's regime was besieged by contrary claims: the ceaseless demands of disgruntled mercenaries and the threats of insurgent conspirators. Gamarra was made dizzy by the vertiginous heights he ascended after he unfettered himself from his *cholo* constituency and then allied himself to the aristocratic sensibilities of elites, who more than all others [in Latin America] inherited the traditional superciliousness of Spain. In the many desperate political situations in which he found himself, he depended on luck and not infrequently on the accomplishments of a woman, his own wife. This kindly and heroic amazon would often strap on a sword and leap fully spurred on horseback to the reviewing field or into the most heated battles, where she galvanized the astonished commanders and the wavering troops with her charming presence. One really could not demand of such a president, whose life was in so many ways perturbing and romantic, to focus his attention on tedious administrative tasks.

* Peru's first president.

† A president of Chile who allied with Gamarra to unravel the Peruvian-Bolivian confederacy. With Bolivia defeated, Gamarra became president of Peru.

But let us follow him for purely artistic reasons, as one might trace the unfolding plot of an imaginative novel rife with alarming and dramatic episodes, until we come to the final denouement of our protagonist in a glorious and useless sacrifice, as he succumbed before the furious charge of Bolivian lancers on the plains of Viacho.*

But returning to these pages, this, surprisingly leaps out at us:

> The citizen Augustín Gamarra, the Grand Marshal, Restorer of Peru, the Heroic, the Excellent, etc. etc. etc.
>
> Considering that to provide steam navigation on the Amazon River and its tributaries, it is necessary to provide facilities and incentives to stimulate the impresarios . . .
>
> Decree 1: We concede to the Brazilian citizen Antonio Marcelino Ribeiro the exclusive privilege to provide navigation on the Amazon River in the part that corresponds to Peru and its affluents . . .

The relevant bits of this decree burnish this hapless caudillo with all the vain elegance of his numerous titles and reveal him as the first ruler who traced for his countrymen the revitalizing march to the east, to the Oriente. But we do not repeat this point to illustrate the contradictory aspects of Peruvian history but rather to emphasize the figure of a *Brazilian*, which would be irrelevant if it did not constitute the first of a series of our unknown compatriots wandering out of our own historical annals and electing themselves, by memorable acts, as the best servants of a neighboring nation.

In fact, insofar as the traces of Peruvian expansion to the east is revealed in the swamp of regulations, degrees, circulars, official letters (these, alas, are the supreme military, political, and administrative obsessions of Peru), one can discern in them the obligatory and incisive presence of Brazilians in yet another unsung but vigorous advance that met with Brazil's own energetic forays toward the west. One could refurbish an entire chapter of our own history, one lost or disjointed, one invisible to the benumbed gaze of the chroniclers, now resurgent in sparse but surprising fragments in between the lines of the history of another people.

And this is revealed in other unrenowned cases. We reveal a few glimpses of them:

In the period embraced by the austere Marshal Ramón Castilla,† the Amazon explorations continued. De Castelnau descended from the headwaters of the

* This was a decisive battle in the War of the Pacific.
† Castilla was president of Peru four times.

Urubamba to the banks of the Amazon. Faustino Maldonado was immortalized by discovering, in an extremely daring excursion, a new route to the Atlantic linked to an as yet unmeasured channel of the Madre de Dios. Raimondi disclosed the "treasures of Mesopotamia" in the 16,000 square leagues of exuberant lands crossed by the watercourses of the Huallaga and the Ucayali. Finally, Montferrier rigorously calculated the riches of that vast Canaan: 50,000,000 ha worth at least half a billion pesos. The arithmetic was lyrical.

The governmental measures of the great marshal soon had the encouragement of the most energetic patriotic stimuli, equaled only by the greed of the most unprincipled adventurers.

The Peruvians, for so long wedded to the sterile littoral, saw the New World for the first time. And its conquest in one of its keenest phases, unfolding in all its vastness. The administration then began a disheartening phase of brilliant but abortive projects. The planned colonies, quickly scattered over the lost and lonely corners in a sort of phantasm of artificial progress, soon flickered out. By 1854 the government of Loreto, an obscure little village whose name now extends throughout those places, informing of the status of two successive colonizations that were established in that department but centralized in Caballo-Cocha near the frontier with Brazil, indicated that they were completely extinct. The same mishaps were generalized throughout the entire region.

This was natural. The limits of human occupation in those outposts were not so simply defined. The first stages were characterized by the instability imposed by life itself, acquired in the movement of the march. A reconnaissance of their new habitat focused on the fruits of immediate riches and how these would provide for recent arrivals in the wandering life of collectors, of gold miners, of herders. These would precede the steadiness of agriculture, before a place was chosen to alight and to root.

This is the eternal social function of nomadism, but in Peru, the devastating turmoil of the *cascarilheiros*—the collectors of quinine bark—was already unfolding, unveiling the obscure backlands stretching from the hills of Carabaya to the most distant scarps of the Beni. This incentive was, however, unraveling.

During that time a tenacious explorer, Clements Markham, commissioned by the English government, traveled in the regions of *Quina calysaia* and quickly managed to transplant to India that key element of Peruvian fortune. By 1862 more than four million trees in Darjeeling were producing an extraordinary 370 tons of quinine, and thus was initiated the triumphant assault on the monopoly of quinine. The anxiously coveted Peruvian backlands were soon shorn, at least for the new inhabitants, of these resources that everywhere had been so purposefully, so lyrically depicted that one could not deflate the always exaggerated hopes of those who immigrated. The *bombanajes*, collectors of palm fibers destined to the gracious industry of "panama" hats woven by the women of Moyobamba, and the gold-bearing gravels of the slopes of the Pastaza, guarded by the bellicose Humbizas, were not enough for a regional economy.

All the acts, magnificent decrees, lucid regulations, generous land concessions of the last government of Castilla would have dissolved in the most abject failure in the final part of his presidency, the year (1862) when the cultivation of quinine in India snatched from the wilderness its greatest allurement, had not an anonymous, humble immortal, invisible to our history but eclipsing in one blow the most imposing administrative thrusts, offered the Peruvians the energetic reagent that encouraged them forward on the route to Amazonia.

A *Brazilian* discovered *caucho*, or at least began its industry of extraction.

In reconstructing this chapter of our history, which if developed later by another historian would merit the title "The Brazilian Expansion in Amazonia," we are hardly alone. A reliable narrator reports:

> Before the year 1862, the incalculable riches of latexes had yet to be explored. . . . After the entrance of some Brazilians into the department [of Loreto], especially the hardworking José Joaquim Ribeiro, this valuable product began to figure in the list of exports to Brazil. The first lot exported was 2,088 kg, the result of some experiments of that Brazilian, who would have contributed so much to the development of that industry had he not, at the beginning, encountered difficulties born of the greed and envy of some petty officials who opposed him at every step.[38]

We will not comment on the antagonism of Peruvian authorities. It was of long standing. Since 1811, don Manuel Ijurra had accused Brazilians in these terms:

> Those Brazilians who live closest to Peru have the barbarous custom of arming military expeditions with the object of hunting down Indians of the Maynas,[39] ignoring the authorities . . .

Or describing us like this:

> Absolute monopolizers of the commerce of imports and exports.

Five years later in an alarming notice, the subprefect of the Maynas urgently requested what actions he should take "that that the Brazilian inhabitants of Caballo-Cocha should depart this province, if not peacefully then by force," and described them, staining them with the most disgraceful libels. Finally the governor general of the missions (1849) decided to demand that Brazilians who entered that country carry passports, stammering in his clumsy Spanish, "We experience no benefit whatsoever from these Brazilian traders, nor are there bayonets enough to restrain them: they do what they want, entering rivers, extracting sarsaparilla, turtle oil, salt, and other goods."

We desist. But one can see from these lines, of which many more could be included, a formidable invasion that stretched out, dominating the regions in the west, challenging the hatred against the foreigner: installing itself through the valley of the great river through Loreto, Caballo-Cocha, Moremote, Pernante, Iquitos, and all the way to the mouth of the Ucayali and up that river past the Pachitea, and leaving signs in the most varied points, in the numerous little farms, the sinuous forest trails, and even the customs that persist to this day, the indelible traces of its passing.

If we were writing this history, we would counterpose the crescendo of invective of petty bureaucrats, ever more stridently derogatory, to the mute invasion that swelled over Peruvian soil, with the concepts of Antonio Raimondi.[40] But the remarkable Joaquim Ribeiro, whom Peru's greatest naturalist encountered on the margins of the Itaya River, master of the region's best farms, materialized an irrefutable answer. He was not hindered by such trivialities. After 1871, rubber emerged as the most important export of Loreto. And bands of extractors without any government support, spontaneous migrants from everywhere, taking on the remotest parts of the wilderness, surpassed in a short time almost a century of efforts so fraught with reversals.

The Oriente was unveiled.

But this picture is not perfect.

The exploitation of *caucho* as it is practiced by Peruvians, by cutting down trees and moving always at random to new, still unclaimed *manchals* [groves] in an endless professional nomadism, brought out in these men all types of brutality in their inevitable encounters with the natives—bringing a systematic disorganization of society. The *caucheiro*, the eternal hunter of new territory, never attaches to the land. In this primitive activity, he perfects only the attributes of slyness, agility, and violence. In the end, it is a barbarous individualism. There is a lamentable involution in a man forever exiled from settlements, wandering from river to river, from forest to forest, always looking for some virgin forest where he hides or takes refuge like a fugitive from civilization.

His passage was devastating. At the end of 30 years of populating the banks of the Ucayali, formerly so ennobled by the self-abnegating works of the missionaries of Sarayaco, today their degraded little villages exhibit an indescribable decadence. Colonel Pedro Portillo, the current mayor of Loreto, who passed through there in 1899, raged: "There is no law there. . . . The strongest, who has the most rifles, is the master of justice . . ." He then denounced the scandalous traffic in slaves. In a similar tone, other travelers without number, a list of whom would be far too tedious to cite, detailed in expressive narratives the regime of manhunts that have become usual in those lands, following the spoor of men traveling these forests, whose sole effect is to barbarize the barbarians.

In the anticipation of the evils of this form of exploitation, which, however, had completely determined the development of its dominion over the region, the Peruvian government never renounced its earlier ambitions for intensive colonization. And at the same time to guarantee the use of the best route to Amazonia via the Ucayali, which travels from the Oroya terminus to the principal tributaries of the Pachitea, it established in 1857, at the margins of the Pozo River, a German colony and monopolized its care and uninterrupted solicitude.

The site itself was admirable. Halfway to Iquitos, near the navigable affluents of the Ucayali, on fertile soil, the settlement was in military and administrative terms the most solid, strategic point in that struggle with the wilderness, justifying the efforts and extraordinary expenses that were invested for rapid development, which the best natural conditions would favor. But things did not work out as planned. As happened in Loreto, the new colonists, even the most persistent, found their efforts fruitless. The colony was paralyzed, immobilized amid the splendors of the forest. It was reduced to subsistence production that hardly satisfied its own needs. The demographic increase was almost imperceptible: the offspring of the tough Prussian stock regressed to the constrained capacities of the Quechua. Visiting in 1870, the prefect of Huánuco, Colonel Vizcarra, was left stunned and sorrowed by the colonists, who appeared to him in rags and starving. The romantic don Manuel Pinzás described the trip and the "heart-rending scene" in lengthy, lachrymose sentences. Five years later, Dr. Santiago Tavara used the same dispiriting tones as he described Admiral John Randolph Tucker's first travels.[41] After thirty years, Colonel Pedro Portillo,* in his Ucayali jaunt, received reliable news of that settlement: it was a miserable shambles. The settlers and their degenerate brood had become victims of an irredeemable fanaticism, lost in painful necessities of penances, endless praying, continuous rosaries, litanies, and lamentations that vied shamefully with the wails of howler monkeys.

Making the disappointment even more painful, our intrepid traveler, who today is one of the most lucid Peruvian politicians, had passed just a few days before through Puerto Victoria at the confluence of the Pichis-Palcazu, affluents of the Pachitea, and witnessed a completely different scene. Puerto Victoria had

* Pedro Portillo was instrumental in the politics of extending Peruvian control on the Purús, Putumayo, and Juruá (see chap. 10).

developed into one of the most animated and opulent settlements of the region, without the government's being the least bit aware of its emergence.

It had never even thought to colonize that area.

It was thought to be ill favored. It was surrounded by the most hostile and wildest of native South Americans, the Campas of the Pajonal to the south, and to the north, the indomitable Cashibos, who had at Chonta Isla (island of the peach palms) brutally murdered the naval officers Tavara and West. The prefect, Benito Arana (father of the more famous king of *caucho*, César Arana), who had passed there that same year, mounted a full-fledged military operation with two launches and an artillery vessel to revenge that bloody affront. They entered the forest, engaged in skirmishes and shootouts, and returned in a singular "triumph" with the savages at their heels in a hail of arrows; they sailed off "victoriously" firing their cannons furiously at the riverbanks, and leaving the name "Playa del Castigo"* as a novelistic remembrance of that embarrassing episode.

For the next three decades the menacing region remained in the most complete isolation as a result of the terror the episode had inspired. Until, one day, coming from the west, there appeared some fearless adventurers, paddling their light canoes, the *ubás*, up the powerful currents of the Pachitea, passing from one difficulty to the next to the next until they arrived at the confluence of the Pichis.

They were tough *caboclos* with dark mestizo coloring and sere, powerful bodies. They were not *caucheiros*. Their language did not vibrate with noisy fanfare. Instead of a "tambo" they put up a "tejupar." Instead of "cuchillas," they carried, thrust in their belts, the "facas," machetes, long as swords.

They quietly prepared themselves for the business at hand and serenely entered the forest. We don't know the details of their feats, which were certainly exceptionally dramatic. The Cashibo Indians have incarnated in their very name the legend of their ferocity: *cashi*, bat; *bo*, similar to. Figuratively: the drinkers of blood. Even in their rare moments of joviality these natives are frightening, because when they laugh one glimpses their teeth tinted black with the juice of the peach palm, or face down on the ground with their mouths near the earth, they ululate the slow notes of a savage refrain. They passed through three hundred years of attempted catechism, indifferent to every brutality, and are still the fiercest tribe on the Ucayali. But they did not seem to impede the vigor of the new pioneers. What faced the bloodthirsty savage, crushing him, was an even more fearful adversary: the *jagunço*.

Those recent arrivals were Brazilians from the north; their patron, their master, Pedro C. de Oliveira, was yet another of those unknown fighters who emerge in the fecund initiatives rounding out the events of extraordinary history.

* This would be translated as "Beach of Punishment."

Figure 2.3. Cashibo Indian, "drinker of blood."

To appreciate his worth, it is enough to note in passing that in 1900, in spite of his ethnicity, he was named governor of the entire territory of which his trading post was the center.

Colonel Pedro Portillo,[42] who received there the warm hospitality so characteristic of our humbler people, so lacking in the ostentatious pomp and false generosity of many others, carried this enchantment throughout his report from his first day to his farewell to the "estimable familia del señor Oliveira." He described the delight he took in the vibrant settlement surrounded by a bountiful agriculture, its dwellings intelligently spread along the top of the riverbank, which one arrived at by a sturdy stairway from the river. He was especially captivated by the calm *sertanejos*, who were unassuming in the face of their complete triumph over wild land and savage. Finally, his clear vision could not help but note that that these outsiders, with neither decree nor subsidy, had solved the problem that had vexed his own government, founding in the most suitable place a station that guaranteed dominion of the Central Route of Amazonia. He was frank about it: Puerto Victoria was the place most appropriate for a military garrison and a customs post that would administer the imports and exports of the colony of Chanchamayo, the northern Pajonal, Tarma, and the forests of Palcazu, Matro, and Pozuzu. And he concluded:

> the house of the Oliveiras should be taken by the supreme government as the obvious site for the administrative offices of the captaincy, customs, and the military.

The recommendation was accepted. A decree from President Nicolás de Pierola ordered the demarcation of Puerto Victoria for the establishment of a *comisaría* destined to protect the colonists of those lands, and in great envy of the excellent situation of the Oliveira estate, revealed the intention of taking exclusive possession of the area by not permitting any settler within the radius of one kilometer.

Peru acquired a really admirable river property. And the Brazilians withdrew.

Five years passed. In 1905 a Parisian tourist, J. Delbeque, descended the Pachitea on his travels to Amazonas and wouldn't have even noticed the formerly flourishing *estancia* had he not been accompanied by some tame Indians who knew the area well. There at the top of the river bluff, which the floods were eroding, one could see just some collapsed roofs and the remains of the agriculture choked by wild brush.

It was a ruin.

Travelers would stay for a few hours in order to dry their soaked clothing in the heat of a fire fed with rotting doors and buckled frames of the old houses, a regular practice of all who passed there on the way to Iquitos. Our voyager mused sorrowfully that if things continued like this, Puerto Victoria would soon be only a memory.

He then left, rowing as fast as possible, fleeing the outpost that had been left in the most complete abandonment.

21

Everyday Forms of Empire

The Tropicalist Ethnography of Euclides da Cunha

Euclides da Cunha, Amazon Tropicalist

Da Cunha's influence on Brazilian nationalist and imperial thought was durable. The Brazilian sociologist Gilberto Freyre pronounced da Cunha "the First Tropicalist." Freyre could claim early usage of the terms *tropicality* (essentialisms about the tropics), *tropicalism* (ideologies that informed the practices of tropical imperialism), and *tropicology* (studies about the tropics).[1] For Freyre, a major mid-twentieth-century international public intellectual, Luso-imperialism was the ideal form of colonial practice because it could integrate autochthonous cultures into global economic systems in a mercantilist, miscegenated, and multicultural way. It was based on the concept of hybridity in all its forms, well before that term had become a cliché in colonial studies. Freyre used these ideas to promote António de Oliveira Salazar's vicious dictatorship (1932–74) and its Portuguese colonialism in one of its especially brutal phases. This stance (and his support for Brazil's authoritarian regime) did much to discredit him, but his insights were perhaps better than his politics.[2]

Freyre, a Northeasterner, yearned, like da Cunha, for a Brazilian national identity rooted in racial blending, as opposed to southern Brazil's proclivity for European preeminence in blood and culture. Like da Cunha, he saw the adaptation of the Luso-temperament and organism to multicultural hybrid equatorial environments as part of an emergent non-European civilization, a "New World in the Tropics." Freyre took a great deal of inspiration from the way da Cunha deployed frameworks for the biosocial dynamics of place, temperament, and culture that produced a uniquely adapted tropical Man. Da Cunha's views on miscegenation and syncretism, on creolizing cultures in producing "tropical citizens," as well as his devotion to the idea of progress, were hallmarks for justifying elements of Brazilian Amazonian imperialism. These approaches were marshaled by Freyre in his writings on Luso-imperialism.[3]

Da Cunha's culminating essays comparing Peruvian and Brazilian tropical imperialisms framed a substantive, "scientific," and "moral" landscape for Brazil's assertions of territorial rights and a rhetorical intervention that would link Amazonia to a mythical collective Brazilian past, a "people," and an imagined project of the future. The Amazon question was not strictly a matter of who was able to "tame the heathen and the wild" but part of an ethical and political ecology. What emerged from this exercise were "homegrown" tropicalities and tropicalisms that challenged both European and Peruvian colonial practices in concrete, discursive, and politically powerful ways, ways that reinforced the Brazilian presence on the ground, as architects of civilization rather than "builders of ruins."

Tropicalism and Orientalism

The tropics themselves are a set of material facts—they are places on the planet at particular latitudes and inhabited by a complex biota that includes people. Ideas about the tropics are not in themselves necessarily tropicality or tropicalism, any more than ideas about the Orient are simply Orientalism. Rather, tropicality questions revolve around systems of thought where ideas of nature, human nature, history, culture, and civilization are played out in the creation of images, institutions, sciences, and discourses that shaped and continue to shape the logics of colonial enterprises, modern development, and even today's environmental strategies. Tropicality is about "essences" and "others," and also about nature and place. These ideas inform and structure the ideologies that underpinned tropicalism—the suite of imperial practices engaging the region.

Recent and increasing use of the term *tropicality* in colonial studies reflects a conceptual borrowing from cultural critic Edward Said's "Orientalism." Tropicality asserts a kind of tropical "Orient" as an embodiment of forms of essentialism: the "beingness" of existing in the tropics as in a particular biogeographic and cultural space, but it is understood less as "place" than as the imaginative terrain whose hold on explanatory sciences, arts, and cultures affected the manner in which enterprises and political endeavors were carried forward in tropical worlds—the essences as translated into ideologies used to justify micro as well as macro politics of power in imperial practices.

Orientalism emphasized the "Known World" and focused on the decadent and the exotic as outcomes of the slow ruin of former civilizations, a world of excessive and depraved sexuality of all kinds, a place that was more or less running down, picking through the rubble of its various collapsed empires, Islam, and despotic states. In contrast, the New World tropics were seen as the "last unfinished page of Genesis." The question for the Orient remained one of culture, Islam, and history, where the European presence would be *liberational*. In the tropics, pagan, wild, and emergent, the colonial project would be *civilizational*. In addition, questions of nature are much more salient. As historian of science David Arnold has observed, tropicality engages the pervasive "alienness" of tropical landscapes as biotically distinctive for Northern Europeans and decisively ecologically foreign to the temperate zone. This "Arnoldian" framing weights nature as at least as powerful, and as much an actor in the "colonial" question, as the agents and the subjects of colonialism themselves.

The tropics and their inhabitants were not unknown to Europe by any means: the Mediterranean world had benefited from trading circuits that extended into the tropical realms of Africa and Asia, returning with the slaves, spices, medicines, dyes, jewels, animals, perfumes, and incenses that had defined elite status and consumption since before the Roman Empire. The torrid zones also figured in ancient medicinal and social theories of the environment and nature's impact on humankind: the humidity and the heat were seen as enhancing nature's creative power such that classification systems could not entirely account for the endless spontaneous generation that occurred there.[4] Travelers' accounts, natural history, and natural philosophy all positioned the tropics in myth and science as a world of monstrous wonders, dangerous, sensual, and fecund; the transfer of mythical terrain (the land of the Amazon) and medieval imaginaries to the New World was part of the "Fantastic Geography" of the Age of Exploration.

Paradise and Pestilence, Edens and Evil Latitudes

Northern tropicality and tropicalisms invoked an imagined Tropics where man and nature were in primal, protean states, outside of history, a "prehistory" of a place *sem fei, sem rei. sem lei*—without creed, king, or law, basically a scrim on which imaginaries, virtues, vices, and dreams of new societies could be projected. This perception framed the tropics, and especially Amazonia, as inchoate space, *terra incognita* as imagined on the first maps, a world ripe for physical and scientific exploration and as *terra nullius*, a land without rulers or rules, a tabula rasa for the practical application of great social and economic experiments. Tropicalism is complex in part because of the diversity of empires, imperial ambitions, divergent methods of claiming, and cultural matrices that engaged the equatorial zone, and how interventions were structured in light of "white" and "black" legends about the tropics and their inhabitants.

The "white legend" of tropicality is an Edenic, discourse: tropical exuberance, abundance, free and innocent sexuality and lots of it, pleasant climate, biosocial harmony, beauty: "nature never sackt," in the words of Walter Raleigh. Derived initially from Renaissance travelers like Jean de Léry, the prelapsarian streak of lost Eden runs through the New World tropical canon from contact to the present day.[5] Its people are charmingly ingenuous, "well formed," noble savages, the result of adapting to an opulent and generous nature, far from the meddling and corrupting insertions of the state. This noble model, historically most associated with de Léry, Raleigh, and later Rousseau and Condorcet, infused the critiques of the European absolutist states of the eighteenth century and exalted less hierarchical, less repressed, secular social formations, not to mention a more unlaced sexuality. The subtext involves an uplifting "natural" morality, part of an anticlerical as well as political critique of European polities. It is not accidental that so many of the early "literary" utopias are located in the tropics,[6] since they take their inspiration from the Age of Exploration and its early anthropologies that contrasted so profoundly with Northern European cultures and seemed to offer up endless sociopolitical imaginaries and possibilities.[7] These utopian writings themselves had afterlives as "colonization manuals" in many cases.[8]

The "black legend," by contrast, evokes ecological intractability, disease, moral corrosion, too much sex (often perverse, or so described by the Inquisition),[9] cannibalism, savagery, an unpleasant and catastrophic climate, laziness and obdurate resistance to the allurements of civilization, or corruption by civilization. The "true nature" of the tropics in this view is a green hell, a festival of horrors. The "green luxuriance" is a deceiving mantle for the essence of the tropics: a place of treachery, falsity, a counterfeit paradise nursing a fundamental debility—agriculture fails, forests don't grow back well or grow back too well, the soil degrades, and people quickly die. The imagery of rot and corruption, whether biological or moral, underpins what is seen initially as fertility. The deep roots of this idea lay in the work of Georges Louis le Clerc, Comte de Buffon, and became widely popularized by this natural historian and encyclopedist. These savage tropics and their inhabitants express a stubborn resistance to higher civilizational possibilities, a stillborn fecundity masking a deeper narrative of peril.

The climate seems agreeable but exhales fatal miasmas or produced lethal storms; and if it is not, as medicine had it at the time, sucking oxygen out from one's lungs, it hastens depravity and moral erosion. Elsworth Huntington in 1905 echoed the earlier views of the Inquisition in Brazil: "Any young man with red blood in his veins is in more danger of deteriorating in character and efficiency because of the women of the tropics than any single cause." And indeed there was the broader "moral climatology" embraced in the urgency of being temperate.[10]

This issue of essence, of what the tropics *are*, shapes "tropicalism" and the set of "development" practices directed at them. For temperate-zone analysts of da Cunha's day there were intrinsic biotic and socio-racial limits to the autochthonous development of tropical "civilization" that inhered in both the "white' and the "black" legends: populations too innocent or too evil (and too dark in color), too degenerate or too frail and undeveloped to support colonial modernities—"land that could hardly bear the weight of civilization." These problems of human nature were compounded by "opulent nature" that quelled the thirst for enterprise, or "intractable nature" that simply overwhelmed. Tropical nature was insufficiently worked by the local populations in their nomadic circuits, or dominated the indolent locals or wilted under the practices of plunder. Thus the equatorial world was an appropriate arena for colonial action, based on other, better, economies, comportments, and rationalities—and the persons—of the Northern *mission civilizatrice*.

Northern and Mediterranean Claiming

The ideas of tropicality developed in the North evolved from a different cultural matrix from that of the Iberian world—the Mediterranean—where cultural familiarity with tropical peoples and products was far deeper and of greater antiquity, stretching back to the cultural links of Roman and Islamic empires with North and sub-Saharan Africa and the tropical realms of the East. Portugal was a creolized culture even before it took to the seas. Later, fifteenth-century slave and evangelical efforts on the West African coast were pioneered by the Portuguese and African Atlantic creoles along with sugar production, as they cast their nets of "Empire on the Ocean Sea" along the coasts of Asia, Africa, and Brazil.[11] As a mercantile empire Portugal was, on the whole, relatively indifferent to how its subject or trading societies delivered goods into commerce; its interests always focused on how local populations could be integrated into the supply chain through trading entrepôts, 'economies of affection," and war booty, rather than rigorous administration of land.[12] Imperial interest for the Portuguese centered on the inhabited world—the labor and ecclesiastic conversions that justified their ventures. Since sugar and slaves were key commodities, the inhabited tropics rather than the "empty equator" held the central allure. The interests of the Portuguese were mercantilist, and for them the terms of trade, rather than the deep structures that delivered the commodities, were of most, and usually monopolist, interest.

Wastelands and Habitus

Nineteenth-century Northern European imperialisms reflected the rise and development of a different colonial model, one more engaged with the pressures of industrialization and mass consumption, more concerned about the organization of production, bureaucracies, and the development of new markets than with the traditional mercantile circuits of preciosities, the spices, sugar, silks, parrots, and porcelains that had traveled on the airy top decks of Portuguese slave crafts. For Northern tropicalists, it was "wasteland"—terrains not cultivated in the European manner—that allowed dispossession, reorganization, and integration into global economies under new regimes of labor management and sovereignty.[13] This invisibility to outsiders of the human hand in tropical landscapes—whether through ignorance of successional agricultures, domesticated landscapes, or a legacy of disease—framed a terrain of imperial opportunity.

"Mediterranean" versions of tropicality were rooted in the imperial aims of the Spanish, Portuguese, and ecclesiastics and in the experience of the inhabited tropics, places of complex cultures, vassals, and dark races. Portuguese interactions with West African states from the 1400s to the 1800s certainly involved engagement with empires, long-distance trade, lively cultures, and plentiful populations, even if the slave trade eventually profoundly destabilized them.[14] The centuries of cultural and economic ties to the tropics (south and east) as a consequence of Islamic conquest made the South Atlantic tropics seem less "foreign" than they otherwise might have. The Iberian history on the West African coast had also created a colonial class of tropical "hands" who would not and did not find the tropics overarchingly strange, due to their substantial familiarity with tropical ecosystems based on the fifteenth-century experience of enslaving Bantus in West Africa, organizing sugar production, the significant African presence in Iberian (and more generally Mediterranean) capitals, and the tropical botany of many crops and ornamentals.[15] Some authors, such as historians John Thornton and Linda Heywood, suggest that early Capuchin Catholic evangelism in Bantu areas, and Islamic cultures in the "Mina" zones, created a kind of Luso-African institutional familiarity (both Islamic and Christian) within the triangle of Brazil, West Africa, and Portugal.[16] This approach was formed from Mediterranean engagement with the entire Southern Atlantic, one that emphasized "fullness" and human populations (necessary to deliver global trade goods) and thus was at variance with Northern European constructions of tropical "emptiness" and its potential for colonial territorial appropriation.

The nineteenth-century naturalists' purchase on the public imagination through their writings, their lectures, and their positions within or relations to modern scientific and imperial institutions, their ideas and enthusiasms—"the improving of the world"—became widely known even as the Iberian sources from which they drew their itineraries faded from public view, and as they traveled from missions to Iberian forts and settlements. The Northern discourse of tropicality—primitivism, emptiness, primal harmony, space for great experiments—paradoxically used the circuits and spaces of "Mediterranean" tropicality (with its people, boats, and farms), to construct a new "wild" tropicality.

"We have met the 'other' and they are us"

Many of the writers and scientists who had introduced the marvels of Amazonia to popular audiences, such as Agassiz, Herndon, Bates, Wallace, and Maury, as well as countless less distinguished commercial travelers—yearned to see the place, in von Humboldt's words, "flooded with Civilization" rather than continuing in the "lassitude" of its situation as they perceived it. They saw the place as empty, marvelous, but with inhabitants who, though the charming in many ways, lacked the qualities of enterprise; this was attributed to climatic environmental determinism and racial deficiencies.

The three pillars of scientific racism (environmentalism, scientific anthropology, and social Darwinism) had justified slavery in Brazil and supported the racial inequality that prevailed after abolition. This body of theory also positioned Brazil's miscegenated gene pool as an essential obstacle to national development and a problem for national identity at the international level. The solution for this racial quandary for development, for da Cunha, was a socioenvironmental analysis with roots in Darwin, as well as the Brazilian "tropicalist" literatti and folklorists—Veríssimo, Henrique Coelho Neto—who wanted to create a national literature and aesthetic to express the culture of an emerging tropical civilization, not some temperate-zone overlay on tropical latitudes. Da Cunha did not much care for the whitening discourses that were widely popular at the time and underpinned the massive migration of European immigrants to Brazil. He would argue that through hybridization and acclimatization Brazilians had evolved the "race" best adapted to the tropics, the one that offered the most promise of an advanced tropical civilization, one superior to that of European colonials.

Clash of Empires of the Amazon: Peru versus Brazil

There were many reasons why Peru might have prevailed in the boundary negotiations with Brazil, as we have seen. Brazil did seem more revanchist—it had only just become a republic, one with a strong authoritarian streak, and had been the last nation in the hemisphere to give up slavery. It was full of mixed-race populations with all the degraded subtext they implied; it barely hung together as a country. Both Peru and Brazil were in the tropical latex trade in an Amazonia divvied up with dusty treaties and questionable maps and generally egregious labor relations. What da Cunha would do was pit the two "tropicalisms"—the widely known white and black legends—against each other in order to assert the Brazilian "moral landscape" as well as its notion of progress. Da Cunha framed his tropicalist arguments in forms of national character and what we might now call political ecology: an early use of comparative economic logics, labor deployment, resource use, occupation patterns, ecologies, and environmental destruction. He also invoked the "black legend": brutalities of Peruvian Amazonia in the most atavistic and savage of tropicalities, nomadism, primitiveness, moral corrosion, cruelty, sexual profligacy, ephemeral settlement, and underpinning it all, the counterfeit paradise—what seems fecund and productive, generating riches, is illusory in the arc of Peruvian plunder. All this explains why the Peruvians could never usher in a civilization in the tropics: they were only "builders of ruins." Their "primitivism" reflected the DNA of conquistadors, looters par excellence who were satisfied with pillage and never engaged in a destiny greater than plunder. At a time when development questions were infused with assumptions about progress and hierarchical social evolution, situating Peruvians in a kind of cultural and genetic stasis or nomadic retrogression was a powerful critique, especially in light of the "fact" that even immigrants who might be possible sources of cultural progression were captured by the primitive Peruvian character and customs.[17] This did not bode well for the future of the huge continental mass being claimed by Peru.

Da Cunha's next argument was racial: "Any street in Lima displays the most varied ethnographic gallery on earth." Thus, if the arguments against Brazil's capacities in the Amazon were to be made on racial grounds, similar arguments would have to be applied to Peru, with its "uncountable forms of mestizos: from the mulattos of all bloods . . . These also find their alienated expression in the pillage of history and landscape."[18] This all produced a "perennially incipient national culture . . . a mingling of races without forming a People." That is, they would have no capacity for a civilizational project. In Brazil, by contrast, the *sertanejos* convert foreigners "to our language, our ways, and our destiny," and to the "dominant components of our nationality," and in their hybrid vigor are "the bedrock of our race." He would explain this in socioenvironmentalist terms: the form of occupation interacting with environment would generate the vibrant pioneer.

Da Cunha was also writing about relations with nature. While Amazonia could provide a new cultural and historical phase for Peruvians, perhaps transforming their culture through the discipline of cultivation, the Peruvian *caucheiros* were "reanimating an apprenticeship to calamity" because they never formed a "brotherhood with nature nor ennobled it with cultivation." When finally Peruvians arrived in the Amazon, according to da Cunha, they dedicated themselves to nomadic extraction, expressing their inherent indifference to place through killing caucho trees and perpetrating the most violent form of terror slavery on Indians. Their cultural regression to nomadism was fueled by an economic logic wed only to private gain—unlike, in da Cunha's *seringueiro* version, where even a questionable form of settlement led to a husbandry that was the first step of a consolidating nationhood.

In the essay "Among the *Caucheiros*" da Cunha articulates all the classic elements of "black legend" duplicity: the enslaved Indians, the *caucho* mansion with a veneer of civilization overlaying its fundamental barbarity, the casual betrayal of women, and the fleeting nature of settlement, with people staying just long enough to plunder a place and then moving on. The driving force behind the *caucho* barons is not some noble civilizing mission but greed for wealth that is to be hedonistically squandered in European capitals. Then there is the moral decay: a novel evolution of a type where the elements of character do not blend but exist in disruptive unity side by side, elegance and brutality inhering in the same character, and a complete indifference to any project other than their own individual gain. Sexual immorality is also commented upon: women as trade goods and chattel, multiple wives and lovers.

In short, someone who would elsewhere be an admirable bourgeois becomes in the Peruvian Amazon a lascivious reprobate.

This vision of a chaotic, plundering culture is contrasted with turbulent yet ultimately stable occupation by mixed-blood *caboclos* and the emergence of the Brazilian city of Lábrea on the far reaches of the Purús. The Brazilian occupation, besides being extensive and progressive, involves the "improvement of the world." Rather than rape and nomadic pillage, it is husbandry and settlement, the "civilizing molecules," the aspirations that inhered in its place-names, that define Brazilian domination.

Even the lamentable social system of Brazilian rubber extraction required at least the tappers' attachment to their little hovels as they set out on daily rounds to provide the latex demanded by industry, however degrading that toil might be. From such modest enterprise they would give birth to the emergent national life of a great tropical republic and a new civilization in the remote forests of the world. These would naturally thrive even more with government's active efforts in infrastructure and telegraphs. The gentle hand of Luso-Brazilian colonialism engages the transformative powers of mestizo adaptation, not part of either the "black" or the "white" tropicality but rather a "third way" rooted in its mixed inhabitants.

"Civilization" actually precedes the state and only awaits its harmonizing support, as, da Cunha was at pains to note, had occurred in the United States, thus equating Brazil with a modern, developing imperial republic clambering onto the global stage and not with the unfortunate caudillismo and tropical nomadism of Peru, whose decadence reflected that of the moribund Spanish empire, incarnating a nomadism that ruptured bonds of place, Those who could make better use of the lands, defined in the moral terms of those transforming them into settled patterns of livelihood, had more right to them. Da Cunha thus echoed the centuries of travelers with imperial ambitions.

"What I had first seen as a desperate crawl . . ."

Da Cunha uses the double ideologies of tropicality to undermine the Peruvian claims but also constructs an alternative, native anti-European colonial tropicality to promote the goals of Brazil in the Amazon Scramble. His approach, basically "we have met the 'other' and they are us," recasts the terms of the debate. Da Cunha's version of the "white legend" transforms the passive ahistoric harmony of the native-born to one of national epic and vital human agency shaped by encounters with nature, disillusion, and greed, a place of biological and "moral"selection that produces the new tropical man, regardless of race. He sees an emergent dynamic harmony, the creation of humanized landscapes (urban and rural). and a "tropicalized" civilization in Amazonia. These were outcomes of the actions of his adapted *seringueiros*, in a place also full of other races and histories, exemplifying the multiculturalism of Luso-Brazilian tropicalism. This was an extremely radical position, given the more general ideas of the time about tropical capacities.

In "Clima caluniado," a masterpiece of irony, da Cunha compares the careful, "hygienic" colonial occupation carried out by the French and British with the tumultuous desperation of the Brazilian Northeasterners. Tropical medicine and colonial administration were burgeoning fields at the time, stimulated by the emerging European empires. Colonial agencies developed a vast array of protocols ranging from civil construction to detailed prescriptions of suitable, salubrious activities and diets.[19] This new colonist becomes a protégé of the state, which strictly regulates his life. Da Cunha remarks, "What is chastening is not so much the prodigious labors of colonial medicine to adapt the colonizer to the environment, but rather the slow unraveling of the most tenacious efforts." He contrasts European colonial states with completely disorganized and chaotic migrations to Acre. "Here was no crisis of growth, nor an excess of population flooding toward the frontiers, marching to fresh horizons, embodiment of the triumphant March of the Races. Rather, it reflected dearth, and utter defeat by natural catastrophe." Da Cunha describes the coast as haunted by "terrifying starvelings, burning with fevers and pox," and a government whose only concern was "to free themselves as soon as possible of the invasion of moribund savages. . . . So they sent them off to Amazonia—that vast, empty unknown—to exile in their own country. . . . No government agent, no doctor accompanied the exiles, whose sole and painful mission was to disappear." And amazingly, "100,000 men, risen from the dead, sprang from nowhere and reclaimed their national heritage in a novel and heroic way, extending the fatherland to the new territories that they occupied."

In sharp contrast to racialist debates of the day, where mixed bloods were pictured as "ugly, indolent, slothful, and inert,"[20] and to commentators who viewed Amazonia's only reasonable future as one based on white occupation, da Cunha describes prosperous yeomen communities with "well-tended orchards and prudent husbandry." He turns the racial environmental determinist discourse on its head: it is the adaptation of the Brazilian mestizos, the "civilizing molecule," accompanied by other ethnicities and histories—the Syrians, the Germans, the Italians—who shape the region. With discussions of the failures of European colonization, he inverts the conventional social Darwinist racial order.

This diverges from the primitive culture of plunder, the nomadism, and all the other elements of tropicality that da Cunha invokes against violent Peruvian occupation and the pathetic arrogance and ineffectiveness of white colonialism. For him, the tropics are a place of habitation, a place of history and civilization. His texts on the Amazon, and especially these on comparative colonialisms, could be seen as elaborated texts on Luso tropicalism, an imperial ideology that sprang from a tropical hearth, from the practices of unknown "bronzed titans" rather than the mandates of proconsuls. The triumph achieved by the mestizoized "bedrock of our race" was perhaps the story of the "lost paradise" he meant to tell in his companion volume to *Os Sertões* through their (and his) odyssey through a land as rich in travails and triumphs as any Homeric epic.

"Among the *Caucheiros*"[21]

On the right margin of the River Ucayali, that softly undulating land that embraces the headwaters of the Javari, the Juruá, and the Purús, a new society appeared about 50 years ago. It developed covertly. Lost in the depths of the forest, for a long time its existence was known only to a few traders in Pará, when, in 1862, from the most remote outposts there began to arrive dark brown slabs formed from another elastic gum, a competitor with rubber for the needs of industry.

It was *caucho*, and those who worked this gum called themselves *caucheiros*, fearless pioneers who boldly explored those forgotten backlands. They came from the West, crossing the Andes and tolerating all the climates on earth—from the burning littoral of the Pacific to the chill *punos* of the cordillera. Between them and our native clay were two soaring ramparts in excess of 6,000 m and a long valley that gave onto the abyss. They faced the Amazon plains, a stretch of hundreds of miles to the northeast, vast enough to lose oneself in forever, facing toward the distant Atlantic without a scarp of a mountain to serve as a guidepost in this immensity. Never have such imposing landscapes been confronted by such mediocre actors.

Naturally, the frontiersmen wandered for many years—few, invisible, fumbling in the perpetual dusk of far-off forests where other difficulties, graver than the unmeasured distances, the thickening wildness, began to dog their uncertain steps.

The entire area where one now provisionally traces the demarcation lines of the Brazil-Peru borders, where the creeks spread out that eventually form the Purús and Juruá, the northernmost falls of the Urubamba and the farthest branches of the Madre de Dios, figure among the most unknown areas of America, less a consequence of their exceptional physical conditions (which were surmounted in 1844 by F. Castelnau)[22] than by the terrifying reputation of the tribes that populate and traverse these regions—they are known by the generic name *chunchos* and incarnate the deepest dread of boldest pioneer.

These tribes are countless. Who ascends the Purús ponders the distance to the outskirts of Cachoeira, where the Panemís people are becoming rarer, little resembling the noble masters of the floodplains they once were. From there, continuing upriver, the inoffensive Ipurinanas; on leaving the Iaco, there are those Tucurinos who seem born old, so much do they reflect in their foolish expressions the decrepitude of their race. At the head of the river, where these unusual forest dwellers reside, is the greatest surprise. There the tribes, diverse in their habits and provenance, are composed of forced groupings—the tame Amahuacas, who congregate at trading posts of *caucho* extractors; the indomitable Coronaus, masters of the headwaters of the Curanjá; the coppery Piros, their teeth tinted with a dark stain that when they smile gives their faces indefinable traces of grim menace; the bearded Cashillos, threatened with extermination by two hundred years of manhunts, chased, as they have been for centuries, over the ruins of the missions of Pachitea; the Conibos with their deformed craniums and chests dazzlingly striped in red and blue; the Setebos, Sipibos, and Yurimauás, the corpulent Mashcos from the Manú—evoking in their aberrant and imposing physiognomy the fabled giants of the early cartographers of Amazonia—and above all, exceeding everyone in fame and valor, the warlike Campa of the Urubamba.

The variety of tribes in so reduced an area implies a strange coercion: this gathering is forced. They have obviously fled to their last redoubt. This is the denouement of the crusade that began with the campaigns of Jesuits in the forests of Maynas, that vast ecclesiastical domain, and extends to our modern expeditions, and whose final episodes are yet murky. The narrator of these last days arrives at the end of a drama, confused and disoriented, contemplates the very end of the last scene, "Civilization," savagely armed with repeating rifles, thoroughly besieges this savagery at the end of the world: the Peruvians from the west and the south, Brazilians from the northeast and southeast, and finally, closing in from the valley of the Madre de Dios, the Bolivians.

The *caucheiros* appear to be the most advantaged warriors of that sinister catechism of fire; they advanced step by step, slaughtering the most interesting aborigines of South America in those most remote backwoods.

They owe their historic mission to the fragility of a tree. The *caucheiro* is obliged to be a nomad and a devotee of combat, of destruction, and of a wandering and chaotic life because *Castilla elastica*, which provides desirable latex, does not permit, as do the Brazilian *Heveas*, a stable exploitation through periodic renovation of the vital tree saps. The *caucho* tree is extremely sensitive, and once pierced it will die or remain useless for a very long time. Thus, to profit from it, the extractor fells the tree once and for all. He drops it, then, meter by meter from the roots to the last leafy branches, and makes a trench beside the fallen tree. He digs out rectangular cavities that correspond to sections of the trunk, and from these he takes out at the end of the week the valuable *planchas* or planks of rubber. Meanwhile, the remains that stick to the bark or on the outside of cuts or that drop on the ground are joined together in the *sernambi* of inferior quality.

The process, as one can see, is rough and ready. The richest *caucho* groves are soon depleted, and because *Castilla* is not evenly distributed in the forests but spreads out in stands quite separated from one another, the *caucheiros* move to new groves, repeating with little variation the dramas of the nomadic life of the hunters of trees. In this way, nomadism is imposed on them, and it is the inviolable condition for success. They plunge boldly into the wilderness; they isolate themselves in successive outposts and never revisit their former paths. They develop a passion for uncharted, completely new landscapes. They reach them, they abandon them. They continue on and never return to those arduously conquered outposts.

When an area of *caucho* trees is discovered, they build their first *tambo*, their hut of Paxiuba palm in a clearing their alongside the grove, and plunge into the tasks at hand. Their primary work implements are Winchester carbines, cocked at the ready for chance meetings in the brush; next, the machete for cutting the unraveling lianas; and the portable compass to orient them through the tangled pathways. They position themselves and then launch a careful review of the surroundings. Then they go in search of the natives whom they will either annihilate or enslave, so that in the same sortie they complete the security of their new outpost and procure the hands they require.

Those who dare to do this necessary, risky task are few. A half-dozen men, dispersed and immersed silently in the jungle, research and survey every recess, encountering the most suspicious lairs, in an exhaustive topographic appraisal (committed to memory) of the most varied terrain. At each step, with ears and eyes alert to the most subtle aspects and vaguest rumors of the murmuring forest airs, they cautiously advance with the prudence and ardor demanded in that shadowy duel of the Spaniard with the wildness.

Some never return. Others return unharmed after useless peregrinations. Some, at the end of this exhausting research, catch a shadowy glimpse, half-distinct, in the foliage of the first native huts. They restrain their cries of triumph and waste no time informing their companions of their discovery.

They further refine their extraordinary cunning. They cling to the ground and, by careful tracking, get as close as they can to their hapless enemy. In this assault there is a poignant trace of heroism—Man, lost in absolute solitude, seeks out the barbarian, taking with him as escort only the eighteen shots of his loaded rifle.

It is a tortuous, long, and slow job of tracking, in which he takes advantage of every roughness of terrain, concealing himself behind trunks. or inserting himself in the sheltering angles of buttressed trees, sliding soundlessly over the litter of decomposing leaves, insinuating himself in the joined stalks of *Heliconias* with their large protective fronds until, with silent and anxious investigation, they inspect, practically in full view, their inexpert adversaries, unaware of the sinister civilized one who spies on them, counts them, and observes their every habit, evaluates their resources—and returns after this minute review to his waiting companions with all the intelligence necessary for the conquest.

Conquest is the preferred term—it's used as a kind of atavistic reminder of the patois of the expeditions of Pizarro. But the *caucheiros* do not proceed with weapons without first exhausting the influence of the rudimentary diplomacy of gifts coveted by the natives.

We heard this from one: "We attract them to the outpost with gifts—clothes, rifles, machetes, etc.—without making them work. We let them go where they like so they can report to their companions how they are treated by *caucheiros*, that we don't make them work, although we advise them to work a little to pay for what we give them . . ."

"Peaceful" means of conquest were not to be trusted. The rule was the unholy hunt, with guns. This was the "heroic" side of the business: a sordid band daring to subjugate a multitude.

There is no lack of examples.

These bands were guided by an unvarying tactic: maximum audacity, maximum intensity of fire. Such were the guarantees of triumph. There were an incalculable number of little battles throughout the forests in which well-armed posses supplanted entire tribes, who in their naive charges against the rifles were immolated by the carbines' thunder. We cite here but one instance. Carlos Fitzcarraldo arrived in 1892 at the headwaters of the Madre de Dios by way of the Ucayali. Between the rivers he had opened the trail that still bears his name. He planned to capture the indomitable Mashco Indians, who were the masters of the region. Among his Piro Indian bearers and captives was a loyal and intelligent translator. Fitzcarraldo was thus able to meet and converse with the native head man.

This meeting was curious and brusque.

The eminent explorer was introduced to the "heathen" and showed them his weapons and other equipment as well as his small army, in which were mingled the varied physiognomies of the tribes he had subdued. Then he tried

Figure 21.1. Muchachos: vicious child warriors of the Upper Amazon.

to demonstrate advantageous alternatives to the inconvenience of a disastrous battle. The sole response of the Mashco was to inquire what arrows Fitzcarraldo carried. Smiling, the explorer passed him a bullet from his Winchester. The native examined it for a long time, absorbed by the small projectile. He tried to wound himself with it, dragging the bullet across his chest. Then he took one of his own arrows and, breaking it, thrust it into his own arm. Smiling and indifferent to the pain, he proudly contemplated the flowing blood that covered the point. Without another word he turned his back on the surprised adventurer, returning to his village with the illusion of a superiority that in a short time, would be entirely discounted.

And indeed, half an hour later roughly one hundred Mashcos, including their recalcitrant chief, lay murdered, stretched out on the riverbank, which to this day bears the name Playa Mashco in memory of that bloody episode.

Thus they mastered this wild region. The *caucheiros* acted with feverish haste. They ransacked the surroundings, killing or enslaving in a radius of several leagues. In a few months, at the side of the initial hut (*tambo*) others multiplied, the little solitary shack transformed itself into an ample barracks or rough *embarcadero*—the houses became more dense—the examples of Cocama and Curanjá at the margins of the Purús mirrored the mirage of surging progress at the edge of the wilderness, but one that would develop and decline in a decade. The *caucheiros* would stay until the last *Castilla* fell. They came, they ravaged, and they left. In general they asked nothing of the earth with the exception of a few manioc or banana plants with which the tame Indians occupied themselves. The only regular agriculture one observes on the Upper Purús, albeit quite diminished, is that of the cotton grown by the neighboring Campa Indians, who in this activity reveal their native independence—collecting, carding, spinning, weaving, and painting the *cushmas* in which they are clad, which cover them from shoulder to foot like a coarse toga.

Those strange white conquerors came in hurried rounds of slaughter of both men and trees, staying just long enough to utterly extinguish both, before seeking other paths where they would unleash the same chaos, passing like a destructive wave and leaving the wilderness yet more wild, more disordered.

The contrast is marked. Proceeding from the Campa village of Tingo Leales to the Peruvian hamlet of Shamboyaco near the source of the Manoel Urbano river, the traveler doesn't pass, as one might expect, from the more primitive to the more advanced stage of human evolution. There is a greater surprise. One goes from straightforward barbarism to a sort of decrepit transitory culture in which all the vices of civilization resurge more markedly from the imprints of progress.

Approach the Peruvian *caucho* station: in the first hours one is enchanted by the picture of a rough and busy existence. The main house and its outbuildings are surrounded sometimes by small hamlets that are always situated on a well-chosen point on a river bluff. In spite of being constructed entirely of the leaves and poles of Paxiuba palms—that providential palm of the Amazon—they are usually composed of two stories and have an elegance of line. In the airy verandas that surround them, they have an appearance totally at odds with the sorry aspect of the dull *barracões* of our tappers.

On the ample patio topping the crest of the river terrace, falling in a sharp slope to the river, there is the pleasing bustle of activity: the powerful bearers passing in long successive lines bent under the slabs of *caucho*, active administrators tearing open the doors of the ground floor and running everywhere, to the warehouses of provisions or to the sparkling shop where hammers and anvils resound, repairing axes and machetes. Below the pier, from the fleets of swift canoes that converge where the slim canoes harpoon the air, come the cackle and babble of the helmsmen, who slap in the water rafts made only of *caucho*, forming a moving pathway: "the merchandise that transports the transporters." Throughout, running up the stairs that wind to the top of the bluff, the red skirts and white bodies of the delicious *cholas* of Iquitos pass and cross each other in festive entanglement.

The traveler passes excited groups and the surprises never cease. Clambering up the stairs that lead to the front veranda, one encounters the main living quarters. On top, the *caucho* baron, a jovial conqueror, standing straight over the stiff heels of his backwoods boots, receives his guest boisterously, throwing open for him the doors of a frank and spectacular generosity. And this completes the charm. One loses all notion of time and space, forgets the thousands of empty kilometers of portage and channel through isolated rivers required to arrive at that remote *estancia*. The visiting explorer is oblivious and feels as if he were at some commercial entrepôt on the coast. The deception lacks for nothing: the long cabinet blocking the main room with its shelves groaning with merchandise; the solicitous, obedient servants at the service of the extremely correct bookkeeper, whom one greeted on entering and who returned quickly to his numbers, bending over his slanted desk; the glass of beer that is offered instead of the more traditional *chicha*,* the artistic paper marking each day; the newspapers from Manaus and Lima—and what is most implausible, the pleasant and refined torture of a phonograph obstinately gabbling the favorite aria of a famous tenor, here at the far end of the wilderness.

But all this surprising exterior cloaks a reality that the garrulous host did not choose to reveal. The disillusion is sudden and impressive. This reflection of a superior life does not go beyond a niggardly strip of land of less than a hectare, constrained between the threatening forest on the sides and behind and the cliff hurtling down to the river in front.

Outside this false scenario, the real, almost unimaginable drama unfolds. Below the exalted state of our opulent baron is, at a deplorable level, the mestizo from Loreto who goes to the jungles in search of his fortune, or the gloomy Quechua brought in from the Andes . . . indeed, there is an unlimited procession of the dispossessed. To discover them, you must travel into the obscure recesses of pathless forest and seek them in their solitary hovels, completely alone, accompanied by their constant companion, their rifle, which guarantees their survival with the haphazard resources of hunting. There they might live uncountably long years, but eventually they sicken, devoured by pests and afflictions, and finally die in utter abandonment. Up to four hundred men, whom no one sees, are dispersed up creeks and gullies, rarely appearing at that glorious thatched palace of the imperious baron who enslaved them. The conqueror doesn't police them;

* *Chicha* is a fermented corn, manioc, or sweet potato beverage.

he knows they can't escape. A radius of six leagues around is all his domain. The region, swarming with "infidels"—Indians—cannot be traversed. The wilderness itself is always vigilant and guards his numerous slaves. Even the stately Campas, whom he captured by wrangling a magisterial hoax on the ingenuous courage of the savage, will no longer leave him, fearing their own wild brethren who will never forgive them their transitory submission.

This genial adventurer, who two years earlier in Lima or Arequipa exercised only the most gracious comportment, feels himself entirely free of the pressures and the infinite correctives of social life. At the same time he acquires the consciousness of limitless command, and the feeling of impunity infuses each caprice or crime; he embraces a most original savagery, one he has entered without having had time to lose the superior attributes of the environment where he was born.

The *caucho* baron is really an original in history. He is antinomian and paradoxical. In the most elaborated ethnographic framework there is no place for him. At first one situates him as yet another vulgar case of the civilized person who "goes native" in a sudden cultural rejection where the superior characteristics are extinguished in primitive forms of activity. But this is wrong. These contradictory characteristics are not joined in a hybrid venture that is defined and stable. Rather, they coexist without fusing. What we see is a case of a psychic mimicry of men who ape savagery to conquer savages. The *caucheiro* is both a gentleman and a brute, depending on circumstances. The curious dualism of one who tries to conserve the highest moral precepts next to a morality especially structured for the wilderness reappears in all the actions of his tumultuous existence. The same man who with enviable rectitude makes all efforts to cover his debts—which sometimes exceed thousands of pounds with the exporters of Iquitos or Manaus—does not hesitate to cheat a miserable peon who works for him of some kilos of ordinary *sernambi*; on occasion he shifts from the most refined gallantry to the maximum brutality, stopping in the midst of a captivating smile and impeccable bow to leap with a bellow, a shining dagger in his fist, onto the disobedient *cholo* who affronted him.

Savagery is a mask he puts on or takes off at will.

He in no way conforms to the incomparable mold of our *bandeirantes*. Antonio Raposo, for example, has an admirable prestige among South American conquistadors. His heroism was brutal, massive, direct, without disguises, without detours. He advanced unintelligently, mechanically, inflexibly, like an unleashed force of nature. He tracked through a diagonal of 1,500 leagues from São Paulo to the Pacific, crossing all of South America through river, mesas, stagnant swamps, jungles, mountains, snowy *páramos*—the tropical alpine grasslands—and desiccated coasts. Amid the surprise and demise of a hundred displaced tribes, he is a terrifying figure of epic dimensions. One feels in that daring individual the marvelous, violent concentration of an entire epoch.

The *bandeirante* was brutal, inexorable, but logical. He was the superman of the wilderness.

The *caucho* baron is irritatingly absurd in his elegant brutality, in his blood-stained gallantry and his vagrant heroism. He is the homunculus of civilization.

But one must understand his logic. The adventurer comes with the exclusive interest of enrichment and return—the sooner the better to flee that depressing and swampy land that doesn't even appear to have enough solidity to tolerate the actual material weight of a society. Accompanying him in all the junctures of his enervated and impetuous activity is the vision of vast cities where one day, transformed into sterling, the black gold of *caucho* will gleam.

Dominated in all by incurable nostalgia for his native land, which he left precisely to prepare the resources that facilitate greater sums of felicity, he is attracted to the forests, buries or subjugates the natives, resists malaria and fevers; he acts madly during four, five, six years, accumulating a few hundred thousand soles, and then suddenly disappears.

He reappears in Paris. He passes in full splendor among the noisy salons and theaters for six months of delirium without people ever discovering the divergence between the impeccable manners and cut of his suits and the slightest vestige of his professional nomadism. He ruins himself quickly, elegantly frittering away his fortune, and then returns . . . he enters into the old routine another four or six years of forced labor, new riches ready to be acquired, then another leap over the ocean, and almost always another return, to recuperate once again his lost affluence, in a stupendous oscillation between the crowded boulevards and the lonely forests.

This pattern has many embellishments among the most notorious *caucho* barons of Manaus.[23]

In this oscillating life he infuses everything with a provisional character, from the land that he devastates and despoils, the house he builds in ten days to last a few years, to the most affectionate long-standing liaisons, which he destroys in a day. On this point especially his inconstancy is unrivaled. One of the *caucheiros* we interviewed had married an extremely genteel Amahuaca woman, who had helped him for the last decade with all the dedication of an exemplary wife. He answered us: "They gave her to me as a present in Pachitea" (Me han hecho regalo en Pachitea). Regalo—a present, a trifle that he would abandon at the first opportunity without worry or care.

The main trader in that dilapidated village, who in Iquitos or Lima would be a sterling example of the quiet and abstemious bourgeois, was a voluptuary of the first rank and presented to his friends and the random traveler his scandalous harem, where particularly notable were the interesting doe-eyed Mercedes, who had cost a battle with the Coronaua tribe, and the enchanting Facunda of the large, untamed, and distrustful eyes, who had cost him one hundred soles. He expounds on this criminal traffic, laughing, absolutely brazen, utterly feckless.

There are no laws. Each carries a penal code with his rifle and exercises justice at his own whim. One day, in July 1905, when the frontier reconnaissance mission composed of Brazilians and Peruvians arrived at the last *caucho* post of the Purús, they saw a naked corpse, horribly mutilated, tossed out from the left bank of the river in a clearing. It was the cadaver of an Amahuaca women. She had been killed for vengeance, it was later vaguely explained. No one thought about the incident, which was trivial and insignificant in the remote camps of these people who only cross but never populate this land, leaving it even more dismal with ruins of their forsaken trading posts .

These abound on the Upper Purús, appearing depressingly in all their varied aspects, which range from the humble shacks of the peons to the once august abodes of the barons. A bit above the Shamboyaca, one particularly impressed us when we descended the river. It had been a trading post of the first rank. We leaped out to examine it, and barely had we scrambled up the scrubby bluff when we discovered, above the old road invaded by brush, the yard where now impenetrable thickets were enshrouding mounds of garbage and debris, cloaking bits of machinery and tools left by the other, departed inhabitants.

The main house was half rotted, the roof caving in, the walls collapsed and about to pull down the moldering beams, so that it appeared to be held up only by the lianas that had penetrated every point, poking through the roof, wrapping around the sagging supports, and lashing them to the closest trees, which thus impeded the total collapse. The nearby outbuildings, embowered in exuberant floriferous vines, were decaying, disappearing bit by bit in an irresistible constriction as the forest reconquered its original terrain.

We did not pay much attention, however, to the magnificent regenerating thrust of the flora that with corollas and garish bunting festooned that deplorable relic. This ruin was not entirely uninhabited.

In one of the most maintained of the outbuildings, the last occupant awaited us. Piro, Amahuaca, or Campa, one couldn't distinguish the native origin. The actual features of the human species were transmuted by the repulsive apparition—a deformed torso swollen by malaria seemed to take up his entire figure, in sharp contrast to the thin arms and bent and twisted legs, like those of a monstrous fetus.

He cringed in a corner and gazed at us impassively. At one side he had all his belongings—a large bunch of green bananas.

This indefinable thing, which by a cruel analogy suggested by the circumstances seemed less a man than a forgotten ball of *caucho* thrown in the corner by the *caucheiros*, answered our questions in a hoarse and fading voice in a completely incomprehensible language. Finally, with enormous will, he lifted an arm and extended it forward as though to indicate something he had been following for a long time, something beyond all those forests and rivers. He babbled and, letting his arm fall heavily as if he had lifted to a great height a heavy burden, murmured, "Amigos."[24]

This: friends, companions, comrades of the busy days of harvests, who had left for other places, forsaking him to absolute solitude.

Of the Spanish words he learned, there remained only that, and thus the forlorn one murmured it with a touching gesture of longing. With poignant sarcasm he unknowingly castigated those vile adventurers, who even then continued in their routine devastation—opening new areas with carbine shots and slashes of machetes, regions that they would leave as they had left here. The final record of their tumultuous works would be scrawled in the crumbling shacks or written on the pathetic figure of the brutalized native. These were the monuments of those builders of ruins.

In contrast to the *caucheiros*, da Cunha's *sertanejos* come onto the historical stage as architects of a Luso-tropical civilization.

"Maligned Climate"

The moral disintegration of those who move to the tropics is a remarkable and necessary component of any description of the climate of the new territories claimed under the Treaty of Petrópolis. It is an element that makes the most obtuse psychologist outshine Hann or other kindred experts in climatology. From the day of their departure these migrants brood about a return home in the shortest time possible. They become a new sort of exile—the exile who asks for banishment, sometimes battles to achieve it, repelling other competitors. At the same time the most mournful images inhabit their fantasies, prefiguring the troubled paradise that attracts them. When they depart, they enter a fretful emotional state that leaves them open to all possible maladies.

At the end of the fifteen endless days that it takes to follow the coast, they enter the Amazon. Revived briefly by that impressive landscape, they are soon oppressed by the disheartening immensity. Their eyes glaze as they take in vistas, certainly vast, but somehow blank and reduced to the hazy outlines of the distant margins. As they press upriver, the idle days pass in the strange immobility of landscapes of only one color, one height, and one form. They have the nerve-wracking sensation that life has stopped. Stupefied by the dulling impressions, they find that even the notion of time is extinguished. The soul retreats into a nostalgia that is a yearning not just for one's native land but for the earth itself, for those natural vistas and perspectives to which we are habituated, but which the eye cannot uncover in the monotony of those vast plains.

They enter one of the great tributaries—the Juruá or the Purús. They arrive at their remote outpost, and their despair is heightened. The land is graceless and sad, because it is new. It is immanent. The forest mantle lacks the artistry of human labor.

There are domesticated landscapes we recognize as echoes of subconscious and ancestral recollections: rolling hills, valleys, coasts fretted with inlets, Even a scorched wilderness is familiar to us to the degree that it provokes some kind of primal reminiscence. Seeing landscapes for the first time, we are programmed with an imaginative lexicon that leaves us primed for the enchantments of instant "recognition" when nature offers us a feature that we had previously known in only an ideal form.

In the Amazon, no. The topographic forms most associated with human presence simply vanish. There is something primordial in this amphibian nature, a mixture of water and earth that veils and entirely obscures its own grandeur. And one has the sense that absent the continuity and consistency of cultures, it will always basically be impenetrable, if it isn't rent apart in the plunder of its riches. Those who live there are molded by its ferocity. They don't cultivate the wilderness, they don't embellish it, they merely try to restrain it. The folks from Ceará and Paraiba, and the other pioneers from the Northeast who were discarded there, unknowingly fulfill one of the greatest enterprises of our innocent and heroic age: they are taming the wilderness. And their simple souls, tempered by a thousand mishaps, ensure even more than do their powerful bodies the success of this formidable endeavor.

Recent arrivals from the south are completely thrown off balance by the tumult and commotion, and ordinarily they collapse. Dazed by the unrecognizable landscape and the society of these titanic pioneers and nation builders, the novice feels dislocated in space and time, removed not just from his country but from human society, led astray into some wasteland in an obscure backwater of history. He can't stand it. His reserves of energy are devoted to the simple effort of sticking it out, useless and inert in the post assigned to him, bumbling through the simplest task. Longingly he eyes the launches heading downstream, and he dreams of his distant home in melancholy and fearful reveries. Then unexpectedly, in the full light of day he is seized by an abrupt and icy shiver running through his body, and he delightedly welcomes the salvation of fever. It is a happy surprise, and in that sudden, close nudge of death, he revives. Malaria means for him a liberating medical certificate. And thus he beats his honorable retreat graced with the justifiable flight, the lawful desertion. Nonchalantly he tells of the fevers and kindred maladies of those forest hellholes, seasoning the evocation of its fears with the impregnable heroism of the survivor.

Retreat must be justified: each nameless creek becomes a drear and pestilent Ganges. The flooded forests or the lakes stretching over the bottomlands are transmuted into endless Pontine marshes. Our revenant limns a terrifying picture of diseases set in an entirely fictitious epic of unending harshness. Amid such unremitting slander and fantasies, nature becomes only a synonym for suffering and death in those remote outposts.

This is, of course, an exaggeration. Acre, which has the most widely vaunted record of mortality in the newly opened outback, gives the lie to this myth.

By their titles the colonial schools of medicine of England and France disclose the care with which the transplanting of people to new habitats is to be undertaken. Evident in this approach and colonial history is a certain nobility in how modern imperialist enterprises gather the strength for those assaults across the world. The most brilliant generals are mere scouts; cadres are later transformed into anonymous battalions of engineers and doctors. A thousand military sorties are but preludes to the final campaign: against the climate. And it is the domination of the "incompetent races" that triggers the development of colonial territories in a magnificent arc from Tonkin to India, Egypt to Tunisia, to the Sudan, to Cuba and the Philippines. The marvelous endeavor of improvement of men and of land is the universal task.

The task is a double one. For these tranquil conquistadors it is not enough to analyze meteorological or telluric causes of the endemic maladies of their recently conquered terrains on the undefined scale that stretches from summer anemia to polymorphic fevers. Their greater task is to adapt their pioneers to their new environments by correcting their temperaments, breaking them of their old unsuited habits or instilling others. The end result of this uplifting and invigorating process is the creation of the entirely adapted individual, sometimes so different in physical and psychic features that he becomes a sort of "native" configured by the strictures of tropical medicine. And this new colonist, this new colonial, becomes essentially a protégé of the state, which strictly regulates his life from the day of his departure (set for the most propitious season) to the specifics of food and clothing. Within the overall regimen dictated by the tropical climate to which these persons are being sent are detailed protocols for each individual: all the contingencies are foreseen, all the possible contretemps (including the inevitable frailty in the early physiological phases of adaptation to this environment) whose dispiriting influence on the newly arrived European erodes his spirit and frays his very musculature. The vast array of prophylactic measures initially inspired by physical factors often concludes in with a lengthy and elaborate code of personal conduct. These instructions are permeated with vulgar ideas about how to deal with to the high heat and humidity so detrimental to blood pressure, which blocks pores and exhausts hearts and nerves.

In the end, this engenders a brooding morbidity that embraces the full range of afflictions, from malaria that drains their very life to the suppurating dermatitis that ravages their skin. Gloomy reveries decisively blunt the efforts of those who brace themselves against the despair, the bitter yearning, the melancholy of a monotonous and primitive existence, the waves of irritability seemingly generated from air itself, electrified and glittering. And above all, their will weakens in a swift, profound spiritual decadence that is the true disease of those outposts—the others are mere symptoms.

Open any book on colonial sanitation. Even the most superficial reading discloses the matchless efforts of today's modern missions and their apostles. Unlike the zealots of old, they no longer seize the transformed savage as a prize for civilization, but instead transplant civilization itself to the savage heart of these barbarian realms. What is chastening in these pages is not so much the prodigious labors of colonial medicine to adapt the colonial to the environment but rather the unraveling of the most tenacious efforts.

In Indochina's almost temperate climate, France applied 15 years of ceaseless labor to overcome its dreadful mortality. This experience and the advice of their best scientists persuaded them not to take up the arduous task of systematic occupation of equatorial Africa. It was the same with the English, German, and Belgian colonies. A three-year tour of duty by colonial district officers became the rule. The return home was necessary to restore their enfeebled bodies. In spite of the great sacrifices, vast expenditures, and monuments of civil engineering that transformed the rough topography of new localities into a truly artful geography, the final result was frail colonies inhabited by a society of eternal convalescents, living only by maintaining exacting medical and dietary regimens.

Compare this pattern of rigorous but ineffective colonial supervision with the turbulent and random occupation of Acre—so surprising in its outcome. It does not require a detailed review to see that this regularly denounced style of occupation is startlingly superior not only when set against most other areas of colonial expansion but even when compared to the great majority of countries that have been inhabited by their own native populations.

There is no more striking historical example of such anarchic immigration, and certainly not one so disrupting to the vulgar clichés of climatic determinism. From 1879 to today, in successive waves backlands populations between Paraíba and Ceará were thrust to that far-flung corner of Amazonia. The very antithesis of ordered and dignified migrations, these departures lacked the most minimal administrative supervision. The occupation of Acre is a historic case, entirely "accidental" and far removed from the orderly aegis of Brazil's watchword, "Progress." Its origins were borne of Calamity: the periodic droughts of our Northeastern backlands that occasioned the exodus en masse of the *flagelados*—the scourged ones. Here was no crisis of growth, no excess of population flooding toward the frontiers, marching to fresh horizons, no embodiment of the triumphant March of the Races. Rather, it reflected dearth and utter defeat by natural catastrophe. Its trajectory was one of stumbling, disordered flight. In the very inverse of natural selection, all the weak, the useless, the worn-out, the sick, and the suffering were sent off willy-nilly to that wilderness.

When the great droughts of 1879–80, 1889–90, and 1900–1901 blasted over the scorched backlands, the coastal cities of the littoral were soon flooded by a new population of terrifying starvelings, burning with fevers and pox. The sole concern of the authorities was to free themselves as quickly as possible of the invasion of these moribund savages clogging roads and waterways with the oppressive mien of the doomed. So they sent them off to Amazonia—vast, empty, largely unknown, an exile in their own country. The martyred multitudes lost their rights, even the comforts of kinship, shattered in the hasty departures, leaving for those frontiers with only a letter "to whom it may concern." They left, famished, febrile, and poxed, thus liable to infect and corrupt the most salubrious place in the world. But once having carried out this purge, the government took no further interest. No government agent, no doctor accompanied the exiles, whose sole and painful mission was to disappear.

But they didn't disappear. On the contrary, in less than thirty years the State of Acre, which had been nothing more than a vague geographic label, a swampy wilderness stretching interminably to the southwest, defined itself out of the blue, surging forward in economic stature. The capital, a city ten years old but established on residue of two centuries, transformed itself into the most important inland navigational metropole in South America. And in the mysterious extreme southwest of the Amazonas, where the remarkable William Chandless penetrated more than 3,200 miles without arriving at its limit, 100,000 frontiersmen, yes, a legion of 100,000 men, risen from the dead, sprang from nowhere and reclaimed their national heritage in a novel and heroic way, extending the fatherland to the new territories that they occupied.

Open the last municipal reports of Acre. In its pages we are dazzled less with the transformations that are documented there than by the inchoate and utterly informal manner in which the occupation of the region is still carried out. Today as a generation ago, absent traumatic droughts and disasters, we see the immigrant's advance without the least vestige of official assistance. The transplanted populations settled, sank their roots to the soil, in an astounding demographic advance—from the headwaters of the Juruá to the confluence of the Abunã—and extended the promised lands of the northern frontier, more sought after with each day that passes.

What a contrast to the traditional colonial pattern! What a blow to the myths of its fatal climate! The story of Acre, a story that is still unfolding albeit at a slower tempo, is that telluric selection that Kirchhoff speaks of, a sort of natural triage imposed by nature herself on those who seek her out, and conceding the right to live only on those she favors. But there is a more general process. In all latitudes, the original elective affinity between man and nature has been vital. Those who survive are those who best balance their personal characteristics with climatic factors. The subtle etiology of adaptation is combined in physical and moral attributes, that reaction between the most tangible thermal, barometric, and hydrographic components of environment to the most subjective impressions of landscape—from actual resistance of cells and muscles to strength of character in all its complex refinement. Prior to hereditary transfer of these qualities of resistance, the acquired ones guarantee individual integrity with the racial adaptation itself, and the process of selection is advertised in statistics of life expectancy and ineluctable, necessary death. All acclimatization of this type is a permanent plebiscite imposed on outsiders: Who will live? Who will die? In the tropics, it is natural that biological scrutiny takes on a more severe character.

There are no ways to cheat. All fall equally under the same incorruptible inspection: the consumptive gulping for air that is caustic, too poor in oxygen, the outcome of unbridled wantonness; the heart felled by the drop in arterial pressure; the alcoholic as a perpetual candidate for all sorts of edemas; the lymphatic immediately snapped up by anemias; along with the glutton, the sleepwalker in the winding sheets of his own insomnia; the indolent, rotting in enervating naps. And the choleric, the neurasthenics, tremble immoderately in our electrified airs and strangely refracted light until the paroxysms of tropical dementia suddenly explode in a sort of sunstroke of the soul.

Each physiological or moral collapse is offset by a corrective physical reaction. What one calls insalubrious is but a purification, a sweeping elimination of the unfit. In the end, we come to the conclusion that it is not the climate that is bad but man himself.

This is what transpired in Acre. The groups occupying the area—ignorant of what awaited them and traveling in deplorable conditions—suddenly arrived only to be plunged into a social order that would do nothing but exacerbate instability and weakness. What always awaits them, and what always will await them, is the vilest organization of labor ever conceived of by human egoism and greed. The *sertanejo* incarnates a social paradox that we cannot emphasize enough: man works there to enslave himself.

When the Italian immigrant relocates from Genoa to São Paulo and then to the most remote coffee plantation, he does so under the paternal eye of our social agencies. The exile from the Northeast has no helping hand and is basically forsaken by one and all. His is a far more arduous voyage, and he survives only on money advanced to him by rapacious labor contractors who cheat him with fictional land parcels and chain him to debts for the rest of his days.

From the first slash of his tapping knife, the *seringueiro* is trapped within an unendurable cycle: the exhausting struggles to undo a debt that constantly expands, always equaling the energy of his toils. In the dreary round of his existence he is completely alone. In this respect, the exploitation of rubber is worse than that of *caucho*: it enforces exile. Dostoyevsky's darkest narratives of Siberia could scarcely capture the tapper's torment: a man confined to the same trail, tethered to the same trees, for his entire life, setting off each day from the same point along the prison of his dark and narrow passage. For him, the task of Sisyphus is to move not a stone but his own body along the cramped arcs of the endless circuit of this wall-less dungeon. It takes but an hour to learn the task that will consume him the rest of his life. Without the most robust inner resources, the tapper soon becomes a mere shell, an dullard drained of all hope, illusions, and the taste for adventure that prompted him to make that fatal bid for fortune. Mental decline is matched by physical decadence. All precepts of diet, which is the key to health in the tropics, are abandoned: poverty rations of suspect and nauseating canned goods plus the luck of the hunt make up his meals.

But the worst is the solitude. By the nature of his trade, the tapper is obliged to be a hermit. Even in Acre, where the greater density of trees permits 16 rubber trails in a square league (such an area could sustain fifty small farming families), has but eight men who are widely dispersed and rarely ever see each other. A medium-sized rubber estate with two hundred *estradas* is about fifteen square leagues; this *latifundia*, which could easily be populated by three thousand active inhabitants, carries on with a scattered and invisible population of one hundred workers. This is the systematic conservation of wilderness, the prison cell for men in an unmeasured amplitude of land.

Despite the haphazard and brutish manner of occupation and of life, the newborn society adapts and progresses. Its evolution, slow and continuous, is apparent to even the least curious traveler in the Purús. At first indifferent to the land, our pioneer settles in. Houses emerge in forest clearings and on the river flats, and on the solid banks above, one can see the first areas of farming. The gloomy barracks covered with palm branches are transmuted into regular homes or ample houses of rock and mortar. In Sebastopol, Canocory, São Luis de Cassauan, Itaituba, Realizam, and dozens of other settlements in the Lower Purús, in Liberdade, Concordia, even the most remote sites, their numerous houses cluster around the small churches and expand into real villages. These are the material images of control and definitive possession, tangible evidence of social evolution.

The place-names are telling. Some are elaborate, but all, from the oldest to the new, are eloquent. In the land without history, such names proffer the first fragmentary intimations of our national epic. In the initial agonizing phase of settlement, the names evoke sadness, martyrdom, cries of despair, and appeals for help. On boards tacked up on the houses of these hamlets the traveler can read: "God Save Us." "Saudade" [Yearning]. "St. John of Mercy." "Damned." "Hell." Other names record a more hopeful aspect and the cheer of the redeemed: "New Enchantment," "Triumph," "Let's See," "Liberty," "Paradise."

The farther up the river, the more pronounced the optimism. After the confluence of the Acre River one sees along some stretches that the various estates fronting the river show signs of well-cultivated homesteads, settled a long time ago. Nothing remains of the uncouth or brutish outposts of the pioneers. Past Cátiana, past Macapá far upriver, until Sobral, with its tiny plantations of coffee which provide for their own needs, one sees the whole story: from the small-scale agriculture, now common, to the well-tended orchards, the prudent husbandry of the settler who created this land, and now will never abandon it.

And the men themselves are admirable. We spoke with them and saw them in action. They had all kept their outlandish names and nicknames, from the opulent "King Caboclo" of Cachoeira to the garrulous "Clumsy" in the hinterlands of the Chandless, to the old "Yellow John" who founded Cataí and still works with a vengeance through the twisted rubber trails, and has for the last seventy years; the bold "Golden Antonio" of Terra Alta, an unerring marksman whose audacious attacks in the skirmishes of 1903 with the *caucheiros* have become a page of our history, vibrant with courage.

What stands out most about these men is the physical integrity inscribed in their taut muscles, and there is also the moral beauty of manly souls who have defeated the wilderness. When we keep in mind the pitiable circumstances in the early days of colonization, which still afflict them though to a lesser degree today, this physical and moral resilience cannot be explained by the exacting demands of a climate so malign and as brutal as that for which Acre is renowned.

The argument that the northern *sertanejos* (or more crudely, *jagunços*), endowed with the austere and warrior nature of Arabs, were preadapted to this new habitat by the discipline imposed by droughts and by relocating themselves along roughly the same latitude as their native soil, is also not convincing. The frontiersmen who opened up the Purús and the Juruá were a diverse lot: the Syrian, arriving from Beirut, gradually displaced the Portuguese in river commerce; the adventurous, artistic Italian rambled along the banks for months on end with his camera for capturing the essence of forest dwellers and their wild countryside; the phlegmatic Saxon, trading his northern mists for the splendors of the tropical air. The great majority still live there, working and prospering, and end up long-lived. Just one example: In 1872, Barrington Brown and Lidstone traversed the Lower Purús to Huytanahan on the launch *Guajara*, under the command of a Captain Hoefner, "a German speaking both English and Portuguese," explained the two travelers in the interesting book they wrote. Thirty-five years later . . . and Captain Hoefner is there! the eternal commander of the launch, toiling away without rest on those damn waters where blood gnats swarm and every mosquito bite presages fevers. There spread out and embellishing the current are the *mururés*—the water hyacinths—floating, with their purple flowers reminiscent of funeral wreaths. But they are not evil portents for the German.

We saw him at the end of 1904, at the confluence of the Acre River. He is an old guy, lively and esteemed, diligent and active, with a face that is open and rosy, framed by hair that is entirely white. If he appeared in Berlin, only the slight tan of his skin would suggest the dark stigma of the tropics. Multiply the cases and a myth evaporates.

Finally there remains, because so insistently repeated, yet another argument: those robust *caboclos* and that exceptional Saxon are not results of the climate but emerged in spite of it; they triumphed in a decisive battle in which all the rest succumbed, all those not armed with the same requisites of energy, abstinence, and hardiness.

Here we must we cast aside sterile sentimentality and acknowledge that the climate has a higher role. Amid the horrific circumstances that prompted and propelled the occupation of Acre, for so many years prey to all the vices and maladies fostered by the indifference of our public administrators, the climate exerted an incorruptible supervision, cleansing that land of the disgraced and the lawless who would otherwise be far more numerous than they are today. It policed and imposed a moral order. It chose and it still chooses for life those most worthy. It eliminates the inept and the unfit by exile or by death.

Surely we must admire a climate that prepares abodes for the strong, the enduring, the Good.

PART 5

Abyss and Oblivion

22

Killing Dr. da Cunha

All tragedies end with the stage strewn with corpses, wailing women, sorrowful comrades yearning for a different ending—perhaps the bourgeois one with its accommodation to circumstances—the wish that the protagonist had perhaps not been so attached to his own nature, so melded by conditions. Most of what we know about Euclides da Cunha's end days, besides the rancorous explanations offered up in revisionist histories by the da Cunha and de Assis factions, is derived from the testimony and depositions from the trial of twenty-two-year-old Dilermando de Assis,[1] along with de Assis's memoir of events, *A trágedia de Piedade* (The tragedy at Piedade), and description of the domestic situation published as part of his formal argument in another Euclides da Cunha murder trial, that of da Cunha's son.[2] There are also the hindsight accounts of various people who saw da Cunha and the other victims in the days before he took the streetcar from his house on the beach at Copacabana on a drizzly morning to the modest working-class neighborhood of Piedade, to the house at 240 Rua Sta. Cruz, home of Dilermando and Dinorah de Assis.[3]

After four years of insecurity, intrigue, and the stresses of cartographic production and political writing for the Baron Rio Branco, da Cunha sought a sinecure, a chair in logic at the Pedro II Gymnasium, the equivalent today of a Brazilian national university. With it he would no longer have to suffer through the formalisms and increasingly dandified atmosphere of diplomatic salons and the endless annoyance of seeing those he considered dishonest mediocrities vaulted into higher, more secure positions through the intercessions of their cronies.[4] He generally stuck to his maps and his studies whenever he could. The impact of his books meant that foreign visitors often liked to see him, and da Cunha, though very prim, was often sloppily attired and if irritated, not particularly polite. Rio Branco, reared in a diplomatic household and having spent considerable time in France, was reflexively courteous and needed da Cunha to be soigné and charming to international visitors, given the delicate politics surrounding Brazil at the time with the Peru-Bolivia adjudication in Argentina and the Peru-Brazil boundary negotiations. Itamaratí was switching gears into a different international diplomacy of trade, treaty connections, and international legations. The kind of international arbitration in which da Cunha had been engaged was on its way out: his efforts were the "last sigh" of this type of chancellery politics, with its maps and historical arguments.

With the end of his diplomatic career in sight, da Cunha began an arduous campaign for the academic post. The trauma of the competition for the chair in logic is described in detail by others and is perhaps not of general interest.[5] The process took place over several months with numerous public presentations on topics such as "The Idea of Being." The contest nourished his paranoia, was extremely exhausting, and made him even more impossible in his domestic life. Da Cunha, in the end, placed second according to the examiners, but final determination of the outcome was to be made by the president of the country. Rio Branco made the decisive intervention to ensure that da Cunha captured the prized chair. Ana reported that he threatened to kill either himself or the examiners if he didn't prevail.[6]

The stressful months of competition leading up to the selection had caused da Cunha's maladies to erupt in hallucinating form. His "Woman in White," the spectral vision that haunted him in his Amazon travels, floated around him so frequently that he hardly dared to sleep because of the panic the apparition caused him.[7] In this period he vacillated between mania, paranoia, and despair. His health was extremely precarious, with frequent crises due to the combination of tuberculosis and malaria coupled with domestic and professional worries. He wrote his friend Oliveira about arriving home trembling with chills when outdoors it was close to 100°F, and he was hemorrhaging from his lungs and spitting up blood.[8] His doctor was coming twice a day. His health was not improved by distress over his dying father, Ana's infidelity, and the frequent visits of the Ratto sisters, whom he detested for a number of reasons, not the least being their family connections to the de Assis cadets and unpleasant meddling. These kinswomen remained Ana's closest friends, confidantes, and most intimate traitors. Other visitors included the rowdy pals of Euclides and Ana's oldest son, Solon, including Dinorah de Assis, the younger brother of Dilermando. In this crowd, whose main pursuits seemed to involve music, intrigues, and gossip, the intellectualism and frailty of da Cunha made him an easy target of ridicule.

Things disappeared from his office: his favorite books, his small portrait of Queen Amélia of Portugal, who incarnated for him the ideal of female beauty. Her resemblance to Ana was actually quite striking: opulent body, dark hair, lively features. Quite a bit of money was lifted from his wallet as well.[9] While we cannot know who the real pilferers were, Lucinda Ratto helpfully pointed out in her deposition about the domestic affairs of the da Cunhas that Ana was diverting her household monies to help support Dinorah and Dilermando.[10] and then there were extra demands from Dilermando. Ana enjoyed giving him presents, as her lover and the father of her most adored child. Fights over money became the norm with Euclides, who often asserted to his wife (regardless of who might be in the room) that he was supporting those that he shouldn't have to, referring in some instances to very cute, very blond little Luis, at other times to more ambiguous dependents.[11]

Among the things disappearing from the household, besides banknotes, treasures, and papers, was Ana herself, who increasingly found the ambience impossible. Even when she encountered her husband on the street he was liable to berate her. She described a scene where after having been insulted by her husband in front of his friend Henrique Coelho, she took the streetcar in order to go shopping, Da Cunha bounded onto the car, seated himself in one of the back rows, and proceeded to publicly harangue her. On their arrival at home da Cunha continued to harass and revile her, saying that she was contemptible and that she should go live with her "big Sergeant" (o Sargentão).[12] Given this domestic milieu. Ana fled to Piedade to escape Euclides' unceasing abuse. She frequently visited the de Assis brothers with the four-year-old Luis

Figure 22.1. Luis da Cunha, who would eventually change his surname to de Assis to reflect his paternity.

and teenaged Solon, and the other children occasionally went as well. Neighbors of de Assis brothers noted her frequent presence and assumed she was a relative.

By July 1909, Euclides's father was extremely ill, perhaps with a stroke. Euclides's own health was such that he couldn't travel: he'd been coughing up clots of blood from his hemorrhaging lungs—what is known in the medical literature and in a more tubercular time as hemoptysis. On July 3 he wrote Otaviano, his brother-in-law, from his bed. He was unable to travel but was sending Ana and Solon, with the idea that if his father improved he could return with them.[13] On July 5 he expressed his distress: "The doctors tell me nothing positive, and advise me to vacation in Madeira.[14] It's not too difficult to figure out the prognosis. . . . Is it true? I don't care, I've given what I have to give. Among my few preoccupations, only one really remains—that my father pass his last days at my side."[15] Apparently the conditions of Manuel and his son were equally precarious. A few days later he had even more reason to be discouraged.

On July 9, in a truly Solomonic decision by the Argentine adjudication, the contested territories between Bolivia and Peru were split. Of the 6,432 leagues at stake, Peru got 3,322 while Bolivia received 3,100.[16] The court recognized some of the Peruvian claims, invoking the 1810 holdings of the Viceroyalty and of the *audiencia* of Charcas based on *uti possedetis*. Bolivia had tried to extend its claims well into the Andes and almost to Cuzco with the

1810 agreements at the moment of revolutionary transformation, as Spain's empire began to crumple. Da Cunha's widely read, dramatic, and controversial rendering of Bolivian rights in his *Peru versus Bolivia* had been countered by the drier legalistic efforts of the Peruvians and the Ecuadorian precedent. La Paz erupted in riots. Rio Branco was furious, although Brazilian territories remained secure. Most of Bolivia's Amazon holdings were upheld, thanks in no small measure to the efforts of the House of Suárez and its tappers.[17] Still, Rio Branco was unhappy about the adjudication, especially since one of his archrivals, Estanislao Severo Zebellos, was foreign minister of Argentina and perhaps had persuaded the plenipotentiaries and taken his revenge for Rio Branco's earlier triumph in the Misiones adjudication. Clearly the decision must have increased general anxiety about the ongoing negotiations between Brazil and Peru.[18] The disappointment cannot have been very salutary for da Cunha, given the turbulent state of his mind and body.

Ana and Solon spent time in São Paulo addressing family concerns and returned. But a later crisis ensued, and on July 31 Ana and Solon went back to São Paulo, this time in the company of Dilermando. Ana charged Angelica and Lucinda Ratto with management of her household during this absence, since both Euclides and his youngest son, Manoel Afonso, were ill and in bed.

Someone must have informed Euclides that Ana was traveling with her lover, and Dilermando's relatives noted that they were in São Paulo and dining with various Assis relatives.[19] A terse telegram to Otaviano Vieira on August 3 indicated that Euclides was too sick to travel: Ana needed to return home at once, and Euclides asked Otaviano to advance the money for the return trip.[20] By August 6 the trio was back in Rio at their respective abodes, but the da Cunha domestic situation was spiraling out of control. Once Solon and Ana returned, Euclides's vituperation was vicious and constant. "Discord reigned in that house," and they snarled at each other on any pretext.[21] This rancor was inevitably witnessed by the Rattos, who, Ana was to testify later, went out of their way to inflame Euclides's jealousy.[22]

Expenses from the trip that Ana and Dilermando took together to São Paulo in early August so she could attend to da Cunha family matters included a bill from Dilermando that exceeded her entire monthly household allowance.[23] As with Emma Bovary, for Ana money was a linchpin in the dynamics of deception and the unraveling of her domestic life. Angelica (Neném) Ratto later reported that Ana showed her invoices for clothing and miscellaneous expenses that arrived with a note from Dilermando urging her to promptly pay his creditors.[24] The number of bills and her inability to pay them perhaps led Ana to hope for some assistance from her kinswomen; a very late confessional evening ensued on Wednesday August 11, when she and the

Ratto sisters stayed up talking until three in the morning. Later, both Neném and Lucinda Ratto reported on it in their depositions. In testimony at the trial of Dilermando, Lucinda explained that she and Neném had witnessed a marital explosion over money, so they asked Ana why she was supporting their nephews given that it caused so much conflict with her husband.[25] The sisters testified that Ana then revealed to them her compelling passion for Dilermando.[26] It would seem that after four years in close proximity with the clandestine couple, this could hardly have been a revelation to Dilermando's interfering aunts. But many reputations were under scrutiny at the trial, and certainly the Ratto sisters, omnipresent as instigators and observers of this affair from the beginning, and hardly innocent, needed to feign a modicum of propriety.

In the Rattos' view (and statements), their nephew Dilermando was affectionate and kind, while Ana's husband was cold and lived only for his books. The ardor of the illicit couple, especially in light of the overheated romantic ideals of the day, was imbued with its own passionate, if disastrous, logic. In her testimony, Neném sanctimoniously suggested that even if Ana were unable to rectify the situation on the moral plane, she should not have deprived her children in favor of the de Assis men. The predawn conversation apparently became increasingly histrionic: Ana averred that she could never forget Dilermando, nor could she ever leave him, because, she asserted, he had promised bitter retaliation if she ever discarded him.[27] Moreover, when she couldn't help him financially, Dilermando threatened her (or blackmailed her) in reference to Solon,[28] who was not exactly unaware of his mother's relations of with the older de Assis brother but whose larger understanding was blurred by the ambiguous oedipal terrain.

Armed with such intimate knowledge, and as witnesses to ugly scenes that played out at every encounter and meal, Angelica and Lucinda maintained their malign presence throughout the family's final days—a pointed interjection here, a nasty crumb of information there, a perverse chorus luring and prodding the various protagonists to calamity.

For the most part, Euclides directed his venom at Ana. Dilermando told of encountering Ana, Solon, and Euclides a week before the events at Piedade in the Carioca Plaza (in Rio's theater district) at midnight, when the shows were letting out. Perhaps Dilermando ached for a glimpse of his lover after their São Paulo tryst. Euclides passed by him, and Dilermando feared that some enormous public accusation would ensue—a concern that suggests the affair had become dangerously overt.[29] But Euclides at that moment chose to berate his wife and Solon, though later he left them in the crowd, announcing that he intended to find, insult, and assault Dilermando.[30] According to Ana, as of that night Euclides became frantically jealous. Even though he had long known that two of Ana's children (the unfortunate Mauro and Luis) had been fathered by Dilermando, in his present state the situation had become explosive. Dilermando had assumed that da Cunha was indifferent to the situation, a complaisant husband; Ana reported that she kept most of the family discord from him.[31] But clearly something had shifted.

After Ana's return from the second São Paulo trip, interactions in the da Cunha household became fraught, to put it mildly. For example, as reported by Ana, one day Euclides asked Neném whether she thought little Luis resembled anyone in his (the da Cunha) family, which had no blonds. Neném answered that her family did include blonds, such as Dilermando. Ana asked the Rattos not to say such things to her husband, who at this particular juncture was becoming ever more, in Ana's words, "irascible." Wasn't it enough, she said, with the melodrama that became the hallmark of this household, that her life was a living hell?[32] At this, Euclides said to Lucinda and Neném that since they were single young women, it was really improper for them to stay or be seen in the house of someone like Ana.[33]

That Thursday, August 12, Euclides told Ana that he knew the truth and that she should show some decorum and courage and—his words dripping irony—seek out someone "worthy" of her, the "Sargentão." Of course Euclides had known "the truth" about little Mauro and knew also that Luis was not his child,[34] so the drama of "discovery" seems unwarranted. The "truth" he knew now, though, seems to have been that the São Paulo trip had been too obvious, had been noticed too publicly. Solon was also involved in it, and acquiescent.

Perhaps in earlier times Euclides had hoped the passion would fade, or at least remain clandestine or socially invisible. Perhaps he had just been too busy to heed it. In any case he did not much care for Ana and probably had little interest in women more generally. But as the affair became more widely known, as it inevitably would in the gossipy circles they lived in, Euclides was increasingly incensed and jealous for reasons of his own honor and public esteem, as well as resentful of the shifting affections of his oldest son. Dilermando believed Euclides was unbothered by the affair, and perhaps he had been for a time. Dilermando, at least, could imagine a domestic détente. After the final São Paulo visit there seems to have been a sea change within Euclides, though. Most of the principals hadn't quite taken the measure of his emotional swing and were surprised and distressed at its rancor. Solon's and Dinorah's panicky calls for Ana to return home immediately on the night before the catastrophe convey this dread.

On that same Thursday, Ana went to Dilermando's house after Dinorah delivered a note stating that his brother felt ill, wanted her to come to him, and suggested the pretext of hunting for a bigger house (necessary for when Euclides's father would come). She took Luis with her, and as their absence extended into evening, Euclides became increasingly upset, stomping about and asking where she was and what was taking so long.[35] Meanwhile, at Dilermando's, Ana told him about the conversation with his aunts and averred that Lucinda and Neném would "upset their happiness."[36] Life for Ana in any of the households where she might land (that of her natal family, her lover, or her husband) was impossible in different ways. At length she proceeded to her mother's house and revealed her situation. She spent Thursday night there.

The following day, Friday August 13, another highly unpleasant altercation ensued at lunch. Euclides asserted that Ana was completely brazen, he knew the truth, and had she any concept of shame, she would no longer be under his roof: they should divorce. He described her as "the treacherous offspring of a *tamboleiro* father"—making derogatory reference to the military coterie of Solon Ribeiro. She pointed out that he hadn't wanted a divorce when it should have been done, and that she would not be alone after she left his house. For the moment Ana removed herself to her mother's with little Luis.[37]

Euclides came later to speak to Ana's mother, Tulia, and these two proceeded to the parlor. Augusta, Ana's sister-in-law, reported that Euclides wanted to "return" Ana to her family's household—an extremely embarrassing turn of events that would reflect badly on the Ribeiros.[38] Ana and her brother Adroaldo (no great fan of Ana) stayed outside the parlor, perhaps ears to the door. Finally, after hearing the term *unfaithful*, Ana swept into the room and, swearing her truthfulness in front of the portrait of her dead father Solon, reported the unpleasant details of her married life and challenged Euclides to state "if her words, sworn on her father's memory, weren't true."[39] After this scene, Euclides and the ailing dona Tulia left the parlor, and soon Tulia entered into to a "nervous crisis." Adroaldo urged Ana to return to her husband in order to avoid further devastating their mother, who was very unwell. Ana indicated that she would think it over. The family honor was quite publicly at stake.

That evening Ana proceeded to Dilermando's house, where, after he returned from horseback riding, they dined together, and Ana, according to her testimony, revealed the extensive rancor and the discussions of divorce that now were part of her home life. De Assis was not yet twenty-one; he lived on a student stipend and whatever Ana could provide. He could certainly not give her the kind of life she had enjoyed with Euclides. Strapped as they always were, the da Cunhas still resided in a good neighborhood, in a big house with servants. Dilermando urged Ana to seek out the Sisters of Charity while he finished his degree, and then he would seek a posting to that refuge of the socially exiled, Mato Grosso.[40] Given Ana's obviously sexual nature, sending her off to a nunnery would have served as a means of both penance and rehabilitation, as well as of stashing her out of sight until public attention had moved onto other titillating scandals. Ana spent Friday night at Dilermando's home.

Young Euclides Jr.—Quidinho—who was residing in a local boarding school, was also ill (most of the various personages in the drama seem to have been afflicted with fevers of one sort and another). On Saturday August 14, Ana proceeded to visit him and then went back to her mother's house. There, during a heart-to-heart discussion with her sister-in-law, Augusta, she wrote Dilermando's name on a slip of paper so as not to speak it out loud (one has the impression of servants, family members, everyone lurking in corners to catch snippets of the next conversation and figure out how to enlarge on this extremely juicy gossip). Augusta, seeing the name, said there was already a great deal of talk about him and that learning the truth fueled her and Adroaldo's growing sense that Ana should not return to the family hearth, where her disgrace would fall on everyone, but neither should she live in the company of Dilermando, as that would only bring further humiliation upon her and the Ribeiro clan. Ana, in her view, should seek some kind of asylum, a job somewhere where she could support herself and little Luis, perhaps among nuns.[41]

All these events as recounted by their protagonists have the air of gothic melodrama, and certainly a scandal of this type was riveting to households, friends, and acquaintances. And it was quite a small world that was chattering about the da Cunha domestic arrangements. Infidelity was hardly unheard of in Brazil's upper classes, but such overt passions between a very handsome young man and a matron seventeen (or twenty-one) years his senior would raise eyebrows (and maybe envy among the female cohort), even among the most libertine of Rio's cosmopolitan *salonistes*. For men, such dalliances would hardly merit comment, but bourgeois ideologies of virtuous womanhood made Ana's trespass socially intolerable.[42] Moreover, at the margins of the elite, among those without large fortunes like the Ribeiros and da Cunhas, who depended on patronage, reputation was everything. This kind of scandal could have serious repercussions, not just for Ana (who would have been socially dead in any case) but also for the other members of the family, for their vocational and social prospects.

Euclides was famous, well connected to the national elites. Divorce at the time could be achieved only through papal annulment, a difficult strategy in this case given that Euclides and Ana had three children between them, in addition to Dilermando's Luis and the dead Mauro.[43] But Euclides was also *cornudo*, a cuckold publicly betrayed by his wife, and his name was at stake. The general view at the time was that tarnished honor could be cleansed only by blood. Given Euclides's own concern with correctness, the profoundly incorrect circumstances of his intimate relations must have become personally and professionally unbearable. One of his last letters to Otaviano describes his frustration at the rising "mediocracy," social and political hypocrisy.[44] He also felt that he was dying, and probably actually was—as he wrote to his father. "The truth is I've abused my resistance and now I must stop."[45] He had little to lose in seeking vengeance and his reputation to gain, no matter what the outcome.

On Saturday August 14, Euclides made his way back to dona Tulia's house to find Ana. Augusta informed him that Ana had not spent the night there. In rage and despair, Euclides went to pick up the ailing Quidinho and took him home for the weekend. Back at home, Euclides informed his children, Euclides Jr., Solon, and Manoel Afonso, that their mother would no longer be living with them.[46]

Solon had been party to the entire affair without exactly realizing it, although some, like the police reviewer[47] and the Ratto sisters, believed that he was not innocent of the nature of the relations between Ana and Dilermando. He spent too much time with the couple not to have been aware of affections that went well beyond customary cousinly exchanges, but the romance itself was perhaps too much of a transgression for him to even acknowledge.[48] Given his temperament, he probably preferred the hearty masculinity of Dilermando to his father's neurasthenic intellectualism, but when it came to deciding which "father" he preferred, Solon was caught, naive if not ingenuous, in a tangle of silences and public secrets. His unwitting role had been to divert attention and legitimize the illicit.

That night Solon had a bitter fight with Euclides. In the early evening, Euclides had called him into the study and said, according to Solon's later testimony, "Your mother is an adulteress; she's no longer at her mother's house, that is to say, she didn't sleep there, and since she didn't sleep here, she has to be somewhere else."[49] In a fury of anguished loyalty and adolescent rage, Solon swore he would never step foot in the da Cunha house again, that he and his father were completely different types of people and completely incompatible.[50] The two had never gotten along that well; the studious little Quidinho was understood to be his father's favorite, while Ana doted on her handsome boys, Solon and Luis. Solon said that he would go and look for Ana. He knew, of course, exactly where she was. In his deposition he reported that Neném had told Euclides that Ana was at the de Assis home that night.[51]

The Rattos were in a position to know, since they had meant to go shopping with Ana that afternoon but had left messages with Dinorah and Solon at the Piedade household that they would be unable to join her. In Quidinho's version of that evening, he reported that he thought that Solon was going to avenge the family honor, and that was why he took da Cunha's revolver with him when he left the house for Piedade.[52]

Earlier in the evening, Ana had urged Dinorah to go to the da Cunha house and see to her other young son, Manoel Afonso, also ailing (and under the care of the Ratto sisters) in the midst of this crisis. Dinorah carried out this charge, but when he heard Euclides's angry voice in the da Cunha house insulting Ana in the most explicit terms, he chose not to visit the obviously enraged Euclides, and returned to the house in Piedade extremely upset. He pleaded with Ana to return to Copacabana that night.[53]

Solon left the da Cunha house with some money given to him by an anguished Euclides and, armed with his father's pistol, set off to Piedade. On arriving he sought his mother in the parlor, where she sat with Dilermando. Solon reported to Ana that his father was extremely agitated and said that she must immediately return home with him. Ana refused to do so, noting that the night was extremely stormy. The panicked Solon was soon embroiled in a heated argument with Dilermando and Ana, demanding that Ana promise that she would return to the da Cunha house and that she swear that she was at the de Assis home as though it were the domicile of a son.[54]

It is not clear exactly when Solon pulled out the gun and aimed it at the lovers. According to Ana's testimony,[55] he announced that if what his father had said was true, he would have to kill Dilermando—clearly that was why he had come to the house armed. In his own testimony Solon reported that he pulled out the gun because Dilermando said he would throw Solon out of the house if he kept insisting that his mother traipse home that night in such foul weather. Solon replied, "If you eject me from this house, I'll dispatch you from this world."[56]

Somehow this escalating argument was defused and this particular oedipal gun battle averted. Ana agreed that she would leave very early the next the morning with Solon. The calming genie of the de Assis household, Dinorah, succeeded in distracting Solon by inviting him to play chess. Solon, comforted, slept in Dinorah's room, but they talked until very late at night.[57]

In what state was Euclides? Ana was gone: she had left him, he had thrown her out, and regardless of his contempt for her, his wounded pride, his own madness, he knew the humiliation that attended her romance would turn the private sniggers about him into unbearable public ridicule. He had to deal with his children. His wife and oldest son had both departed on the bitterest possible terms. "I have to make a clean slate of things," he declared that evening as he smoked and paced. According toEném, he vowed to kill Dilermando the next day.[58] At this point, with his health ruined, his family ravaged, his best professional days apparently behind him, he had little to preserve but his honor.

Euclides had gone to a movie the week before with Henrique Coelho Neto—a western where the betrayed husband had killed the lover—and perhaps this plot element echoed into his distress. The nearest witnesses to his state of mind were the Ratto sisters, looking after Afonso and Quidinho, perhaps reveling in the intrigue, and stoking the engines of his fury. That evening Euclides was bordering on the hysterical: "Angelica, I've gone mad, I no longer care to live, my son Solon has said he'll never enter this house again, that we're completely irreconcilable . . ." He spent the rest of the evening chain-smoking and striding around furiously in his office. The Rattos retired to the upper rooms.[59]

Descending the next morning, the Rattos greeted Euclides: "How did you sleep?" "You must be joking," he said, pointing to a floor awash in cigarettes. He'd been up all night. He'd asked the housemaid whether Solon had returned and received a heartbreaking, but unsurprising, no.[60]

Breakfast that morning included the Ratto sisters plus Euclides, Quidinho, and Afonso. Young Quidinho asked one of the ladies, "What does a woman deserve who betrayed her husband?" To this Eném promptly answered, "The husband should kill the wife, and afterward spit in her face."[61] This kind of mealtime conversation, it appears, had been regular fare over the previous few days.

At this moment Euclides once again requested the address of Dilermando. Ana had taken the precaution of lying about where the de Assis brothers lived, so Euclides's ignorance was understandable. Eném sharply retorted: "Those who should know are your sons Euclides and Solon, who have the habit of visiting the house of the de Assis brothers."[62]

Loyal Quidinho, perhaps for self-protection, perhaps to shield his father, replied that this wasn't the case—although, in fact, Solon *was* a regular visitor and had even been photographed at the house a few days before. Then Euclides turned again to Eném, imploring her on the ashes of her deceased mother to tell him the address. And she did: it was Estrada Real de Sta. Cruz 240: "Who knows if Ana isn't there too?"[63]

Da Cunha plummeted into a state of hysterical despair. Attempting to calm him, the Rattos pointed out that Solon was certainly fine and must have spent the night with his great friend Dinorah—a thought that perhaps was not all that comforting. Now Euclides asked Lucinda where Dinorah lived. She

Figure 22.2. House at Piedade: Dinorah at the gate, Solon da Cunha at the window.

again responded, "In Piedade, on Sta. Cruz 204." She provided this crucial bit of knowledge "only," as she said in her sworn testimony, "so that Dr. Euclides could meet up with his son, and to calm him down, because absolutely, Dr. Euclides did not tell *me* he had the intention of murdering anyone. . . . After arranging some papers and getting formally dressed, he left about 9:30. I and my sister, after lunching lightly, returned to our pension on Rua Silveira Martins that afternoon. We only heard of the tragedy later that night."[64]

That morning, his agitation largely passed and his decision made, after straightening his papers for Rio Branco and changing his clothes, Euclides went to Botafogo, where his dearest relative, his cousin Arnaldo, lived with his brother Nestor. Arnaldo had been Euclides's companion in the hard Amazon days and his compatriot in the diplomatic corps of the Baron. Using the excuse of a rabid dog in his neighborhood, Euclides asked Nestor to locate a gun since Solon had made off with his. He chose a Smith and Wesson .22-caliber pistol. He drank a cup of coffee, and everyone later remarked that he was very calm. Then, having taken a streetcar to the suburb of Piedade, Euclides wandered around the neighborhood, asking where the two cadets lived.[65]

Constantino Fontainha, who worked at the *Jornal do Comércio*, happened to be on his way to the butcher shop to select some choice meats for the Sunday lunch that is so much a part of Brazilian family life. At the storefront, with gleaming carcasses in the background, he encountered a man who asked whether he knew where the cadets lived. Constantino cheerily remarked, “Why, they’re my neighbors!” and pointed the way to the de Assis house.[66]

It was a quiet morning, still drizzly and hot as Rio’s turgid air tends to be after a night of storms. Solon was at the bottom of the garden, Ana and Dilermando were breakfasting with little “Lulu” in the dishabille normal to intimates. Then Dinorah, going to the front of the house for some cigarettes, heard a knock, looked through a window, and saw Euclides. Dinorah fled to the rear of the house and informed the couple that Euclides was at the door.[67]

At first Dilermando thought this was a joke in remarkably bad taste, but soon he realized the gravity of the situation and told Dinorah not to let Euclides in just yet. He directed Ana, the cook (a former da Cunha family retainer), and Luis to hide in a pantry that had been modified into a photographic darkroom. He told Dinorah to stall Euclides because he needed to get into more formal attire to receive him—and to get the .38 on his dresser. Dilermando would ultimately become Brazil’s sharpshooting champion; he was rarely far from his firearm.[68]

Dilermando’s idea was that Solon would escort his father through the house and see that Ana wasn’t there, and then Dilermando and Euclides could have a man-to-man conversation. But Euclides barged into the parlor, pulled a gun from his jacket, and shouted, “I’ve come to kill or be killed!” He lunged past Dinorah into the hall and stopped at the first door, which happened to be to Dilermando’s room.

All the witnesses reported that the first shot was loud and powerful, probably from Dilermando’s .38. Dinorah reported that Euclides kicked open the door of Dilermando’s room and fired in two shots, and that after the third shot Dinorah grabbed him from behind and pushed him toward the dining room in the back of the house while Euclides took random shots behind him. That third shot grazed Dinorah, as did the fourth. Dinorah said he let go of Euclides, turned to race to his own room and get his weapon, and was hit in the nape of the neck with the fifth bullet fired by Euclides. In the midst of the shootout, Solon, who reported Dinorah yelling to his brother to “grab the other gun and finish him off,” also picked up his pistol with the intent of avenging his father by killing Dilermando, shouting, “You’ve killed my father!” as he raced to the front of the house, but he was quickly disarmed by Dinorah. Dinorah made it to his room, where he found neither his gun nor the strength to return to the gunfight.[69]

There was the roar of powerful discharges, and then a blasting silence. Dinorah went out of his room to see his brother at the top of the entry stairs and Euclides a few meters away on the concrete walkway in the garden, reflexively, impotently, moving his finger on the trigger of his little .22 pistol, bleeding from two shots to his chest.[70] Solon then rushed to his dying father, begging forgiveness.[71] The central question in the trial was whether Euclides was leaving because he had run out of bullets, and thus Dilermando shot a fleeing man, or whether he had turned to renew his attack.

Euclides's shots had also hit home: Dilermando had three wounds. But Euclides's pistol lacked the force necessary for mortal damage. The bullets fired by Dilermando in the house had hit Euclides in his arm and wrist—the intention had been to disarm him. While interpretations differ, Dilermando reported later that he didn't know whether Euclides would continue to attack or not.[72] Euclides may not have had any shells left in his pistol. As he fled out to the front garden, he turned on the porch stairs to face Dilermando, who then fired the two deadly shots. Euclides fell bleeding with one bullet to his right lung and another under his clavicle.[73] Fontainha, emerging from the butcher shop, heard shots and ran down the road to see the person he had so helpfully directed collapsing, dying in the front garden.

With everyone bleeding, Euclides was lifted from the garden and carried inside to Dilermando's bed, where some pointless first aid was applied, including vigorous chest rubbing and a glass of port. The wounded cadets stood by. "What insanity was this, Dr Euclides?" asked Dinorah. The final words as Euclides da Cunha's fading gaze took in his wife, his son, and his detested rival: "I despise you . . . honor . . . forgiveness . . . intrigues . . . calumnies . . ."[74]

Dilermando spoke to Ana: "I've killed your husband, but I and Dinorah will also die here"—he had three wounds and Dinorah two, including a bullet lodged in his spine, and there was every reason to imagine that infection and clumsy tropical surgery would extinguish the young men as well.[75]

A newspaper copy editor, Mario Hora, lived nearby and was a casual acquaintance of the two cadets; in happier times, he carried out diversions with them like shooting tossed limes out of the air. Dilermando never missed. In fact, Dilermando practiced his aim regularly with such gunplay, so the neighbors were not quite as alarmed as one might have expected with such a barrage of bullets on an early Sunday morning. But Hora's mother, looking out her kitchen window, saw Dilermando standing on his porch covered in blood; she yelled at the sleeping Mario to run over to help his hemorrhaging friend. Mario later described the situation this way: "Dilermando told me to get a doctor: he had shot someone in order not to be killed."[76]

Mario returned with Dr. Capanema, only to find that Euclides had already expired. When told that the corpse was Euclides da Cunha, of *Os Sertões*, the doctor glared at Dilermando, who then opened his bloody shirt to reveal his

Figure 22.3. Corpse of Euclides da Cunha at the police morgue.

own bullet wounds. Mario reported, "It was only at this time that I noticed the woman sobbing uncontrollably on the chest of Dilermando, even as he held a blood-soaked cloth to stanch the flowing wound. And impressively immobile, with the pallor of death already on his strong features, with his arms extended along his body, lay Euclides. It was only later that I realized the woman, the daughter of the celebrated Solon Ribeiro, was the wife of the lifeless man. Dilermando asked that I telegraph the Baron of Rio Branco and Coelho Neto. Piedade had no telegraph at the time, so I ran to Cascadura to send the message to Itamaratí and then over to the house of Coelho Neto."[77]

Coelho Neto was the first of da Cunha's intimates to hear the news. The message asked him to bring Afrânio Peixoto, a celebrated symbolist writer, who was also a medical doctor, and the two men hurried to the crime scene, where the cadaver of da Cunha was laid out on Dilermando's bed. Dilermando himself was already in custody, and a wounded Dinorah was circulating around the house in shock. An anguished Solon swore vengeance. Of the two brothers, Dinorah was in better shape. Before she returned to the da Cunha house in Copacabana, Ana asked him to send this telegram to Angelica

Figure 22.4. Solon and Euclides II after the death of their father.

(Neném) Ratto: "Your will is done: Euclides is dead, Dilermando dead or condemned."[78]

The corpse of da Cunha was taken in a police coach to the morgue. Reporters and "the curious" as well as intimates waited for the autopsy, and among the last visitors to the da Cunha remains was the Baron Rio Branco. Afrânio Peixoto, as the chief medical officer, performed the autopsy of his friend, eviscerating the body of the most prominent writer of his day. As part of the necrology, da Cunha's brain was removed and carefully studied, just as Antonio Conselheiro's had been. But while Conselheiro's brain autopsy revealed no morphological abnormalities that could explain his behaviors, that of da Cunha revealed some brain lesions, which, given the frequency of his malarial relapses and history of tuberculosis, are not surprising. These lesions, according to a review of the autopsy report by Dr. Walter Guerra,[79] perhaps explain the virulence of his temper, the sudden shifts in mood, the hallucinations, the paranoia. His body lay in state at the National Academy. The country mourned his death and was dismayed by the means of it, although the scandal sold many newspapers.

Da Cunha's brain was placed in formalin and deposited for viewing at the National Museum (eventually it was removed to the town of his birth, and then buried in 1988).

Dilermando de Assis's trial pivoted on whether after the wounded da Cunha had fired all his bullets, he was fleeing the scene and no longer posed a threat but was being pursued. De Assis fired the lethal shots that pierced the lung and clavicle and killed da Cunha from the vantage of the porch. Evaristo de Moraes, de Assis's lawyer, maintained that in the commotion of the moment, the confusion over whether da Cunha might continue his attack justified de Assis's actions. In the case of "crimes of passion" (which were also crimes of honor), Brazilian jurisprudence seldom condemned. De Assis also had powerful protectors in his military faction.

The fates were not quite done with da Cunha/de Assis households; the sons especially lived in a strange oedipal anguish. The photo taken of them with a journalist shortly after Euclides's death surely reveals profound torment (see fig. 22.4).

Three weeks later, September 8, 1909, the boundaries between Brazil and Peru were redrawn largely based on the reports and maps of Buenaño and da Cunha.

The jury absolved Dilermando de Assis of the crime of murder on May 5, 1911.

He and Ana married a week later.

23

Hamlet's Lament

Fatherless, and with their mother in disgrace, the da Cunha children were sent away from her immediately after the shoot-out to become the wards of various relatives and protectors. Certainly life in Rio was less alluring, and quite emotionally distressing, especially after their mother married Dilermando. The tutor of the all the boys, da Cunha's cousin Nestor, had nothing good to say about the domestic arrangements and the children's apparent accommodation to it.[1] Dilermando was closer in age to the boys than to their mother and had been a sort of an older playmate for them. The murderous eroticism that clung to the couple, who, for obvious reasons, were socially isolated, and the enormity of the transgression, infused the family circle. This overtly sexualized oedipal situation was certainly immensely complicated for the teenage boys.

Solon, fifteen at the time, was placed under the care of Cândido Rondon, the great Brazilian explorer and almost president, while Euclides II (Quidinho) became a ward of José

Carlos Rodriguez. Manoel Afonso was the ward of Ana's sister Alquimena, who had broken all relations and communication with her.

As an older teenager, once his studies were finished, Solon soon found himself on expeditions slogging through the southwestern Amazon toward Acre Territory to maintain and repair the telegraph line between Cuiaba and points north in a famous expedition under the direction of Rondon.[2] He was in fact actualizing a policy—telegraph lines—suggested by his father for integrating the remote Amazon into national life. According to Dilermando, Solon broke off all relations with his mother and her lover from 1910 to 1915.[3] Once Solon had acquired a post in Acre he returned to Rio to collect a few mementos of his father (like the proofs of *Os Sertões*), and initially he seemed warm and forgiving, but by the end of his stay he was cold and remote. "He departed," said Dilermando, "saying nothing, without even seeking his mother's blessing. He left to go North, to his independence, and . . . to his death." Dilermando indicated that the malign gossip and the larger question of vengeance constantly agitated the sons and turned them against himself and Ana.[4] The internal psychological distress created by his relationship with their mother seems not to have occurred to him.

Perhaps it is not too surprising that Solon finally opted to stay in Amazonia, in the Purús valley in Acre Territory, not too far from the terrain mapped and finally formally claimed through the intercessions of da Cunha. There was probably a great deal still unresolved about his father in this young man's heart, and his failure to avenge him perhaps shadowed him through his very short life. It was a time of transition in Amazonia. The rubber monopoly was failing, but there was the new presence of the nation: an extension of order, of civilization, where the conflicts were increasingly supposed to be mediated by local state officials rather than the *seringalista* militias and gunslingers of the frontier heyday.

Solon lived in the village of Tarauacá, in the core of Acre Territory, just as the boom was winding down. He was a type of sheriff's deputy and was much liked in the little town. He apparently had a fiancée, friends, and "tender esteem," as the scribe who accompanied him to his death put it—a world well apart from the intrigues he had been living. My account is taken from the reportage of this scribe, Sancho Pinto Ferreira Gomes.[5]

In April 1916, Solon was sent to open an inquiry and to take into custody those accused of three murders in a backwoods vendetta, henchmen of the powerful Correa-Lima family, an influential trading and estate owning clan. According to the weekly journal of Tarauacá, *O Jornal Oficial*, three *seringalistas* from the estates of Sant'anna, Mira Flores, and Sta. Cruz organized an ambush of the police delegation in order to protect their gunslingers. As reported in the journal by the police scribe, the group managed to capture two of the perpetrators and took them prisoner. They arrived at a *barracão* called Ambrosio. A quick reconnaissance revealed four men hidden inside, waiting with rifles poised, "ready to resist whoever might arrive." The men who had earlier been taken prisoner by the sheriff's delegation were in fact part of a ruse arranged by the ambushers to warn them of the police arrival and to interfere with the strike force. As the deputies entered the open patio in front of the trading post, a cry would be given, and the men inside would have their rifles at the ready and the hapless lawmen in their sights.

Solon went up the stairs of the trading post, announcing that he was a delegate of the police, at which time two of the men opened fire. The volley from Francisco Leandro hit Solon in the middle of his belly. Though wounded, Solon blasted his opponent with a shot to the chest, killing him, and ordered the army men with him to fire, which they did, taking down Solon's other assailant. The other two men inside the post fled the scene. Less than an hour away by foot, another group of nine men, armed and ready, awaited this delegation in ambush, at least according to one of their prisoners.

Outnumbered and with the mortally wounded Solon to attend to, the police resolved to leave as quickly as possible. A couple of men were dispatched to Mira Flores Estate to seek medical help. Solon was placed in a hammock and carried through the rubber forests for several hours toward the trading post known as Revolta. After three hours, he asked the men carrying him to stop. He said he was almost blind, and then with a long sigh said, "Aí, meu pai"—a cry of pain and a whimper for the comfort of his father. But *aí* has another meaning: "here," or "there you have it, it's done"—his death, an offering, or perhaps both. The men walked until one in the morning through, as Ferreira Gomes put it, "horrible *varadouros*." At dawn, near the post of Carneiro de Tal, they buried Euclides da Cunha's oldest son.

The police scribe, in the same mournful article, reported that as Solon said his good-byes embarking from Tarauacá, he had been certain he would die on the expedition and so had left final letters for his brother and fiancée.[6] The good citizens of Tarauacá erected a monument to Solon, a mark of affection and of his prestige within the local society. Where his actual remains now lie is not known.

We do not know what was in the letter from Solon to his brother Quidinho, who would outlive him by only two and a half months. We do know that

Euclides Jr. nurtured a profound hatred for his stepfather. Partly this was explained by the obvious affection and compatibility between Euclides I and II. Quidinho's life with Nestor, whose loathing for Ana and Dilermando was unmuted and implacable, cannot have helped. Even Quidinho's grandmother was repulsed by the situation: as she wrote to Euclides's sister, "You mourn a brother, but I lament that I ever gave life to my daughter."

With Such Cozenage

When Ana married Dilermando upon his release two years after the incidents at Piedade, Quidinho (Euclides II) was sent to boarding schools, where the scandal was so widely known that his mother could not visit him lest he become the brunt of even greater abuse. Ana was seen as a person of questionable morals, with a taste for adolescents, so there were even some reservations about the propriety of Ana's calling at a boy's school. While Quidinho was tutored by Nestor, the venom regarding "the situation" was palpable. At the heart of his condition was Hamlet's lament:

> [He] hath kill'd my king and whored my mother;
> . . . And with such cozenage—'tis not perfect conscience,
> To quit him with this arm? And is't not to be damn'd
> To let this canker of our nature come in further evil?[7]

How then, did it happen that on July 4, 1916, Euclides da Cunha II went to the Vara dos Órfãos (the social service administration for orphans) and sought out Dilermando de Assis, who astonishingly managed to survive four shots from the younger da Cunha and, as happened with the father, was able even with multiple wounds to aim true enough to kill him?

Ever since the dreadful Saturday that preceded his father's death, the younger Euclides had lived in torment. Letters he wrote later from boarding school to his mother suggest how painful his life had become and how torn he was in his filial affections:

> In reference to what you wrote about coming to visit me—I was very happy about it but I have an objection to make: as you know tongues can be treacherous and hurtful. Your coming here could prompt the meaner spirits to scandal. They'll say that I'm partial to the murderer of my father, because you are married to him.
>
> I don't want to meet you here in Juiz da Fora,* because as you know, I won't

* Juiz da Fora is in the northwestern part of the state of Rio de Janeiro.

> have very much freedom. And as I already said, in this town (as in all of them) there are many people who like only to gossip. In any case, set the date, the place, and which train to take and indicate [to the school authorities] that I must go . . .[8]

Manoel Afonso, meanwhile, was probably not having such a wonderful time either. He was being educated in a Silesian school in São Paulo and was very unhappy, the most orphaned of all the da Cunha boys. He wrote to his mother about his extreme dejection and said he was losing his mind in despair, a situation also noted by the school's director. The de Assis household tried to contact Ana's sister for help, but apparently they lacked an address for her. At fifteen, Manoel Afonso fled his boarding school, determined to go back to Rio de Janeiro, where he was returned to the care of his uncle Nestor.[9]

Ana was in the eighth month of her latest pregnancy when on June 13 Manoel Afonso arrived at the farm where she and Dilermando resided. He stated that he no longer cared to live with Nestor, nor did he want him as his tutor, because during a meal Nestor had accused Ana of "murdering his father and his brother Solon," apparently now a regular topic at that da Cunha table.[10] The young runaway precipitated a series of disastrous events.

Born in 1901, Manoel Afonso had been very young during the pre-Amazon times and had been largely superseded in Ana's affections by the much-doted-upon Luis. The catastrophic last few years of the da Cunha marriage could hardly have been a recipe for childhood happiness. Now the new de Assis family was growing by a new member almost every year: João (1910), Laura (1912), Judith (1913), and Carlos (1914)—and Manoel Afonso's mother was again in late pregnancy in June 1916. Who knows what yearnings for family normality animated him? The de Assis farm with its babies, rural pleasures, and general fecundity must have seemed like an idyll compared with the ascetic dormitories of religious boarding schools in São Paulo, where probably the same taunts that plagued Quidinho haunted him. And his da Cunha relatives were austere and accusing.

What transpired in the following weeks is less clear. Nestor was adamant that Manoel Afonso must not continue in the de Assis household, since Ana and Dilermando had initially shoved the child into foster care with his aunt and Nestor for his education and welfare. He was their ward. The couple had been largely out of contact with all the boys for years. The de Assises actually retained few legal rights over the teenager, whatever moral or emotional privileges might be invoked. Clearly the boy did want to stay in a more domestic setting, but soon he was part of a larger vendetta.

At the behest of Nestor and his lawyer, the police were sent out to retrieve the boy, who apparently did not want to return to school or his uncle. This first attempt at removing him was thwarted, but a few days later a warrant, a scribe, and a policeman arrived. This effort to take away the boy was also unsuccessful. Finally, once again another scribe, warrant, and policeman, with some backup from the army , surrounded the house. Once again Dilermando and Ana were embroiled in public scandal.

Ana begged this delegation to wait for the arrival of Dilermando at the end of that day. When he finally appeared, the situation was reviewed, and the child was deemed of the age of reason and thus able to decide on his own where he preferred to reside. But the policeman insisted that Dilermando and Manoel Afonso appear the next day in front of a judge of child services (*juiz de órfões*) and explain the situation.[11]

To get in the middle of a da Cunha–de Assis controversy was asking for trouble. Nestor's lawyer, Rodrigo São Paulo, appeared before the judge and argued that the child had been more or less abducted and that the moral terrain of de Assis household was questionable and unfit for him. Other allegations were made that the couple was illegally holding da Cunha estate materials that rightfully belonged elsewhere.[12]

The judge, however, decided that for the time being the child should stay with the de Assis household until another guardian could be arranged, in this case General Emídio Dantas Barreto, in some ways a döppelganger for Euclides. He was a former war minister under President Hermes Fonseca, former governor of Pernambuco, and veteran of the great Brazilian battles of Paraguay, Canudos and the 1910 millenarian revolt, the Contestado in Santa Caterina. He was also a man of literature in his way. He had written, like da Cunha, an account of the Canudos War[13] (Dantas Barreto's is largely forgotten), and was an essayist of some repute. He held the chair previously occupied by Joaquim Nabuco at the Brazilian Academy of Letters. Dantas Barreto was a person of the military-political class rather than of the intelligentsia but had enough Euclides "analogues" to act as the child's guardian. He was also an important behind-the-scenes protector of de Assis.[14]

At the meeting with the judge, Nestor's lawyer argued that Euclides II, for his part, was extremely upset about the situation, the "ignominious" circumstances of his brother's living under the roof of his father's assassin, the dishonor of the lack of revenge, and so on. For Dilermando, the problem was the deeper social rancor against him, the resentment about his release from prison, and the large da Cunha clan's longing for a more active form of justice, given the absence of any harsh punishment by the military tribunals. After leaving his boarding school, Euclides Jr. had returned to Rio for naval training but had commanded his mother not to visit him or seek him out.

He sent her a letter denouncing her for "having infringed on the concept of Dignity, betrayed her filial relations, and had been made craven by gossip," and he ordered her not to write him because "he wasn't sure where he would reside."[15]

Dilermando regarded the next phase of the feud as a kind of social conspiracy, where, stimulated both by the military concept of honor imbued in young cadets and by the da Cunha clan's yearning for revenge, Euclides Jr.'s friends and the broader public opinion inflamed the young man's utter loathing of his stepfather. His sense of his own (and his brothers') failure to settle things and his ideas about distortions of justice cannot have helped. Quidinho was likely manipulated into becoming an angel of vengeance. In the normal day to day, he would have had no idea of Dilermando's movements; someone had to inform him, after all, of where Dilermando would be on July 4, and that someone was probably Nestor.

After hearing testimony from Manoel Afonso and Dilermando, the judge had requested that Dilermando go to the Varas do Órfões—the Orphans' Services Administration—and fill out some forms and declarations in reference to this case so that the child could have a new tutor. It was there, on July 4, 1916, that Quidinho and Dilermando met for the last time. Quidinho had left a farewell letter: "I'm going to do this, whatever I do, from sentiments of honor. I need do nothing more than act!!"[16] Hamlet, indeed.

It was shortly after noon, and Dilermando had just started reading the documents in the Vara dos Órfões when he heard an explosion, felt a bullet, and saw Quidinho about two meters away, dressed in his formal uniform. Quidinho fired again; both shots entered Dilermando's lungs. The others in the room—a scribe and a bureaucrat—were paralyzed, terrified. Dilermando moved away from his assailant, opened his military tunic, and took out his revolver, which contained only three bullets. As he fell, another bullet perforated his liver and diaphragm. He was seriously wounded, and the Vara was now the scene, as the papers put it, of a "duel to the death."

Dilermando stated later that he thought he was probably going to die and that he at least had the right to defend himself and his name. His military discipline had taught him never to flee. Quidinho, a good shot, had perhaps lowered his guard in light of his accurate shots into Dilermando's torso, and probably did not expect him to rise and start his own attack. The first two bullets from Dilermando hit Quidinho's arm, and then, when he turned in surprise, a last volley from the Brazilian sharpshooting champion smashed into his brain. He fell dead to the floor. Quidinho, like his father, had badly wounded Dilermando but not survived him.

On hearing the news of the murder of her son by her husband, Ana went into premature labor, and her child, almost at term, was stillborn.

A military tribunal absolved Dilermando de Assis for killing Euclides da Cunha II on grounds of self-defense.

Rosencrantz and Guildenstern Are Dead

There was another victim, an ancillary player who was unfortunately situated in the drama. Euclides da Cunha had not just wounded Dilermando, remember, but also hit Dinorah in the neck. Dinorah de Assis, athletic and quite sweet in disposition, was an accomplished footballer who played for one of Rio's most important teams, Botafogo. Fans used to the modern professionalism of Brazilian soccer may be surprised to hear of a cadet rising to such heights, but he was apparently extremely talented, a football star of the first rank, and very well known in Brazilian soccer and sport circles. Alongside his companion Pullen, Dinorah played both fullback and defensive midfielder for this esteemed team.

With the championship against the team's traditional rival, Fluminense, coming up, Botafogo was in despair because Dinorah had sustained injuries in the first da Cunha shootout and would not be able to play. The rivalry between Rio's home teams is intense, as anyone who has spent any time in a Rio bar will know, and this competition between teams stretches back more than a century. But a week after the Piedade shootings, against all imaginable odds, Dinorah was once again on the field, a fact that caused no little commentary even among the trial lawyers. Although Botafogo lost, the game was close. The next year, once again, Dinorah with his teammate Pullen helped the team win the championship of Rio.

But the following year Dinorah was gone. The bullet in his neck had migrated (probably not helped by the professional level of soccer that he played), and he became partially paralyzed. He was not even twenty when both his military and his sports careers were definitively over.

The journalist Acélio Dauat wrote of Dinorah in his "Forgotten Victim of Euclides da Cunha."[17] Dauat describes the profound resentment that Dinorah nurtured against a destiny that had exiled him from his life without extinguishing it. He came to despise "the simple and innocent pleasures of youth." His afflictions separated him from more "humble happiness," leaving to him only the easy pleasures of inebriation and the reprobate habits of a life in brothels. Dauat narrates his spiraling downfall: "With his organism ruined by his orgiastic nights, compromised by alcohol toxicity, and finally ravaged by syphilis, Dinorah very shortly fell into dementia. In his madness, Dinorah entered into conflict with his brother and came to loathe the one person he had most loved, the one person who suffered the bitterness of Dantesque contrition for being the inadvertent cause of his brother's disability."[18] Dinorah despised his brother and his dependence on him. Even though the bullet was finally removed from his spine in 1913, the neural (not to mention psychic) damage had been done and was irreversible. Dinorah rejected all charity from Dilermando and, increasingly homeless and destitute, refusing all family help, came to beg on the streets. Finally by 1921 he had had enough and, finding himself on a quay in Porto Alegre, threw himself into the sea.

The gods were done playing. The carnage was over.

Dilermando de Assis rose to the rank of general but always had to live up to, or live down, da Cunha. De Assis wasn't a bad writer, and in addition to his volumes that argued his defense (*Um conselho de guerra: A morte de aspirante da Marinha Euclides da Cunha Filho* [To the military tribunal: The story of the death of cadet Euclides da Cunha II] and *A tragédia da Piedade* [The tragedy of Piedade]), he produced an autobiography (*Um nome, uma vida, uma obra* [A name, a life, a labor]), a war history, and a saga of unrelenting tedium on transport engineering in Sao Paulo.[19]

He left Ana in 1926.

The events of de Assis's mature life were eclipsed by the early Euclidean times, but he did have a role in other historical events. By kinship, training, and political alliance, he was part of a powerful military coterie from Rio Grande de Sul that would become the incubator for the ideologues and army men who would shape the Brazilian military state that would rule the country for much of the twentieth century. De Assis was the commander of the crucial São Paulo garrison that supported the Vargas coup in 1930 and vanquished its opposition by 1932. This critical victory entrenched the military coalition in governing the country. De Assis thus helped usher in Brazil's period of Authoritarian Modernism, which lasted on and off for the next fifty years.

These military modernists and their "March to the West" and "National Integration" programs transformed the Brazilian economy, and they set their sights on the Amazon as a central national project. Dilermando died in 1950, just as Generals Golbery de Couto e Silva, Emilio Garrastazu Medici, and Humberto Castelo-Branco would look at their maps and see a very different Amazonia from that of Euclides, a place usefully “empty,” ripe for a new geopolitics and a different kind of nation building, a place for a modernizing nation to make its mark, to “flood the land with civilization.” By 1964, they ruled. They turned their sights to the north and the Amazon and unleashed an apocalypse.[20]

24

Illusions and Oblivion

Today the Upper Purús River is considered one of the most remote and pristine forests in all the Amazon basin. Classified as part of the Juruá-Purús moist forest ecoregion, the area is now embraced by the Alto Purús Park, extending over some 2.5 million hectares (about 5 million acres). It links to the 1.7-million-hectare Manú National Park and the Chandless State Park in Acre. These three protected areas alone are almost the size of Costa Rica.

Today, as mostly it was in Euclides da Cunha's time, the area is largely forested, with trees almost a hundred feet in height and emergents stretching up another fifty feet. The diversity of flora is exceptionally high, with forests of more than 250 species per hectare. There are many hardwoods at low densities. Extensive mahogany forests, characteristic of older-growth secondary vegetation, have made it an area of rampant timber poaching; this lovely veneer wood has been pushed close to extinction almost everywhere else in the basin.

The fauna is also diverse and endemic. There are about 170

mammals and more than 554 avian species.[1] It is a region where naturalists thrill to the presence of very rare species such as the giant Amazon river otter, the short-eared dog, and the Amazon river turtle, which in other parts of its range has had its eggs consumed and been itself hunted to proto-extinction. Top predators such the jaguar and prized prey items like peccaries are also part of charismatic fauna of the remote and pristine Purús.

The Upper Purús, according to modern assessments, is a refuge of tribes that "run from man." Indeed, part of the justification for the preservation was to protect the traditional territorial rights of the Masco-Piro, who flee at the sight of missionaries, loggers, and environmentalists in equal panic.[2] Humans there, in this version of the Purús, like nature, are seen at their most primal. Portrayed as "the last nomadic indigenous people on earth,"[3] the people are herded into an imaginary Eden, a setting somehow outside tides of any history. These modern descriptions echo Father João Daniel's assessment, written from his inquisitional cell at the end of the eighteenth century: "There is nothing between the Madeira and the Javari." Yet the Masco-Piro had a long and fabled history in the realms of rubber, as da Cunha was at pains to let us know, as two of the ethnicities of the thousands of native slaves who were lured by Carlos Scharff and the Fitzcarraldos. Even *über-caucheiro* Arana had some men roaming in this outback, as da Cunha reported in his confidential report to Rio Branco.

It is very difficult for us to imagine what transpired in those forgotten forests in now forgotten times. For Amazonia, the death of the metastasizing rubber economy had been foretold in the seeds stolen for Clements Markham in 1876 by Henry Wickham a full thirty years before da Cunha's travels. Markham structured the institutional mechanisms that underpinned the "empire of botany," and as quinine had before, rubber would now shape the global economy. These shifts, as da Cunha had noted, had "stolen the future," or at least one kind of future, from the Amazon.

The seeds that had been planted out with varying success were now trees swaying in the breezes of Ceylon plantations. In 1906, as da Cunha was poring over his maps and reports for Baron Rio Branco, Sir Henry Blake hosted the first ever rubber exhibition in Ceylon, with products from India and Java as well. Planting and processing techniques were discussed in detail, and within the decade, Asian rubber would eclipse the wild product, collapsing the world that da Cunha so movingly described. The language and ideologies of nationalism that were developed for a particularly Amazonian context by Rio Branco's mandarins would, by 1912, have lost their salience as Malaysian rubber flooded world markets and the New World latex economy faded into torpid obscurity.

In any case, the rest of the world was having other scrambles: World War I would reinforce the dissolution of the Russian, Hapsburg, and Ottoman empires, modern and civilized Europe would descend into its own barbarity in the trenches of Flanders and Verdun with its mustard gas and bomb-dropping airplanes, the last war where horses were used in pathetic cavalry charges into machine-gun fire. In this noisy and desperate world, with Europe's own imperial reconfiguration, the plights of some debt peons on a far-flung tropical tributary and battles over Creole empires lost their fascination. Yet the maps we have today would look very different without the definitive Brazilian triumph in the Amazon Scramble, one that, at the end of the day (and certainly compared with the blood lust of Europe), seemed stately and civilized in its quiet adjudications.

The area would depopulate and revert to the simple rural producers who had been its slaves in earlier times. A hundred years later, the history and drama of the place would be excised and forgotten. The Upper Purús was inscribed in the twenty-first century as the Empire of Nature, a land without history.

Da Cunha's assessments of Amazonia were eclipsed by the durable triumph of *Os Sertões* and the frisson of scandal that surrounded his death. What is clear is that the fates of millions all over the globe had been inextricably woven into the tropical biologies of latex trees, whose influences had already figured in complex ways in Latin American history.

The great estates that flourished along the Purús, the vast commodity chain stretched the from the Rothschild banking houses to oozing saps that had ignited the avarice of British, French, US, and Belgian imperial ambitions and galvanized traditional New World rivalries into warfare in the Amazon, all forgotten and glossed with a knowing glance backward: how silly the boom had been, with the estate owners' pathetic ambitions for opera in the tropics, how profligate the *arriviste* extravagance of sending laundry to be washed by nuns in Lisbon while bathing one's mistress in champagne. Really, what could one expect from such short-sighted economic actors?

But the "boom," the latex industry, lasted a century. Few other enterprises were so lucrative, and given the capitalisms of the time, few boasted of social relations that were much better anywhere in the world. The collapse brought down tropical tycoons and banking houses of many stripes. Da Cunha's dream of a defiant new Amazonian Civilization was over. The inhabitants would not be the vanguard of novel tropical society, at least not just then. It would take another seventy years, and the rubber tapper movements would have to be imbued with a completely different logic, for that part of the Amazon to re-emerge as a symbol of a global trajectory, this time in the arc of populist and environmentalist politics. The late twentieth century would see the region's invisible inhabitants writing a new history in the last days of the military republics set into motion by da Cunha and his cohorts a hundred years before.

Da Cunha's dreams of empires in the tropics were restless ones. Amazonian dwellers and developers would again clash over the "lost paradise" for the last half of the twentieth century, and well into the twenty-first. The Scramble never really ended but reflected the profound changes in the nation and the state of nature, and new forms of globalization. But da Cunha had foretold this:

> Such is the River, such is its history: always turbulent, always insurgent . . .

A Note on the Text

Fragments, Translation, and Photos

The Lost Chapters

Leandro Tocantins, one of the great Amazon analysts and himself a distinguished scholar of da Cunha's Amazon time, pulled together many of the fragments of da Cunha's Amazon writings for the centenary of his birth in 1966. Through this compendium many Amazon researchers and readers have been exposed to da Cunha's Amazoniana. But most readers still know him only through the disturbing prose of *Os Sertões*. The rebel story dominated his "literary life," but his Amazon phase involved much more field time and employed and obsessed him until the last. Even recent biographies have scanted this part of his life.[1]

The fragments of da Cunha's Amazon writing were mostly published in newspapers, as reports, and as part of the posthumous collection *À margem da história*.[2] Most of these writings were meditations and drafts meant to become elements of his unachieved project *Paraíso Perdido*. The writings are a mixed bag: they repeat and rework. Insightful, beautiful, sometimes ranting, homesick, even overwritten: they are what we have of his Amazon oeuvre. Some of his writings were for specialists; for example, his analysis of the Purús, "Rio em abandono," was initially published in the *Revista de Instituto Histórico e Geográfico* in 1905 and later reworked and published in the eclectic posthumous *À margem da história*. The greater volume of his Amazon work, however, was part of the efforts of a writer for hire, and he was very specifically classified as an "ideologue." I have used the Scramble for the Amazon to situate him because he was thrust by Baron Rio Branco right in the middle of it. I base my analysis not on the essays alone but also on the imperial work and "nation narration" these texts were meant to do. I also treat his cartography as texts. This was the logic I used in carrying out and editing my translations of his work.

This book relies on his letters, his formal correspondence with the Baron, his explicitly political Amazon writings from *Contrastes e confrontos*, his publications in newspapers, *Peru versus Bolivia*, many of his letters to Rio Branco as well as friends and family, some of his the essays from *À margem da história*, and his reports from the Purús expedition itself. However, these are placed in a larger context of geopolitics, travel genres, and emerging elements of both natural and colonial science and ethnography. I wanted da Cunha to have his own say about what was transpiring, but without the unifying idea of the Scramble for the Amazon and bitter Peru-Brazil-Bolivia controversies, much of the writing seems a strange blend of national and naturalist lyricism, screed, and policy recommendations, and this is more or less the position taken by a recent translation and biographers.[3] Also, much of the oeuvre seems disconnected if it is disembodied text.

I was greatly helped by insights on translation from my UCLA colleague Efraim Kristal's commentary on Jorge Luis Borges, the Argentinean writer who also had an illustrious career as a translator.[4] I have taken some of my inspiration from him. First, Borges tended to treat texts, even final texts, as drafts; he points out that just as there are no perfect translations, there are no perfect texts. Since for him texts are not sacred, Borges's ambition in translation was to create a convincing work of literature, through several strategies. First, remove the padding—writers can fall in love with their words and include passages that are redundant and superfluous. This is the case with many of da Cunha's Amazon writings. He reworked short pieces for introductions and speeches, recycled newspaper articles, sent them off in letters to friends, and reworked structures and images from the *Os Sertões* into a humid tropical counterpart. Compilers of his Amazon work often included everything or concentrated on one collection. There is certainly a justification for this. Borges also removed textual distractions—he chose to smooth or change the text if he felt the literary effect would be more convincing, and he was bold when it came to creative translations. I have followed his lead and taken liberties with the fragments in an attempt to mold them into a narrative, yet I seek to frame them within the debates they were addressing, events of the time, and places where they were transpiring.

Borges often restructured a translation in light of another work. He thought that since an author's output is a totality, sometimes a more effective structure can be partially transposed to make the narrative stronger. Da Cunha used the campaigns of the war at Canudos as a framework on which to hang his reflections on place, history, and human progress. I have used his own Amazon odyssey as the apparatus for connecting various fragments that, of course, are also meditations on place, race, history, Brazilian nationalism, and human progress. In this I have used themes to link the various fragments together. As he puts it: "I'll write a kind of 'Lost Paradise,' for example, or something else whose amplitude frees me from the precise definitions."[5] But his thought was evolving, and his analyses had to be focused on very specific products—articles, reports—in the short term. These were less literary but deeply connected to the territorial and narrative construction of Brazil's nationalist/imperial claims, phrased in historical, national, and moral languages and in the idea of "moral landscapes."

I part ways with Borges too. He felt that great art stands on its own. He said that when younger, he believed context was necessary, but in his later years he concluded that it makes no difference: the work is good or it isn't. Not all da Cunha's Amazonian writings can be considered "magisterial." When he was called upon to contribute to an imperial venture and political agenda, his goal was not purely artistic, and our understanding of his beautiful words is enriched when we grasp something of his context, one unfamiliar to most readers in English and even to most Latin Americans. He has many interesting things to say about this period in Amazonia, whether every phrase rose to the pinnacles of high art or not.

In taking these positions I would not only annoy Borges but really irritate that other supreme translator, Vladimir Nabokov.[6] In contrast to Borges, Nabokov embraces text as definitive, and any attempt to impose "what the translator thinks the author wanted" merits total condemnation, probably involving a purgatory in elevators with the most emollient Muzak. The problem here is that da Cunha's fragments can be repetitive and his perfectly honed "text" doesn't exist. So I've decided to take the risk of both giving him context and, where necessary, editing the text, thereby alienating both Borges and Nabokov.

I invoke the great translator Gregory Rebassa as my defense. His question: "Can we ever feel what an author is feeling as he wrote the words we are transforming?" His answer: "Only if we know what the book is about."[7] Rebassa's position informs my approach to this translation and semibiography. I feel the biggest stumbling block for translators who approached da Cunha's Amazon writing has been not really knowing what it's about—the Amazon: the time, the history, and what else was afoot. While there are good translations of some of da Cunha's Amazon work in *À margem da história*, the limited textual focus of that volume curbs the understanding of "what it's about." My translations of da Cunha could offer something that most literary translators would not have: decades of Amazonian field experience in the area he actually traveled; grounded experience of its social history, da Cunha's terrain, and his forest dwellers; and an understanding of Amazonia from its pre-Columbian history to its modern integration into global commodity circuits. In the popular mind Amazonia is still "a land without history," although da Cunha marked out its past in the scattered tribes, in the place-names of little upriver settlements, and in the archives and maps of the epic, centuries-long Scramble for Amazonia.

I also thought that da Cunha's world could be understood with photography of the place, the people, and the expedition itself. The boundary survey of the Purús and Juruá (1905) is among the earliest uses of expeditionary photography in Amazonia. There were also other roughly contemporaneous sources of appropriate photographs. These include small book of photos that the Manaus Chamber of Commerce prepared for the 1893 Chicago World's Fair. There were the photographs from expedition reports by experts on the rubber economy, like Pearson, who photographed as they traveled. There were the works of commercial photographers: for example, the Paraense studio photographer Emilio Facão produced a volume with photos of all the great *seringais* in the region, a kind of *Estates of Beverly Hills* of its time, showing all the conventions of bourgeois representation of wealth: the workers, the numerous *pelles* (giant globes of rubber) that often cram the foreground, the ladies riding sidesaddle on white mules, the portraits of private steamships. Such photographs were displays of wealth as obvious as the jewelry and blood stock in the portraiture of European and American elites and their estates.

I ended up writing more of a biographical essay than I intended. Like da Cunha's own Amazon fragments, this work is part biography, part social history, part nature writing, part geographic translation. The last part of my book relies extensively on the court documents and depositions of the Dilermando de Assis trials, along with Amazonian materials produced by the police scribe who accompanied Solon da Cunha on his fatal trip. For this reason my version of the final days of Euclides da Cunha is different from those found in the biographies of Roberto Ventura and Frederic Amory, who emphasized different sources, including Euclides da Cunha II and Euclides I's close friend Henrique Coelho Neto.

This book meant to introduce English-speaking as well as Brazilian readers to another dimension of Euclides da Cunha rather than stranding him and his backlanders forever on the bloody banks of the Vassa Barris River watching the last few vanquished dwellers of Canudos stagger through a corpse-filled landscape on their way to Salvador. He had a far more triumphal history to record for his *sertanejos*.

His own epic goes far beyond that time, one that is actually inscribed in every map of Brazil, and in his history of a land that is always seen to be without one.

List of Translations

Chapter 3

Selections from "Maréchal de Ferro"

Chapter 11

"Contrastes e confrontos" (Contrasts and Comparisons)
"Conflito inevitable" (Inevitable Conflict)
"Contra os *caucheiros*" (Against the Caucho Hunters)
"Entre o Madeira e o Javari" (Between the Madeira and the Javari)

Chapter 12

"Amazônia: A gestação de um mundo" (Amazonia: A Nascent World)

Chapter 13

"Impressões gerais," part 1 (Such Is the River)
"Impressões gerais," part 2 (At the Threshold of a Marvelous World)

Chapter 16

Relatório de Comissão Mixta Brasileiro-Peruana de Reconhecimento do Alto Purús (Report of the Joint Commission to the Governments of Peru and Brazil)

Chapter 17

Relatório confidencial ao Barão do Rio Branco (Confidential report to Baron Rio Branco)

Fragments:
“Uma entrevista” (An Interview)
“A minha terra é retilinea e alta como as palmeiras” (My Land Is as Upright as the Palms)
“As promessa divinas de esperança” (Divine Promises of Hope)

Chapter 19

“Entre os seringais” (In the Rubber Estates)
“Judas-Asvero” (The Wandering Jew)

Chapter 20

Mapping the Purús: Between the Madeira and the Javari
“Geografía real e a mitológia: De foz as cabeceiras” (Mapping the Purús: Real and Mythological Geographies from the Mouth to the Headwaters—my title for “De foz as cabeceiras,” the first section of “Observações sobre a história da geografia do Purús” [Observations on the History of the Geography of the Purús])
“Nas cabeceiras” (At the Headwaters)
“Os varadouros: A TransAcreana” (Forest trails: Acrean integration)
“O povoamento: Da foz as cabeceiras” (Populating the Purús from Its Mouth to the Headwaters)
“At the Headwaters”
“Brasileiros” (Brazilians)

Chapter 21

“Os caucheiros” (The Caucheiros)
“Um clima caluniado” (Maligned Climate)

Glossary

aldeia. Missionized Indian settlement.
aniga. Giant aquatic phildendron.
aturizais. Thickets of shrubby tropical legumes, *Machaerium*.
audiencia. Court. In the New World, *audiencias* were given a consultative and quasi-legislative role in administration of the territories, but the viceroyalty, like that of Lima, was the ultimate seat of power over the *audiencias* under Spanish colonial law.
bandeirante. Brazilian flag–bearing explorer or slaver.
barracão, pl. **barracões.** Trading post.
batelão. Small vessel, usually with multiple rowers or small engines.
brejos. Wetlands.
buchita (*Calliandra trinerva*). An aquatic vine common in the Upper Amazon that often clogs the surface of small rivers.
caboclos. Acculturated Indians (early meaning); backwoods populations, hicks (later).
cafuz. Person of mixed black and Indian ancestry.
cascarilheiro. Quinine gatherer.
caucheiro. Extractor of latex from *Castilla* trees. The term can refer to either the workers themselves or the barons.
caudillo. Military strongman.

cholas. Mixed-blood populations of white and native from Peru.
chuncho. Uncatechized, "wild," or savage Indian.
Code Noir. Code defining the conditions of slavery—its rights, obligations, and sanctions—in the French colonial empire. It defined the terms of liberty of various kinds of slaves, restricted the activities of free blacks, required Roman Catholicism as the defining religion, and ordered all Jews out of France's colonies. See *Wikipedia*'s helpful checklist in its "Code Noir" entry.
comísariá. Checkpoint, customs house, and police station.
comissário. Head of a commission or chief of party. Da Cunha uses this term often when referring to Alexandre Buenaño.
cordel. Illustrated chapbook.
coronais. Landowners and militia leaders.
currais. Literally, corrals or cattle camps, but applied during the Northeastern droughts to refugee camps.
cushma. Woven cotton robe, similar to a Roman toga, sewn at the shoulders and sides; widely used by the Campa Indians and Shipibo of the Upper Amazon.
Directoria. Indian labor depot village.
drogas do sertão. Amazon backland essences and animal products—for example, sarsaparilla (thought to be a cure for syphilis), turtle butter, manioc, dried fish, oils, and herbs.
estirão, pl. **estirões.** Bayou.
estrada. Rubber tapping trail.
feitoria. Processing warehouse, factory.
flagelados. Literally "scourged ones"—used in reference to refugees from the Northeast.
furo. Side channel.
garimpo. Placer gold mine.
Heliconias. Widespread tropical genus related to bananas, best known for the bird of paradise flower.
igapó. Flooded forest.
infieles. "Wild" Indians, called infidels because they hadn't converted to Christianity.
itaúba. Heavy wood highly resistant to water damage and insect attack, widely used for large canoes in Amazonia.
jagunços. Fighters of the Northeastern outback.
juquiri. Semiaquatic plant of the floodplains.
latifundia. Large estate.
mameluco. Brazilian racial category: a person born to a white father and an Indian mother.
manchal. Dense stand of a given tree species, here *caucho* trees.
mateiro. Skilled backwoodsman.
mita. Tribute of eight years' labor supplied by native villages.
montaña. Forest.
montária. Small canoe, the "horse" or mount of the Amazon, because it was so necessary for getting around.
muchachos. Arana's child warriors—the "boys."
palenque. Spanish term for *quilombo*, or runaway slave community.
papiri. Palm-thatched hut or lean-to.

paramo. Vegetation formation between 3000 and about 5000 m altitude. Largely formed by glacier retreat, with a mix of lakes, peat bogs, and wet grasslands intermingled with shrublands and forest patches.
puno. High-altitude grassland vegetation of the Andes.
selva. Tropical forest.
seringa. *Hevea* rubber (we get our English term *syringe* from this Portuguese word).
seringal. Rubber estate.
seringalista. Owner of a rubber estate.
seringueiro. Rubber tapper.
sernambi. Low-quality latex scrapings.
sesmaria. Royal land grant.
tambo. Lean-to or hut of latex workers.
tapuyas. "Tame" Indians.
terra nullius. No-man's-land. This category, originating in Roman law, applies to a territory that has never been subject to any state and can be acquired through occupation or political decree.
tuxauá. Headman, usually one with shamanic powers.
ubá. Light small canoe, usually less than ten feet in length with a very shallow draw.
uti possedetis. Roman legal tenet: who has, keeps.
varadouro. Forest trail connecting watersheds, headwaters, or tributaries of the Amazon. Some were pre-Columbian, and many were developed in the rubber period to improve the speed of movement of products, to avoid custom houses, and for military purposes during the many guerrilla wars.
varzea. Floodplain ecosystem.
zambo. Mixed blood.

Notes

Chapter 1

1. Positivist scholar, professor at the Praia Vermelho Military Academy, and framer of the new Republic's first constitution.

2. See Bartelt, *Nation Gegen Hinterland*. Galvão, *No calor da hora*.

3. De Queiroz, *Messianismo e conflito social*; Diacon, "Bringing the Countryside Back In"; Diacon, *Millenarian Vision, Capitalist Reality*; Giumbelli, "Religion and Social (Dis)Order."

4. Introduction to the 1905 edition of *Os Sertões*. Also Borges, "Puffy, Ugly, Slothful and Inert"; Burns, "Destruction of a Folk Past."

5. The best summary of this "Tropicalist" school of Brazilian literature is Ventura, *EstiloTropical*.

6. Needell, *Tropical Belle Époque*; Aguiar and Leite, *Civilização e exclusão*; Bello, *Inteligência do Brasil*.

7. Da Cunha, *"Plano de uma crusada."*

8. Davis, *Late Victorian Holocausts*; Quinn, Neal, and Demayolo, "El Niño Occurrences over the Past 4½ Centuries."

9. In this regard see Markham, *"Fifty Years Work of the Royal Geographical Society"*; and Herdon and Gibbon, *Exploration of the Valley of the Amazon*, among many others.

10. Conrad, "Geography and Some Explorers."

11. Betts, *"Scramble" for Africa*; Comaroff and Comaroff, *Of Revelation*

and Revolution; Gifford and Louis, *France and Britain in Africa*; Pakenham, *Scramble for Africa*; Wesseling and Pomerans, *Divide and Rule*.

12. da Rocha, *O Acre*; Méndez and Tribunal Arbitral Boliviano-Brasileño, *Defensa de los derechos de Bolivia*; Ourique, *O Amazonas e o Acre*.

13. Burns, *Unwritten Alliance*; Pontes, *Euclides da Cunha*.

14. Ferreira, *A epopéia bandeirante*; Monteiro, *Negros da terra*; Morse, *Bandeirantes*.

15. Da Cunha will describe Urbano, who appears as guide and muse in the travels of Agassiz and Agassiz: *Journey in Brazil*; in Chandless, "Ascent of the River Purus"; and in James, *Brazil through the Eyes of William James*. Da Cunha also describes the Manaus administrator Silva Coutinho: Daniel, *Tesouro descoberto no máximo Rio Amazonas*; Rodrigues da Cunha, *O naturalista Alexandre Rodrigues Ferreira*.

16. Da Cunha, letter to Veríssimo, January 12, 1905.

17. Seed, *American Pentimento*; Seed, *Ceremonies of Possession*. Also see Pratt, *Imperial Eyes*.

18. See Weinstein, *Amazon Rubber Boom*; Pearson, *Rubber Country of the Amazon*; Woodroffe, *Upper Reaches of the Amazon*.

19. See da Cunha, "Land without History," chap. 13 below.

20. Burnett, *Masters of All They Surveyed*; Rivière, *Absent-Minded Imperialism*.

21. Da Cunha, "Impressions Completely New to Me," chap. 12 below, p. ΔΔΔ.

22. Da Cunha to Veríssimo, March 18, 1905, Manaus.

23. Some of the Amazon essays in *À margem* have been recently translated by Ronald Sousa under the title *The Amazon: Land without History*.

24. See chapter 13 below.

25. Da Cunha, "Inferno Verde," in *Paraíso Perdido*, ed. Tocantins.

26. Abdala Júnior and Alexandre, *Canudos*; Amory, *Euclides da Cunha*; Bernucci and da Cunha, *A imitação dos sentidos*; Bernucci da Cunha, *"Os Sertões"*; Fernandes and Gaudenzi, *O clarim e a oração*; de Santana, *Ciência e arte*; do Nascimento, *"Os Sertões" de Euclides da Cunha*; Rabello, *Euclides da Cunha*; Ventura, de Santana, and Carvalho, *Retrato interrompido da vida de Euclides da Cunha*.

Chapter 2

1. Rabello, *Euclides da Cunha*.

2. Stepan, *Rise of the Brazilian Sciences*.

3. Da Costa, *Da monarquía à república*; de Carvalho, *A formação das almas*; de la Vega, *Evolucionismo versus positivismo*.

4. 4 Barman, *Citizen Emperor*; Schwarcz, *Emperor's Beard*. Also see da Costa, *Brazilian Empire*; da Costa, *Da monarquía à república*; Needell, *Party of Order*.

5. See Schultz, *Tropical Versailles*.

6. Barman, *Citizen Emperor.*

7. Bethell, *Brazil*; Graham, *Patronage and Politics*.

8. Da Costa, *Brazilian Empire*.

9. Graham, *Patronage and Politics*.

10. Ibid.

11. Needell, *Tropical Belle Époque*.

12. Da Costa, *Da monarquía à república*.

13. See Needell, *Tropical Belle Époque*.

14. Schultz, *Tropical Versailles*.

15. See Needell, *Party of Order*.

16. See for example Agassiz and Agassiz, *Journey in Brazil*.

17. Hartt and Agassiz, *Thayer Expedition*; see also James, *Brazil through the Eyes of William*

James, Agassiz, *Journey in Brazil*, and Pedro, *O imperador do Brasil e os seus amigos de Novo Inglaterra*.

18. See Barman, *Princess Isabel of Brazil*.

19. There is an immense literature that addresses Brazilian slavery, but see Bastide, *African Religions of Brazil*; Butler, *Freedoms Given, Freedoms Won*; Naro, *Blacks, Coloureds and National Identity*; Bethell, *Abolition of the Brazilian Slave Trade*; Reis and Gomes, *Liberdade por um Fio*; Schwartz, "Black Latin America"; and the many Portuguese references cited in the bibliography, especially those of Reis and Gomes.

20. Eltis et al., *Trans-Atlantic Slave Trade*; Eltis, Lewis, and Sokoloff, *Slavery in the Development of the Americas*.

21. Lovejoy, *Transformations in Slavery*.

22. See Conrad, *World of Sorrow*; Eltis, *Rise of African Slavery in the Americas*; Heywood, *Central Africans and Cultural Transformations*; Lovejoy and Trotman, *Trans-Atlantic Dimensions of Ethnicity*; Freyre, *The Masters and the Slaves*; Karasch, *Slave Life in Rio de Janeiro*; Mattoso, *To Be a Slave in Brazil*; Monteiro, *Negros daTerra*; Reis and Gomes, *Liberdade por um fio*; Vainfas, *A heresia dos índios* and *Ideologia e escravidão*.

23. Costigan et al., *Diálogos da conversão*; Hemming, *Amazon Frontier*; Hemming, *Red Gold*; Langfur, "Myths of Pacification"; Lavalle, "Frontiers, Colonization and Indian Manpower"; MacLachlan, "Indian Labor Structure" in Alden, *Colonial Roots of Modern Brazil*; Restall, *Beyond Black and Red*; Schwartz and Langfur, "Tapahuns, Negros da Terra and Curibocas" in Restall, *Beyond Black and Red*.

24. Santos Granero, *Vital Enemies*.

25. The most comprehensive is Hemming, *Red Gold*. Recent historiography places this colonial process more explicitly in resistance and negotiation. See Langfur, *Forbidden Lands*; Restall, *Beyond Black and Red*; Treece, *Exiles, Allies and Rebels*.

26. Denevan, "Native Population of Amazonia"; also see *The Native Population of the Americas in 1492*; Mann, *1491*.

27. For details of this resistance, no better chronicler can be found than Hemming, *Amazon Frontier*. Also see Fausto, *Inimigos fiéis*; Langfur, *Forbidden Lands*; Langfur, "Myths of Pacification"; Whitehead, "Native Peoples Confront Colonial Regimes," in Salomon and Schwartz, eds., *Cambridge History of Native Peoples of the Americas*; Wright and da Cunha, "Destruction, Resistance, and Transformation," in Salomon and Schwartz, eds., *Cambridge History of Native Peoples of the Americas*. See also references in n30.

28. See Eltis et al., *Trans-Atlantic Slave Trade*; Eltis and Richardson, *Routes to Slavery*; Falola and Childs, *Yoruba Diaspora*; Mattoso, *To Be a Slave in Brazil*; Sweet, *Recreating Africa*.

29. Linebaugh and Rediker, *Many-Headed Hydra*; Price, *First-Time*; Sweet, *Recreating Africa*.

30. Barickman, *Bahian Counterpoint*; Brandão, *O escravo na formação social do Piauí*; Brazil, *Fronteira negra*; Carney and Rosomoff, *In the Shadow of Slavery*; dos Reis, *Histórias de vida familiar e afetiva de escravos*; Heywood, *Central Africans and Cultural Transformations*; Karasch, *Slave Life in Rio de Janeiro*; Lovejoy, *Transformations in Slavery*; Lovejoy and Trotman, *Trans-Atlantic Dimensions of Ethnicity*; Silva, *Um rio chamado Atlántico*.

31. Carney, *Black Rice*; Voeks, *Sacred Leaves of Candomblé*.

32. Falola and Childs, *Yoruba Diaspora*; Reis, *Rebelião escrava no Brasil*; Reis, "Revolution of the Ganhadores"; Sweet, *Recreating Africa*.

33. Hecht, "Quilombos and the Hidden History of Amazonia."

34. Reis, "Revolution of the Ganhadores"; Gomes, "Outras paisagens coloniais," in Gomes, *Nas terras do Cabo Norte*; Gomes, "'Safe Haven.'"

35. Brazil, *Fronteira negra*; Fiabani, *Mato, palhoça e pilão*; Graden, *From Slavery to Freedom in Brazil*; Huggins, *From Slavery to Vagrancy in Brazil*; Johnson, *The Chattel Principle*; Lavergne,

"Quilombismo—the Case of Cafundó,"; Mendonça, *Cenas da abolição*; Reis and Gomes, *Liberdade por um fio*; Santos, *A Balaiada e a insurreição de escravos no Maranhão*.

36. Bethell, *Abolition of the Brazilian Slave Trade*.

37. Schwarcz, *Spectacle of the Races*.

38. Beattie, *Tribute of Blood*; Castro and de Araújo, *Militares e política na nova República*; Kraay, *Race, State, and Armed Forces*; McCann, *A nação armada*; Sodré, *História militar do Brasil*.

39. Gomes, "Entre fronteiras e limites"; Gomes, "Outras paisagens coloniais"; Gomes, "Safe Haven."

40. Beattie, *Tribute of Blood*.

41. Ibid.

42. De Moraes, Costa, and de Oliveira, *A tutela militar*; Hayes, *Armed Nation*; Smallman, *Fear and Memory*.

43. Kraay, *Race, State, and Armed Forces*; Kraay and Whigham, *I Die with My Country*; Leuchars, *To the Bitter End*; Marques and Bethell, *A Guerra do Paraguai*; Salles, *Guerra do Paraguai*.

44. See Kraay and Whigham, *I Die with My Country*. See also Marques and Bethell, *A Guerra do Paraguai*; Whigham, *The Paraguayan War*.

45. Bastide and Fernandes, *Brancos e negros em São Paulo*; Fernandes, *A Integração do negro à sociedade de classes*; Fernandes, *Os Militares como categoria social*.

46. Sarmiento and Ross, *Facundo*.

47. Marques and Bethell, *A Guerra do Paraguai*.

48. See Kraay and Whigham, *I Die with My Country*. and Leuchars, *To the Bitter End*.

49. Cerqueira, *Reminiscências da Campanha do Paraguai*.

50. Burton, *Letters from the Battlefields of Paraguay*.

51. dos Santos, *Guerreiros e Jesuítas na Utopía do Prata*; Ganson, *Guaraní under Spanish Rule*; Kern, *Missões: Uma utopia política*.

52. See description in Leuchars, *To the Bitter End*.

53. Fragoso, *História da Guerra entre a Tríplice Aliança e o Paraguai*.

54. See Whigham, *Paraguayan War*; Marques and Bethell, *A Guerra do Paraguai*.

55. Hemming, *Red Gold*.

56. See the debate in *Latin American Research Review* 34, no. 1 (1999): 174–86; Reber, "Demographics of Paraguay."

57. Fragoso, *História da Guerra entre a Tríplice Aliança e o Paraguai*.

58. See Whigham and Potthast, "Paraguayan Rosetta Stone."

59. See Beattie, *Tribute of Blood*; also Kraay and Whigham, *I Die with My Country*.

60. Lemos, "Benjamin Constant," in Kraay and Whigham, *I Die with My Country*..

61. Conrad, *Struggle for the Abolition of the Brazilian Slave Trade*; Toplin, *Abolition of Slavery in Brazil*; Soares, *O militarismo na República*.

62. Beattie, *Tribute of Blood*.

63. Castro and Soares, *Militares e política na Nova República*, report that Spencer and Haeckel were broadly studied in the military curricula of the 1880s and the influence of these thinkers was widespread.

64. See Beattie, Kraay, Stepan, Hahner, among others, for a more detailed discussion of the internal dynamics of the military units.

65. Amory, "Euclides da Cunha and Brazilian Positivism."; Beattie, *Tribute of Blood*.

66. De Carvalho, *A Formação das almas*; Romero, *O evolucionismo e positivismo no Brasil*.

67. See Amory, "Euclides da Cunha and Brazilian Positivism." Also see Amory, *Euclides da Cunha*; Hale, *Transformation of Liberalism*; Lemos, *Benjamin Constant*.

68. Hahner, *Civilian-Military Relations in Brazil*.

69. Da Cunha, "Em Tourno da Vida de Euclides da Cunha."

70. Rondon and de Viveiros, *Rondon conta sua vida.*

71. Ventura, de Santana, and Carvalho, *Retrato interrompido da vida de Euclides da Cunha.*

72. Pontes, *Vida dramática de Euclides da Cunha.* Also Rabello, *Euclides da Cunha.*

73. Ventura, Santana, and Carvalho, *Retrato interrompido da vida de Euclides da Cunha.*

Chapter 3

1. Da Cunha, "Maréchal de Ferro," in *Obra completa.*

2. Da Cunha, letter to Lucio de Mendonça, 1904.

3. Rabello, *Euclides da Cunha..*

4. Da Cunha, "Maréchal de Ferro."

5. Rabello, *Euclides da Cunha.*

6. Da Cunha, letter to Tulia Ribeiro, January 7, 1894, Rio de Janeiro.

7. Da Cunha, letter to Solon Ribeiro, January 10, 1895, São Paulo.

8. Da Cunha, letter to Tulia Ribeiro, January 7, 1894.

9. Della Cava, *Miracle at Juazeiro*; Pessar, *From Fanatics to Folk*; Pessar, "Millenarian Movements in Rural Brazil."

10. De Queiroz, "Messianic Myths and Movements"; de Queiroz, *Messianismo no Brasil*; de Queiróz, *Radicais da República*; Metcalf, "Millenarian Slaves?"; Pessar, "Millenarian Movements in Rural Brazil"; Risério, *Utopía brasileira*; Urbano and Beascoechea, *Utopía, messianismo y milenarismo*; Walker, "Canudos Revisited."

11. Da Cunha, *Os Sertões.*

12. Galvão, *No calor da hora.*

13. Ventura, Barreto de Santana, and Carvalho, *Retrato interrompido.*

14. Auzel, "The Vendée in History"; Clemanceau, *Histoire de la Guerre*; Paret, *Internal War and Pacification*; Ventura, "Nossa Vendéia"; Woell, *Small-Town Martyrs and Murderers.*

15. Sampaio, a geographer, early ethnographer, and geologist of the Northeast, had enormous impact on the American geologists Hartt and Derby, and on da Cunha. Barreto de Santana, *Ciência e arte*; Barreto de Santana, "Natural Science and Brazilian Nationality"; Sampaio, *Rio São Francisco*; Ventura, Barreto de Santana, and Carvalho, *Retrato interrompido.*

16. Da Cunha, *A nossa Vendéia.*

Chapter 4

1. The name *favela* as a term for an urban settlement comes from the Canudos soldiers who camped on the hills opposite the Ministry of War in Rio de Janeiro waiting for their very tardy pay, mimicking their unhappy vigil over Canudos.

2. Da Cunha, *Rebellion in the Backlands*, 83.

3. Hastenrath, "Circulation and Teleconnection Mechanisms of Northeast Brazil Droughts"; Davis, *Late Victorian Holocausts.* For the contemporary impacts of drought and migration, see Scheper-Hughes, *Death without Weeping.* Also see Arons, *Waiting for Rain.*

4. Arons, *Waiting for Rain*; Quinn, Neal, and Demayolo, "El Niño Occurrences over the Past 4½ Centuries."

5. Barickman, *Bahian Counterpoint.*

6. Von Spix, von Martius, and Lloyd, *Travels in Brazil.*

7. See Loureiro, *"Gazeta do Purús," scenas de uma épocha.*

8. See Barickman, *Bahian Counterpoint.*

9. Sampaio, *Rio de S. Francisco.*

10. Greenfield, *Realities of Images.*

11. Ibid.

12. Arons, *Waiting for Rain.*

13. Graham, "Another Middle Passage?" and Slenes, "Brazilian Internal Slave Trade," in Johnson and Brown, eds., *Chattel Principle*.

14. Slenes, "Brazilian Internal Slave Trade"; Slenes,"Comments on Slavery in a Non- export Economy."

15. Slenes, "Brazilian Internal Slave Trade." Also Conrad, *World of Sorrow*.

16. Slenes, "Comments on Slavery in a Non-export Economy."

17. Slenes, "Brazilian Internal Slave Trade."

18. Bruno, *Café e negro*; dos Reis, *Histórias de vida familiar e afetiva de escravos*; Johnson and Brown, eds., *Chattel Principle*.

19. Letter to Baron of Jeremoabo from Antero de Cirqueira Galo, January 23, 1897, in Sampaio, ed., *Canudos: Cartas para o Barão*.

20. Arruda, *Canudos: Messianismo e conflito social*; Braga-Pinto, *Promessas da história*; Burns, "Destruction of a Folk Past"; Della Cava, "Brazilian Messianism and National Institutions"; Pessar, *From Fanatics to Folk*; Pessar, "Millenarian Movements in Rural Brazil."

21. Rough chapbooks, a popular written form in the Sertão.

22. Hermann, ed., *Sebastianismo e sedição*; Serrão, *Do Sebastianismo ao socialismo*.

23. Clastres, *Land without Evil*; Del Giudice and Porter, *Imagined States*; Mesters and Suess, *Utopía cativa*; Pontes, *Doce como o Diabo*; Risério, *Utopía brasileira e os movimentos negros*.

24. Falola and Childs, *Yoruba Diaspora*; French, "Buried Alive" and *Legalizing Identities*; Lienhard, "Kalunga"; Metcalf, "Millenarian Slaves?"; Sweet, *Recreating Africa*; Vainfas, *Herésia dos Indios*.

25. This literature has been mostly treated as case studies, but syncretic messiahs abound. See for example Vanderwood's study of Tomochic, *The Power of God against the Guns of Government*; Brown and Fernández, *War of Shadows*; de Queiroz, "Messianic Myths and Movements"; Lanternari, *Religiões dos oprimidos*; Lindoso, *Utopía armada*; Milhou, "Messianic and Utopian Currents"; Risério, *Utopía brasileira*.

26. Giménez and Coelho, *Bahia indígena*; Wright, "Prophetic Traditions," in Hill and Granero, eds., *Comparative Arawakan Histories*; Wright-Rios, "Indian Saints and Nation-States."

27. Vanderwood, *Power of God against the Guns of Government*.

28. Mooney, *Ghost-Dance Religion*; Smoak, *Ghost Dances and Identity*.

29. Andrade, *História de interpretação de "Os Sertões"*; Arruda, *Canudos*; do Nascimento, *"Os Sertões" de Euclides da Cunha*; Facó, *Cangaceiros e fanáticos*; Moniz, *Canudos: Guerra social*; Moniz, *Canudos, a luta pela terra*; Silva, *Antônio Conselheiro*.

30. See Davis, *Late Victorian Holocausts*; Brown and Fernández, *War of Shadows*; Lindoso, *Utopía armada*; Metcalf, "Millenarian Slaves?"; Milhou, "Messianic and Utopian Currents," in Schaer, Claeys, and Sargent, eds., *Utopía*; Pessar, *From Fanatics to Folk*; Pessar, "Millenarian Movements in Rural Brazil"; Risério, *Utopía brasileira*; Vainfas, *Confissões da Bahia*; Vainfas, *Herésia dos Indios*.

31. Barman, *Princess Isabel of Brazil*; Daibert Junior, *Isabel*.

32. Galvão, *No calor da hora*.

33. Especially important here are the studies of the *cordel*, which carried striking political commentary: Arons, *Waiting for Rain*; Calasans da Silva, *Canudos na literatura de cordel* and *No tempo de Antônio Conselheiro*; Facó, *Cangaceiros e fanático*; Pontes, *Doce como o Diabo*; Slater, "Representations of Power in Pilgrim Tales of the Northeast."

34. Da Cunha, *Rebellion in the Backlands*, 143.

35. Barickman, *Bahian Counterpoint*; Barman, "Brazilian Peasantry Re-examined"; Joffily, *Quebra-Quilo*; Kraay, "As Terrifying as Unexpected"; Millet, *Quebra-Quilos e a crise de lavoura*; Souza, *Sabinada*.

36. Caldeira, *Mutirão*; Galvão, *Mutirão no Nordeste*; Wagner de Almeida, "Terra do Preto."

37. Dantas makes the point that agrarian crisis in the Northeast was probably much more

extreme than is usually recognized, especially with drought. Also see Levine's *Vale of Tears*. Chandler, *Feitosas and the Sertão dos Inhamuns*; de Mello, *Norte agrário e o império*; Monteiro, *Nordeste insurgente*.

38. Brito, *Abolição na Bahia*; Cardoso, *Escravo ou camponês?*; de Azevedo, *Onda negra, medo branco*; Gomes, *Experiências atlânticas*; Graden, *From Slavery to Freedom in Brazil*; Schwartz, *Slaves, Peasants, and Rebels*.

39. Levine, "Canudos in the National Context."

40. Arons, *Waiting for Rain*; Greenfield, *Realities of Images*; Pessar, *From Fanatics to Folk*.

41. Letter to Dantas Martins from Aristides Borges, February 4, 1894, in Sampaio, ed., *Canudos: Cartas para o Barão*.

42. Jose Américo to Baron of Jeremoabo, February 10, 1894, in Sampaio, ed., *Canudos: Cartas para o Barão*.

43. Lovejoy and Trotman, *Trans-Atlantic Dimensions of Ethnicity*.

44. Barickman, *Bahian Counterpoint*.

45. Barbosa, *Negros e quilombos em Minas Gerais*; Bergad, *Slavery*; Higgins, *"Licentious Liberty"*; Machado Filho, *Negro e o garimpo em Minas Gerais*; Reis, "Revolution of the Ganhadores."

46. Mott, "Indios e a pecuaria nas fazendas de gado no piauí."

47. See Hecht, "Cattle Ranching in the Amazon"; Rivière, *Forgotten Frontier*; Cabral, *Caminhos do gado*; Wilcox, "Cattle and Environment"; Wilcox, "Law of the Least Effort."

48. Barros, *Derradeira gesta*; Facó, *Cangaceiros e fanáticos*; Fontes, *Lampião na Bahia*; Freixinho and Olinto, *Sertão arcaíco do Nordeste do Brasil*; Machado, *Táticas de guerra dos cangaceiros*; Johnson, "Subalternizing Canudos."

49. Scrub thorn forests.

50. It is not my intention to review the debates over social banditry in "prepolitical" forms of resistance in different historical moments. But see Blok, "The Peasant and the Brigand"; de Carvalho, *Lampião e a sociologia do cangaço*; Singelman, "Political Structure and Social Banditry Northeast"; and the classic on this topic, Hobsbawm, *Social Bandits and Primitive Rebels*.

51. Da Cunha, *Rebellion in the Backlands*, 153.

52. Langfur, "Moved by Terror"; Schwartz and Langfur, "Tapahuns, Negros da Terra and Curibocas," in Restall, *Beyond Black and Red*.

53. See Hemming, *Red Gold*; Barickman, "Tame Indians, Wild Heathens and Settlers"; da Cunha, *História dos indios no Brazil*; Puntoni, *Guerra dos bárbaros*.

54. De Queiroz, *Messianismo no Brasil*; Giumbelli, "Religion and Social (Dis)Order"; Vainfas, *Herésia dos Indios*. See also Ferrari, *Os Kariri*; Lowie, "The Cariri," in *Handbook of South American Indians*; Siqueira, *Cariris do Nordeste*.

55. Barickman, "Bit of Land Which They Call Roça"; Facó, *Cangaceiros e fanáticos*; Pires, *Guerra dos bárbaros*; Schwartz, "Indian Labor and New World Plantations."

56. See Hemming, *Die If You Must*; *Red Gold*; Langfur, "Myths of Pacification."

57. Costigan et al., *Diálogos da conversão*; Eisenberg, "Cultural Encounters, Theoretical Adventures"; Neves, *Vieira e a imaginação social jesuítica*; Cohen, *Fire of Tongues*.

58. David, *Inimigo invisível*.

59. See Ferrari, *Os Kariri*; Mascarenhas, "Toda nação em Canudos." The Kariris had been allying themselves with anti-Brazilian movements for most of a century by the time of the last days of Canudos.

60. Reis, *Slave Rebellion in Brazil*.

61. Mascarenhas, "Toda nação em Canudos."

62. Abdala and Alexandre, *Canudos*; Della Cava, "Brazilian Messianism and National Institutions"; Levine, "Mud-Hut Jerusalem"; Levine, *Vale of Tears*; Moniz, *Canudos: Guerra social*; Moniz, *Canudos, a luta pela terra*; Nogueira, *Antônio Conselheiro e Canudos*; Sola, *Canudos*.

63. Miller, *Way of Death*; Thornton, *Warfare in Atlantic Africa*.

64. Thornton, *Warfare in Atlantic Africa.*

65. This literature is extensive, but see Anderson, "*Quilombo* of Palmares"; Péret and Henriquez, *Quilombo de Palmares*; Schwartz, "Black Slaves in Palmares"; Fiabani, *Mato, Palhoça e Pilão*; Gomes, *Hidra e os pântanos*; Moura, *Quilombos e a rebelião negra.*

66. Hill, *History, Power, and Identity*; Restall, *Beyond Black and Red*; Schwartz, "New Peoples and New Kinds of People," in Salomon and Schwartz, eds., *Cambridge History of Native Peoples of the Americas*; Schwartz and Langfur, "Tapahuns, Negros da Terra and Curibocas"; Whitehead, "Black Read as Red," in Restall, ed., *Beyond Black and Red.*

67. Schwartz, "Cantos e quilombos numa conspiração de escravos haussas," in Reis, ed., *Liberdade por um Fio.*

68. Why people revolt (or do not) is one of the central questions of Latin American social history. Slave uprisings and *quilombos* are very dynamic research areas within Brazilian historiography. See Reis, *Slave Rebellion in Brazil*; Gomes, *Hidra e os pântanos*; Moura, *Quilombos e a rebelião negra*; Moura, *Quilombos*; Silberling, "Displacement and *Quilombos.*"

69. Scott, *Moral Economy of the Peasant*; Scott, *Weapons of the Weak.*

70. Price, *First-Time.*

71. Barickman, "Bit of Land Which They Call Roça"; da Cunha and Gomes, *Quase-cidadão*; Gomes, *Histórias de quilombolas.*

72. Landes, *City of Women*; Merrell, *Capoeira and Candomblé*; Voeks, *Sacred Leaves of Candomblé.*

73. Araújo dos Anjos, *Territórios das communidades*; Barbosa, Silva, and Silvério, *De preto a afro-descendente*; Barbosa, *Negros e quilombos em Minas Gerais*; French, "Buried Alive in the Northeast"; Gomes, *Hidra e os pântanos*; Moura, *Quilombos na dinâmica social do Brasil*; Reis and Gomes, *Liberdade por um fio.*

74. Wagner de Almeida, "Terra do Preto."

75. Scott, *Moral Economy of the Peasant.*

76. Da Cunha notes that after mass the *sertanejos* would go off to a *terreiro* for observance of more rousing Candomblé rituals. See *Rebellion in the Backlands*, 112.

77. Treccani, *Territórios quilombolas.*

78. Marin and Castro, *Negros do Trombetas*; Coudreau, *Voyage au Trombetas.*

79. Silberling, "Displacement and *Quilombos.*"

80. Dantas, *Fronteiras moveçidas.*

81. Sampaio, *Canudos: Cartas para o Barão.*

82. Da Cunha, *Rebellion in the Backlands*, 143.

83. Ibid., 161.

84. See Sampaio, *Canudos: Cartas para o Barão*, especially the letter from Jemoabo's cousin José Americo (February 28, 1894): "those miserable people—everyone who was a slave, all the criminals from every province . . . it's he (Antonio) who makes the laws, creates armies and does what ever he pleases."

85. Barbosa, Silva, and Silvério, *De preto a afro-descendente*; Fiabani, *Mato, palhoça e pilão;* Gomes, *Hidra e os pântanos*; Moura, *Quilombos na dinâmica social do Brasil.*

86. Acevedo Marin and Castro, *No caminho de Pedras de Abacatal*; Treccani, "Diferentes caminhos para a resgate dos territórios quilombolas"; Treccani, *Territórios quilombolas.*

87. Della Cava points out the institutional differences and compromising positions that were taken by Padre Cicero at Joazeiro to conform to regional oligarchies. Also see the transformation of backland "saints' from radicalizing icons to expressions of "folklore." Pessar, *From Fanatics to Folk*; Pessar, "Millenarian Movements in Rural Brazil."

88. See Barreto, *Última expedição a Canudos*; Galvão, *No calor da hora*; and of course *Os Sertões.*

89. Letters to the Baron of Jeremoabo routinely complain about the "Judas Iscariots" in

their midst and the surprising identities of those Canudos sympathizers. Sampaio, *Canudos: Cartas para o Barão*.

90. Da Cunha, *Rebellion in the Backlands*, 148.

91. Price, *Maroon Societies*; Stedman, Price, and Price, *Narrative of a Five Years Expedition*; Gomes, *Hidra e os pântanos*; Gomes, "Safe Haven"; Karasch, "Quilombos do ouro na Capitania de Goias," in Reis and Gomes, eds., *Liberdade por um fio*.

Chapter 5

1. The lore of the time and the very humorous *urubú* songs were captured in collections by Bahian folklorist and historian Calasans Brandão da Silva: *Canudos na literatura de cordel* and *Quase biografias de jagunços*.

2. The number of actual men involved in the attack is somewhat confused. Impressments were yielding up young children, and most of the men were not trained or ready for combat. See Araripe, *Expedições militares contra Canudos*; Galvão, *No calor da hora*; Levine, *Vale of Tears*; Macedo, *Belo Monte*; Neto, "Canudos na Boca do Povo"; Villela Júnior, *Canudos*.

3. Coordenação do Novo Movimento Histórico de Canudos, *Noventa anos depois*; Galvão, *No calor da hora*; Néry and da Silva, *Quarta expedição contra Canudos*.

4. Da Cunha, *Rebellion in the Backlands*, 444.

5. Da Silva, "Jagunçinho de Euclides da Cunha," cited in Ventura, *Retrato interompida;* da Cunha, *Rebellion in the Backlands*, 407–8.

6. Da Cunha, *Rebellion in the Backlands*, 426.

7. Walnice Galvão's collection of newspaper articles that covered the fourth expedition is indispensable for contextualizing da Cunha's writings. See Galvão, *No calor da hora*.

8. Da Cunha, *Rebellion in the Backlands*, 464.

9. Ibid., 473.

10. Ibid., 464.

11. Ibid., 439.

12. Ibid., 444.

13. Ibid., 475.

14. Ibid., ΔΔΔ??.

Chapter 6

1. The most important analyst of these forays remains Reis, *Amazônia e a cobiça internacional* and *Limites e demarcações na Amazônia Brasileira*.

2. See for example Rio Branco, *Questões de limites*.

3. Betts, *"Scramble" for Africa*; Pakenham, *Scramble for Africa*; Wesseling and Pomerans, *Divide and Rule*.

4. In his comparison of de la Condamine and Rodrigues Ferreira, Mauro Coelho points out that for the latter Iberian scientific exploration was explicitly about its political ends, while de la Condamine (at least publicly) was more concerned with Enlightenment models of abstract science. See Coelho, "Viagens filosóphicas de Charles-Marie de la Condamine e Alexandre Rodrigues Ferreira." On European scientific exploitation, see for example Drayton, *Nature's Government*; Gifford and Louis, *France and Britain in Africa*; Miller and Reill, *Visions of Empire*; Mills and Barton, *Drugs and Empires*.

5. Daniel, *Tesouro descoberto no máximo Rio Amazonas*.

6. See da Cunha, "The Real and Mythological Geography of the Purús," chap. 20 below.

7. The British Royal Geographical Society ran a training school in survey and fielded numerous mappers and surveyors, including the Schomburgk brothers, Chandless, and Fawcett. One of the central functions of military schools in Latin America (and especially Praia Vermelha) was to develop a cadre of geographers and mappers. See Driver, *Geography Militant*.

8. The importance of Amazonian products like quinine certainly facilitated imperial expansion elsewhere. See Webb, *Humanity's Burden*; McNeill, *Mosquito Empires*.

9. Burnett, *Masters of All They Surveyed*; Cortesão, *História do Brasil nos velhos mapas*. Cortesão masterfully shows the territorial construction of Brazil through its imaginative cartographers who progressively pushed the Tordesillas line farther to the west. This idea is also developed in Costa, *História de um país inexistente*.

10. Burnett, *Masters of All They Surveyed*. Also Rivière, *Absent-Minded Imperialism*.

11. See Cleary, "Lost Altogether to the Civilised World."

12. Cortesão, *História do Brasil nos velhos mapas*; Davidson, "How the Brazilian West Was Won."

13. Cabeça de Vaca, *Naufragios*.

14. See Cortesão, *História do Brasil nos velhos mapas*.

15. Monteiro, *Negros da terra*; Morse, *Bandeirantes*.

16. Heckenberger, Petersen, and Neves, "Of Lost Civilizations and Primitive Tribes." Also see Buarque de Holanda, *Monsões*.

17. Karasch, "Quilombos do Ouro"; Toral, "Indios Negros ou os Carijó de Goiás."

18. See Block, *Mission Culture on the Upper Amazon*.

19. Davidson, "How the Brazilian West Was Won."

20. Cortesão, *História do Brasil nos velhos mapas*; de Gusmão and Cortesão, *Alexandre de Gusmão*.

21. Cortesão , *Missão dos padres matemáticos*.

22. Botelho Gosálvez, *Proceso del imperialismo del Brasil*; Davidson, "How the Brazilian West Was Won"; Varela Marcos, Vernet Ginés, *Tratado de Tordesillas*.

23. See Tambs, "Rubber, Rebels, and Rio Branco."

24. Fifer, "Bolivia's Boundary with Brazil"; Tocantins, *Formação histórica do Acre*.

25. Da Cunha's writings on this topic involve eight newspaper articles from 1906 that were later republished (and almost instantly translated into Spanish): da Cunha, *Peru versus Bolívia*.

26. Benton, *Law and Colonial Cultures*.

27. Moore, *Principles of American Diplomacy*.

28. Burns, *Unwritten Alliance*.

29. Quevedo dos Santos, *Guerreiros e Jesuítas*; Ganson, *Guaraní under Spanish Rule*.

30. Seed, *Ceremonies of Possession*.

31. Ibid.

32. Davidson, "How the Brazilian West Was Won."

33. These regions were famously contested for centuries. See Benton, *Law and Colonial Cultures*; Besouchet, *Rio-Branco e as relações entre o Brasil e a Républica Argentina*; Dominguez, Schmidel, and Cabeza de Vaca, *Conquest of the River Plate (1535–1555)*; Severo Zeballos, *Diplomacia desarmada*.

34. Burns, *Unwritten Alliance*.

35. These included Brazilian conflicts with Peru, Bolivia, Paraguay, Guyana, French Guyana, Colombia; Suriname with French Guyana, Venezuela with Colombia; Colombia with Ecuador and Peru; Peru with Colombia, Ecuador, Bolivia, Brazil.

36. Barboza, *Cartografia política do Barão*; Burns, *Unwritten Alliance*; de Abranches, *Rio Branco e a política exterior*.

Chapter 7

1. Boomert, "Between the Mainland and the Islands"; Boomert, "Gifts of the Amazons"; Ferguson and Whitehead, *War in the Tribal Zone*; Whitehead, "Native Peoples Confront Colonial Regimes"; Whitehead, "Recent Research."

2. Boomert, "Arawak Indians"; Boomert, "Gifts of the Amazons"; Nimuendaju, *In Search of a Lost Amazon;* Rostain, "Archaeology of the Guianas"; Whitehead, *Lords of the Tiger Spirit.*

3. Vidal, "Kuwe Duwakalumi"; also "Amerindian Cartography."

4. Roosevelt, *Mound Builders of the Amazon*; Schaan, "Recent Investigations on Culture, Marajó Island, Brazil"; Meggers and Evans, *Archaeological Investigations at the Mouth of the Amazon*; McEwan, *Unknown Amazon.*

5. Whitehead, "Carib Ethnic Soldiering"; Whitehead, "Native Peoples Confront Colonial Regimes."

6. The most famous "swamp rebels" were the Cottica rebels of the mangroves of Suriname who regularly threatened Paramaibo, its capital, in the 1760s. Also see Gomes, "Safe Haven," and *A hidra e os pântanos.*

7. Von Humboldt, *Personal Narrative of Travels to the Equinoctial Regions.*

8. McNeill, *Mosquito Empires.*

9. Ferguson and Whitehead, *War in the Tribal Zone*; Vidal, "Amerindian Cartography"; Whitehead, "Carib Ethnic Soldiering"; Whitehead, "Native Peoples Confront Colonial Regimes."

10. Galvin, *Patterns of Pillage*; Lane, *Pillaging the Empire*; Latimer, *Buccaneers of the Caribbean*; Rediker, *Villains of All Nations.*

11. Williams, *Brazil and French Guiana.*

12. Boxer, *Dutch in Brazil*; Emmer, *Dutch in the Atlantic Economy*; Maurits et al., *Johan Maurits Van Nassau-Siegen.*

13. Tyacke, "English Charting of the River Amazon"; Reis, *Território do Amapá.*

14. Reis, *Amapá: Perfil histórico*; Gomes and de Queiroz, "Between Frontiers and Limits."

15. Freyre, *Luso e o Tropico*; Schwartz and Langfur, "Taphuans, Negros da Terra and Curibocas"; Cavasini et al.,"Duffy Blood Group Gene Polymorphisms"; McNeill, *Mosquito Empires*; Mann, *1492.*

16. Costigan et al., *Diálogos da conversão*; Monteiro, *Negros da terra*; Schwartz and Langfur, "Tapahuns, Negros da Terra and Curibocas."

17. Figueroa and de Acuña, *Informes de Jesuitas en el Amazonas*; Hoornaert and Comisión de Estudios de História, *História da Igreja na Amazônia.*

18. Touchet, *Botanique et colonisation en Guyane Française*; Spary, "Of Nutmegs and Botanists."

19. Ellis, *Eighteenth-Century Coffee-House Culture*; Pendergrast, *Uncommon Grounds*; Hemming, *River of Trees.*

20. Instituto Brasileiro do Café, *O café no Brasil.*

21. Reis, *Portugueses e Brasileiros na Guiana Francesa*; Reis, *Território do Amapá*; Cardoso, *Guiana Francesa.*

22. Capistrano de Abreu. *Chapters of Brazil's Colonial History.*

23. Williams, "Brazil and French Guiana."

24. Capistrano de Abreu, *Chapters of Brazil's Colonial History.*

25. Bruleaux et al., *Deux siècles d'esclavage en Guyane Française*; Mam Lam Fouck, *Guyane Français*; Soublin, *Cayenne 1809.*

26. Gomes, *Terras de Cabo Norte;* Reis, *Portugueses e brasileiros na Guiana Francesa.*

27. Eltis and Richardson, *Extending the Frontiers*; Eltis and Richardson, *Atlas of the Transatlantic Slave Trade.*

28. Funes, "Mocambos do Trombetas"; Gomes, "Safe Haven"; Gomes and de Queiroz, "Between Frontiers and Limits."

29. Funes, "Mocambos do Trombetas"; Gomes, "Entre fronteiras e limites"; Gomes, "Fronteiras e mocambos;" Gomes, *Terras do Cabo Norte*; Gomes,"Outras paisagens coloniais"; Cardoso, *Economia e sociedad.*

30. The main Demerara uprising occurred in 1823, but there had been previous "troubles." See da Costa, *Crowns of Glory, Tears of Blood.*

31. See Stedman. *Expedition to Surinam*; Naipaul, *Loss of El Dorado*; da Costa, *Crowns of Glory, Tears of Blood.*

32. Hoogbergen, *Boni Maroon Wars in Suriname*; Rothschild, "Horrible Tragedy in the French Atlantic."

33. da Costa, *Crowns of Glory, Tears of Blood.*

34. Gomes, "Safe Haven."

35. Debbasch, "Crime d'empoisonnement"; Prance, "Poisons and Narcotics of the Amazonian Indians"; Savage, "'Black Magic' and White Terror." Poisoning by natives and slaves was emphasized in the Inquisitional visits to Pará. See Lapa, *Livro da visitação*; Stedman, *Narrative of a Five Years Expedition*; Gomes, *Terras do Cabo Norte*; Gomes, "Fronteiras e mocambos."

36. Stedman, *Narrative of a Five Years Expedition;* Gomes, *Nas terras do Cabo Norte*; Gomes, "Fronteiras E Mocambos".

37. Darrell Posey described ways that Kayapo Indians provided trekking foods, including planting in tree falls, along pathways, and around old camps. It is likely that similar strategies were used by maroons. See also Rival, *Trekking through History*; Funes, "Nasci na Matas nuca tive Senhor."

38. Guerreiro et al., "Genetical-Demographic Data"; Perna, Cardoso, and Guerreiro, "Duffy Blood Group Genotypes."

39. Kenneth Maxwell argues that the expulsion of Jesuits from the Amazon was in fact the "spark" that resulted in the more general expulsion of religious orders from European politics as well as colonial life. See Maxwell, "Spark."

40. Mukerji, "Dominion, Demonstration and Domination"; Spary, "Of Nutmegs and Botanists."

41. This family included Joseph, who traveled with de la Condamine and wrote the first definitive monograph on quinine. Other siblings managed the Jardin du Roi, the national botanical garden. See Plotkin, Boom, and Allison, "Ethnobotany of Aublet's *Histoire des plantes*."

42. Spary, "Of Nutmegs and Botanists"; Touchet, *Botanique et Colonisation.*

43. Plotkin, Boom, and Allison, "Ethnobotany of Aublet's *Histoire des plantes*."

44. Bruleaux et al., *Deux siècles d'esclavage en Guyane Française*; Lowenthal, "Colonial Experiments in French Guiana"; Mam Lam Fouck, *Guyane Français*; Michel, *Guyane sous l'Ancien Régime.*

45. Abbé Raynal in his *Histoire des Deux Indes* described the Kourou area as the Land of Candide. Choiseul corresponded with Voltaire on the topic of Kourou, and from this interaction perhaps formed his imaginary of an Enlightenment utopia in contrast to the grisly slave society observed by Candide. See Raynal, *Histoire philosophique et politique*; Voltaire and Raffel, *Candide, or Optimism*; Rothschild, "Horrible Tragedy in the French Atlantic."

46. Lowenthal, "Colonial Experiments in French Guiana."

47. Daubigny, *Choiseul et la France d'outre mer*, cited in ibid.

48. Quoted in Michel, *Guyane sous l'Ancien Régime.*

49. Lowenthal, "Colonial Experiments in French Guiana."

50. Exact numbers remain uncertain. See McNeill, *Mosquito Empires*; Michel, *Guyane sous l'Ancien Régime.*

51. Rothschild, "Horrible Tragedy in the French Atlantic."

52. Lowenthal, " Colonial Experiments"; Coudreau, *Français en Amazonie.*

53. MacLaughlin, "Indian Directorate"; Harris, *Rebellion in the Amazon.*

54. Cleary, "Lost Altogether to the Civilized World."

55. Hemming, *Amazon Frontier*; MacLachlan, "Indian Directorate"; Mendonça, *Amazônia na era Pombalina.*

56. Mendonça, *Amazônia na era Pombelina.*

57. Fernandes, *Missionários jesuítas no Brasil no tempo de Pombal*; Maxwell, "Spark"; Mendonça, *Amazônia na era Pombalina*; Sommer, "Cracking Down on the Cunhamenas"; Harris, *Rebellion in the Amazon.*

58. MacLachlan, "Indian Directorate" and "Indian Labor Structure in the Portuguese Amazon"; Harris, *Rebellion on the Amazon*; Bruleaux et al., *Deux siècles d'esclavage en Guyane Française*; Geggus, *Impact of the Haitian Revolution*; Martin and Favre, *De la Guyane à la diaspora africaine*; Mendonça, *Amazônia na era Pombalina*; Sommer, "Cracking Down on the Cunhamenas."

59. The colony formally lasted from 1769 to 1783.

60. Vidal, *Mazagão, la ville qui traversa l'Atlantique.*

61. Schiebinger, *Plants and Empire.*

62. Bertani et al., "Evaluation of French Guiana Traditional Antimalarial Remedies"; Jullian et al., "Validation of Use of a Traditional Antimalarial Remedy"; Lopes et al., "Antimalarial Use of Volatile Oil"; Vigneron et al., "Antimalarial Remedies in French Guiana."

63. Linebaugh and Rediker, *Many-Headed Hydra*; Price, *To Slay the Hydra.*

64. Geggus and Fiering, *World of the Haitian Revolution*; Linebaugh and Rediker, *Many-Headed Hydra*; Reis, *Invenção de liberdade*; Reis and Gomes, *Liberdade por um fio.*

65. Reis and Gomes, "Repercussions of the Haitian Revolution in Brazil." Also see Gomes, *Hidra e os pântanos*; Gomes and de Queiroz, "Between Frontiers and Limits"; Soublin, *Cayenne 1809.*

66. Bénot, *Guyane sous la Revolution Française*; Dubois, *Colony of Citizens*; Adelman, *Sovereignty and Revolution in the Iberian Atlantic*; Adélaïde-Merlande, *Caraïbe et la Guyane.*

67. See discussions in da Costa, *Crowns of Glory, Tears of Blood*, as well as the complaints of the Fructidor exiles: Anon., *History of the Revolution of the 18th Fructidor.*

68. Reis, *Território do Amapá.*

69. Bénot, *Guyane sous la Révolution Française*; Bruleaux et al., *Deux siècles d'esclavage en Guyane Française.*

70. García and Camacho, *Comunidades afrodescendientes*; Guedes, "L'insurrection negre de Coro"; Perez, "Journey to Freedom."

71. Bénot, *Guyane sous la Révolution Française*; Anon., *History of the Revolution of the 18th Fructidor*; Coggiola, *A Revolução Francesa e seu impacto na América Latina.*

72. Adélaïde-Merlande, *Caraïbe et la Guyane*; Bénot, *Guyane sous la Révolution Française*; Larue, *Histoire du dix-huit Fructidor*; Michelot, *Guillotine sèche.*

73. Anon., *History of the Revolution of the 18th Fructidor.*

74. Redfield, *Space in the Tropics.*

75. Amis des Archives d'Outre-Mer, *Terres de Bagne*; Redfield, *Space in the Tropics.*

76. Gomes, "Entre fronteiras e limites."

77. Marin and Castro, *Negros do Trombetas*; Coudreau, *Voyage au Trombetas*; Funes, "Mocambos do Trombetas"; Gomes, "Entre fronteiras e limites."

78. Castro, *Escravos e senhores de Bragança.*

79. Bediaga, "Joining Pleasure and Work."

80. Bénot, *Guyane sous la Revolution Française*; Soublin, *Cayenne 1809.*

81. Reis and Gomes, "Repercussions of the Haitian Revolution in Brazil."

82. For much more detail see Cleary, "Lost Altogether to the Civilised World"; Di Paolo, *Cabanagem*; Pinheiro, *Visões da Cabanagem*; Salles, *Memorial da Cabanagem*; Thorlby, *Cabanagem na fala do povo*; Harris, *Rebellion in the Amazon.*

83. The tension between local elites and sovereignty is discussed widely (although not for the Amazon) in Adelman, *Sovereignty and Revolution.*

84. Gomes, *Terras de Cabo Norte.*

85. Wallace, *Narrative of Travels.*

86. Petot, *L'or de Guyane*; Stroebel, *Gens de l'or.*

87. Price, "Liberdade, Fronteiras, e Deuses," in da Cunha and Gomes, *Quase-cidadão.*

88. Stroebel, *Gens de l'Or.*

89. Mam Lam Fouck, *Guyane Français.*

90. Reis, *Amazônia e a cobiça internacional.*

91. Coudreau, *France equinoxiale*; Coudreau, *Français en Amazonie.*

92. Thouar, *Explorations dans l'Amérique du Sud.*

93. Coudreau, *Chez nos Indiens.*

94. Coudreau, *France equinoxiale.*

95. Ibid.

96. Ibid.

97. Ibid.

98. Reis, *Território do Amapá.*

99. Coudreau, *France equinoxiale.*

100. Meira, *Fronteiras sangrentas*; Reis, *Território do Amapá.*

101. Reis, *Amazônia e a cobiça internacional.*

102. Goeldi, "Excavações archeológicas em 1895."

103. Cited in Reis, *Território do Amapá.*

104. Ibid.

105. Hentenryk, "Leopold II et la question de l'Acre."

106. Moore, "Franco-Brazilian Boundary Dispute"; Burr, "Guiana Boundary."

Chapter 8

1. Hanna and Hanna, *Confederate Exiles to Venezuela.*

2. United States et al., *Counter Case.*

3. Martin, "Influence of the United States"; Tavares Bastos, *Valle do Amazonas.*

4. Bueno, *Política externa da Primeira República*; Bueno, *República e sua política exterior*; Cervo and Bueno, *História da política exterior do Brasil*; Luz, *Amazônia para os negros americanos*; Reis, *Amazônia e a cobiça internacional*; Tocantins, *Formação histórica do Acre.*

5. See Comaroff and Comaroff, *Of Revelation and Revolution*; Driver and Martins, *Tropical Visions*; Hobsbawm, *Age of Empire*; Johannsen, Haynes, and Morris, *Manifest Destiny and Empire.*

6. The Christmas Bird Count is a long-standing program of the National Audubon Society. See www.audubon.org/Bird/cbc/.

7. Burnett, "Matthew Fontaine Maury's 'Sea of Fire,'" in Driver and Martins, eds., *Tropical Vision in an Age of Empire*, 113–36.

8. Maury was knighted in Russia, Denmark, Portugal, and Belgium; he received national medals of merit from Prussia, France, Austria, the Netherlands, Sweden, Norway, and Sardinia, and honorary doctorates from Cambridge, North Carolina, and Columbia. His Austrian awards and friendships proved especially useful for him later in life.

9. Herndon and Gibbon, *Exploration of the Valley of the Amazon.*

10. Dozer, "Matthew Maury's Letter of Instruction."

11. Ibid.

12. See Horne, *Deepest South.*

13. Emerson, "Young America"; Widmer, *Young America.*

14. Wilentz, *Rise of American Democracy.*

15. Harrison, "Young Americans"; Widmer, *Young America.* The political ramifications are described in Eyal, *Young America Movement*; Hettle, *Peculiar Democracy.*

16. See chapter 13 in this volume, "Land without History."

17. Da Cunha, see chapter 13.

18. Maury, *Amazon and the Atlantic Slopes of South America.*

19. Maury, "Commercial Prospects for the South"; Maury, "Great Commercial Advantages"; Maury, "Our Gulf States and the Amazon."

20. Herndon and Gibbon, *Exploration of the Valley of the Amazon.*

21. Graham Burnett reviews the meaning of the tropics as a physical engine for the movement of currents. Burnett, "Matthew Fontaine Maury's 'Sea of Fire,'" in Driver and Martins, *Tropical Vision in an Age of Empire.*

22. DeBow, "South American States."

23. May, *Manifest Destiny's Underworld*; May, *Southern Dream of a Caribbean Empire.*

24. Fredrickson, *Black Image in the White Mind.*

25. Maury, "Letter to Mrs William Blackford," in *Maury Papers VI.* Cited in Bell, "Relation of Herndon and Gibbon's Exploration."

26. Maury, "Letter to Ann Maury."

27. Maury, "Letter to Mrs William Blackford."

28. Dozer, "Matthew Maury's Letter of Instruction."

29. Ibid

30. Maury, *Amazon and the Atlantic Slopes of South America.*

31. Dozer, "Matthew Maury's Letter of Instruction."

32. Ibid.

33. Guyot, *Earth and Man*; Maury, *Amazon and the Atlantic Slopes of South America.*

34. Guyot, *Creation*; Guyot, *Earth and Man.*

35. Maury, "Our Gulf States and the Amazon."

36. But see Fredrickson, *Black Image in the White Mind*; Johnson, *Soul by Soul.*

37. Bates, *Naturalist on the River Amazons.*

38. Mansfield, *Paraguay, Brazil and the Plate.*

39. Pitts, *Turn to Empire*; Seed, *American Pentimento*; Vattell, Chitty, and Ingraham, "Law of Nations."

40. Maury, "Commercial Prospects for the South."

41. Dozer, "Matthew Maury's Letter of Instruction."

42. Maury used a pseudonym, "Inca," for a series of letters and articles published in the *National Intelligencer* and Union newspapers.

43. Herndon and Gibbon, *Exploration of the Valley of the Amazon.*

44. Dozer, "Matthew Maury's Letter of Instruction"; Maury, *Amazon and the Atlantic Slopes of South America.*

45. Herndon and Gibbon, *Exploration of the Valley of the Amazon.*

46. Your author ascribes her safety and survival in adventures in the eastern Amazon to a Grulla mule with tiger stripes named Comprida, who also eventually had to be returned to muleteers.

47. Many explorer accounts dwell on the formidable rapids and falls that impeded their travels, but no one has captured the drama as well as novelist Marcio Souza, *Emperor of the Amazon.*

48. Gibbon was at pains to note the importance of his black boatmen in getting through the rapids. The region, as later twentieth-century experts of Amazonian colonization in Rondonia would ultimately note, was famously malarial. The prevalence of black boatmen and black colonization on the Madeira may well have reflected the benefits of the sickle cell gene. See Sawyer and Sawyer, *Malaria on the Amazon Frontier.* For a more general discussion also see McNeill, *Mosquito Empires.*

49. Herndon and Gibbon, *Exploration of the Valley of the Amazon*, 234.

50. Block, *Mission Culture on the Upper Amazon*; Denevan, *Aboriginal Cultural Geography*; Erickson, "Artificial Landscape-Scale Fishery in the Bolivian Amazon"; Erickson, "Domesticated

Landscapes," in Balée and Erickson, eds., *Time and Complexity in Historical Ecology*; F. E. Mayle et al., "Long-Term Forest-Savannah Dynamics."

51. Block, *Mission Culture on the Upper Amazon*.

52. The confusion over boundary lines meant that many fugitive blacks moved into areas not technically free from slavery. The slaves who fled into to the west from the Madeira may well have thought they were in Bolivian territory.

53. See da Cunha, *Peru versus Bolívia*.

54. See Church, *Route to Bolivia*; Craig, *Recollections of an Ill-Fated Expedition*; Keller-Leuzinger, *Amazon and Madeira River*. Also see Foot-Hartmann, *Trem Fantasma*.

55. Hill, "Confederates in Middle America."

56. Exemplary in this regard was Dunn, *Brazil, the Home for Southerners*—a kind of travelogue about places and forests and reviewing terrains for Southern colonists.

57. The Dawseys devote chapters to the Baptists and Methodists as significant organizing elements of the immigrant communities: Dawsey and Dawsey, *Confederados*; Weaver, "Confederate Immigrants." The earlier proselytizer was Daniel Kidder, and his volume also had considerable influence on migrants. See Kidder, *Brazil and the Brazilians*.

58. The entrance and settlement of US citizens has periodically provided a justification for invading a country. See Graham-Yooll, *Imperial Skirmishes*; Hill, *Roosevelt and the Caribbean*; Leonard, *United States–Latin American Relations*; Rivas, *Missionary Capitalist*.

59. Dawsey and Dawsey, *Confederados*; Hill, "Confederates in Middle America"; Rolle, *The Lost Cause*.

60. Dunn, *Brazil, the Home for Southerners*.

61. Hastings, *Immigrants Guide to Oregon and California*; Hastings, *New Description of Oregon and California*.

62. The Donner party was a group of immigrants who were stranded by sets of blizzards in the Sierra. Those who survived did so by cannibalism. Mark Twain provided a blackly humorous description of the event in *Roughing It*.

63. Hastings was not entirely foolish. In siting the colony at Santarém, he had settled his party on one of the largest outcroppings of *terra preta*, the highly fertile, anthropogenic Amazon dark earths. His descriptions of high yields may not have been entirely a function of his own hyperbole. See Petersen, Neves, and Heckenberger, "Gift from the Past," in McEwan, Barreto, and Neves, eds., *Unknown Amazon*; Guilhon, *Confederados em Santarém*; Harter, *Lost Colony of the Confederacy*. Guilhon states that the first and largest rubber plantations were planted there by *confederados* and that the performance of this plantation (some 10,000 trees) helped convince Henry Ford of the viability of rubber for the creation of his huge rubber estate, Fordlândia. Dean, *Brazil and the Struggle for Rubber*; Grandin, *Fordlândia*.

64. Dawsey and Dawsey, *Confederados*; Guilhon, *Confederados em Santarém*.

65. Grandin, *Fordlândia*; Guilhon, *Confederados em Santarém*; Hecht and Cockburn, *Fate of the Forest*.

66. See for example Hartt and Agassiz, *Thayer Expedition*; James, Agassiz, and Pedro, *Imperador do Brasil*.

67. Dean, *Brazil and the Struggle for Rubber*; Jackson, *Thief at the End of the World*.

68. Werlich, *Admiral of the Amazon*.

69. Orton, *Andes and the Amazon*.

70. Twain, "Turning Point of My Life," in Paine, *Favorite Works of Mark Twain*.

71. Quoted in Richardson, *Messages and Papers of the Presidents*, 6:54–55.

72. Horne, *Deepest South*. Also see Bennett, *Forced into Glory*.

73. Bennett, *Forced into Glory*.

74. Francis Blair cited in Horne, *Deepest South*.

75. Cleven, "Some Plans for Colonizing."

76. May, *Manifest Destiny's Underworld.*

77. Yrisarri, who was acting foreign minister of El Salvador and Guatemala, was quite stern in his rejection, seeing in this colonization "the evils . . . in this form of colonization." See Cleven, "Some Plans for Colonizing."

78. Pomeroy, "Information."

79. Espino, *How Wall Street Created a Nation*; LaRosa and Mejía, *United States Discovers Panama*; Missal, *Seaway to the Future.*

80. Carter, *Road to Botany Bay*; Forster, *France and Botany*; Redfield, *Space in the Tropics.*

81. Webb, *Slavery and Its Tendencies.*

82. Eyal, *Young America Movement*; Fredrickson, *Black Image in the White Mind.*

83. See Eyal, *Young America Movement.*

84. These kinds of arrangements were quite common for state colonization in Brazil and in indentured labor more generally.

85. Webb, *On the Necessity of Supplying Brazil*, cited in Horne, *Deepest South.*

86. Guilhon, *Confederados em Santarém.* Also see comments in Smith, *Brazil, the Amazons and the Coast.*

87. See Markham's preface to Church's posthumously published *Aborigines of South America.*

88. See for example Church, *Interoceanic Communication on the Western Continent*; Church, *Route to Bolivia.*

89. Church, *Desiderata in Exploration.*

90. This number uses the 2009 GDP value of the 1870 pound relative to that of 2009.

91. Cited in Craig, *Recollections of an Ill-Fated Expedition.*

92. Church's own reports here are of great interest. The most entertaining description of the behavior of Bolivia in international markets is supplied by the first chapters of Neville Craig's memoir *Recollections of an Ill-Fated Expedition.*

93. Ibid.

94. Stols, "Belges au Mato Grosso et en Amazonie," in Dumoulin and Stols, *Belgique et l'Étranger.*

95. Tambs reported this, but Piper does not appear in the more thorough and probably more accurate list of Purús explorers elaborated by da Cunha.

96. In terms of the navigation rights on the Madeira, it appears that both Piper and Church had been given the same "monopoly." Selling overlapping rights seemed to be a habit, as we will see in the next chapter.

97. Colonization and Commercial Company of Bolivia and Church, *Bolivian Colonization.*

Chapter 9

1. Or, as Bolivians called it, Puerto Alonso.

2. Tocantins, *Formação histórica do Acre.*

3. Balf, *Darkest Jungle*; Ovidio Diaz, *J. P. Morgan, Teddy Roosevelt and the Panama Canal*; Espino, *How Wall Street Created a Nation*; Freehoff, *America and the Canal*; Lindsay-Poland, *Emperors in the Jungle*; Collin, *Theodore Roosevelt's Caribbean.*

4. *Provincia do Pará*, March 17, 1899.

5. The full text of the accord is in vol. 1 of Tocantins, *Formação histórica do Acre.*

6. Tocantins, *Formação histórica do Acre.*

7. Ibid.; Santos-Granero and Barclay, *Tamed Frontiers.*

8. Barbosa, *Imprensa*, June 10, 1899.

9. *Provincia do Pará*, June 5, 1899.

10. Cited in Tocantins, *Formação histórica do Acre.*

11. Hay generally held the view that South American diplomats were, as he so colorfully put

it, "Dago ambassadors from powerless, insignificant countries." See Espino, *How Wall Street Created a Nation*.

12. Hecht and Cockburn, *Fate of the Forest*. A novelized version of the events is described by Amazonian writer Marcio Souza, *Emperor of the Amazon*.

13. Fifer, "Bolivia's Boundary with Brazil"; Fifer, "Empire Builders"; Pacheco, *Integración económica y fragmentación social*.

14. Tambs, "Rubber, Rebels and Rio Branco."

15. Tocantins, *Formação histórica do Acre*.

16. See Cashman, *America in the Gilded Age*.

17. Espino, *How Wall Street Created a Nation*; Lindsay-Poland, *Emperors in the Jungle*.

18. Bandeira, "O Barão de Rothschild."

19. De Wiart, *Leopold II*.

20. Stols, *Les belges au Mato Grosso*.

21. Ascherson, *King Incorporated*; Hochschild, *King Leopold's Ghost*.

22. Panama involved a commercial contract for construction; the country was later taken over by the US government, which then more or less had sovereignty for a hundred years. See Collin, *Theodore Roosevelt's Caribbean*; Espino, *How Wall Street Created a Nation*; Lindsay-Poland, *Emperors in the Jungle*.

23. Costa du Rels, *Félix Avelino Aramayo y su época*.

24. Stols, "Investiments belges aux Brasil."

25. Hentenryk, "Leopold II et la question.de Acre."

26. Bandeira, "Barão de Rothschild e a questão de Acre"; Tocantins, *Formação histórica do Acre*.

27. Hay to Bridgman, letter, cited in vol. 2 of Tocantins, *Formação histórica do Acre*.

28. Moore, *Principles of American Diplomacy*.

29. Rio Branco, communiqué to Olinto de Margalhães, June 12, 1902.

30. August Plane, *L'Amazonie*.

31. Bridgman. Documents pertaining to Bolivian Syndicate: The Acre Territory. Documents concerning the Controversy between Brazil and Bolivia over a contract made with American Citizens.

32. De Assis-Brasil, "Report to Olinto Magalhães," cited in Tocantins, *Formação histórica do Acre*.

33. Bryan to Hay, May 6, 1902, cited in Bradford Burns, *Unwritten Alliance*.

34. Tambs, "Rubber, Rebels and Rio Branco." Also see Bueno, *Política externa da Primeira República*; Cervo and Bueno, *História da política exterior*; Posada Carbó, *Wars, Parties and Nationalism*.

35. Cabral, *Plácido de Castro*; de Castro, *Estado independente do Acre*.

36. Tocantins, *Formação histórica do Acre*.

37. Cabral, *Plácido de Castro*; da Rocha, *Acre*; Nobre, *Epopéia acreana*; Ribeiro, *Acre e os seus heroes*; Ricardo, Brazil, and Bolivia, *Tratado de Petrópolis*.

38. Aguirre Achá, *De los Andes al Amazonas*; de Castro, *Estado independente do Acre*; Fernandez, *Campaña del Acre*.

39. Moore, *Brazil and Peru Boundary Question*; Moore, *Principles of American Diplomacy*.

40. See Bandeira, "Barão de Rothschild."

41. This would be equivalent to almost 200 million pounds in today's currency, and a very large sum at the time. The region was, however, generating hundreds of millions in annual revenue.

42. Ibid.; Ricardo, Brazil, and Bolivia, *Tratado de Petrópolis*; Tambs, "Rubber, Rebels and Rio Branco"; Tocantins, *Formação histórica do Acre*.

43. Fawcett's self-publicized exploits until his death in the "wilds of Mato Grosso" have

been a durable source of Amazonian best-sellers. The most recent entry into this pantheon is Grann, *Lost City of Z*. A delightful antidote to this genre is Fleming, *Amazon Adventure*, whose "tutelary deity was Burlesque."

44. Farquar's fiscal disaster is discussed in detail in Hecht and Cockburn, *Fate of the Forest*.

Chapter 10

1. See chapter 14.
2. Huber, "Observações sobre as árvores de borracha"; Schultes, "Amazon Indian and Evolution in *Hevea* and Related Genera."
3. Huber, "Observações sobre as árvores de borracha."
4. Weinstein, *Amazon Rubber Boom*.
5. Serier, *Les barons du caoutchouc*; Domingues and Gomez, *Economia extractiva en la Amazonia colombiana*.
6. Santos-Granero and Barclay, *Selva central*.
7. Gow, *Of Mixed Blood*; Hill, *Rethinking History and Myth*; Murphy, *Rubber Trade and the Mundurucú Village*.
8. Pineda Camacho, *Holocausto en el Amazonas*; Taussig, *Shamanism, Colonialism, and the Wild Man*; Valcárcel, *Proceso de Putumayo y sus secretos inauditos*.
9. Casaverde, *Viajeros al infierno verde*.
10. Collier, *The River That God Forgot*; Pineda Camacho, *Holocausto en el Amazonas*; Rey de Castro, *Escándalos del Putumayo*; Santos-Granero and Barclay, *Tamed Frontiers*; Taussig, *Shamanism, Colonialism, and the Wild Man*; Woodroffe, *Upper Reaches of the Amazon*.
11. Casement and Mitchell, *Amazon Journal of Roger Casement*.
12. Arana and Select Committee on Putumayo, *Cuestiones del Putumayo*; Gómez, Lesmes, and Rocha, *Caucherías y conflicto colombo-peruano*; Lagos, *Arana, rey del caucho*; Larrabure y Correa, *Perú y Colombia en el Putumayo*; Rey de Castro, Gray, and Chirif, *Defensa de los caucheros*.
13. Casement, *Amazon Diaries*.
14. Iribertegui, *Amazonas, el hombre y el caucho*; Lagos, *Arana, rey del caucho*; Linden, *Por las tierras del caucho*; Pennano A., *Economía del caucho*.
15. García Jordán, *Cruz y arado, fusiles y discursos*.
16. Da Cunha, *Peru versus Bolívia*.
17. Da Cunha, *Cuestión de límites entre Bolivia y el Perú*.
18. Moore, *Brazil and Peru Boundary Question*.
19. Treaty of Commerce Navigation and Limits, signed in Lima, 1852.
20. Moore, *Brazil and Peru Boundary Question*.
21. See Santos Graneros, *Tamed Frontiers*; Pennano, *Economía del caucho*; Casement, *Amazon Journals*; Gómez, Lesmes, and Rocha, *Caucherías y conflicto colombo-peruano*.
22. Loureiro, *Gazeta do Purús*; Tocantins, *Formação histórica do Acre*.
23. Tocantins, *Formação histórica do Acre*.
24. Gow, "Canção Purús."
25. Scharff has several cameo appearances in the tales of the rubber economy of the Madre de Dios. See Huertas Castillo, *Indigenous Peoples in Isolation*; Casaverde, *Viajeros del infierno verde*; and of course the Amazon writings of Euclides da Cunha.
26. Flores-Martin, *Explotación de caucho en el Perú*. Scharff was eventually killed when his native workers rebelled against him.
27. Loureiro *Gazeta do Purús*
28. Da Cunha also describes the capture of natives. See chapter 21, "Among the *Caucheiros*."
29. Lange, *In the Amazon Jungle*.
30. Tocantins, *Formação histórica do Acre*.
31. Da Cunha to Domicio da Gama, November 16, 1907, Rio de Janeiro.

Chapter 11

1. Da Cunha to Machado Assis, Santos, Februrary 15, 1904.

2. Pontes, *Vida dramática de Euclides da Cunha*.

3. See Ventura, Barreto de Santana, and Carvalho, *Retrato interrompido*.

4. Ibid.

5. These include the essays "Contrastes e confrontos," "Conflito inevitável," "Contra os caucheiros," and "Entre a Maderia e a Javari" in da Cunha, *Contrastes e confrontos*.

6. Euclides had made comments about the tropical climate in Mato Grosso and Pará in *Os Sertões*, but these were by way of a regional geographic background for the Sertão.

7. Rio Branco pressured for rapid publication essays in a small volume that could be easily distributed. See da Cunha to da Gama, November 18, 1905.

8. This conflict persisted throughout the twentieth century and was finally resolved only in 1998.

9. Needell, *Tropical Belle Époque*.

10. Ibid.

11. Burns, *Nationalism in Brazil*; Burns, *Unwritten Alliance*.

12. These included Getulio Vargas; see Garfield, "Continent Apart from the Rest of the World"; Golbery de Couto e Silva, *Geopolítica do Brasil* and *Geopolítica e poder*; Freyre, *Perfil de Euclydes e otros perfís*.

13. Freyre notes that several of the Baron's top diplomatic posts were given to prominent writers: Oliveira Lima, Assis-Brasil, da Gama, and Joaquim Nabuco, as well as da Cunha.

14. Hecht, "Last Unfinished Page of Genesis."

15. Da Cunha is referring to forests of the Amazon.

16. Freyre, *Order and Progress*; Freyre, *Perfil de Euclydes e otros perfís*.

17. In nineteenth- and twentieth-century colonial models, nomadism was more primitive than settled agriculture, and an emblem of barbarism. The idea of extraction and nomadism thus places the Peruvian national character at a lower order than the settled agriculture of Brazilian development. Historical ecologists have critiqued this theme: Cormier, "Between the Ship and the Bulldozer"; Rival, "Amazonian Historical Ecologies"; Rival, *Trekking through History*. Also: Whitehead, *Histories and Historicities in Amazonia*.

18. C. Weiner was a diplomat to Chile in the 1880s who intimated the existence of Machu Picchu before its discovery by Hiram Bingham. Markham, *Incas of Peru*.

19. Maldonado met his end in the rapids of the Rio Madeira in 1861.

20. Breman et al., *Imperial Monkey Business*; Dickens, *Social Darwinism*.

21. By 1905, about half the population of São Paulo was composed of white immigrants, whose culture (as well as genomes) would, in da Cunha's view, swamp the true Brazilian culture. See Graham et al., *Idea of Race in Latin America*; Schwarcz, *Spectacle of the Races*; Skidmore, *Black into White*; Stepan, *Hour of Eugenics*.

22. Tavares Bastos helped a great deal with the logistics of the Agassiz and Morgan expeditions to the Amazon.

23. Maury never traveled in the Amazon but based his book on Herndon. Maury, *Amazon and the Atlantic Slopes of South America*.

24. The Brazilian names are those of the governors and administrators. Silva Coutinho is especially important because of his travels on the Purús.

25. Viscount James Bryce was a liberal British diplomat who was posted in the United States and had written in 1888 a well-known book, *The American Commonwealth*, about US development.

26. Garfield, "Continent Apart from the Rest of the World"; Craig, *Recollections of an Ill-Fated Expedition*; Ferreira, *Ferrovía do diabo*.

27. Da Cunha to Luis Cruls, Lorena, February 20, 1903.
28. Da Cunha to Vicente de Carvalho, Guarujá, April 24, 1904.
29. Da Cunha to Coelho Neto, Rio de Janeiro, April 22, 1904.
30. Da Cunha to Veríssimo, Guarujá, June 24, 1904.
31. Burns, *Unwritten Alliance*.
32. Da Gama, "Euclides da Cunha."
33. Da Cunha to Veríssimo, Rio de Janeiro, June 6, 1904.
34. Da Cunha to Manuel da Cunha, Rio de Janeiro, October 10, 1904.

Chapter 12

1. Da Cunha to Manuel da Cunha, Manaus, December 30, 1904.
2. Da Cunha, "Amazônia: A gestão do mundo" (fragment of his speech given at his inception into the Brazilian Academy of Letters); da Cunha and Tocantins, *Um paraíso perdido*.
3. He is referring here to the endless portraits of Alexander von Humboldt.
4. A suburb of Rio de Janeiro.
5. The Araceae are probably best known to temperate zone readers through the house plant known as philodendron.
6. The local names that da Cunha uses in the text are unfamiliar to most temperate zone readers but he felt the metropole should learn the language of the tropics. In that spirit I have retained them, but a glossary at the end of the book will also help clarify. His use of the popular plant names and his attention to them echo his use of plant names as metaphors in *Os Sertões* and creates a more general evocation of place and authenticity. These local names are more rhythmic and beautiful words than their translations:

aninga: *Philodendron speciosum*
canaranas: aquatic grasses, mainly a native Amazonian *Paspalum,* often forming floating islands
aturias: the leguminous shrubs *Machaerium lunatum* and *Drepanocarpus ferox*
burití: *Mauritia flexuosa*, one of the most common and widely used palms throughout Amazonia

7. Da Cunha to Veríssimo, Manaus, January 13, 1905.
8. The Christie Question refers to the 1864 blockading of Rio de Janeiro by Britain as a means of interfering with Brazil's slave trade. This came at an especially inconvenient moment of the Paraguayan War.
9. Da Cunha to Veríssimo, Manaus, February 2, 1905.
10. Da Cunha to Artur Lemos, Manaus, no month/day, 1905.
11. Da Cunha to Veríssimo, Manaus, March 10, 1905
12. Da Cunha to Oliveira Lima, Manaus, January 16, 1905.
13. The governor of Amazonas, Eduardo Ribeiro.
14. Da Cunha to Coelho Neto, Manaus, March 10, 1905.
15. Letter to Veríssimo, Manaus, March 2, 1905.
16. See chapter 17.
17. The activist bishop of Belém, who supported justice for native populations and laborers. In many ways he was a liberation theologian *avant la lettre*.
18. A geographer of the Upper Amazon. See da Cunha, *Peru versus Bolívia*. One of the great explorers who for over forty years in the late eighteenth century explored the upper Xingu and the Guaporé.
19. Lacerda e Almeida (1750–97) formed part of the demarcation team established by the Treaty of Idelfonso. An astronomer, a geographer, and possibly Brazil's first hydrologist, he

provided some of the earliest precise location measurements of the Amazon River. He died in Mozambique.

20. A fashionable Rio neighborhood.

21. Letter to Rangel, March 20, 1905, Manaus.

22. Angelica Ratto testimony in Galvão and Silva Neto, *Crónica de uma tragédia*, 68–72.

23. Ana Solon da Cunha in ibid., 126.

24. As part of urban sanitation efforts and the modernization of Rio de Janeiro, the republic had embarked on a major public health vaccination initiative. Resistance to the measures involved some questions about the vaccination itself, but also the idea that women inoculated in their homes might have to endure a breach of modesty and would be somehow be dishonored—a kind of Freudian interpretation of vaccination. See Needell, "Revolta contra Vacina of 1904."

25. See chaps. 20 and 21 below.

26. The annals of tropical exploration do not lack for examples of gay or bisexual men who fled the constraints of their societies. Iconographic in this regard are Alexander von Humboldt and the famously notorious Roger Casement. See Casement and Mitchell, *Amazon Journal of Roger Casement*; Casement and Sawyer, *Roger Casement's Diaries*; Singleton-Gates and Girodias, *Black Diaries*. Though not so "out," Clements Markham, Orville Derby, and Richard Spruce, along with the bisexual Richard Burton, add to this distinguished list. Certainly the tropics have provided a liberational exploration of the senses in imagery and practice for every orientation. After all, the durable image of the tropical vacation is a beach and people with almost no clothing. See Littlewood, *Sultry Climates*; McClintock, *Imperial Leather*; Mott, *Homossexuais da Bahia*.

Chapter 13

1. Da Cunha, "General Observations," in *À margem da historia*.

2. Da Cunha, "Um rio abandonado."

3. Two contradictory essays and other incoherent elements in *À margem*, organized in the last months of his life, suggests that he was unraveling.

4. Orme, "American Geomorphology at the Dawn of the 20th Century"; Orme, "Rise and Fall of the Davisian Cycle of Erosion."

5. Orme, "American Geomorphology at the Dawn of the 20th Century."

6. Barreto de Santana, *Ciência e arte*.

7. Anderson, "Charles Lyell, Uniformitarianism, and Interpretive Principles"; Gould, *Time's Arrow, Time's Cycle*; Virgili, "Charles Lyell and Scientific Thinking in Geology"; Wilkinson, "Ecology before Ecology."

8. Katzer, *Grundzüge der Geologie*.

9. Kalliola et al., "Upper Amazon Channel Migration"; Rasanen et al., "Recent and Ancient Fluvial Deposition Systems."

10. Campbell, Frailey, and Romero-Pittman, "Pan-Amazonian Ucayali Peneplain"; Kalliola et al., "Upper Amazon Channel Migration"; Neller, Salo, and Rasanen, "On the Formation of Blocked Valley Lakes"; Pitman et al., "Tree Species Distributions"; Toivonen, Mäki, and Kalliola, "Riverscape of Western Amazonia."

11. Godoy, Petts, and Salo, "Riparian Flooded Forests"; Kalliola et al., "Upper Amazon Channel Migration"; Rasanen et al., "Recent and Ancient Fluvial Deposition Systems."

12. Da Cunha, "Relatorio de Comissão Mixta Brasileiro-Peruana."

13. Gould, *Structure of Evolutionary Theory*; Hecht, "Last Unfinished Page of Genesis."

14. Gilbert, "New Light on Isostasy."

15. See map in chapter 16.

16. Fragment of the essay from the introduction to Alberto Rangel's Book *Inferno verde*, published in 1907.

17. Katzer, *Grundziige der Geologie.*

18. The Agassiz expedition, funded by J. P. Morgan.

19. Maspero (1846–1916) was the preeminent Egyptologist of his time, and if Herodotus took on the East in his review of civilizations, Maspero can be seen as the historical ethnographer of the southern Mediterranean.

20. Da Cunha's observations are supported by modern research: Rossetti and de Toledo, "Biodiversity from a Historical Geology Perspective"; Rossetti et al., "Quaternary Tectonics in a Passive Margin"; Rossetti, Valeriano, and Thales, "Abandoned Estuary within Marajó Island."

21. See Cohen et al., "Wetland Dynamics of Marajó Island"; Rossetti et al., "Quaternary Tectonics in a Passive Margin."

22. Smith, *Brazil, the Amazons and the Coast.*

23. The town of Obidos is situated at the narrowest part of the Amazon's channel.

24. This is the second part of "General Observations" from *À margem da história.*

25. Here da Cunha is making reference to the "mathematical" Jesuits who were meant to determine with celestial as well as more earthbound techniques the boundaries for the Treaty of Madrid. See chap. 7 above.

26. Buscalione was a botanist who developed one of the major European collections of Amazon flora and is well known for the first major botanical description of *Castilloa*, the *caucho* tree. He was an early promoter of an international research institute for the central Amazon, an analogue to the Goeldi museum, rather like today's Institute for Amazonian Research (INPA). See Daly and Millozza, "'Lost' Plant Collections from the Amazon."

27. Gaspar Barleus (1584–1648) was the biographer of Johan Mauritz (1630–54) and chronicler of his administration in Brazil. His observations of the colonial period, including the endless slave and native wars, prompted his cynical aphorism.

Chapter 14

1. Barham and Coomes, *Prosperity's Promise*; Evans, Buckland, and Lefer, *They Made America*; Kadir, "Natural Rubber."

2. Huber, "Observações sobre as arvores de borracha da região Amazônica."

3. Barham and Coomes, *Prosperity's Promise*, point out that there was quite a bit of mobility in many tapping systems, a position also argued by Weinstein, *Amazon Rubber Boom*. For what it was like to sit out the rainy season, few authors are more evocative than Lange: *In the Amazon Jungle.*

4. These numbers are based on customs house export data from Loreto, Mollendo, Manaus, Putumayo, and Belém. There was quite a bit of "leakage" in these systems, so total volume may well be underestimated.

5. For a broader discussion of de la Condamine, see Safier, *Measuring the New World.*

6. De la Condamine's tropical travels were meant to discover the circumference of the earth but were famously ill starred. While his own movements were not especially fraught, the rest of his team suffered innumerable travails. His cartographer was swept down the Amazon and ended up in penury in the French colony of Amapá until rescued by his loyal and courageous wife. The story was novelized in 2004 in Whitaker, *The Mapmaker's Wife*. Condamine's botanist, Jussieu, wrote only one monograph on quinine before losing both his collections and his mind. See the travel documents: de la Condamine, *Voyage sur l'Amazone*; de la Condamine and Académie des sciences, *Journal du voyage.*

7. Fresneau described the latex of the "Mapá" tree (hence Amapá), "from which the

Portuguese make all kinds of useful and curious articles." But his real find was what he called the "Seriengue" tree, "which the Spanish call Cauchuc." This seems to have been its first description: the seeds were used for oil, and the sap, "which was copious," made waterproof items. Memorandum of François Fresneau (1747), cited in Schidrowitz and Dawson, *History of the Rubber Industry*.

8. See Clavijero and Cuevas, *Historia antigua de México*; Fernández de Oviedo y Valdés and López, *De la natural historia de las Indias*; López Austin, *Textos de medicina náhuatl*.

9. De las Casas and Griffin, *Short Account of the Destruction of the Indies*; de las Casas, O'Gorman, and Manrique, *Los indios de México y Nueva España*.

10. Anglerius, "De orbo novo petri anglerii decades octo." Also see Fernández de Oviedo y Valdés, *Historia general y natural de las Indias*.

11. D'Orbigny, *Voyage pittoresque dans les deux Ameriques*; Fernández de Oviedo y Valdés, *Historia general y natural de las Indias*; Myers, Scott, and Fernández de Oviedo y Valdés, *Fernández de Oviedo's Chronicle of America*.

12. Hosler, Burkett, and Tarkanian, "Prehistoric Polymers."

13. Tarkanian and Hosler, "America's First Polymer Scientists."

14. Personnal communication, Michael Heckenberger, July 14, 2006.

15. Fernández de Oviedo y Valdés, *Historia general y natural de las Indias*.

16. The ubiquity and antiquity of these games are reviewed in Stern, *Rubber Ball Games of the Americas*.

17. Sinkholes in Karst landscapes; most associated with Yucatán history.

18. Tarkanian and Hosler, "America's First Polymer Scientists."

19. De Landa, *Relación de cosas de la Yucatan,* 1566.

20. Terraciano, *Mixtecs of Colonial Oaxaca*. Other medicinal uses were noted by one of the earliest observers of native medicinal practices, Fra Bernardo de Sahagún: it was used as a curative incense or mixed with cacao and taken for dysentery. López Austin, *Textos de medicina náhuatl*.

21. Stone, "Spirals, Ropes, and Feathers."

22. Carreón Blaine, *El olli en la plástica mexica*; Stone, "Spirals, Ropes, and Feathers."

23. Tarkanian and Hosler, "America's First Polymer Scientists."

24. Stone, "Spirals, Ropes, and Feathers"; Whittington,, *Sport of Life and Death*.

25. Clavijero and Cuevas, *História antígua de Mexico*.

26. Santos, *História econômica da Amazônia*.

27. See Edwards, *Voyage up the River Amazon*.

28. *Heteroploidy* refers to the multiples of the basic number of chromosomes of a group. *Polyploidy* often produces larger fruits or unusual characteristics, for example in tomatoes.

29. A review by one of South America's experts on domestication hardly mentions tree crops at all: Pickering, "Domestication of Plants in the Americas."

30. Piperno and Pearsall, *Origins of Agriculture in the Lowland Tropics*.

31. Clement, "1492 and the Loss of Amazonian Crop Genetic Resources."

32. McKey et al., "Pre-Columbian Agricultural Landscapes, Ecosystem Engineers."

33. Arce-Nazario, "Human Landscapes Have Complex Trajectories"; Denevan, *Cultivated Landscapes*; Erickson, "Domesticated Landscapes"; Posey and Plenderleith, *Kayapó Ethnoecology and Culture*.

34. See Anderson, May, and Balick, *Subsidy from Nature*; Hecht, "Kayapó Savanna Management"; Peters et al., "Oligarchic Forests of Economic Plants in Amazonia"; Porro, "Palms, Pastures, and Swidden Fields"; Rival, "Amazonian Historical Ecologies."

35. Clement, "1492 and the Loss of Amazonian Crop Genetic Resources"; Clement, "Center of Crop Genetic Diversity."

36. Descola, *In the Society of Nature*.

37. Huber, "Observações sobre as árvores de borracha"; Seibert, "Study of *Hevea*"; Seibert, "Uses of *Hevea* for Food."

38. Plotkin, Boom, and Allison, "Ethnobotany of Aublet's *Histoire des plantes*."

39. Schultes, "Amazon Indian and Evolution in *Hevea* and Related Genera."

40. Cited in Seibert, "Uses of *Hevea* for Food."

41. These techniques are described in many publications: Clement, "1492 and the Loss of Amazonian Crop Genetic Resources"; Coomes and Burt, "Indigenous Market-Oriented Agroforestry"; Denevan et al., "Indigenous Agroforestry in the Peruvian Amazon"; Posey and Balée, *Resource Management in Amazonia*; Posey and Balick, *Human Impacts on Amazonia*; Whitehead, *Histories and Historicities in Amazonia*.

42. Schultes, "Amazon Indian and Evolution in *Hevea* and Related Genera."

43. Rival, "Amazonian Historical Ecologies"; Rival, *Trekking through History*. See also Posey and Balée, *Resource Management in Amazonia*.

44. Posey and Balick, *Human Impacts on Amazonia*.

45. Parssinen, Schaan, and Ranzi, "Pre-Columbian Geometric Earthworks in the Upper Purus."

46. De Carvajal et al., *Discovery of the Amazon*.

47. Simón, Bollaert, and Markham, *Expedition of Pedro de Ursúa & Lope de Aguirre*.

48. Heckenberger, Petersen, and Neves, "Of Lost Civilizations and Primitive Tribes, Amazonia"; Heckenberger et al., "Legacy of Cultural Landscapes in the Brazilian Amazon"; Lathrap, *Upper Amazon*; McEwan, *Unknown Amazon*; Schaan et al., "Geoglifos da Amazônia occidental."

49. Balée and Erickson, *Time and Complexity in Historical*; Hecht, "Indigenous Management"; Heckenberger et al., "Legacy of Cultural Landscapes in the Brazilian Amazon."

50. Miles Silman, personal communication.

51. Mann, "Ancient Earthmovers of the Amazon"; Fernando Santos-Granero, "Boundaries Are Made to Be Crossed."

52. Payne, *Amazonian Linguistics*.

53. Santa-Anna Néry, *Le pays des Amazones*.

54. Schidrowitz and Dawson, *History of the Rubber Industry*.

55. Edwards, *Voyage up the River Amazon*.

56. Ibid.

57. See Schidrowitz and Dawson, *History of the Rubber Industry*.

58. See Dean, *Brazil and the Struggle for Rubber*. See Santa-Anna Néry, *Le pays des Amazones*, where he cites Amazon references before de la Condamine noted the use of *caucho* (Gallicized as "caoutchouk"). "Father Manoel da Esperança . . . had found it in use among the Cambebas Indians" and he "called it they say by the singular name of 'seringa,' having remarked that these intelligent savages used it to make bottles and bowls in the form of a syringe. Thence came the name of '*syringers*' or '*seringueros*' by which the extractors of the milky sap are still known in Amazonia, and that of *seringais* given to the places where they extract this product by incision." Also see Edwards, *Voyage up the River Amazon*; Keller-Leuzinger, *Amazon and Madeira River*.

59. See chap. 20 below.

60. Santos, *História econômica da Amazônia*.

61. While the price collapsed, rubber production itself remained high until the 1920s.

62. Those involved in the upper reaches of the *Hevea* economy worried a great deal about the problems of overtapping. See Dean, *Brazil and the Struggle for Rubber*.

63. Hecht, "Last Unfinished Page of Genesis."

64. See Weinstein, *Amazon Rubber Boom*, and Nugent, *Amazonian Caboclo Society*.

65. Barham and Coomes, *Prosperity's Promise*; Barham and Coomes, "Wild Rubber"; Domínguez and Gómez, *Economía extractiva*; Pacheco B., *Integración económica y fragmentación*.

66. Nugent, *Amazonian Caboclo Society*; Taussig, *Shamanism, Colonialism, and the Wild Man*.

67. Barham and Coomes, "Wild Rubber"; Nugent, *Amazonian Caboclo Society*.

68. The classic study of the Mundurukú is Murphy, *Rubber Trade and the Mundurucú Village*. Bates and William Chandless, whose steps da Cunha followed up the Purús, noted that natives there were integrated into the rubber economy even in the 1860s. See Chandless, "Ascent of the River Purús."

69. See chap. 21.

70. Lagos, *Arana, rey del caucho*.

71. Bates reports how insalubrious the river became later in the 1850s: Bates, *Naturalist on the River Amazons*. Anderson points out that the 1850s and 60s saw numerous epidemics of yellow fever, cholera, and measles. Anderson, *Colonization as Exploitation*. See also McNeill, *Mosquito Empires*.

72. Stanfield, *Red Rubber, Bleeding Trees*.

73. Lange, *In the Amazon Jungle*.

74. Casement, *Amazon Diaries*.

75. Coudreau, *Voyages au Tocantins Araguaia*.

76. Pacheco B., *Integración económica y fragmentación social*.

77. See Edelman, "Central American Genocide."

78. Pearson, *Rubber Country of the Amazon*.

79. Rocha, *Memorandum de viaje*.

80. Taussig, *Shamanism, Colonialism, and the Wild Man*.

81. Robuchon, *En el Putumayo y sus afluentes*.

82. Domínguez and Gómez, *Economía extractiva*.

83. Pineda Camacho, *Holocausto en el Amazonas*; Taussig, "Culture of Terror"; Taussig, *Shamanism, Colonialism, and the Wild Man*.

84. Arana and Select Committee on Putumayo, *Cuestiones del Putumayo*.

85. See Casement and Sawyer, *Roger Casement's Diaries*; Goodman, *Devil and Mr. Casement*.

86. The so-called Black Diaries, which sealed Casement's fate in many ways, hardly read as racy texts today. As the scandal of gayness has waned, Casement's considerable importance in indigenous rights has emerged strongly, and the building of global campaigns against the brutality of rubber regimes has made him a hero. Vargas Llosa, *El sueño del celta*; Goodman, *Devil and Mr. Casement*.

87. The research was conducted from 1995 to 2005 with a focus on biomass and carbon studies.

88. Hecht and Saatchi, "Rubber Boom and a Deforestation Pulse."

89. Domínguez and Gómez, *Economía extractiva*.

90. Hecht and Saatchi, "Rubber Boom and a Deforestation Pulse."

91. Assies, "From Rubber Estate to Simple Commodity Production"; Schroth, Moraes, and da Mota, "Increasing the Profitability of Traditional, Planted Rubber."

92. Silva Coutinho, who carried out his own exploration on the Purús at the behest of the imperial administrators in Manaus, was the first to call attention to the potential problems of demand outstripping supply. The more general crisis is described both by Dean, *Brazil and the Struggle*, and by Weinstein, *Rubber Boom*.

93. See Dean, *Brazil and the Struggle for Rubber*; Grandin, *Fordlândia*; Hecht and Cockburn, *Fate of the Forest*; Kennedy and Lucks, "Rubber, Blight, and Mosquitoes"; Santos-Granero and Barclay, *Tamed Frontiers*. Also see Hemming, *Tree of Rivers*.

94. Brass, *Towards a Comparative Political Economy of Unfree Labour*; Eltis, *Coerced and Free Migration*; Emmer and Boogaart, *Colonialism and Migration*; Look Lai, *Indentured Labor, Caribbean Sugar*; Northrup, *Indentured Labor in the Age of Imperialism*.

95. See chap. 19 below.

96. Agassiz and Agassiz, *Journey in Brazil.*

97. Cf. Hecht and Cockburn, *Fate of the Forest*; Cleary, "Lost Altogether to the Civilised World."

98. Questions of gender and families remain problematic in studies of extractive economies. For a review see Hecht, "Factories, Forests, Fields and Family." For the new historiography of women in Amazonia, see Alvares and d'Inção, *A mulher existe?*; Gomes, "Gênero, etnicidade e memoria na Amazônia"; Martinelli, *Mulheres da Amazônia.*

99. Padoch, *Várzea*; Nugent, *Amazonian Caboclo Society*; Raffles, *In Amazonia.*

100. At the height of the boom, 1900–1910, the Purús drainage supplied a great deal of the Amazon's latex, but after the collapse, when there was still substantial latex production (averaging about 29,344 tons per year), the islands and Acre were the largest regional suppliers and produced about 54,170 tons—almost one-quarter of the basin's total of 234,756 tons exported from Belém in 1916–23. The islands supplied a slightly greater proportion (51 percent). These calculations are based on export statistics from the port of Belém.

101. Wolff, *Mulheres da floresta.*

102. Facão, *Album do Acre.*

103. Weinstein, *Amazon Rubber Boom*; Nugent, *Amazon Caboclo*; Hecht and Cockburn, *Fate of the Forest.*

104. Grandin, *Fordlândia.*

105. Araújo dos Anjos, *Territórios das communidades.*

106. Da Cunha and Gomes, *Quase-cidadão*; Gomes, *Hidra e os pântanos*; Gomes, "Safe Haven."

107. Acevedo Marin and de Castro, *Negros do Trombetas.*

108. Coudreau, *Voyage au Cuminá*; Gomes, "Fronteiras e mocambos"; Gomes, *Hidra e os Pântanos.*

109. De Rivière, "Explorations of the rubber districts of Bolivia."

110. Torres and Martine, *Amazonian Extractivism.*

Chapter 15

1. Da Cunha, "Relatorio de Comissão Mixta Brasileira-Peruana."

2. Agassiz and Agassiz, *Journey in Brazil*; Bates, *Naturalist on the River Amazon*; Markham and Blanchard, *Markham in Peru*; Mathews, *Up the Amazon and Madeira Rivers*; Orton, *Andes and the Amazon*; Pöppig, *Viaje al Perú y al Río Amazonas*; von Humboldt, Bonpland, and Williams, *Personal Narrative*; von Spix and von Martius, *Viagem pelo Brasil*; Wallace, *Narrative of Travels.*

3. And womanly courage: see Whitaker, *The Mapmaker's Wife.*

4. Safier, *Measuring the New World.*

5. Von Humboldt, Bonpland, and Williams, *Personal Narrative;* Weigl, "Between Technique and Eroticism."

6. Ibid.

7. Dettelbach, "Humboldtian Science"; Fagot, "World Consciousness"; Helferich, *Humboldt's Cosmos*; Miller and Reill, *Visions of Empire*; Zimmerer, "Humboldt's Nodes and Modes." The sources of von Humboldt's imaginary are also discussed in Cañizares-Esguerra, *Nature, Empire, and Nation.*

8. Coelho, "As viagens filosóphicas"; Falcão et al., *Viagem filosófica*; Faulhaber and de Toledo, *Conhecimento e fronteira.*

9. Driver, *Geography Militant.*

10. Schultes, "Richard Spruce"; Schultes, "Several Unpublished Ethnobotanical Notes."

11. But see Rivière, *Absent-Minded Imperialism.*

12. Von Humboldt, Bonpland, and Williams, *Personal Narrative.* Von Martius, very interested in native rights and ethnobotany, invoked a racial hierarchy as the best means to develop

Brazil: whites would serve as civilizers and indigenes must do what was necessary to retrieve their native dignity to move toward civilization, while blacks, primitives, though capable of civilization, were stumbling blocks to Brazil's development. This essay was in response to a contest sponsored by the Instituto Histórico on "how to write the history of Brazil." Lisboa, *Nova Atlântida*; von Martius, "Como se deve escrever."

13. Park worked for the British Africa Company, Stanley for King Leopold, and Brazza for the French crown.

14. The eighteenth-century imperial effort relied on economic botany and massive collection. Later nineteenth-century efforts, though still engaged in extensive collection, sought to theorize the links in more complex geographical ways. Schiebinger and Swan, *Colonial Botany*.

15. See Endersby, *Imperial Nature*.

16. The Jesuit Serafim Leite has compiled many of these documents in his monumental study *História de Companhia de Jesus no Brasil*.

17. Hoornaert and Comisión de Estudios de História, *História da Igreja na Amazônia*.

18. See Herndon and Gibbon, *Exploration of the Valley of the Amazon*.

19. See Lubrich, "Egypt Everywhere."

20. Muir and Branch, *John Muir's Last Journey*.

21. See *Amazon Adventure* by Peter Fleming (brother of 007's inventor Ian Fleming), who wrote about the attempt to find Colonel Fawcett and his El Dorado. Mostly it's a tale of a bunch of Oxford hearties let loose in the tropics, but the entire story is saved by the good humor and superior writing of its protagonist.

22. Wallace, *Narrative of Travels*.

23. See Pratt, *Imperial Eyes*; Nicolson, "Humboldtian Science." In Cañizares-Esguerra, *Nature, Empire, and Nation*, especially see "How Derivative Was Humboldt? Microcosmic Narratives in Early Modern Spanish America and the (Other) Origins of Humboldt's Ecological Sensibilities" (112–29). Also see Browne, *Secular Ark*. In Amazonia, von Humboldt's "convivencia" at the mouth of the Orinoco focused on the local knowledge of his native expeditionaries. See von Humboldt, Bonpland, and Williams, *Personal Narrative*.

24. See Burnett, *Masters of All They Surveyed*; Raffles, *In Amazonia*.

25. Bates, *Naturalist on the River Amazon*; Burnett, "It Is Impossible to Make a Step without the Indians"; Coudreau, *Chez nos Indiens*; Davis, *One River*; Davis and Schultes, *Lost Amazon*; von Humboldt, Bonpland, and Williams, *Personal Narrative*.

26. Coudreau, *Chez nos Indiens*; Coudreau, *Voyage au Tocantins-Araguaya*; Coudreau, *Voyage au Xingú*; Herndon and Gibbon, *Exploration of the Valley of the Amazon*.

27. Barham and Coomes, "Wild Rubber"; Collier, *River That God Forgot*; Hemming, *Amazon Frontier*; Santos-Granero and Barclay, *Tamed Frontiers*; Stanfield, *Red Rubber, Bleeding Trees*; Taussig, *Shamanism, Colonialism, and the Wild Man*.

28. Benchimol, *Eretz Amazônia*; Harris and Nugent, *Some Other Amazonians*.

29. Ibid. In the town of Lábrea, Christians, Muslims, and Jews could all be found. Lábrea had an active Kardist sect, as well as native practices and candomblé. See Loureiro, *"Gazeta do Purús."*

30. See Drayton, *Nature's Government*; Richard Grove, *Green Imperialism*; Burnett, *Masters of All They Surveyed*; Driver and Martins, *Tropical Visions*; Cosgrove and Daniels, *Iconography of Landscape*; Arnold, *Problem of Nature*; Miller and Reill, *Visions of Empire*.

Chapter 16

1. As part of the demilitarization, Peru and Brazil had set up separate regional administrative posts.

2. The beriberi outbreak at the administrative outpost of Novo Lugar, referred to earlier, did not afflict the Joint Commission.

3. Da Cunha describes here a phenomenon known locally as a *friagem*, a cold snap. While unexpected by many travelers, they are hardly uncommon. Anyone who spends the summer months in Amazonia will experience these sharp extensions of Antarctic air in the tropics, which are extremely unpleasant, although short lived.

Chapter 17

1. Da Cunha, "A minha terra e retilinea e alto como as palmeiras," in da Cunha and Tocantins, *O Paraíso Perdido,* 2nd ed., 208–9.
2. Da Cunha, "As promessas divinas de esperança," in da Cunha and Tocantins, *O Paraíso Perdido*, 2nd ed., 210–12.
3. Da Cunha, "Uma entrevista," in da Cunha and Tocantins, *O Paraíso Perdido*, 1st ed., 274–78.

Chapter 18

1. Da Cunha telegram to Rio Branco, Manaus, October 19.
2. Da Cunha to Jose Veríssimo, November 8, 1905.
3. Da Cunha to Rio Branco, November 10, 1905.
4. Ibid.
5. The term "love nest" is Dilermando's. Dilermando de Assis, *Conselho de guerra*.
6. Ibid.
7. Judith de Assis and de Andrade, *Ana de Assis*.
8. This debate unfolded between the de Assis faction, including Dilermando de Assis (*A tragédia da Piedade*) and later Judith de Assis, daughter of Ana and Dilermando, in her 1987 memoir (de Assis and de Andrade, *Ana de Assis*) and the da Cunha side through a descendant of Manuel Afonso da Cunha, Joel Bicalho Tostes, who, in collaboration with the da Cunha scholar Adelino Brandão, produced the 1990 *Aguas de amargura* as a counterpoint to the narrative supplied by Judith. They supplied some damning documents, including Ana's birth certificate showing she was four years older than she claimed (and with a lover seventeen years her junior a little fudging on age is certainly understandable) and accounts from the Bazaar America.
9. De Assis, *Conselho de guerra*.
10. J. de Assis and de Andrade, *Ana de Assis*.
11. Ana de Solon da Cunha, in Galvão and Silva Neto, *Crónica*, 126.
12. Ibid. 127.
13. Ibid.
14. Manuel da Cunha to Euclides da Cunha, June 26, 1905.
15. Euclides da Cunha to Manuel da Cunha, Rio, February 14, 1906, Rio.
16. De Assis, *Tragédia da Piedade*.
17. Tostes and Brandão, *Aguas de amargura*.
18. Da Cunha to Dilermando de Assis, undated 1906, Rio.
19. De Assis, *Tragédia da Piedade*. This is corroborated in the deposition given by Ana.
20. Ana Solon da Cunha in Galvão and Silva Neto, *Crónica*, 127.
21. De Assis, *Conselho de guerra*.
22. Ana Solon da Cunha in Galvão and Silva Neto, *Crónica*, 127.
23. Ibid.
24. Ibid., 127–28.
25. Ibid., 128.
26. Ibid.
27. Ibid.
28. Ibid.

29. De Assis and de Andrade, *Ana de Assis*, 53–54.

30. Tostes and Brandão, *Aguas de amargura*.

31. Da Cunha to Henrique Coelho, Rio, July 30, 1906.

32. Ana Solon da Cunha in Galvão and Silva Neto, *Crónica*, 129.

33. Da Cunha to Firmo Dutra, Rio, July 9, 1906.

34. Da Cunha to Manuel da Cunha, Rio, July 24, 1906.

35. The letters to Ana from Dilermando are transcribed in de Assis and de Andrade, *Ana de Assis*, 47–53.

36. Da Cunha to H. Coelho, Rio, June 30, 1906.

37. See Hecht and Cockburn, *Fate of the Forest*.

38. For a more detailed analysis of the events surrounding the railroad and its tycoons, see ibid.

39. Da Silva, *Madeira–Mamoré–Martírios*; Ferreira, *Ferrovía do diabo*.

40. Da Cunha to Oliveira Lima, Rio, May 23, 1906.

41. Da Cunha to Firmo Dutra, Rio, September 30, 1906.

42. Da Cunha to Francisco Escobar, Rio, December 31, 1906.

43. Ana Solon da Cunha in Galvão and Silva Neto, *Crónica*, 129.

44. Ibid.

45. Da Cunha to Otaviano Vieira, Rio, July 4, 1907.

46. Ana Solon da Cunha in Galvão and Silva Neto, *Crónica*, 129.

Chapter 19

1. Peru and Maurtua, *Juicio de límites entre el Perú y Bolivia*; Peru, Ministerio de Relaciones Exteriores, *Perú*; Santos-Granero and Barclay, *Selva central*.

2. Santos-Granero and Barclay, *Tamed Frontiers*.

3. Renan, *Qu'est-ce qu'une nation?*, in Forest 1991.

4. Cañizares-Esguerra, *How to Write the History of the New World*, 204–6.

5. Burns, *Unwritten Alliance*; Dennison, *Joaquim Nabuco*; Rivière, *Absent-Minded Imperialism*. Rio Branco had placed the abolitionist Joaquim Nabuco in charge of this adjudication, and though he was antislavery he was also anti-Indian. Since the British argument was based on human rights and fair treatment of the natives (and the excellent mapping work of Schomburgk), that adjudication was mishandled.

6. Da Cunha to Francisco Escobar, Rio, June 13, 1906.

7. See Driver, *Geography Militant*; Edney, *Mapping an Empire*.

8. Harley, "Maps, Knowledge and Power"; Harley and Woodward, *History of Cartography*; Perkins, "Cultures of Map Use."

9. See, for example, Harley, "Historical Geography and the Cartographic Illusion"; Harley, "Rereading the Maps of the Columbian Encounter."

10. Harley, "Maps, Knowledge, and Power."

11. Carter, *Road to Botany Bay*.

12. Safier, *Measuring the New World*.

13. Da Cunha, "Plano de uma cruzada," 153.

14. Cortesão, *História do Brasil nos velhos mapas*.

15. Da Cunha discusses this in great deal in "Real and Mythical Geographies of the Purús" (see chap. 20).

16. See, for example, Davidson, "How the Brazilian West Was Won"; Farage, *Muralhas dos sertões*; Volpato, *Conquista da terra*.

17. Burnett, *Masters of All They Surveyed*; Farage, *Muralhas dos sertões*; Hemming, *Roraima*; Rivière, *Absent-Minded Imperialism*.

18. Peru did have a navy outpost in Iquitos, but it was poorly supplied and was highly dependent on tax revenues from the latex trade.

19. Da Cunha letter to Firmo Dutra, Rio, September 30, 1906.

20. *Varadouros* reflected many indigenous practices and polities. Wallace describes some on the upper Rio Negro. Also see la Combe, von Hassel, and Pesche, *El Istmo de Fitscarrald.*

21. This is a fragment of a larger piece, "Transacreana," in *À margem da história.*

22. This is not the case. Indigenous mapping exercises report extensive cuts between meanders.

23. Da Cunha, "Transacreana."

24. Ibid.

25. See Heckenberger 2008.

26. For a history of the reserves, see Hecht and Cockburn, *Fate of the Forest.*

27. Da Cunha, "Entre os seringais."

28. See Harley, "Deconstructing the Map."

29. Da Cunha, "Entre os seringais."

Chapter 20

1. Coutinho and Urbano would both serve as guides for the Agassiz expedition. Coutinho was one of the unsung heroes of the Upper Amazon and one of the keenest observers of the rubber economy. By the 1880s he was to note that the production of wild rubber would reach its limits and be replaced by plantations. See Dean, *Brazil and the Struggle for Rubber.*

2. Chandless, "Ascent of the River Purús."

3. Seed, *Ceremonies of Possession.*

4. Ibid.

5. Hadfield, *Brazil and the River Plate.*

6. Orton, *Andes and the Amazon.*

7. For an overview of Haenke, see William Denevan's appreciation, "Tadeo Haenke,."

8. Da Cunha derives virtually all his discussions of Urbano from the Chandless report, which included a synopsis of the Coutinho travels as well. Chandless, "Ascent of the River Purús."

9. "From the rising of the sun he formed a much better estimate of the general course than I should have thought possible in so tortuous a river, and not a bad one of the distance of leagues." See ibid.

10. Wallis's attempts on the river were marred by an unending series of catastrophes, including drowning.

11. This litany of hardship was clearly experienced by da Cunha.

12. This was more or less the same methodology used by the Da Cunha / Buenaño survey.

13. Chandless, "Ascent of the River Purús."

14. Fifer, "Empire Builders." Herzog generally seems to have had a horrible time in Amazonia filming *Aguirre* but most especially filming *Fitzcarraldo.* See Herzog and Winston, *Conquest of the Useless.*

15. Davis and Schultes, *Lost Amazon.*

16. Heath, "Exploration of the River Beni."

17. His essay in *À margem* called "A Transacreana" is more or less a meditation on an infrastructure development program based on the *varadouros* (see Souza, *Land without History*). Da Cunha's more interesting meditations on the meaning of the land connections are explored in chapter 19 under the section "*Varadouros.*"

18. To clarify the locations, see da Cunha map 11 in chapter 19.

19. See the already cited studies by Hemming, Langfur, Garfield, and Manuela da Cunha.

20. Rondon would for a short time become the guardian of da Cunha's sons after his death.

21. Rondon is an underappreciated figure in Amazonian history. He founded the Indian protection service and was one of the Amazon's great surveyors and explorers (including acting as a guide for Teddy Roosevelt's trip on the "River of Doubt"). See Coutinho, *Rondon*; Diacon, *Stringing Together a Nation*; Millard, *River of Doubt*.

22. De Carvajal et al., *Discovery of the Amazon*.

23. Hill and Santos-Granero, *Comparative Arawakan Histories*; Santos-Granero, "Boundaries Are Made to Be Crossed."

24. Lathrap, *Upper Amazon*.

25. The anthropogenic landscapes of the Purus are described in chapter 14, "In the Realm of Rubber."

26. See Balée and Erickson, *Time and Complexity*; Chirif, *Etnicidad y ecología*; Denevan, *Aboriginal Cultural Geography*; Denevan, *Cultivated Landscapes*.

27. Xerez, *Verdadera relación de la conquista del Perú*.

28. See de Acuña and Kane, *Nuevo descubrimiento del Gran Rio de las Amazonas*. The interactive area between Andean cultures and Amazonian ones was also noted by the mercenary soldier Ulrich Schmeidl.

29. For overviews of regional indigenous politics that emphasize the Mura and their conflicts, see Manuela da Cunha, *História dos índios no Brasil*; Hemming, *Amazon Frontier*; Porro, *O povo das águas*.

30. See Hemming, *Red Gold*. The explorers von Spix and von Martius also saw them after their capitulation.

31. Hemming, *Red Gold*; MacLachlan, "Indian Directorate."

32. Leonel de Almeida was also one of the heroes in the Peruvian skirmishes. Some of this early history is captured in a selection of newspaper articles from the Lábrea newspaper *Gazeta do Purús*.

33. Venâncio was a significant character on the Purús and one of its important political headmen. An Ashinika leader who worked as a contractor for Carlos and Delfim Fitzcarraldo, he and his kinsmen migrated from the Ucayali over to the Purús in the dry season to collect *caucho* on the Cujar. Venâncio also worked for Carlos Scharff; the elaborately attired chief with the towering headdress shown in figure 17.2 may be Venâncio. Peter Gow argues that the ability to mobilize and move his kinsmen set Venâncio up as a privileged labor supplier who delivered his affines into the rubber economy through the bonds of blood rather than iron shackles, and himself became a *patron* of the Madre de Dios and Manú Rivers. Da Cunha met Venâncio, and indeed the native leader was regularly sought out by authorities of both Peru and Brazil as well as ecclesiastics of various stripes. Gow, "Gringos and Wild Indians"; Santos-Granero and Barclay, *Tamed Frontiers*.

34. The South American history of science and ethnobiology is enjoying a boom. In part this reflects the increased attention to indigenous knowledge systems that began in the mid-1980s, the biological turn in colonial studies and the promotion of pharmaceuticals as means of saving rain forests through bioprospecting. Iconic for ethnobotany in this regard are Balick, Elisabetsky, and Laird, *Medicinal Resources of the Tropical Forest*; Markham, *Peruvian Bark*; Renner, *History of Botanical Exploration in Amazonian Ecuador*.

35. Curtin, *Disease and Empire*; Curtin, *Migration and Mortality in Africa and the Atlantic World*; Faust, *This Republic of Suffering*; McNeill, *Mosquito Empires*.

36. Gramiccia, *Life of Charles Ledger*; Rocco, *Miraculous Fever Tree*. A revisionist history now holds that Markham's mission ultimately failed through a lack of indigenous knowledge of the effectiveness of local varietals.

37. These ports, especially Mollendo, were important export platforms for the Ucayali and Madre de Dios latexes, especially *caucho*.

38. Da Cunha here quotes from José Wilkens de Matos, *Dicionário topográfico do departamento de Loreto*.

39. This region was an important area of ecclesiastic settlements, and here the pattern from southern Brazil of raiding missions for slaves was clearly carried out.

40. Raimondi was an encyclopedic naturalist and collected material on geology, botany, ethnography, zoology, etc. He was Italian by birth but naturalized as Peruvian and was perhaps the greatest Peruvian scientist of the nineteenth century. He took part in boundary demarcations and hydrology in the Peruvian Amazon. Da Cunha is referring quite specifically to Raimondi's stance against abuse and exploitation of Chinese immigrants and post-emancipation Afro-Peruvians, positions not unlike da Cunha's own. Raimondi was also very impressed by native peoples, indigenous knowledge, and local ecologies. Raimondi, *Notas de viajes para su obra "El Perú"*; Raimondi, Balta, and Sociedad Geográfica de Lima, *El Perú*.

41. Tucker, as noted in earlier chapters, was a transplanted US southern seaman who worked as a river surveyor on the Ucayali and its tributaries.

42. Portillo was concerned to both extend Peruvian terrains and limit Brazilian incursions. Roger Casement complained bitterly about his collusion with *caucheiros*, especially the Aranas. Casement and Sawyer, *Roger Casement's Diaries*.

Chapter 21

1. Freyre, *Perfil de Euclydes e otros perfís*. Also see Burke and Pallares-Burke, *Gilberto Freyre*.

2. See Burke and Pallares Burke, *Gilberto Freyre*.

3. Freyre's writings on Brazilian identity were durable and are still widely commented upon in Brazil as well as his historiography, which, as E. P. Thompson noted, emphasizes the "common people" as the true bearers of culture.

4. Haas and Mansfeld, *Aristotle*; Lang, *Order of Nature in Aristotle's Physics*.

5. Slater, *Entangled Edens*.

6. See More, *Utopia*; Voltaire, *Candide*; Campanella and Donno, *City of the Sun*; Baptiste and de las Casas, *Bartolomé de Las Casas and Thomas More's "Utopia"*; Bertrand, "Utopian Dreamers in Hispanic America in the 16th Century"; Brown and Fernández, *War of Shadows*; Folguera, *Construcción de la utopía*; Milhou, "Messianic and Utopian Currents."

7. See, for example, Baudot, *Utopia and History in Mexico*; Cañizares-Esguerra, *Puritan Conquistadors*; Greenblatt, *New World Encounters*.

8. Baudot, *Utopia and History in Mexico*; Bertrand, "Utopian Dreamers in Hispanic America in the 16th Century"; Folguera, *Construcción de la utopía*; Gassent, *Utopía y derecho en la conquista de América*; Greene, *Science, Ideology, and World View*; Milhou, "Messianic and Utopian Currents"; Urbano and Beascoechea, *Utopía, mesianismo y milenarismo*.

9. Vainfas, *Heresia dos índios*; Vainfas, *Trópico dos pecados*.

10. Huntington, *Civilization and Climate*.

11. Heywood, "Portuguese into African"; Heywood and Thornton, *Central Africans, Atlantic Creoles*.

12. Brooks, *Eurafricans in Western Africa*; Curtin, *Image of Africa*; Falola and Childs, *Yoruba Diaspora*; Florentino, *Em costas negras*; Heywood and Thornton, *Central Africans, Atlantic Creoles*; Russell-Wood, *Portuguese Empire*.

13. Beinart, "Men, Science, Travel and Nature"; Pratt, *Imperial Eyes*.

14. Haenger, Shaffer, and Lovejoy, *Slaves and Slave Holders*; Heywood and Thornton, *Central Africans, Atlantic Creoles*; Klieman, *"The Pygmies Were Our Compass"*; Thornton, "Warfare in Atlantic Africa."

15. Stepan, *Picturing Tropical Nature*; Schiebinger and Swan; *Colonial Botany*.

16. See, for example, Thornton, *Congolese Saint Anthony*; Heywood, "Portuguese into African"; Fernandez-Armesto, *Before Columbus*.

17. See essays in chapter 11, above.

18. Da Cunha "Contrastes e confrontos."

19. Arnold, *Problem of Nature*; Arnold, *Warm Climates and Western Medicine*; Stepan, *Idea of Race in Science*.

20. Borges, "Puffy, Ugly, Slothful and Inert."

21. From da Cunha's *À margems da historia*.

22. François Louis de Caumont LaPorte, Comte de Castelnau (1810–80), was a French naturalist and diplomat. Greatly influenced by Cook's travel stories as a child, he studied natural science in Paris under Cuvier (Agassiz was also a protégé of Cuvier) and Geoffroy Saint-Hilaire. Castelnau became the French consul in Bahia after a four-year expedition from Rio to Lima and into the Amazon to Pará. Castelnau, *Expedítion dans les parties centrales de l'Amerique du Sud*.

23. Although hardly the only caucho barons, certainly the most famous were the Aranas, whose trading house was based in Manaus and already very active.

24. It is possible that da Cunha's Indian was referring and pointing to the Amigos River, a tributary of the Madre de Dios, on which many "friendly Indians" had been taken by *caucheiros* into the Purús.

Chapter 22

1. Galvão and Silva Neto, *Crônica de uma tragédia inesquecível*, is a compilation of the records and testimony (*autos*) of the murder trial of Dilermando. This chapter relies heavily on these documents, which I will refer to as *Crônica*, followed by the page numbers.

2. De Assis, *Conselho de guerra*.

3. See especially the moving book by his closest friend then: Coelho Neto, *Livro de prata*.

4. His letters are remarkably explicit on this point. See his letter to Francisco Escobar, Rio, April 10, 1908.

5. Rabello, *Euclides da Cunha*; Ventura, Barreto de Santana, and Carvalho, *Retrato interrompido*; Amory, *Euclides da Cunha*. His letters reveal some of this despair as well. See da Cunha to Oliveira Lima, May 5, 1909.

6. Ana Solon da Cunha in *Crônica*, 130.

7. This type of fear is actually quite common and increasingly is listed among the symptoms of posttraumatic stress, but it also has roots in other afflictions. In essence a person wakes up paralyzed and perceives an evil or upsetting presence and a crushing sensation in the chest. This is called sleep paralysis. Two brain systems contribute to sleep paralysis. One involves the inner brain structures that monitor surroundings for threats and launch responses to perceived dangers. REM-based activation of this system triggers a sense of an ominous entity nearby. Other neural areas that contribute to dream imagery may draw on personal or cultural knowledge to flesh out the evil presence.

8. Da Cunha to Oliveira Lima, Rio, May 5, 1909.

9. Coelho Neto, *Livro de prata*.

10. Lucinda Ratto in *Crônica*, 79.

11. Ana Solon da Cunha in ibid., 130.

12. Ibid.

13. Letter to Otaviano Vieira, Rio, July 3, 1909.

14. Before antibiotics the "cure" for tuberculosis was geographical—sanitariums or clinics in arid mountainous areas were deemed most healthful. But these were often the last steps before the final one into the abyss. See, for example, Mann, *Magic Mountain*.

15. Letter to Vieira, July 5, 1909.

16. A league here is about 18 km^2.

17. Fifer, "Bolivia's Boundary with Brazil"; Fifer, "Empire Builders."

18. De Abranches, *Rio Branco e a política exterior*; de la Pedraja Tomán, *Wars of Latin America*; Pontes, *Euclides da Cunha*; Zeballos, *Diplomacia desarmada*.

19. Anna Almeida Lima in *Crónica*, 72.

20. Da Cunha telegram to Vieira, Rio, August 3, 1909.

21. Lucinda Ratto in *Crônica*, 75.

22. Ana Solon da Cunha in ibid., 67.

23. Angelica Ratto in ibid., 69.

24. Ibid.

25. Ibid.

26. Lucinda Ratto in ibid., 80.

27. Ibid.

28. Ibid.

29. De Assis, *Conselho de guerra*. Though there is no paper trail, the degree to which everyone was on edge suggests that the Dilermando-Ana romance somehow became much more public and dangerous after this trip. The correspondence between the two had heated up. Judith de Assis, in her family memoir, *Ana de Assis*, reports that Ana's mother's precarious health gave her hope for the end of her marriage to Euclides, because upon Tulia's death she would inherit a useful pension that would free her economically from him.

30. De Assis, *Conselho de guerra*.

31. Ana Solon da Cunha in *Crônica*, 131.

32. Ibid.

33. Ibid.

34. Later in life Luis changed his surname from da Cunha to de Assis to reflect his true paternity.

35. Lucinda Ratto in *Crônica*, 80.

36. Ana Solon da Cunha in ibid., 132.

37. Ibid.

38. The term used here is *entregar*, which in this sense means the return of damaged goods. Typically it was applied to daughters who married but turned out not to have been virgins or were defective in some other way. In Iberian cultures such an *entrega* involved extreme shame and dishonor. For a novelistic rendering, see García Márquez's *Chronicle of a Death Foretold*.

39. Ana Solon da Cunha in *Crônica*, 132.

40. Ibid.

41. Augusta Solon Ribeiro in ibid., 82; also Ana Solon da Cunha, 133.

42. Judith Ribeiro de Assis paints her mother as a kind of protofeminist, and the Brazilian serial made about Ana and Dilermando (*Desejo*—Desire) expands on this theme. In descriptions of Ana by Judith, Dilermando, Dinorah, and others, however, there is nothing other than this calamitous affair to suggest feminist undercurrents.

43. Judith Ribeiro de Assis claimed that Ana was just about to sail to Rome to procure such an annulment. This seems quite unlikely in light of Ana's financial circumstances.

44. Da Cunha to Vieira, Rio, August 8, 1909.

45. Da Cunha to Manuel da Cunha, Rio, August 9, 1909.

46. Ana Solon da Cunha in *Crônica*, 135.

47. Police inquiry in ibid., 104.

48. Solon's insistence that night that his mother swear that she would stay the night as though it were the house of a son speaks to his psychological turmoil and need to deny the true state of affairs.

49. Ana Solon da Cunha in *Crônica*, 38.

50. Ibid.; also Lucinda Ratto, 80.

51. Ana Solon da Cunha in ibid., 139.

52. Da Cunha Filho, "Verdade sobre a morte."
53. Dinorah de Assis in *Crônica*, 76.
54. Ana Solon da Cunha in ibid., 134.
55. Ibid.
56. Solon da Cunha in ibid., 97.
57. Ibid., 54; Ana Solon da Cunha, 134.
58. Angelica Ratto in ibid., 71; also Ana Solon da Cunha, 78.
59. Angelica Ratto in ibid., 70.
60. Ibid., 70–71.
61. Ana Solon da Cunha in ibid., 135.
62. Ibid., 135–36.
63. Angelica Ratto in ibid., 72.
64. Ibid., 71, 72.
65. Quidinho, who as a child was perhaps not able to appreciate everything that was unfolding, eventually wrote a memoir of these days that surely captures the anguish. See da Cunha Filho, "Verdade sobre a morte."
66. Constantino Fontainha in *Crônica*, 85.
67. Dinorah de Assis in ibid., 121.
68. Dilermando de Assis in *Crônica*, 100.
69. Dinorah de Assis in ibid., 122.
70. Ibid., 122.
71. Ibid., 101.
72. De Assis, *Tragédia da Piedade*.
73. Ibid.; Dilermando de Assis in *Crônica*.
74. Dinorah de Assis in *Crônica*, 101.
75. Ana Solon da Cunha in ibid., 67.
76. Hora cited in de Assis, *Conselho de guerra*.
77. Ibid.
78. Ana Solon da Cunha in *Crônica*, 135.
79. W. Guerra, *Revista Paulista de Medicina*, cited in de Assis and de Andrade, *Ana de Assis*.

Chapter 23

1. De Assis and de Andrade, *Ana de Assis*.
2. Diacon, *Stringing Together a Nation*.
3. De Assis, *Conselho de guerra*.
4. Ibid.
5. Ferreira Gomes, "Como se deu o lamentável acontecimento" (How the unhappy event unfolded).
6. Ibid.
7. Shakespeare, *Hamlet*, act 5, scene 2.
8. Euclides II, letter to Ana Ribeiro de Assis, cited in de Assis, *Conselho de guerra*.
9. Ibid.
10. De Assis and de Andrade, *Ana de Assis*.
11. De Assis, *Conselho de guerra*.
12. Ibid.
13. Barreto, *Última expedição a Canudos*.
14. His works weirdly echoed those of da Cunha: a memoir of Canudos, a history of an expedition to Mato Grosso, and various commentaries on the politics of the day. It is a measure of the prestige of the Ribeiro, da Cunha, and de Assis families that a person of this rank would take on the scandal-plagued child.

15. De Assis, *Conselho de guerra.*

16. See the anonymous piece "Pelo honra da nome."

17. Dauat, "Vítima esquecido de Euclides da Cunha." Cited in de Assis and de Andrade, *Ana de Assis.*

18. Ibid.

19. De Assis, *Um nome, uma vida, uma obra.*

20. The twentieth-century history of Amazonia is covered in Hecht and Cockburn, *Fate of the Forest.*

Chapter 24

1. Fagan and Shoobridge, "Investigation of Illegal Mahogany Logging."

2. Ibid.

3. Ibid.

A Note on the Text

1. Amory, *Euclides Da Cunha*; Ventura, Barreto de Santana, and Carvalho, *Retrato interrompido.*

2. Three of the five Amazon pieces in *À margem da história* had actually been published earlier in academic journals or in newspapers.

3. Amory, *Euclides Da Cunha*; da Cunha, Sousa, and Sá, *The Amazon.*

4. Kristal, *Invisible Work.*

5. Letter to Henrique Coelho Neto, Manaus, March 10, 1905.

6. I am indebted to George Lovell, William Woods, and Charles Mann for stimulating a useful discussion on questions of translation and especially the Nabokov position in Morelia, Mexico, in 2005. Appel, Newman, and Nabokov, *Nabokov*; Nabokov, *Notes on Prosody.*

7. Rabassa, *If This Be Treason.*

Bibliography

Writings of Euclides da Cunha

Listed here are the writings referred to in the text as well as specific writings that I cite in more detail. They are organized in rough chronological order.

Books and Collections

da Cunha, Euclides. *Os Sertões*. Rio de Janeiro: Laemmert, 1903.

da Cunha, Euclides, and Samuel Putnam. *Rebellion in the Backlands*. Chicago: University of Chicago Press, 1970.

da Cunha, Euclides, and Leopoldo Bernucci. *Os Sertões*. São Paulo: Ateliê, 2002.

da Cunha, Euclides. *Caderneta de Campo*. São Paulo: Editora Cultrix, 1975

———. *Relatorio de Comissão Mixta Brasileiro-Peruana de Reconhecimento do Alto Purús*. Rio de Janeiro: Imprensa Nacional, 1905.

———. *Relatorio de Comissão Mixta Brasileiro-Peruana de Reconhecimento do Alto Purús*. Rio de Janeiro: Superintendencia para Valorizaçã o da Economia Amazonica (SPVEA), 1960.

Selections used in this text (numbers refer to pages):

"Instrucões"

"A viagem"

"Notas complementares: 1o, Observacoes sobre a história da geografía do Purús"

"Da foz às cabeceiras," 63–72

"Nas cabeceiras," 72–74

"Os varadouros," 74–78

"Notas complementares: 2o, O Povoamento"

"Da foz às cabeceiras," 79–86

"Nas cabeceiras," 86–88

———. *Peru versus Bolívia*. Rio de Janeiro: Typ. do Jornal do Commercio, Rodrigues, 1907.

———. *Peru versus Bolívia*. São Paulo: Editora Cultrix, 1975.

———. *Contrastes e confrontos*. Porto: Emprêsa Litteraria e Typographica, 1907.

———. *Contrastes e confrontos*. São Paulo: Editora Cultrix, 1975.

Selections used in this text:

"Maréchal de Ferro," 30–34

"Plano de uma cruzada," 57–61

"Conflito inevitável," 93–97

"Contra os caucheiros," 98–101

"Entre o Madeira e o Javari," 102–6

———. "Entre os seringais." *Kosmos* 3, no. 1 (1905).

———. "Um rio abandonado." *Instituto Histórico e Geográfico Brasileiro* 68, no. 113 (1905).

———. *À margem da história*. Porto: Livraria Chardron, 1909.

———. *À margem da história*. São Paulo: Martins Fontes, 1999.

Selections used in this text:

"Impressões gerais," 1–15

"Rios em abandono," 16–27

"Clima caluniado," 28–40

"Os Caucheros," 40–52

"Judas Asvero," 52–58

"Brasilieros," 58–71

"A Transacreana," 71–84

———. *Obra completa*. Edited by Afrânio Coutinho. 2 vols. Rio de Janeiro: J. Aguilar, 1966.

da Cunha, Euclides, and Leandro Tocantins. *Um paraíso perdido: Ensaios, estudos e pronunciamentos sobre a Amazônia; Ed. comemorativa do 80a. aniversário da presença de Euclides da Cunha na Amazônia (1905–1985)*. Rio de Janeiro: J. Olympio Editora and Fundação Desenvolvimento de Recursos Humanos da Cultura e do Desporto do Governo do Estado do Acre, 1986.

Selections used in this text:

"Amazônia: A gestão do mundo" (fragment of speech given at his reception into the Brazilian Academy of Letters), 3–5

"Inferno verde," 200–207

"A minha terra e retilinea e alto como as palmeiras," 208–9

"As promessas divinas de esperança," in 2nd ed. (1994), 210–12

"Uma entrevista," 274–78

da Cunha, Euclides, Ronald Sousa, and Lúcia Sá. *The Amazon: Land without History*. Oxford: Oxford University Press, 2006.

Euclides da Cunha Correspondence

The place of writing is supplied where available.

da Cunha to Tulia Ribeiro, Rio de Janeiro, January 7, 1894
da Cunha to Solon Ribeiro, São Paulo January 10, 1895
da Cunha to Luis Cruls, Lorena, February 20, 1903
da Cunha to Lucio de Mendonça, undated 1904
da Cunha to Machado Assis, Santos, February 15, 1904
da Cunha to Coelho Neto, Rio de Janeiro, April 22, 1904
da Cunha to Vicente de Carvalho, Guarujá, April 27, 1904
da Cunha to José Veríssimo, Guarujá, June 24, 1904
da Cunha to Veríssimo, Rio de Janeiro, June 6, 1904
da Cunha to Manuel da Cunha, Rio de Janeiro, October 10, 1904
da Cunha to Manuel da Cunha, Manaus, December 30, 1904
da Cunha to Veríssimo, Manaus, January 13, 1905
da Cunha to Oliveira Lima, Manaus, January 16, 1905
da Cunha to Verissimo, Manaus, February 2, 1905
da Cunha to Coelho Neto, Manaus, March 10, 1905
da Cunha to Coelho Neto, Manaus, March 19, 1905
da Cunha to Artur Lemos, Manaus, no date 1905
da Cunha to Alfredo Rangel, Manaus, March 20, 1905
da Cunha to Domício da Gama, Manaus, May 11, 1905 (telegram)
da Cunha to Jose Veríssimo, Manaus, November 8, 1905
da Cunha to Rio Branco, Manaus, November 10, 1905
da Cunha to Rio Branco, Manaus, November 22, 1905
da Cunha to Dilermando de Assis, Rio, undated 1906
da Cunha to Firmo Dutra, Rio, January 16, 1906
da Cunha to Verissimo, Rio, February 2, 1906
da Cunha to Manuel da Cunha, February 14, 1906
da Cunha to Francisco Escobar, Rio, April 18, 1906
da Cunha to Oliveira Lima, Rio, May 23, 1906
da Cunha to Francisco Escobar, Rio, June 13, 1906
Manuel da Cunha to Euclides, Trinidade, June 26, 1906
da Cunha to Henrique Coelho, Rio, June 30, 1906
da Cunha to Firmo Dutra, Rio, July 9, 1906
da Cunha to Manuel da Cunha, Rio, July 24, 1906
da Cunha to Henrique Coelho, Rio, July 30, 1906
da Cunha to Firmo Dutra, Rio, September 30, 1906
Manuel da Cunha to Euclides da Cunha, Trinidade, December 13, 1906
da Cunha to Francisco Escobar, Rio, December 31, 1906
da Cunha to Otaviano Vieira, Rio, July 4, 1907
da Cunha to Domício da Gama, Rio, November 16, 1907
da Cunha to Francisco Escobar, April 10, 1908
da Cunha to Oliveira Lima, Rio, May 5, 1909
da Cunha to Otaviano Vieira, Rio, July 3, 1909

da Cunha to Otaviano Vieira, Rio, July 5, 1909
da Cunha to Otaviano Vieira, Rio, August 3, 1909
da Cunha to Manuel da Cunha, August 8, 1909
da Cunha to Otaviano Vieira, Rio, August 9, 1909

Other Works

Abdala, Benjamin, Júnior, and Isabel Alexandre. *Canudos: Palavra de Deus, sonho da terra*. São Paulo: Boitempo Editorial, 1997.

Acevedo Marin, Rosa, and Edna de Castro. *Negros do Trombetas*. Belém: Universidade Federal do Pará, 1997.

———. *No caminho de pedras de Abacatal: Experiência social de grupos negros no Pará*. Belém: Universidade Federal do Pará, 2004.

Adalberto da Prussia, Príncipe. *Brasil: Amazonas, Xingu*. São Paulo: Itatiaia, 1977 (orig. 1842).

Adélaïde-Merlande, Jacques. *La Caraïbe et la Guyane au temps de la révolution et de l'Empire, 1789–1804*. Paris: Karthala, 1992.

Adelman, Jeremy. *Sovereignty and Revolution in the Iberian Atlantic*. Princeton, NJ: Princeton University Press, 2006.

Agassiz, Louis, and Elizabeth Cabot Agassiz. *A Journey in Brazil*. 6th ed. Boston: Ticknor and Fields, 1869.

Aguiar, Flávio, and Ligia Chiappini Moraes Leite. *Civilização e exclusão: Visões do Brasil em José Veríssimo, Euclides da Cunha, Claude Lévi-Strauss e Darcy Ribeiro*. São Paulo: Boitempo Editorial, 2001.

Aguirre Achá, José. *De los Andes al Amazonas: Recuerdos de la Campaña del Acre*. 3rd. ed. La Paz: Imprenta Superior, 1980.

Alcorn, Janis B. *Huastec Ethnobotany*. Austin: University of Texas, 1982.

Alegre, R., et al. "HLA Genes in Cubans and the Detection of Amerindian Alleles." *Molecular Immunology* 44, no. 9 (2007): 2426–35.

Alvares, Maria Luzia Miranda, and Maria Angela d'Inção. *A mulher existe? Uma contribuição ao estudo da mulher e gênero na Amazônia*. Belém: Gepem: Goeldi, 1995.

Amis des Archives d'Outre-Mer. *Terres de Bagne: Le Bagne en Guyane et en Nouvelle-Calédonie, 1852–1953*. Aix-en-Provence, France: Amis des Archives d'Outre-Mer, 1990.

Amory, Frederic. *Euclides da Cunha: Uma odisseia nos trópicos*. São Paulo: Ateliê Editorial, 2009.

———. "Euclides da Cunha and Brazilian Positivism." *Luso-Brazilian Review* 36, no. 1 (1999): 87–94.

Anderson, Anthony B., Peter H. May, and Michael J. Balick. *The Subsidy from Nature: Palm Forests, Peasantry, and Development on an Amazon Frontier*. New York: Columbia University Press, 1991.

Anderson, Owen. "Charles Lyell, Uniformitarianism, and Interpretive Principles." *Zygon* 42, no. 2 (2007): 449–62.

Anderson, Robert N. "The Quilombo of Palmares: A New Overview of a Maroon State in Seventeenth-Century Brazil." *Journal of Latin American Studies* 28 (1996): 545–66.

Anderson, Robin L. *Colonization as Exploitation in the Amazon Rain Forest, 1758–1911*. Gainesville: University Press of Florida, 1999.

Andrade, Olímpio da Sousa. *História de interpretação de "Os Sertões."* São Paulo: Editora Art, 1960.

Anglerius, Petrus Martyr. "De orbo novo petri anglerii decades octo." Seville: Alcalá, 1530.

Anonymous. "Pelo honra da nome." *Correio da Manha*, May 7, 1916.

Anonymous. *History of the Revolution of the 18th Fructidor (September 4th, 1797): And of the Deporations to Guiana, in Consequence of That Revolution.* London: J. Wright, 1800.

Antas, João Baptista de Campo Moraes. *Breve reposta a memória Baptista de Campo Moraes ao tenente da Armada Americana—Inglese F. Maury*. Rio de Janeiro: M. Barreto, 1854.

Appel, Alfred, Charles Hamilton Newman, and Vladimir Vladimirovich Nabokov. *Nabokov: Criticism, Reminiscences, Translations, and Tributes*. Evanston, IL: Northwestern University Press, 1971.

Appelbaum, Nancy P., Anne S. Macpherson, and Karin Alejandra Rosemblatt. *Race and Nation in Modern Latin America*. Chapel Hill: University of North Carolina Press, 2003.

Aquino, Ivânia Campigotto. *Literatura e história em diálogo: Um olhar sobre Canudos*. Passo Fundo, Brazil: Universidade de Passo Fundo, 2000.

Aragão, R. Batista. *Guerra da honra ofendida: Brasil-Paraguai*. Fortaleza, 1990.

Arana, Julio César, and Great Britain Parliament, House of Commons, Select Committee on Putumayo. *Las cuestiones del Putumayo: Declaraciones prestadas ante el Comité de Investigación de la Cámara de los Comunes y Debidamente Anotadas*. Barcelona: Impr. Viuda de L. Tasso, 1913.

Araripe, Tristão Alencar. *Expedições militares contra Canudos: Seu aspecto marcial*. 2nd ed. Rio de Janeiro: Biblioteca do Exército Editora, 1985.

Araújo dos Anjos, Rafael. *Territórios das communidades remanscentes de antigos quilombos*. Brasilia: Mapas Editora, 2000.

Araujo, Mundinha. *Insurreição de escravos em Viana, 1867*. São Luís, Brazil: Sioge, 1994.

Arber, Edward, and Richard Eden. *The First Three English Books on America, 1511–1555 A.D.* New York: Kraus Reprint, 1971.

Arce-Nazario, Javier A. "Human Landscapes Have Complex Trajectories: Reconstructing Peruvian Amazon Landscape History from 1948 to 2005." *Landscape Ecology* 22 (2007): 89–101.

Arnold, David. "Legible Bodies: Race, Criminality and Colonialism in South Asia." *English Historical Review* 120, no. 486 (2005): 555–56.

———. *The Problem of Nature: Environment, Culture, and European Expansion*. London: Blackwell, 1999.

———. *The Traveler and the Tropical Gaze*. Seattle: University of Washington Press, 2006.

———. *Warm Climates and Western Medicine: The Emergence of Tropical Medicine, 1500–1900*. Amsterdam: Rodopi, 1996.

Arons, Nicholas. *Waiting for Rain: The Politics and Poetry of Drought in Northeast Brazil.* Tucson: University of Arizona, 2004.

Arruda, João. *Canudos: Messianismo e conflito social*. Fortaleza, Brazil: Edições Federal University of Ceara, 1993.

Asad, Talal. *Formations of the Secular: Christianity, Islam, Modernity*. Stanford, CA: Stanford University Press, 2003.

———. *Genealogies of Religion: Discipline and Reasons of Power in Christianity and Islam.* Baltimore: Johns Hopkins University Press, 1993.

Ascherson, Neal. *The King Incorporated: Leopold the Second and the Congo*. London: Granta Books, 1963.

Asner, G. P., D. E. Knapp, E. N. Broadbent, P. J. C. Oliveira, M. Keller, and J. N. Silva. "Selective Logging in the Brazilian Amazon." *Science* 310, no. 5747 (2005): 480-82.

Assies, Willem. "From Rubber Estate to Simple Commodity Production: Agrarian Struggles in the Northern Bolivian Amazon." *Journal of Peasant Studies* 29, no. 3-4 (2002).

Auzel, Jean-Baptiste. "The Vendée in History: Proceedings of the International Colloquium (French)." *Bibliotheque de l'Ecole des Chartes* 153, no. 2 (1995): 579-81.

Bak, Hans, and Walter Hölbling. *"Nature's Nation" Revisited: American Concepts of Nature from Wonder to Ecological Crisis*. Amsterdam: Vrei Universität Press, 2003.

Balée, William, and Clark L. Erickson. *Time and Complexity in Historical Ecology: Studies in the Neotropical Lowlands*. New York: Columbia University Press, 2006.

Balf, Todd. *The Darkest Jungle: The True Story of the Darién Expedition and America's Ill-Fated Race to Connect the Seas*. New York: Crown, 2003.

Balick, Michael J., Elaine Elisabetsky, and Sarah A. Laird. *Medicinal Resources of the Tropical Forest: Biodiversity and Its Importance to Human Health*. New York: Columbia University Press, 1996.

Bandeira, Luis Alberto Moniz. "O Barão de Rothschild e a questão de Acre." *Revista Brasileira de Política Internacional* 43, no. 2 (2000): 150-69.

Baptiste, Victor N., and Bartolomé de las Casas. *Bartolome de Las Casas and Thomas More's "Utopia": Connections and Similarities; A Translation and Study*. Madrid: Labyrinthos, 1990.

Barbosa, H. A., A. R. Huete, and W. E. Baethgen. "A 20-Year Study of NDVI Variability over the Northeast Region of Brazil." *Journal of Arid Environments* 67, no. 2 (2006): 288-307.

Barbosa, Lucia Maria de Assunção, Petronilha Beatriz Gonçalves e Silva, and Valter Roberto Silvério. *De Preto a Afro-Descendente: Trajetos de pesquisa sobre o negro, cultura negra e relações étnico-raciais no Brasil*. São Carlos, Brazil: Editora Universidade San Carlos, 2003.

Barbosa, Waldemar de Almeida. *Negros e quilombos em Minas Gerais*. Belo Horizonte: Editora Federal Universidade de Minas Gerais, 1972.

Barboza, Mário Gibson. *A cartografía política do Barão do Rio-Branco*. Brasília: Ministério das Relações Exteriores, Seção de Publicações, 1970.

Barham, Bradford L., Jean-Paul Chavas, and Oliver T. Coomes. "Sunk Costs and the Natural Resource Extraction Sector: Analytical Models and Historical Examples of Hysteresis and Strategic Behavior in the Americas." *Land Economics* 74, no. 4 (1998): 429-48.

Barham, Bradford L., and Oliver T. Coomes. *Prosperity's Promise: The Amazon Rubber Boom and Distorted Economic Development*. Boulder: Westview, 1996.

———. "Reinterpreting the Amazon Rubber Boom: Investment, the State, and Dutch Disease." *Latin American Research Review* 29, no. 2 (1994): 73-109.

———. "Wild Rubber: Industrial Organization and the Microeconomics of Extraction during the Amazon Rubber Boom (1860-1920)." *Journal of Latin American Studies* 26 (1994): 37-72.

Barickman, Bart J. *A Bahian Counterpoint: Sugar, Tobacco, Cassava, and Slavery in the Recôncavo, 1780-1860*. Stanford, CA: Stanford University Press, 1998.

———. "A Bit of Land Which They Call Roça." *Hispanic American Historical Review* 72, no. 4 (1994).
———. "The Changing Position of Women in the Brazilian Polity and Economy, 1889-1940." *Hispanic American Historical Review* 72, no. 3 (1992): 492-94.
———. "Persistence and Decline: Slave Labour and Sugar Production in the Bahian Recôncavo, 1850-1888." *Journal of Latin American Studies* 28 (1996): 581-633.
———. "Tame Indians, Wild Heathens and Settlers in the Late 18th and Early 19th Centuries." *The Americas* 51, no. 3 (1995): 325-68.
Barman, Roderick J. "The Brazilian Peasantry Re-examined: The Implications of the Quebra-Quilo Revolt." *Hispanic American Historical Review* 57, no. 3 (1977): 401-24.
———. *Citizen Emperor: Pedro II and the Making of Brazil, 1825-1891*. Stanford, CA: Stanford University Press, 1999.
———. *Princess Isabel of Brazil: Gender and Power in the Nineteenth Century*. Wilmington, DE: SR Books, 2002.
Barreto, Benedicto Bastos. *No tempo dos bandeirantes*. 2nd ed. São Paulo: Departamento de Cultura, 1940.
Barreto, Dantas. *Última expedição a Canudos*. Porto Alegre: Franco e Irmão, 1898.
Barretto, Maria Amália Pereira. *Os voduns do Maranhão*. São Luís, Brazil: Fundação Cultural do Maranhão, 1977.
Barros, Luitgarde Oliveira Cavalcanti. *A derradeira gesta: Lampião e nazarenos guerreando no Sertão*. Rio de Janeiro: Fundação de Amparo à Pesquisa do Estado do Rio de Janeiro, 2000.
Bartelt, Dawid Danilo. *Nation Gegen Hinterland: Der Krieg Von Canudos in Brasilien; Ein Diskursives Ereignis (1874-1903)*. Stuttgart: Steiner, 2003.
Bastide, Roger. *The African Religions of Brazil: Toward a Sociology of the Interpenetration of Civilizations*. Baltimore: Johns Hopkins University Press, 1978.
Bastide, Roger, and Florestan Fernandes. *Brancos e negros em São Paulo*. São Paulo: Editora Nacional, 1959.
Bastos, José Augusto Cabral Barretto. *Incompreensível e bárbaro inimigo: A guerra simbólica contra Canudos*. Salvador, Brazil: Editora da Universidade Federal da Bahia, 1995.
Bates, Henry Walter. *The Naturalist on the River Amazons: A Record of Adventures, Habits of Animals, Sketches of Brazilian and Indian Life, and Aspects of Nature under the Equator during Eleven Years of Travel*. London: J. Murray, 1863.
Baudot, Georges. *Utopia and History in Mexico: The First Chroniclers of Mexican Civilization (1520-1569)*. Niwot: University Press of Colorado, 1995.
Beattie, Peter M. *The Tribute of Blood: Army, Honor, Race, and Nation in Brazil, 1864-1945*. Durham, NC: Duke University Press, 2001.
Bediaga, Begonha. "Joining Pleasure and Work in the Making of Science: The Jardim Botânico do Rio de Janeiro, 1808 to 1860." *História, Ciências, Saúde-Manguinhos* 14 (2007): 1131-57.
Beiguelman, Paula. *A prática nacionalista: Dever intrínseco das Forças Armadas Nacionais*. São Paulo: Sindicato dos Escritores no Estado de São Paulo, 1987.
Beinart, William. "Men, Science, Travel and Nature in the Eighteenth and Nineteenth-Century Cape." *Journal of Southern African Studies* 24, no. 4 (1998): 775-99.
Bell, Whitfield J. "The Relation of Herndon and Gibbon's Exploration of the Amazon to

American Slavery, 1850–1855." *Hispanic American Historical Review* 19, no. 4 (1939): 494–503.

Bello, José Maria. *Inteligência do Brasil: Ensaios sobre Machado de Assis, Joaquim Nabuco, Euclides da Cunha e Rui Barbosa; Síntese da evolução literária do Brasil*. 3rd ed. São Paulo: Editora Nacional, 1938.

Benchimol, Samuel. *Eretz Amazônia: Os judeus na Amazônia*. Manaus: Editora Valer, 1998.

Bennett, Lerone. *Forced into Glory: Abraham Lincoln's White Dream*. Chicago: Johnson, 2000.

Bénot, Yves. *La Guyane sous la Révolution Française, ou l''impasse de la Révolution Pacifique*. Kourou, French Guiana: Ibis Rouge Éditions, 1997.

Benton, F. Weber. *Semi-tropic California: The Garden of the World; Including a Concise History of Panama and the Panama Canal, with Map, and the Missions of California, Together with Verses in Accord with the Topic*. 2nd ed. Los Angeles: Benton, 1915.

Benton, Lauren. *Law and Colonial Cultures: Legal Regimes in World History*. New York: Cambridge University Press, 2002.

———. *A Search for Sovereignty: Law and Geography in European Empires, 1400–1900*. Cambridge: Cambridge University Press, 2009.

Bergad, Laird W. *Slavery and the Demographic and Economic History of Minas Gerais, Brazil, 1720-1888*. New York: Cambridge University Press, 1999.

Bernucci, Leopoldo M. *A imitação dos sentidos: Prógonos, contemporâneos e epígonos de Euclides da Cunha*. São Paulo: Editora da Universidade de São Paulo (EDUSP), and Boulder: University of Colorado at Boulder, 1995.

———. Preface and notes to *Os Sertões* by Euclides da Cunha. Cotia, Colecão clássicos comentados. Sao Paulo: Ateliê, 2002.

Bertani, S., G. Bourdy, I. Landau, J. C. Robinson, P. Esterre, and E. Deharo. "Evaluation of French Guiana Traditional Antimalarial Remedies." *Journal of Ethnopharmacology* 98, nos. 1–2 (2005): 45–54.

Bertrand, Michel. "Utopian Dreamers in Hispanic America in the 16th Century." *Caravelle* 78 (2002): 57–66.

Besouchet, Lidia. *Rio-Branco e as relações entre o Brasil e a República Argentina*. Rio de Janeiro: Ministério das Relações Exteriores, 1949.

Besselaar, J. J. van den, and António Vieira. *António Vieira, profecia e polêmica*. Rio de Janeiro: Editora Universidade Estadual Rio de Janeiro, 2002.

Bethell, Leslie. *The Abolition of the Brazilian Slave Trade; Britain, Brazil and the Slave Trade Question, 1807–1869*. Cambridge: Cambridge University Press, 1970.

Betts, Raymond F. *The "Scramble" for Africa: Causes and Dimensions of Empire*. Boston: Heath, 1966.

Blashford-Snell, John, and Richard Snailham. *East to the Amazon: In Search of Great Paititi and the Trade Routes of the Ancients*. London: John Murray, 2002.

Block, David. *Mission Culture on the Upper Amazon: Native Tradition, Jesuit Enterprise and Secular Policy in Moxos, 1660–1880*. Lincoln: University of Nebraska Press, 1994.

Blok, Anton. "The Peasant and the Brigand: Social Banditry Reconsidered." *Comparative Studies in Society and History* 14, no. 4 (1972): 494–503.

Boomert, Arie. "The Arawak Indians of Trinidad and Coastal Guiana, c. 1500–1650." *Journal of Caribbean History* 19, no. 2 (1984): 123-88.

———. "Between the Mainland and the Islands: The Amerindian Cultural Geography of Trinidad." *Bulletin of the Peabody Museum of Natural History* 50, no. 1 (2009): 63–73.

———. "Gifts of the Amazons: Greenstone Pendants and Beads as Items of Ceremonial Exchange in Amazonia and the Caribbean." *Antropologica* 67 (1987): 33–44.

Borges, Dain. "Puffy, Ugly, Slothful and Inert: Degeneraton in Brazilian Social Thought." *Journal of Latin American Studies* 25, no. 2 (1993): 235–56.

Botelho Gosálvez, Raúl. *Proceso del imperialismo del Brasil, de Tordesillas a Roboré*. La Paz: Editorial Universitaria, 1960.

Bowd, Gavin, and Daniel Clayton. "Fieldwork and Tropicality in French Indochina: Reflections on Pierre Gourou's *Les Paysans du Delta Tonkinois*, 1936." *Singapore Journal of Tropical Geography* 24, no. 2 (2003): 147–68.

———. "Tropicality, Orientalism, and French Colonialism in Indochina: The Work of Pierre Gourou, 1927-1982." *French Historical Studies* 28, no. 2 (2005).

Boxer, C. R. *The Dutch in Brazil, 1624–1654*. Oxford: Clarendon, 1957.

Bradford, Phillips Verner, and Harvey Blume. *Ota Benga: The Pygmy in the Zoo*. New York: St. Martin's, 1992.

Braga-Pinto, César. *As promessas da história: Discursos proféticos e assimilação no Brasil colonial (1500-1700)*. São Paulo: Editora da Universidade de São Paulo (EDUSP), 2003.

Brandão, Adelino. *Euclides da Cunha e a questão racial no Brasil: A antropologia de "Os Sertões."* Rio de Janeiro: Presença, 1990.

Brandão, Tanya Maria Pires. *O escravo na formação social do Piauí: Perspectiva histórica do século XVIII*. Teresina, Brazil: Editora Gráfica da Universidade Federal de Piauí, 1999.

Brantlinger, Patrick. *Dark Vanishings: Discourse on the Extinction of Primitive Races, 1800-1930*. Ithaca, NY: Cornell University Press, 2003.

Brass, Tom. *Towards a Comparative Political Economy of Unfree Labour: Case Studies and Debates*. London: Frank Cass, 1999.

Brazil, Maria do Carmo. *Fronteira negra: Dominação, violência e resistência escrava em Mato Grosso, 1718–1888*. Passo Fundo, Brazil: Universidade de Passo Fundo, 2002.

Breman, Jan, Piet De Rooy, Ann Laura Stoler, and Wim F. Wertheim. *Imperial Monkey Business: Racial Supremacy in Social Darwinist Theory and Colonial Practice*. Amsterdam: Vrei Universität Press, 1990.

Bridgman, Frederick. The Acre Territory. Documents concerning the controversy between Brazil and Bolivia over a contract made with American citizens, 1902? LC control number 04019530, Record ID 4005096. UCLA Libraries and Collections, Los Angeles.

Brito, Jailton Lima. *A abolição na Bahia: 1870–1888*. Salvador, Brazil: Centro de Estudos Baianos, 2003.

Brockway, Lucile. *Science and Colonial Expansion: The Role of the British Royal Botanic Gardens*. New York: Academic Press, 1979.

Brooks, George E. *Eurafricans in Western Africa: Commerce, Social Status, Gender, and Religious Observance from the Sixteenth to the Eighteenth Century*. Athens: Ohio University Press, and Oxford: James Currey, 2003.

Brown, Charles Henry. *Agents of Manifest Destiny: The Lives and Times of the Filibusters*. Chapel Hill: University of North Carolina Press, 1980.

Brown, Michael F., and Eduardo Fernández. *The War of Shadows: The Struggle for Utopia in the Peruvian Amazon*. Berkeley: University of California Press, 1991.

Browne, Janet. *Secular Ark: Studies in the History of Biogeography*. New Haven, CT: Yale University Press, 1983.

Bruleaux, Anne-Marie, Régine Calmont, and Serge Mam Lam Fouck. *Deux siècles d'esclavage en Guyane Française, 1652–1848*. Cayenne, French Guiana: Centre Guyanais d'Études et de Recherches; Paris: L'Harmattan, 1986.

Bruneau, Michel. "From a Centred to a Decentered Tropicality: Francophone Colonial and Postcolonial Geography in Monsoon Asia." *Singapore Journal of Tropical Geography* 26, no. 3 (2005): 304-22.

Bruno, Ernani Silva. *Café e negro: Contribuição para o estudo da economia caféeira de São Paulo na fase do trabalho servil*. São Paulo: Atalanta, 2005.

Buarque de Holanda, Sérgio. *Monções*. Rio de Janiero, 1945.

Buckle, Henry Thomas. *History of Cvilization in England*. 3 vols. New York: L. D. Appleton, 1867.

Bueno, Clodoaldo. *A República e sua política exterior (1889 a 1902)*. São Paulo: Editora Universidade de São Paulo, 1995.

———. *Política externa da Primeira República: Os anos de apogeu (de 1902 a 1918)*. São Paulo: Paz e Terra, 2003.

Bueno-Hernandez, A. A., and J. E. Llorente-Bousquets. "The Other Face of Lyell: Historical Biogeography in His Principles of Geology." *Journal of Biogeography* 33, no. 4 (2006): 549–59.

Buffon, Georges Louis Leclerc, and Jean-Antoine-Nicolas de Caritat Condorcet. *Buffon's Natural History: Containing a Theory of the Earth, a General History of Man, of the Brute Creation, and of Vegetables, Minerals, &C, &C*. 10 vols. London: H. D. Symonds, 1807.

Bunker, Stephen G. *Underdeveloping the Amazon*. Chicago: University of Chicago Press, 1985.

Burke, Peter, and Maria Lucia Pallares Burke. *Gilberto Freyre: Social Theory in the Tropics*. Oxford: Peter Lang, 2008.

Burnett, D. Graham. "'It Is Impossible to Make a Step without the Indians': Nineteenth-Century Geographical Exploration and the Amerindians of British Guiana." *Ethnohistory* 49, no. 1 (2002): 3–40.

———. *Masters of All They Surveyed: Exploration, Geography, and a British El Dorado*. Chicago: University of Chicago Press, 2000.

———. "Matthew Fontaine Maury's 'Sea of Fire': Hydrography, Biogeography and Providence in the Tropics." In *Tropical Vision in an Age of Empire*, edited by Felix Driver and Luciana Martins, 113-36. Chicago: University of Chicago Press, 2005.

Burns, E. Bradford. "The Destruction of a Folk Past: Da Cunha and the Cataclysmic Cultural Clash." *Review of Latin American Studies* 3, no. 1 (1990): 17-36.

———. *Nationalism in Brazil: A Historical Survey*. New York: Praeger, 1968.

———. *The Poverty of Progress: Latin America in the Nineteenth Century*. Berkeley: University of California Press, 1983.

———. *The Unwritten Alliance: Rio-Branco and Brazilian-American Relations*. New York: Columbia University Press, 1966.

Burns, E. Bradford, Thomas E. Skidmore, and Virginia Bernhard. *Elites, Masses, and Modernization in Latin America, 1850–1930*. Austin: University of Texas Press, 1979.

Burnside, Madeleine, and Rosemarie Robotham. *Spirits of the Passage: The Transatlantic Slave Trade in the Seventeenth Century*. New York: Simon and Schuster, 1997.

Burr, George Lincoln. "The Guiana Boundary: A Postscript to the Work of the Commission." *Hispanic American Historical Review* 6, no. 1 (1900): 49–64.

Burrieza Sánchez, Javier. *Jesuitas en Indias entre la utopía y el conflicto: Trabajos y misiones de la Compañía de Jesús en la América moderna*. Valladolid, Spain: Universidad de Valladolid, Secretariado de Publicaciones e Intercambio Editorial, 2007.

Burton, Richard Francis. *Letters from the Battlefields of Paraguay*. London: Tinsley Brothers, 1870.

Butler, Kim D. *Freedoms Given, Freedoms Won: Afro-Brazilians in Post-abolition São Paulo and Salvador*. New Brunswick, NJ: Rutgers University Press, 1998.

Cabral, Francisco Pinto. *Plácido de Castro e o Acre Brasileiro*. Brasília: Thesaurus, 1986.

Cabral, Maria do Socorro Coelho. *Caminhos do Gado: Conquista e ocupação do sul do Maranhão*. São Luís, Brazil: Edições Secretaria de Cultura de Maranhão 1992.

Caldeira, Clovis. *Mutirão, formas de ajuda mútua no meio rural*. São Paulo: Companhia Editora Nacional, 1956.

Caldeira, Jorge. *Mauá: Empresário do império*. São Paulo: Companhia das Letras, 1995.

Campbell, K. E., C. D. Frailey, and L. Romero-Pittman. "The Pan-Amazonian Ucayali Peneplain, Late Neogene Sedimentation in Amazonia, and the Birth of the Modern Amazon River System." *Paleogeography Paleoclimatology Paleoecology* 239, nos. 1–2 (2006): 166–219.

Campanella, Tommaso, and Daniel J. Donno. *The City of the Sun: A Poetical Dialogue*. Berkeley: University of California Press, 1981.

Cañizares-Esguerra, Jorge. *How to Write the History of the New World: Histories, Epistemologies, and Identities in the Eighteenth-Century Atlantic World*. Stanford, CA: Stanford University Press, 2001.

———. *Nature, Empire, and Nation: Explorations of the History of Science in the Iberian World*. Stanford, CA: Stanford University Press, 2006.

———. *Puritan Conquistadors: Iberianizing the Atlantic, 1550–1700*. Stanford, CA: Stanford University Press, 2006.

Cardoso, Ciro Flamarion Santana. *Economia e sociedade em áreas colonais periféricas: Guiana Francesa e Pará*. Rio de Janeiro: Graal, 1984.

———. *Escravo ou camponês? O protocampesinato negro nas Américas*. São Paulo: Editora Brasiliense, 1987.

———. *La Guyane Français, 1715–1817: Aspects économiques et sociaux; Contribution à l'étude des sociétés esclavagistes d'Amérique*. Petit-Bourg, Guadeloupe, France: Ibis Rouge, 1999.

Carney, Judith A. *Black Rice: The African Origins of Rice Cultivation in the Americas*. Cambridge, MA: Harvard University Press, 2001.

Carney, Judith A., and Richard Rosomoff. *In the Shadow of Slavery*. Berkeley: University of California Press, 2009.

Carreón Blaine, Emilie A. *El olli en la plástica mexica: El uso del hule en el siglo XVI*. Mexico City: Universidad Nacional Autónoma de México, Instituto de Investigaciones Estéticas, 2006.

Carter, Paul. *The Road to Botany Bay: An Exploration in Landscape and History*. Chicago: University of Chicago Press, 1989.

Casement, Roger, and Angus Mitchell. *The Amazon Journal of Roger Casement*. London: Anaconda Editions, 1997.

Casement, Roger, and Roger Sawyer. *Roger Casement's Diaries: 1910, the Black and the White*. London: Pimlico, 1997.

Cashman, Sean Dennis. *America in the Gilded Age: From the Death of Lincoln to the Rise of Theodore Roosevelt*. New York: New York University Press, 1984.

Castelnau, Francis, Comte de. *Expedítion dans les parties centrales de l'Amérique du Sud, de Rio a Lima et de Lima au Pará, executés par l'order de gouvernement français*. Paris: P. Bertrand, 1850.

Castro, Celso, and Maria Celina Soares de Araújo. *Militares e política na Nova República*. Rio de Janeiro: Fundação Getulio Vargas, 2001.

Castro, Edna Maria Ramos. *Escravos e senhores de Bragança*. Belém: Universidade Federal do Pará, 2006.

Castro, Edna Maria Ramos, Rosa Elizabeth Acevedo Marin, and Universidade Federal do Pará, Nucléo de Altos Estudos Amazônicos. *Amazônias em tempo de transição*. Belém: NAEA, 1989.

Cavalcanti Filho, Sebastião Barbosa. *A questão jesuítica no Maranhão colonial (1622–1759)*. São Luís, Brazil: Serviço de Imprensa e Obras Gráficas do Estado, 1990.

Cavasini, C. E., et al. "Duffy Blood Group Gene Polymorphisms among Malaria Vivax Patients in Four Areas of the Brazilian Amazon Region." *Malaria Journal* 6 (2007).

Cerqueira, Dionísio. *Reminiscências da Campanha do Paraguai, 1865–1870*. Rio de Janeiro: Biblioteca do Exército, 1958.

Cervo, Amado Luiz, and Clodoaldo Bueno. *História da política exterior do Brasil*. São Paulo: Editora Atica, 1992.

Chandler, Billy Jaynes. *The Feitosas and the Sertão dos Inhamuns: The History of a Family and a Community in Northeast Brazil, 1700–1930*. Gainesville: University of Florida Press, 1972.

Chandless, William. "Ascent of the River Purús." *Journal of the Royal Geographical Society* 36 (1866): 86–118.

Chaplin, Joyce E. "Nature and Nation: Natural History in Context." *Transactions of the American Philosophical Society* 93 (2003): 75–96.

Chardon, Jean-Pierre, Dynamique Industrielle S.A., and France, Ministère de l'Industrie, Moyennes et petites industries de la Guadeloupe. *Le cabotage dans la Caraïbe: Vers la création d'une flotte française rattachée à la zone Caraïbe; Étude*. 2 vols. Pointe-à-Pitre, Guadeloupe: Ministère de l'Industrie, Moyennes et petites industries de la Guadeloupe, 1980.

Chirif, Alberto. *Etnicidad y ecología*. Lima: Centro de Investigación y Promoción Amazónica, 1979.

Christopher, Emma, Cassandra Pybus, and Marcus Rediker. *Many Middle Passages: Forced Migration and the Making of the Modern World*. Berkeley: University of California Press, 2007.

Church, George Earl. *Aborigines of South America*. London: Chapman and Hall, 1912.

———. *Desiderata in Exploration*. London: Royal Geographical Society, 1907.

———. *Interoceanic Communication on the Western Continent: A Study in Commercial Geography*. London: William Clowes and Sons, 1902.

———. *The Route to Bolivia via the River Amazon: A Report to the Governments of Bolivia and Brazil*. London, 1877.

Cipolletti, María Susana. "Lacrimabili-Statu, Part 1: Indigenous Slaves in the Northwestern Area of the Amazon (XVII–XIX Centuries)." *Revista de Índias* 55, no. 205 (1995): 551–71.

Clastres, Hélène. *The Land without Evil*. Urbana: University of Illinois Press, 1995.

Clavijero, Francesco Saverio, and Mariano Cuevas. *Historia antigua de México*. Mexico City: Editorial Porrua, S.A., 1958.

Cleary, David. "'Lost Altogether to the Civilised World': Race and the Cabanagem in Northern Brazil, 1750 to 1850." *Comparative Studies in Society and History* 40, no. 1 (1998): 109–35.

———. "Towards an Environmental History of the Amazon." *Latin American Research Review* 36, no. 2 (2001): 64–96.

Clemanceau, Joseph. *Histoire de la Guerre de la Vendée (1793–1815)*. Paris: Nouvelle Librairie Nationale, 1909.

Clement, Charles R. "1492 and the Loss of Amazonian Crop Genetic Resources: Part 2, Crop Biogeography at Contact." *Economic Botany* 53, no. 2 (1999): 203–16.

———. "A Center of Crop Genetic Diversity in Western Amazonia." *Bioscience* 39, no. 9 (1989): 624–31.

Cleven, Andrew N. "Some Plans for Colonizing Liberated Negro Slaves in Hispanic America." *Journal of Negro History* 11, no. 1 (1926): 35–49.

Coelho, Mauro. "As viagens filosóphicas de Charles-Marie de la Condamine e Alexandre Rodrigues Ferreira: Ensaio comparativo." In *Nas terras do Cabo Norte: Fronteiras, colonização e escravidão na Guiana Brasileira*, edited by Flávio dos Santos Gomes, 98–127. Belém: Fundação Estadual de Cultura do Estado Amapá. 1999.

Coelho Neto, Henrique. *Livro de prata*. São Paulo: Livraria Liberdade, 1928.

"Coffee-Houses Vindicated in Answer to the Late Published Character of a Coffee-House Asserting from Reason, Experience, and Good Authours, the Excellent Use and Physical Vertues of That Liquor: With the Grand Conveniency of Such Civil Places of Resort and Ingenious Conversation." London: J. Lock for J. Clarke, 1673.

Coggiola, Osvaldo, *A Revolução Francesa e seu impacto na América Latina*. Brasilia: Nova Stella, 1990.

Cohen, M. C. L., et al. "Wetland Dynamics of Marajó Island, Northern Brazil, during the last 1000 Years." *CATENA* 76, no. 1 (2008): 70–77.

Cohen, Thomas M. *The Fire of Tongues: António Vieira and the Missionary Church in Brazil and Portugal*. Stanford, CA: Stanford University Press, 1998.

Collier, Richard. *The River That God Forgot: The Story of the Amazon Rubber Boom*. London: Collins, 1968.

Collin, Richard H. *Theodore Roosevelt's Caribbean: The Panama Canal, the Monroe Doctrine, and the Latin American Context*. Baton Rouge: Louisiana State University Press, 1990.

Colonization and Commercial Company of Bolivia and George Earl Church. *Bolivian Colonization: Being Prospectus of the Colonization and Commercial Co. Of Bolivia. Incorporated in San Francisco, California, January 25th, A.D. 1870*. San Francisco: Alta California Printing House, 1870.

Comaroff, Jean, and John L. Comaroff. *Of Revelation and Revolution*. Chicago: University of Chicago Press, 1991.

Comaroff, John L. *Ethnicity, Nationalism, and the Politics of Identity in an Age of Revolution*. Chicago: American Bar Foundation, 1994.

Conrad, Joseph. "Geography and Some Explorers." *National Geographic* 45 (1924): 1–31.

———. *Heart of Darkness*. New York: Signet Classics, 1997.

———. *Nostromo*. London: Penguin Books, 1994.

Conrad, Robert Edgar. *World of Sorrow: The African Slave Trade to Brazil*. Baton Rouge: Louisiana State University Press, 1986.

Coomes, Oliver T., and G. J. Burt. "Indigenous Market-Oriented Agroforestry: Dissecting Local Diversity in Western Amazonia." *Agroforestry Systems* 37, no. 1 (1997): 27–44.

Cooper, Frederick, and Ann Laura Stoler. *Tensions of Empire: Colonial Cultures in a Bourgeois World*. Berkeley: University of California Press, 1997.

Coordenação do Novo Movimento Histórico de Canudos. *Noventa anos depois: Canudos de novo*. Nazaré, Brazil: Coordenação do Movimento Histórico de Canudos, 1985.

Cormier, Loretta A. "Between the Ship and the Bulldozer: Guajá Subsistence, Sociality and Symbolism after 1500." In *Time and Complexity in Historical Ecology*, edited by William Balée and Clark L. Erickson. New York: Columbia University Press, 2006.

Corro Ramos, Octaviano. *Los cimarrones en Veracruz y la fundación de Amapa*. Veracruz, Mexico: Imprenta Comercial, 1951.

Cortesão, Jaime. *Alexandre de Gusmão e o Tratado de Madrid (1750)*. Rio de Janeiro: Ministério das Relações Exteriores, Instituto Rio-Branco, 1950.

———. *História do Brasil nos velhos mapas*. Rio de Janeiro: Ministério das Relações Exteriores Instituto Rio Branco, 1965.

———. *A missão dos padres matemáticos no Brasil*. Lisbon: Agência Geral do Ultramar, 1958.

Cortesão, Jaime, and Pedro de Angelis. *Antecedentes do Tratado de Madri: Jesuítas e bandeirantes no Paraguai, 1703–1751*. Rio de Janeiro: Biblioteca Nacional Divisão de Obras Raras e Publicações, 1955.

Cosgrove, Denis E., and Stephen Daniels. *The Iconography of Landscape: Essays on the Symbolic Representation, Design, and Use of Past Environments*. Cambridge: Cambridge University Press, 1988.

Costa, Maria de Fátima G. *História de um país inexistente: O Pantanal entre os séculos XVI e XVIII*. São Paulo: Estação Liberdade, Livraria Kosmos Editora, 1999.

Costa du Rels, Adolfo. *Félix Avelino Aramayo y su época, 1846–1929*. Buenos Aires: D. Viau, 1942.

Costigan, Lúcia Helena, et al. *Diálogos da conversão: Missionários, índios, negros e judeus no contexto ibero-americano de período barroco*. Campinas, Brazil: Editora Universidade de Campinas, 2005.

Coudreau, Henri. *Chez nos Indiens: Quatre années dans la Guyane Française (1887–1891)*. Paris: Hachette et cie, 1893.

———. *Les français en Amazonie*. Paris: A. Picard et Kaan, 1887.

———. *La France equinoxiale*. Paris: Challamel Aine, 1886.

———. *Voyage au Tocantins-Araguaya, 31 décembre 1896–23 mai 1897 . . . illustré de 87 vignettes et d'une carte des rivières "Tocantins-Araguaya."* Paris: A. Lahure, 1897.

———. *Voyage au Xingú, 30 mai 1896–26 octobre 1896*. Paris: A. Lahure, 1897.

Coudreau, Otile. *Voyage au Cuminá, 20 avril 1900–7 septembre 1900*. Paris: A. Lahure, 1901.

———. *Voyage au Trombetas*. Paris: A. Lahare, 1899.

Coutinho, Edilberto. *Rondon, o civilizador da última fronteira*. Rio de Janeiro: Civilização Brasileira, 1975.

Craig, Neville B. *Recollections of an Ill-Fated Expedition to the Head-Waters of the Madeira River in Brazil*. Philadelphia, London: J. B. Lippincott, 1907.

Curtin, Philip D. *Death by Migration: Europe's Encounter with the Tropical World in the Nineteenth Century*. Cambridge: Cambridge University Press, 1989.

———. *Disease and Empire: The Health of European Troops in the Conquest of Africa*. Cambridge: Cambridge University Press, 1998.

———. *The Image of Africa: British Ideas and Action, 1780–1850*. Madison: University of Wisconsin Press, 1964.

———. *Migration and Mortality in Africa and the Atlantic World, 1700–1900*. Aldershot, UK: Ashgate/Variorum, 2001.

Cushner, Nicholas P. *Why Have You Come Here? The Jesuits and the First Evangelization of Native America*. New York: Oxford University Press, 2006.

da Costa, Emília Viotti. *The Brazilian Empire: Myths and Histories*. Rev. ed. Chapel Hill: University of North Carolina Press, 2000.

———. *Crowns of Glory, Tears of Blood: The Demerara Slave Rebellion of 1823*. New York: Oxford University Press, 1994.

———. *Da monarquia à república: Momentos decisivos*. São Paulo: Editorial Grijalbo, 1977.

da Costa, Luiz Monteiro. *Na Bahia colonial: Apontamentos para história militar da Cidade do Salvador*. Salvador: Progresso, 1958.

da Costa e Silva, Alberto *Um rio chamado Atlántico: A Africa no Brasil e o Brasil na África*. Rio de Janeiro: Nova Fronteira, Universidade Federale do Rio de Janeiro (UFRJ), 2003.

da Cunha, Arnaldo Pimenta. "Em tourno da vida de Euclides da Cunha." *Revista do Instituto Geográfico e Histórico de Bahia* 46 (1920): 255–64.

da Cunha Filho, Euclides. "A verdade sobre a morte do meu pai." *Dom Casmurro* 10 (1946): 60–61.

da Cunha, Manuela Carneiro. *História dos índios no Brasil*. São Paulo: Fundação de Amparo à Pesquisa do Estado de São Paulo, Companhia das Letras, Secretaria Municipal de Cultura, 1992.

———. *Negros, estrangeiros: Os escravos libertos e sua volta à África*. São Paulo: Brasiliense, 1985.

da Cunha, Olivia Maria Gomes, and Flávio dos Santos Gomes. *Quase-cidadão: Histórias e antropólogias da pós-emancipação no Brasil*. Rio de Janeiro: Editora Fundacão Getulio Vargas, 2007.

da Cunha, Osvaldo Rodrigues. *O naturalista Alexandre Rodrigues Ferreira: Uma análise comparativa de sua viagem filosófica (1783–1793) pela Amazônia e Mato Grosso com a de outros naturalistas posteriores*. Belém: Museu Paraense Emílio Goeldi, 1991.

da Gama, Domicio. "Euclides da Cunha." *Revista do Gremio Euclides da Cunha*, 1927.

Daibert, Robert, Júnior. *Isabel, a"redentora" dos escravos: Uma historia da princesa entre olhares negros e brancos, 1846–1988*. Bauru, Brazil: Editora da Universidade do Sagrado Coração , 2004.

Daly, D. C., and A. Millozza. "'Lost' Plant Collections from the Amazon, Part 1: The 1899 Expedition of Dr. Luigi Buscalioni." *Taxon* 56, no. 1 (2007): 185–99.

Daniel, João. *Tesouro descoberto no máximo do Rio Amazonas*. 2 vols. Belém: Prefeitura da Cidade, 2004.

d'Anjou, Leo. *Social Movements and Cultural Change: The First Abolition Campaign Revisited*. New York: Aldine de Gruyter, 1996.

Dantas, Beatriz G., José Augusto L. Sampaio, and Maria do Rosario G. Carvalho. "Os povos indigenas do nordeste brasiliero: Um esboço historico." In *História dos índios no Brasil*, edited by Manuela Carneiro da Cunha. São Paulo: Companhia das Letras, 1992.

Dantas, Monica Duarte. *Fronteiras movediças: A comarca de Itapicuru e a formação do arraial de Canudos (Relações sociais na Bahia do século XIX)*. São Paulo: Aderaldo e Rothschild Editores, 2007.

Dantas, Paulo, and Euclides da Cunha. *"Os Sertões" de Euclides e outros Sertões*. São Paulo: Conselho Estadual de Cultura, Comissão de Literatura, 1969.

da Rocha, Júlio Ribeiro. *O Acre, documentos para a historia da sua occupação pelo Brazil*. Lisbon: Minerva Lusitana, 1903.

da Silva, Joaquim Caetano. *L'Oiyapoc et l'Amazon*. 2 vols. Paris: L. Martinet, 1861.

da Silva, José Calasans. *Canudos na literatura de Cordel*. São Paulo: Atica, 1984.

———. *No tempo de Antônio Conselheiro: Figuras e fatos da Campanha de Canudos*. Salvador, Brazil: Aguiar e Souza, Livaria Progresso, 1959.

———. "O jagunçinho de Euclides da Cunha." *Gazeta do Rio Pardo*, 1980.

———. *Quase biografias de jagunços: O séquito de Antonio Conselheiro*. Salvador, Brazil: Universidade Federal da Bahia Centro de Estudos Baianos, 1986.

da Silva, Renato Ignácio. *Madeira–Mamoré–Martírios: Caminhos da ilusão*. São Paulo: RENIG, Editora e Assessoria Publicitária, 1991.

Dauat, Acelio. "A vítima esquecido de Euclides da Cunha." *Folha da Tarde*, 1946.

Daubigny, Eugène Théodore. *Choiseul et la France d'outre-mer après le Traité de Paris: Étude sur la politique coloniale au XVIIIe siècle, avec un appendice sur les origines de la question de Terre-Neuve*. Paris: Hachette, 1892.

David, Onildo Reis. *O inimigo invisível: Epidemia na Bahia no século XIX*. Brasília: Sarah Letras, 1996.

Davidson, David M. "How the Brazilian West Was Won." In *Colonial Roots of Modern Brazil*, edited by Dauril Alden, 61–106. Berkeley: University of California Press, 1972.

Davis, Mike. *Late Victorian Holocausts*. London: Verso, 2001.

Davis, Wade. *One River: Explorations and Discoveries in the Amazon Rain Forest*. New York: Simon and Schuster, 1996.

Davis, Wade, and Richard Evans Schultes. *The Lost Amazon: The Photographic Journey of Richard Evans Schultes*. San Francisco: Chronicle, 2004.

Dawsey, Cyrus B., and James M. Dawsey. *The Confederados: Old South Immigrants in Brazil*. Tuscaloosa: University of Alabama Press, 1995.

de Abranches, Dunshee. *Rio Branco e a política exterior do Brasil, 1902–1912*. Rio de Janeiro: Imprensa Nacional, 1945.

de Abreu, Joao Capistrano. *Chapters of Brazil's Colonial History, 1500–1800*. New York : Oxford University Press, 1997.

de Acosta, José, Edward Grimeston, and Clements R. Markham. *The Natural and Moral History of the Indies*. London: Hakluyt Society, 1880.

de Acuña, Cristóbal, and Grenville Kane. *Nuevo descubrimiento del Gran Rio de las Amazonas*. Madrid: En la imprenta del Reyno, 1641.

Dean, Warren. *Brazil and the Struggle for Rubber: A Study in Environmental History*. New York: Cambridge University Press, 1987.

———. *With Broadax and Firebrand: The Destruction of the Brazilian Atlantic Forest*. Berkeley: University of California Press, 1997.

de Assis, Dilermando. *A tragédia da Piedade, mentiras e calúnias da "A vida dramática de Euclides da Cunha."* 2nd ed. Rio de Janeiro: O Cruzeiro, 1951.

———. *Um conselho de guerra—a morte de aspirante da Marinha, Euclides da Cunha Filho.* Rio de Janeiro: Tipografia dos Anais, 1916.

———. *Um nome, uma vida, uma obra.* Rio de Janeiro: Tipografia Duarte, Neves, 1946.

de Assis, Judith, and Jefferson de Andrade. *Ana de Assis: História de um trágico amor.* Rio de Janeiro: Codecri, 1987.

de Assis-Brasil, Joaquim Francisco. "Report to Olinto Magalhães." Edited by Foreign Ministry. Rio de Janeiro, August 31, 1902.

de Azevedo, Celia Maria Marinho. *Onda negra, medo branco: O negro no imaginário das elites, século XIX.* Rio de Janeiro: Paz e Terra, 1987.

Debbasch, Yvan. "Le crime d'empoisonnement aux îles pendant la période esclavagiste." *Revue d'Histoire d'Outre-mer* (1963).

DeBow, J. B. D. "The South American States." *DeBow's Review* 6 (1848): 3–24.

de Cadornega, António de Oliveira. *História geral das guerras Angolanas, 1680.* 3 vols. Lisbon: Agência-Geral do Ultramar, 1972.

de Carvajal, Gaspar, et al.. *The Discovery of the Amazon According to the Account of Friar Gaspar de Carvajal and Other Documents.* New York: American Geographical Society, 1934.

de Carvalho, José Murilo. *A construção da ordem: A elite política imperial.* Rio de Janeiro: Editora Universidade Federale do Rio de Janeiro (UFRJ), Relume Dumará, 1996.

———. *A formação das almas: O imaginário da República no Brasil.* São Paulo: Companhia das Letras, 1990.

———. "Political Elites and State Building: The Case of 19th-Century Brazil." *Comparative Studies In Society and History* 24, no. 3 (1982): 378–99.

de Carvalho, Rodrigues. *Lampião e a sociologia do cangaço.* Rio de Janeiro: Gráfica Editora do Livro, 1976.

de Castro, Ferreira, and Charles Duff. *Jungle, a Tale of the Amazon Rubber-Tappers.* New York: Viking, 1935.

de Castro, Genesco. *O estado independente do Acre e J. Plácido de Castro.* Rio de Janeiro: Typ. São Benedicto, 1930.

de Fernandes, Rinaldo, and Tripoli Gaudenzi. *O clarim e a oração: Cem anos de "Os Sertões."* São Paulo: Geração Editorial, 2002.Defoe, Daniel. *The Life and Strange Surprizing Adventures of Robinson Crusoe.* London: Oxford University Press, 1972.

de la Condamine, Charles-Marie. *Voyage sur l'Amazone.* Paris: Maspero, 1981.

de la Condamine, Charles-Marie, and Académie des sciences (France). *Journal du voyage fait par ordre du roi, a l'équateur, servant d'introduction historique a la mesure des trois premiers degrés du méridien.* Paris: Imprimerie Royale, 1751.

de Landa, Diego. *Relación de las cosas de Yucatan.* Peabody Museum Papers 18. Cambridge, MA: Peabody Museum of American Archaeology and Ethnology, 1941.

de la Pedraja Tomán, René. *Wars of Latin America, 1899–1941.* London: McFarland, 2006.

de las Casas, Bartolomé, and Nigel Griffin. *A Short Account of the Destruction of the Indies.* London: Penguin, 1992.

de las Casas, Bartolomé, Edmundo O'Gorman, and Jorge Alberto Manrique. *Los indios de México y Nueva España: Antología.* 2nd ed. Mexico City: Editorial Porrúa, 1971.

de la Vega, Marta. *Evolucionismo versus positivismo: Estudio teórico sobre el positivismo y su significación en América Latina.* Caracas: Monte Ávila Editores Latinoamericana, 1998.

de Lavôr, João Conrado Niemeyer. *História do Jardim Botânico do Rio de Janeiro.* Rio de Janeiro: Instituto Brasileiro de Desenvolvimento Florestal, Jardim Botânico do Rio de Janeiro, 1983.

de Léry, Jean. *History of a Voyage to the Land of Brazil, Otherwise Called America.* Orig. 1581; Berkeley: University of California Press, 1992.

Del Giudice, Luisa, and Gerald Porter. *Imagined States: Nationalism, Utopia, and Longing in Oral Cultures.* Logan: Utah State University Press, 2001.

de Lira, José Tavares Correia. "Hidden Meanings: The Mocambo in Recife." *Social Science / Information sur les Sciences Sociales* 38, no. 2 (1999): 297–327.

Della Cava, Ralph. "Brazilian Messianism and National Institutions: Reappraisal of Canudos and Joaseiro." *Hispanic American Historical Review* 48, no. 3 (1968): 402–20.

———. *Miracle at Juazeiro.* New York: Columbia University Press. 1970.

de Matos, José Wilkins. *Diccionário topográfico de Departamento de Loreto.* N.p., 1872.

de Mello, Evaldo Cabral. *O norte agrário e o império, 1871–1889.* Rio de Janeiro: Editora Nova Fronteira, 1984.

de Moraes, João Quartim, Wilma Peres Costa, and Eliézer Rizzo de Oliveira. *A tutela militar.* São Paulo: Edições Vértice, Editora Revista dos Tribunais, 1987.

Denevan, William M. *The Aboriginal Cultural Geography of the Llanos de Mojos of Bolivia.* Berkeley: University of California Press, 1966.

———. *Cultivated Landscapes of Native Amazonia and the Andes.* Oxford: Oxford University Press, 2001.

———. "Haenke Tadeo: His Work in Andes and Bolivian Forest." *Hispanic American Historical Review* 58, no. 2 (1978): 322–23.

———. "The Native Population of Amazonia in 1492 Reconsidered." *Revista de Índias* 62, no. 227 (2003): 175-88.

Denevan, William M., et al. "Indigenous Agroforestry in the Peruvian Amazon: Bora Indian Management of Swidden Fallows." *Interciencia* 9, no. 6 (1984): 346–57.

Denevan, William M., and B. L. Turner. "Forms, Functions and Associations of Raised Fields in Old World Tropics." *Journal of Tropical Geography* 39 (December 1974): 24--33.

Dennison, Stephanie. *Joaquim Nabuco: Monarchism, Panamericanism and Nation-Building in the Brazilian Belle Époque.* Oxford: Peter Lang, 2006.

de Queiroz, Maria Isaura Pereira. "Messianic Myths and Movements." *Diogenes,* no. 90 (1975): 78–99.

———. *O messianismo no Brasil e no mundo.* 2nd ed. São Paulo: Editora Alfa-Omega, 1977.

de Queiroz, Mauricia Vinhas. *Messianismo e conflito social: A guerra sertaneja do contestado, 1913–1916.* Rio de Janeiro: Civilização Brasileira, 1966.

de Queiróz, Suely Robles Reis. *Os radicais da república: Jacobinismo—ideologia e ação, 1893–1897.* São Paulo: Brasiliense, 1986.

Derby, Orville. "Contribução para o estudo da geologia do Vale do R. São Francisco." *Archivos do Museu Nacional* 4 (1879): 87–119.

———. "Contribuções para a geologia da região de Baixo Amazonas." *Archivo do Museu Nacional* 2 (1877): 77–104.

de Rivière, H. Arnous. "Exploration in the Rubber District of Bolivia." *Journal of the American Geographical Society of New York* 32, no. 5. (1900) : 432–40.

de Saint-Hilaire, Auguste, and Clado Ribeiro de Lessa. *Viagem ás nascentes do Rio São Francisco e pela provincia de Goyaz*. São Paulo: Companhia Editora Nacional, 1944.
de Santana, José Carlos Barreto. *Ciência e arte: Euclides da Cunha e as ciências naturais*. Feira de Santana, Brazil: Editora Hucitec, 2001.
———. "Natural Science and Brazilian Nationality: *Os Sertões* by Euclides da Cunha." *Science in Context* 18, no. 2 (2005): 225–47.
Descola, Philippe. *In the Society of Nature: A Native Ecology in Amazonia*. Cambridge: Cambridge University Press, 1994.
de Toledo, M. B., and M. B. Bush. "A Holocene Pollen Record of Savanna Establishment in Coastal Amapá." *Anais da Academia Brasileira de Ciências* 80, no. 2 (2008): 341–51.
———. "A Mid-Holocene Environmental Change in Amazonian Savannas." *Journal of Biogeography* 34, no. 8 (2007): 1313–26.
Dettelbach, Michael. "Humboldtian Science." In *Cultures of Natural History*, edited by Nicholas Jardine, James A. Secord, and Emma C. Spary, 287–305. New York: Cambridge University Press, 1995.
de Wiart, Carton. *Leopold II: Souvenirs des derniers années*. Brussels: Goemaere, 1944.
Di Gregorio, Mario A. "Worldwide Enigmas and Wonders of Life: Ernst Haeckel—His Work, Impact and Legacy." *Journal of the History of Biology* 33, no. 2 (2000): 401–4.
Di Paolo, Pasquale. *Cabanagem: A revolução popular da Amazônia*. 2nd ed. Belém: Edições CEJUP, 1986.
Diacon, Todd A. "Bringing the Countryside Back In: A Case Study of Military Intervention as State-Building in the Brazilian Old Republic." *Journal of Latin American Studies* 27 (1995): 569–92.
———. *Millenarian Vision, Capitalist Reality: Brazil's Contestado Rebellion, 1912–1916*. Durham, NC: Duke University Press, 1991.
———. *Stringing Together a Nation: Cândido Mariano da Silva Rondon and the Construction of a Modern Brazil, 1906–1930*. Durham, NC: Duke University Press, 2004.
Dickens, Peter. *Social Darwinism: Linking Evolutionary Thought to Social Theory*. Buckingham, UK: Open University Press, 2000.
Diegues, Antônio Carlos Sant'Ana, and Brent Millikan. *A dinâmica social do desmatamento na Amazônia: Populações e modos de vida em Rondônia e sudeste do Pará*. São Paulo: United Nations Research Institute for Social Development, Núcleo de Apoio à Pesquisa sobre Populações Humanas e Areas Umidas Brasileiras, 1993.
do Couto e Silva, Golbery. *Geopolítica do Brasil*. Coleção Documentos Brasileiros 126. Rio de Janeiro: J. Olympio, 1967.
———. *Geopolitica e Poder*. Rio de Janeiro: UniverCidade, 2003.
do Nascimento, José Leonardo. *"Os Sertões" de Euclides da Cunha: Releituras e diálogos*. São Paulo: Editora Universidade de São Paulo, 2002.
Domínguez, Camilo, and Augusto Gómez. *La economía extractiva en la Amazonia colombiana, 1850–1930*. Bogotá: Tropenbos, 1990.
Dominguez, Luis L., Ulrich Schmidel, and Alvar Nuñez Cabeza de Vaca. *The Conquest of the River Plate (1535–1555)*. New York: B. Franklin, 1964.
d'Orbigny, Alcide Dessalines. *Voyage pittoresque dans les deux Ameriques; Résumé général de tous les voyages de Colomb, Las-Casas, Oviedo . . . Humboldt . . . Franklin . . . etc.* Paris: L. Tenré, 1836.
dos Reis, Isabel Cristina Ferreira. *Histórias de vida familiar e afetiva de escravos na Bahia*

do século XIX. Salvador, Brazil: Centro de Estudos Baianos, Editora da Universidade Federal da Bahia (EDUFBA), 2001.

dos Santos, Júlio Ricardo Quevedo. *Guerreiros e Jesuítas na utopia do Prata*. Bauru, Brazil: Editora Universidade de Santa Caterina, 2000.

Doyle, Don Harrison, and Marco Antonio Villela Pamplona. *Nationalism in the New World*. Athens: University of Georgia Press, 2006.

Dozer, Donald. "Matthew Maury's Letter of Instruction to William Herndon." *Hispanic American Historical Review* 28 (1948): 212–28.

Drayton, Richard. *Nature's Government: Science, Imperial Britain, and the "Improvement" of the World*. New Haven, CT: Yale University Press, 2000.

Driver, Felix. *Geography Militant: Cultures of Exploration and Empire*. Oxford: Blackwell, 2001.

Driver, Felix, and David Gilbert. "Heart of Empire? Landscape, Space and Performance in Imperial London." *Environment and Planning D: Society and Space* 16, no. 1 (1998): 11–28.

———. *Imperial Cities: Landscape, Display and Identity*. Manchester, UK: Manchester University Press, 1999.

Driver, Felix, and Luciana Martins. *Tropical Visions in an Age of Empire*. Chicago: University of Chicago Press, 2005.

Dubois, Laurent. *Avengers of the New World: The Story of the Haitian Revolution*. Cambridge, MA: Belknap Press of Harvard University Press, 2004.

———. *A Colony of Citizens: Revolution and Slave Emancipation in the French Caribbean, 1787–1804*. Williamsburg: University of North Carolina Press, 2004.

Dunn, Ballard S. *Brazil, the Home for Southerners*. New York: G. B. Richardson, 1866.

Edelman, Marc. "A Central American Genocide: Rubber, Slavery, Nationalism, and the Destruction of the Guatusos-Malekus." *Comparative Studies in Society and History* 40, no. 2 (1998): 356–90.

Edney, Matthew H. *Mapping an Empire: The Geographical Construction of British India, 1765–1843*. Chicago: University of Chicago Press, 1997.

Edwards, William Henry. *A Voyage up the River Amazon, Including a Residence at Pará*. London: J. Murray, 1847.

Eisenberg, José. "Cultural Encounters, Theoretical Adventures: The Jesuit Missions to the New World and the Justification of Voluntary Slavery." *History of Political Thought* 24, no. 3 (2003): 375–96.

Eisenberg, Peter L. *Homens esquecidos: Escravos e trabalhadores livres no Brasil, séculos XVIII e XIX*. Campinas, Brazil: Editora da Universidade de Campinas, 1989.

———. *The Sugar Industry in Pernambuco; Modernization without Change, 1840–1910*. Berkeley: University of California Press, 1974.

Ellen, Roy F., Peter Parkes, and Alan Bicker. *Indigenous Environmental Knowledge and Its Transformations: Critical Anthropological Perspectives*. Amsterdam: Harwood Academic, 2000.

Ellis, Markman. *The Coffee-House: A Cultural History*. London: Weidenfeld and Nicolson, 2004.

———. *Eighteenth-Century Coffee-House Culture*. London: Pickering and Chatto, 2006.

Eltis, David. *Coerced and Free Migration: Global Perspectives*. Stanford, CA: Stanford University Press, 2002.

———. *The Rise of African Slavery in the Americas*. Cambridge: Cambridge University Press, 2000.

Eltis, David, Stephen D. Behrendt, David Richardson, and Herbert S. Klein. *The Trans-Atlantic Slave Trade: A Database on CD-ROM*. Cambridge: Cambridge University Press,1999.

Eltis, David, Frank D. Lewis, and Kenneth Lee Sokoloff. *Slavery in the Development of the Americas*. Cambridge: Cambridge University Press, 2004.

Eltis, David, and David Richardson. *Atlas of the Transatlantic Slave Trade*. New Haven, CT: Yale University Press, 2010.

———. *Extending the Frontiers: Essays on the New Transatlantic Slave Trade Database*. New Haven, CT: Yale University Press, 2008.

———. *Routes to Slavery: Direction, Ethnicity, and Mortality in the Transatlantic Slave Trade*. London: Frank Cass, 1997.

Emerson, Ralph Waldo. "Young America." *Dial* (1844): 484–507.

Emmer, Pieter C. *The Dutch in the Atlantic Economy, 1580–1880: Trade, Slavery and Emancipation*. Aldershot, UK: Ashgate, 1998.

Emmer, Pieter C., and Ernst van den Boogaart. *Colonialism and Migration: Indentured Labour before and after Slavery*. Hingham, MA: Kluwer Boston, 1986.

Endersby, Jim. *Imperial Nature: Joseph Hooker and the Practices of Victorian Science*. Chicago: University of Chicago Press, 2008.

Erickson, Clark L. "An Artificial Landscape-Scale Fishery in the Bolivian Amazon." *Nature* 408, no. 6809 (2000): 190–93.

———. "Domesticated Landscapes of the Bolivian Amazon." In *Time and Complexity in Historical Ecology*, edited by William Balée and Clark Erickson. New York: Columbia University Press, 2006.

Espino, Ovidio Diaz. *How Wall Street Created a Nation: J. P. Morgan, Teddy Roosevelt and the Panama Canal*. New York: Four Walls Eight Windows, 2001.

Estrella, Hernani. *Direitos da mulher: Doutrina, legislação e jurisprudência*. Rio de Janeiro: J. Konfino, 1975.

Evans, Harold, Gail Buckland, and David Lefer. *They Made America: From the Steam Engine to the Search Engine; Two Centuries of Innovators*. New York: Little, Brown, 2004.

Eyal, Yonathan. *The Young America Movement and the Transformation of the Democratic Party, 1828–1861*. New York: Cambridge University Press, 2007.

Facão, Emilio. *Album do Acre*. Belém: Facão, DATE?

Facó, Rui. *Cangaceiros e fanáticos: Gênese e Lutas*. 2nd ed. Rio de Janeiro: Editôra Civilização Brasileira, 1965.

Fagan, Chris, and Diego Shoobridge. 2005. "An Investigation of Illegal Mahogany Logging in Peru's Alto Purús National Park and Its Surroundings." Parks Watch. http://www.upperamazon.org/PDF/2005_Purus_Report_eng.pdf (accessed June 12, 2012).

Fagot, P. "World Consciousness, Alexander von Humboldt and the Unfinished Project of a Different Modernism." *Études Germaniques* 58, no. 1 (2003): 116–17.

Falcão, Edgar de Cerqueira, et al. *Viagem filosófica as capitanias do Grão-Pará, Rio Negro, Mato Grosso e Cuiabá*. São Paulo: Gráficos Brunner, 1970.

Falci, Miridan Britto Knox. *Escravos do Sertão: Demografia, trabalho e relações sociais, Piauí, 1826–1888*. Teresina, Brazil: Fundação Cultural Monsenhor Chaves, 1995.

Falcon, Andrea. *Aristotle and the Science of Nature: Unity without Uniformity*. Cambridge: Cambridge University Press, 2005.
Falola, Toyin, and Matt D. Childs. *The Yoruba Diaspora in the Atlantic World*. Bloomington: Indiana University Press, 2004.
Farage, Nádia. *As muralhas dos Sertões: Os povos indígenas no Rio Branco e a colonização*. São Paulo: Paz e Terra, 1991.
Farinelli, Franco. "Friedrich Ratzel and the Nature of (Political) Geography." *Political Geography* 19, no. 8 (2000): 943–55..
Faulhaber, Priscila, and Peter Mann de Toledo. *Conhecimento e fronteira: História da ciência na Amazônia*. Belém:, Museu Paraense Emílio Goeldi, 2001.
Faust, Drew Gilpin. *This Republic of Suffering: Death and the American Civil War*. New York: Alfred A. Knopf, 2008.
Fausto, Carlos. *Inimigos fiéis: História, guerra e xamanismo na Amazônia*. São Paulo: Editora da Universidade de São Paulo (EDUSP), 2001.
Fausto, Carlos, and Michael Heckenberger. *Time and Memory in Indigenous Amazonia: Anthropological Perspectives*. Gainesville: University Press of Florida, 2007.
Feldpausch, T. R., et al. "When Big Trees Fall: Damage and Carbon Export by Reduced Impact Logging in Southern Amazonia." *Forest Ecology and Management* 219, nos. 2–3 (2005): 199–215.
Ferguson, R. Brian, and Neil L. Whitehead. *War in the Tribal Zone: Expanding States and Indigenous Warfare*. Santa Fe, NM: School of American Research Press; Oxford: James Currey, 2005.
Fernandes, Antônio Paulo Cyriaco. *Missionários Jesuítas no Brasil no tempo de Pombal*. 2nd ed. Porto Alegre: Edição da Livraria do Globo, 1941.
Fernandes, Florestan. *A Integração do negro à sociedade de classes*. São Paulo: Faculdade de Filosofia, Ciências e Letra da Universidade de São Paulo, 1964.
Fernandes, Heloisa Rodrigues. *Os militares como categoria social*. São Paulo: Global Editora, 1978.
Fernández de Oviedo y Valdés, Gonzalo, and Enrique Alvarez López. *De la natural historia de las Indias (sumario de historia natural de las Indias) con un estudio preliminar y notas*. Madrid: Editorial Summa, 1942.
Fernandez, Emilio. *La campaña del Acre (1900–1901)*. Buenos Aires: Imprenta Litografía y Encuadernación de J. Peuser, 1903.
Fernández-Armesto, Filipe. *Before Columbus: Exploration and Colonization from the Mediterranean to the Atlantic*. Philadelphia: University of Pennsylvania Press, 1987.
Ferrari, Alfonso Trujillo. *Os Kariri, o crepúsculo de um povo sem história*. São Paulo: Escola de Sociologia e Politica, 1957.
Ferraz, S. F. D., C. A. Vettorazzi, D. M. Theobald, and M. V. R. Ballester. "Landscape Dynamics of Amazonian Deforestation between 1984 and 2002 in Central Rondonia, Brazil: Assessment and Future Scenarios." *Forest Ecology and Management* 204, no. 1 (2005): 67–83.
Ferreira, Alexandre Rodrigues. *Viagem ao Brasil de Alexandre Rodrigues Ferreira: A expedição philosophica pelas capitanias do Pará, Rio Negro, Mato Grosso e Cuyabá*. 3 vols. Lisbon: Kapa Editorial, 2005.
Ferreira, Antônio Celso. *A epopéia bandeirante: Letrados, instituições, invenção histórica (1870–1940)*. São Paulo: Editora Universidade Estadual Paulista (UNESP), 2002.

Ferreira, Manoel Rodrigues. *A ferrovía do diabo: História de uma estrada de ferro na Amazônia*. São Paulo: Melhoramentos: Secretaria de Estado da Cultura, 1981.

Fiabani, Adelmir. *Mato, Palhoça e Pilão: O quilombo, da escravidão as comunidades remanescentes (1532–2004)*. São Paulo: Expressão Popular, 2005.

Fifer, J. Valerie. "Bolivia's Boundary with Brazil: A Century of Evolution." *Geographical Journal* 132 (1966): 360–82.

———. "The Empire Builders: A History of the Bolivian Rubber Boom and the Rise of the House of Suarez." *Journal of Latin American Studies* 2, no. 2 (1970): 113–46.

Figueroa, Francisco de, and Cristóbal de Acuña. *Informes de Jesuitas en el Amazonas, 1660–1684*. Iquitos, Peru: Peruvian Amazon Research Institute (IIAP), 1986.

Fleming, Peter. *Amazon Adventure*. London: Jonathan Cape, 1933.

Florentino, Manolo. *Em costas negras: Uma história do tráfico atlântico de escravos entre a Africa e o Rio de Janeiro, séculos XVIII e XIX*. Rio de Janeiro: Arquivo Nacional, 1995.

Flores-Martin, Antonio. *La explotación de caucho en el Perú*. Lima: Consejo Nacional de Ciencia y Tecnologia, 1987.

Folguera, Jose Miguel Morales. *La construcción de la utopía: El proyecto de Felipe II (1556–1598) para Hispanoamérica*. Madrid: Biblioteca Nueva, 2001.

Fontes, Oleone Coelho. *Lampião na Bahia*. Petrópolis, Brazil: Vozes, 1988.

Foot-Hartmann, Francisco. *Trem fantasma: A modernidade na selva*. São Paulo: Companhia das Letras, 1988.

Forest, Philippe Renan Ernest, et al. *Qu'est-ce qu'une nation? Ernest Renan (texte intégral), littérature et identité nationale de 1871 à 1914 (textes de Barrès, Daudet, R. de Gourmont, Céline)*. Paris: Pierre Bordas et fils éditeur, 1991.

Forster, Colin. *France and Botany Bay: The Lure of a Penal Colony*. Carlton South, Australia: Melbourne University Press, 1996.

Fragoso, Tasso. *História da guerra entre a Tríplice Aliança e o Paraguai*. Rio de Janeiro: Imprensa do Exército, 1934.

Fredrickson, George. *The Black Image in the White Mind: The Debate on Afro-American Character and Destiny*. Hanover, NH: Wesleyan University Press, 1987.

Freehoff, Joseph C. *America and the Canal Title, or An Examination, Sifting and Interpretation of the Data Bearing on the Wresting of the Province of Panama from the Republic of Colombia by the Roosevelt Administration in 1903 in Order to Secure Title to the Canal Zone*. New York: Author, 1916.

Freixinho, Nilton, and Antônio Olinto. *O Sertão arcaíco do nordeste do Brasil: Uma releitura*. Rio de Janeiro: Imago, 2003.

French, Jan Hoffman. "Buried Alive: Imagining Africa in the Brazilian Northeast." *American Ethnologist* 33, no. 3 (2006): 340–60.

———. *Legalizing Identities: Becoming Black or Indian in Brazil's Northeast*. Chapel Hill: University of North Carolina Press, 2009.

Freyre, Gilberto. *The Masters and the Slaves (Casa-Grande and Senzala): A Study in the Development of Brazilian Civilization*. New York: Alfred A. Knopf, 1946.

———. *O Luso e o Tropico: Sugestões em torno dos métodos portugueses de integração de povos autóctones e de culturas diferentes da Europeia num complexo novo de civilização, o Luso-Tropical—Ensaios*. Lisbon: Comissão Executiva das Comemorações do V Centenário da Morte do Infante D. Henrique, 1961.

———. *Order and Progress: Brazil from Monarchy to Republic*. New York: Alfred A. Knopf, 1970.
———. *Perfil de Euclydes e otros perfís*. Rio de Janeiro: J. Olympio, 1944.
———. *Portuguese Integration in the Tropics: Notes concerning a Possible Lusotropicology Which Would Specialize in the Systematic Study of the Ecological-Social Process of the Integration in Tropical Environments of Portuguese, Descendants of Portuguese and Continuators of Portuguese*. Lisbon: Grafica Silvas, 1961.
Freyre, Gilberto, Fernando Henrique Cardoso, Edson Nery da Fonseca, and Gustavo Henrique Tuna. *Casa-Grande e Senzala: Formação da família brasileira sobre o regime da economia patriarcal*. 47th ed. São Paulo: Global Editora, 2003.
Fry, Peter, and Carlos Vogt. *Cafundó: A Africa no Brasil*. São Paulo: Companhia das Letras, 1996.
Funari, Pedro Paulo de Abreu. "A arquéologia de Palmares: Sua contribução para o conhecimento da história da cultura afro-americana." In *Liberdade por um fio*, edited by João José Reis and Flavio dos Santos Gomes Reis, 26–51. São Paulo: Companhia das Letras, 1996.
Funes, Eurípedes. "Mocambos do Trombetas: História, memoria e identidade." Fortaleza, Brazil: Universidade Federal do Ceará, 2009.
———. "Nasci na Matas Nuca Tive Senhor: Historia e Memoria de Mocambos do Baixo Amazonas." PhD diss., Universidade de São Paulo, 1995.
Galloway, J. H. "The Last Years of Slavery on the Sugar Plantations of Northeastern Brazil." *Hispanic American Historical Review* 51, no. 4 (1971): 586–605.
Galvão, Hélio. *O mutirão no nordeste*. 2nd ed. Brasília: Fundação Centro de Formação do Servidor Público, 1988.
Galvão, Walnice. *No calor da hora: A Guerra de Canudos nos jornais, quarto expedição*. São Paulo: Atica, 1974.
Galvão, Walnice, and Domecio Pacheco da Silva Neto, eds. *Crônica de uma tragédia inesquecível: Autos do processo de Dilermando de Assis, que matou Euclides da Cunha*. São Paulo: Albatroz Loqüi, Terceiro Nome, 2007.
Gálvez, Lucía. *Guaranís y Jesuitas: De la tierra sin mal al paraíso*. Buenos Aires: Editorial Sudamericana, 1995.
Galvin, Peter R. *Patterns of Pillage: A Geography of Caribbean-Based Piracy in Spanish America, 1536–1718*. New York: Peter Lang, 1999.
Ganson, Barbara Anne. *The Guaraní under Spanish Rule in the Río de la Plata*. Stanford, CA: Stanford University Press, 2003.
Garay, Blas. *El comunismo de las misiones: La revolución de la independencia del Paraguay*. Asunción: Instituto Colorado de Cultura, 1975.
García, Jesús Chucho, and Nirva Rosa Camacho. *Comunidades afro-descendientes en Venezuela y América Latina*. Caracas: Red de Organizaciones Afro-venezolanas, 2002.
García Jordán, Pilar. *Cruz y arado, fusiles y discursos: La construcción de los orientes en el Perú y Bolivia, 1820–1940*. Lima: Instituto Francés de Estudios Andinos, Instituto de Estudios Peruanos, 2001.
García, Lorenzo. *Historia de las misiones en la Amazonia ecuatoriana*. Quito: Ediciones Abya-Yala, 1985.
García Márquez, Gabriel. *Chronicle of a Death Foretold*. New York: Ballantine Books, 1984.

Garfield, Seth. "A Continent Apart from the Rest of the World: Geography and Race in the Making of the Brazilian Amazon." Unpublished manuscript, 2008.

———. *Indigenous Struggle at the Heart of Brazil*. Durham, NC: Duke University Press, 2001.

Gasman, Daniel. "Haeckel's Scientific Monism as Theory of History." *Theory in Biosciences* 121, no. 3 (2002): 260–79.

———. *The Scientific Origins of National Socialism*. New Brunswick, NJ: Transaction, 2004.

Gassent, Paz Serrano. *Vasco de Quiroga: Utopía y derecho en la conquista de América*. Madrid: Universidad Nacional, 2001.

Gates, E. Nathaniel. *The Concept of "Race" in Natural and Social Science*. New York: Garland, 1997.

Geggus, David Patrick. *The Impact of the Haitian Revolution in the Atlantic World*. Columbia: University of South Carolina, 2001.

Geggus, David Patrick, and Norman Fiering. *The World of the Haitian Revolution*. Bloomington: Indiana University Press, 2009.

Gifford, Prosser, and William Roger Louis. *France and Britain in Africa: Imperial Rivalry and Colonial Rule*. New Haven, CT: Yale University Press, 1971.

Gilbert, G. K. "New Light on Isostasy." *Journal of Geology* 3, no. 3 (1895): 331–34.

Giménez, Célia Beatriz, and Raimundo dos Santos Coelho. *Bahia Indígena: Encontro de dois mundos; Verdade do descobrimento do Brasil*. Rio de Janeiro: Faculdades do Descobrimento, 2005.

Giumbelli, Emerson. "Religion and Social (Dis)Order: Contestado, Juazeiro, and Canudos in Sociological Studies on Religious Movements." *Dados: Revista de Ciências Sociais* 40, no. 2 (1997): 251–82.

Glacken, Clarence J. *Traces on the Rhodian Shore: Nature and Culture in Western Thought from Ancient Times to the End of the Eighteenth Century*. Berkeley: University of California Press, 1973.

Glaser, Bruno, and William I. Woods. *Amazonian Dark Earths: Explorations in Space and Time*. Berlin: Springer, 2004.

Godoy, J. R., G. Petts, and J. Salo. "Riparian Flooded Forests of the Orinoco and Amazon Basins: A Comparative Review." *Biodiversity and Conservation* 8, no. 4 (1999): 551–86.

Goeldi, Emilio. "Excavações archeológicas em 1895: As Cavernas funerárias artificiaes de indios hoje extinctos no Rio Cunany." *Memorias do Museu Paraense de História Natural e Ethnografia*, no. 1 (1900).

Goldblatt, Peter. *Biological Relationships between Africa and South America*. New Haven, CT: Yale University Press, 1993.

Gomes, Flávio dos Santos. *A hidra e os pântanos: Mocambos, quilombos e comunidades de fugitivos no Brasil (séculos XVII–XIX)*. São Paulo: Universidade Estadual Paulista (UNESP), 2005.

———. "Entre fronteiras e limites: Identitidades e espaços transnacionais na Guiana Brasileira." *Estudos Ibero-Americanos* 28, no. 1 (2002): 21–50.

———. *Experiências atlânticas: Ensaios e pesquisas sobre a escravidão e o pós-emancipação no Brasil*. Passo Fundo, Brazil: Universidade de Passo Fundo (UPF) Editora, 2003.

———. "Fronteiras e mocambos." In *Nas terras do Cabo Norte: Fronteiras, colonização, escravidão na Guyana Brasiliera XVIII–XIX*, edited by Flávio dos Santos Gomes, 225–318. Belém: Universidade Federal do Pará (UFPA), 1999.

———. "Gênero, etnicidade e memoria na Amazônia: Notas de Pesquisas etnográficas em communidades negras." In *Mulher e modernidade na Amazônia*, edited by Maria Luzia Miranda Alvares, Eunice Ferreira dos Santos, and Maria Angela d'Incao, 151–80. Belém: Cejup, 1997.

———. *Histórias de quilombolas: Mocambos e comunidades de senzalas no Rio de Janeiro, século XIX*. Rio de Janeiro: Arquivo Nacional, 1995.

———. "Outras paisagens coloniais: Notas sobre desertores militares na Amazônia setecentista." In *Nas terras do Cabo Norte*, edited by Flávio dos Santos Gomes, 196–223. Belém: Universidade Federal do Pará (UFPA), 1999.

———. "A 'Safe Haven': Runaway Slaves, Mocambos, and Borders in Colonial Amazonia, Brazil." *Hispanic American Historical Review* 82, no. 3 (2002): 469–98.

———. *Nas terras do Cabo Norte: Fronteiras, colonização, escravidão na Guyana Brasiliera XVIII–XIX*. Belém: Universidade Federal do Pará (UFPA), 1999.

Gomes, Flávio dos Santos, and Jonas M. de Queiroz. "Between Frontiers and Limits: Transnational Identities and Space in the Brazilian Guyanas in the 18th and 19th Centuries." *Estudos Ibero-Americanos* 28, no. 1 (2002): 21–50.

Gomes, Gínia Maria. *Euclides da Cunha: Literatura e história*. Porto Alegre: Editora da Universidade Federal do Rio Grande do Sul (UFRGS), 2005

Gomes, Sancho Pinto Ferreira. "Como se deu o lamentável acontecimento." *Journal Official* (Tarauacá, Acre, Brazil), May 21, 1916.

Gómez, Augusto, Ana Cristina Lesmes, and Claudia Rocha. *Caucherías y conflicto colombo-peruano: Testimonios, 1904–1934*. Bogotá: Disloque Editores, 1995.

Goodman, Jordan. *The Devil and Mr. Casement: One Man's Battle for Human Rights in South America's Heart of Darkness*. New York: Farrar, Straus and Giroux, 2010.

Gorender, Jacob. *Brasil em preto e branco: O passado escravista que não passou*. São Paulo: Editora Senac, 2000.

Gould, Stephen Jay. *The Mismeasure of Man*. New York: W. W. Norton, 1981.

———. *The Structure of Evolutionary Theory*. Cambridge, MA: Belknap Press of Harvard University Press, 2002.

———. *Time's Arrow, Time's Cycle: Myth and Metaphor in the Discovery of Geological Time*. Cambridge, MA: Harvard University Press, 1987.

Gow, Peter. "'Canção Purús': Nacionalização e tribalização no sudoeste da Amazônia." *Revista Antropologica* 49, no. 1 (2006): 431–64.

———. "Gringos and Wild Indians: Images of History in Western Amazonian Cultures." *Homme* 33, nos. 2–4 (1993): 327–47.

———. *Of Mixed Blood: Kinship and History in the Peruvian Amazon*. Oxford: Clarendon, 1991.

Graden, Dale. *From Slavery to Freedom in Brazil: Bahia, 1835–1900*. Albuquerque: University of New Mexico Press, 2006.

Graham, Richard. "Another Middle Passage? The Internal Slave Trade in Brazil." In *The Chattel Principle*, edited by Walter Johnson and David Brown, 293–325. New Haven, CT: Yale University Press, 2004.

———. *Patronage and Politics in Nineteenth-Century Brazil*. Stanford, CA: Stanford University Press, 1990.

Graham, Richard, Thomas E. Skidmore, Aline Helg, and Alan Knight. *The Idea of Race in Latin America, 1870–1940*. Austin: University of Texas Press, 1990.

Graham-Yooll, Andrew. *Imperial Skirmishes: War and Gunboat Diplomacy in Latin America*. New York: Olive Branch, 2002.

Gramiccia, Gabriele. *The Life of Charles Ledger (1818–1905): Alpacas and Quinine*. Basingstoke, UK: Macmillan, 1988.

Grandin, Greg. *Fordlândia*. New York: Metropolitan, 2009.

Grann, David. *The Lost City of Z*. New York: Doubleday, 2009.

Green, J. N. "The Abolition of Slavery in Brazil: The 'Liberation' of Africans through the Emancipation of Capital." *Historian* 64, nos. 3–4 (2002): 727–28.

Greenblatt, Stephen. *New World Encounters*. Berkeley: University of California Press, 1993.

Greene, John C. *Science, Ideology, and World View: Essays in the History of Evolutionary Ideas*. Berkeley: University of California Press, 1981.

Greenfield, Gerald Michael. *The Realities of Images: Imperial Brazil and the Great Drought*. Phildelphia: American Philosophical Society, 2001.

Griffin, Roger. *The Nature of Fascism*. London, New York: Routledge, 1993.

Gross, D. G. Eiten, et al. "Ecology and Acculturation among Native Peoples of Central Brazil." *Science* 206 (1979): 104--50.

Grove, Richard. *Green Imperialism*. New York: Cambridge University Press, 1995.

Guedes, Jose Maciel Ramos. "L'insurrection negre de Coro em 1795 au Venezuela." In *De la Revolution Francaise aux revolutions créoles e negres*, edited by Michel Martin and Alain Yacou. Paris: Editions Caribeens, 1989.

Guerreiro, J. F., et al. "Genetical-Demographic Data from Two Amazonian Populations Composed of Descendants of African Slaves: Pacoval and Curiaú." *Genetics and Molecular Biology* 22 (1999): 163–67.

Guilhon, Norma. *Confederados em Santarém: Saga americana na Amazônia*. Rio de Janeiro: Presença, 1987.

Gumplowicz, Ludwig. *Der Rassenkampf: Sociologische Untersuchungen*. Innsbruck, Austria: Wagner'sche Univ.-Buchhandlung, 1883.

Guyot, Arnold. *Creation, or the Biblical Cosmogony in the Light of Modern Science*. New York: Charles Scribner, 1884.

———. *The Earth and Man: Comparative Physical Geography in Relation to the History of Mankind*. Translated by C. Felton. Boston: Gould and Lincoln, 1860.

Haas, Frans A. J. de, and Jaap Mansfeld. *Aristotle: On Generation and Corruption, Book 1*. Oxford: Clarendon, 2004.

Haberly, David T. *Three Sad Races: Racial Identity and National Consciousness in Brazilian Literature*. Cambridge: Cambridge University Press, 1983.

Hadfield, William. *Brazil and the River Plate*. Surrey, UK: Sutton, 1877.

Haeckel, Ernst Heinrich Philipp August. *Anthropogénie, ou Histoire de l'évolution humaine, leçons familières sur les principes de l'embroyologie et de la phylogénie humaines*. Paris: Reinwald, 1877.

Haenger, Peter, J. J. Shaffer, and Paul E. Lovejoy. *Slaves and Slave Holders on the Gold Coast: Towards an Understanding of Social Bondage in West Africa*. English ed. Basel: Schlettwein, 2000.

Hagen, Joshua. "Thomas M. Lekan, Imagining the Nation in Nature: Landscape Preservation and German Identity, 1885–1945." *Journal of Historical Geography* 31, no. 1 (2005): 190–92.

Hahner, June Edith. *Civilian-Military Relations in Brazil, 1889–1898*. Columbia: University of South Carolina Press, 1969.

Hale, Charles A. *The Transformation of Liberalism in Late Nineteenth-Century Mexico*. Princeton, NJ: Princeton University Press, 1989.

Hall, Gwendolyn Midlo. *Slavery and African Ethnicities in the Americas: Restoring the Links*. Chapel Hill: University of North Carolina Press, 2005.

Hanna, Alfred, and Kathryn Hanna. *Confederate Exiles to Venezuela*. Tuscaloosa, AL: Confederate, 1960.

Hardenburg, Walter Ernest, C. Reginald Enock, and Roger Casement. *The Putumayo, the Devil's Paradise: Travels in the Peruvian Amazon Region and an Account of the Atrocities Committed upon the Indians Therein*. London: T. Fisher Unwin, 1913.

Harley, J. B. "Deconstructing the Map." *Cartographica* 26, no. 2 (1989): 1–20.

———. "Historical Geography and the Cartographic Illusion." *Journal of Historical Geography* 15, no. 1 (1989): 80–91.

———. "Maps, Knowledge and Power." In *The Iconography of Landscape*, edited by Denis E. Cosgrove and Stephen Daniels, 277–311. Cambridge: Cambridge University Press, 1988.

———. "Rereading the Maps of the Columbian Encounter." *Annals of the Association of American Geographers* 82, no. 3 (1992): 522–42.

Harley, J. B., and David Woodward. *The History of Cartography*. Chicago: University of Chicago Press, 1987

Harris, Mark. *Rebellion on the Amazon: The Cabanagem, Race, and Popular Culture in the North of Brazil, 1798–1840*. Cambridge: Cambridge University Press.

Harris, Mark, and Stephen Nugent. *Some Other Amazonians: Perspectives on Modern Amazonia*. London: Institute for the Study of the Americas, 2004.

Harrison, Brady. "The Young Americans: Emerson, Walker and the Early Literature of American Empire." *American Studies*, no. 40 (1999): 75–97.

Harter, Eugene C. *Lost Colony of the Confederacy*. New York: University of Mississippi Press, 1985.

Hartt, Charles Frederick, and Louis Agassiz. *Thayer Expedition: Scientific Results of a Journey in Brazil, by Louis Agassiz and His Travelling Companions; Geology and Physical Geography of Brazil*. Boston: Fields Osgood, 1870.

Hastenrath, S. "Circulation and Teleconnection Mechanisms of Northeast Brazil Droughts." *Progress in Oceanography* 70, nos. 2–4 (2006): 407–15.

Hastings, Lansford. *Immigrants Guide to Oregon and California*. Cincinnati: G. Conclin, 1845.

———. *A New Description of Oregon and California and a Vast Amount of Information Relating to the Soil, Climate, Productions and Rivers*. Cincinnati: H. M. Rulison, 1855.

Haubert, Maxime. *L'église et la défense des "sauvages"; Le Père Antoine Vieira au Brésil*. Brussels, 1964.

Hawkins, Mike. *Social Darwinism in European and American Thought, 1860–1945: Nature as Model and Nature as Threat*. Cambridge: Cambridge University Press, 1997.

Hayes, Robert Ames. *The Armed Nation: The Brazilian Corporate Mystique*. Tempe: Center for Latin American Studies, Arizona State University, 1989.

Heath, Edwin R. "Exploration of the River Beni in 1880–1." *Proceedings of the Royal Geographical Society* 5, no. 6 (1883): 327–41.

Hecht, Susanna B. "The Black Amazon." Unpublished manuscript, n.d.

———. "Brazil through the Eyes of William James." *Geographical Review* 97, no. 4 (2007): 565–68.

———. "Cattle Ranching in the Amazon: Analysis of a Development Program." PhD diss., University of California, 1982.

———. "Environment, Development and Politics: Capital Accumulation and the Livestock Sector in Eastern Amazonia." *World Development* 13, no. 6 (1985): 663–84.

———. "Factories, Forests, Fields and Family: Gender and Neoliberalism in Extractive Reserves." *Journal of Agrarian Change* 7, no. 3 (2007): 316–47.

———. "Indigenous Management and the Creation of Amazonian Dark Earths: Implications of Kayapó Practices." In *Amazonian Dark Earths: Origins, Properties Management*, edited by J. Lehman, D. Kern, B. Glaser, and W. Woods, 355–73. Dordrecht: Kluwer, 2003.

———. "Kayapó Savanna Management: Fire, Soils, and Forest Islands in a Threatened Biome." In *Amazonian Dark Earths: Wim Sombroek's Vision*, edited by William I. Woods et al., 143–62. Heidelberg: Springer, 2009.

———. "The Last Unfinished Page of Genesis: Euclides da Cunha and the Amazon." *Historical Geography* 32 (2004): 43–69.

Hecht, Susanna B., and Alexander Cockburn. *The Fate of the Forest: Developers, Destroyers, and Defenders of the Amazon*. London: Verso, 1989.

Hecht, Susanna B, and Sasan Saatchi. "The Rubber Boom and a Deforestation Pulse in the Upper Amazon." Unpublished manuscript, 2011.

Heckenberger, Michael J., et al. "Amazonia 1492: Pristine Forest or Cultural Parkland?" *Science* 301, no. 5640 (2003): 1710–14.

Heckenberger, Michael J., James B. Petersen, and Eduardo Góes Neves. "Of Lost Civilizations and Primitive Tribes, Amazonia: Reply to Meggers." *Latin American Antiquity* 12, no. 3 (2001): 328–33.

———. "Village Size and Permanence in Amazonia: Two Archaeological Examples from Brazil." *Latin American Antiquity* 10, no. 4 (1999): 353–76.

Heckenberger, Michael J., J. Christian Russell, Joshua R. Toney, and Morgan J. Schmidt. "The Legacy of Cultural Landscapes in the Brazilian Amazon: Implications for Biodiversity." *Philosophical Transactions of the Royal Society B–Biological Sciences* 362, no. 1478 (2007): 197–208.

———. "Pre-Columbian Urbanism, Anthropogenic Landscapes, and the Future of the Amazon." *Science* 321, no 5893 (2008): 1214–17.

Helferich, Gerard. *Humboldt's Cosmos: Alexander von Humboldt and the Latin American Journey That Changed the Way We See the World*. New York: Gotham Books, 2005.

Hemming, John. *Amazon Frontier: The Defeat of the Brazilian Indians*. London: Macmillan, 1987.

———. *Die If You Must: Brazilian Indians in the Twentieth Century*. London: Macmillan, 2003.

———. *Red Gold: The Conquest of the Brazilian Indians*. Cambridge, MA: Harvard University Press, 1978.

———. *Roraima: Brazil's Northernmost Frontier*. London: Institute of Latin American Studies, 1990.

———. *The Search for El Dorado*. London: M. Joseph, 1978.

———. *Tree of Rivers: The Story of the Amazon*. London: Thames and Hudson, 2008.

Hentenryk, G. Kurgan. "Leopold II et la question de l'Acre." *Bulletin de l'Academie de l'Outre Mer*, 1975, 339–63.

Herman, Marc. *Searching for El Dorado: A Journey into the South American Rain Forest on the Tail of the World's Largest Gold Rush*. New : Doubleday, 2003.

Hermann, Jacqueline, ed. *Sebastianismo e sedição: Os rebeldes do Rodeador na cidade do paraíso terrestre*. Vol. 6. Rio de Janeiro: Tempo, 2001.

Herndon, William Lewis, and Lardner Gibbon. *Exploration of the Valley of the Amazon*. Washington, DC: R. Armstrong, 1853.

Herzog, Werner, and Krishna Winston. *Conquest of the Useless: Reflections from the Making of "Fitzcarraldo."* New York: Ecco, 2009.

Hettle, Wallace. *The Peculiar Democracy: Southern Democrats in Peace and War*. Athens: University of Georgia Press, 2001.

Heywood, Linda M. *Central Africans and Cultural Transformations in the American Diaspora*. Cambridge: Cambridge University Press, 2002.

———. "Portuguese into African: Eighteenth Century Central African Background to Atlantic Creole Cultures." In *Central Africans and the Cultural Transformations of the American Diaspora*, edited by Linda M. Heywood, 91–117. Cambridge: Cambridge University Press, 2002.

Heywood, Linda M., and John K. Thornton. *Central Africans, Atlantic Creoles, and the Foundation of the Americas, 1585–1660*. New York: Cambridge University Press, 2007.

Higgins, Kathleen J. *"Licentious Liberty" in a Brazilian Gold-Mining Region: Slavery, Gender, and Social Control in Eighteenth-Century Sabará, Minas Gerais*. University Park: Pennsylvania State University Press, 1999.

Hill, Howard C. *Roosevelt and the Caribbean*. Chicago: University of Chicago Press, 1927.

Hill, Jonathan David. *History, Power, and Identity: Ethnogenesis in the Americas, 1492–1992*. Iowa City: University of Iowa Press, 1996.

———. *Rethinking History and Myth: Indigenous South American Perspectives on the Past*. Urbana: University of Illinois Press, 1988.

Hill, Jonathan David, and Fernando Santos-Granero. *Comparative Arawakan Histories: Rethinking Language Family and Culture Area in Amazonia*. Urbana: University of Illinois Press, 2002.

Hill, Lawrence. "Confederates in Middle America." *Southwestern Historical Quarterly* 4 (1936): 309–26.

Hobsbawm, Eric J. *The Age of Empire, 1875–1914*. London: Weidenfeld and Nicolson, 1987.

———. *Primitive Rebels: Studies in Archaic Forms of Social Movement in the 19th and 20th Centuries*. 2nd ed. New York: Praeger, 1963.

Hochschild, Adam. *King Leopold's Ghost: A Story of Greed, Terror, and Heroism in Colonial Africa*. London: Macmillan, 1999.

Hodell, David A., et al. "An 85-Ka Record of Climate Change in Lowland Central America." *Quaternary Science Reviews* 27, nos. 11–12 (2008): 1152–65.

Holmes, Rachel. *African Queen: The Real Life of the Hottentot Venus*. New York: Random House, 2007.

Hoogbergen, Wim S. M. *The Boni Maroon Wars in Suriname*. Leide: Brill, 1990.

Hoornaert, Eduardo, and Comisión de Estudios de História de la Iglesia en Latinoamérica. *História da Igreja na Amazônia*. Petrópolis, Brazil: Vozes, 1992.

Horne, Gerald. *The Deepest South: United States, Brazil and the African Slave Trade*. New York: New York University Press, 2007.

Hosler, Dorothy, Sandra Burkett, and Michael Tarkanian. "Prehistoric Polymers: Rubber Processing in Ancient Mesoamerica." *Science* 284 (1999): 1988–91.

Hossfeld, Uwe. "Friedrich Ratzel (1844–1904): Naturalist, Geographer, Scholar." *History and Philosophy of the Life Sciences* 21, no. 2 (1999): 195–213.

Howse, Derek. *Greenwich Time and the Discovery of the Longitude*. Oxford: Oxford University Press, 1980.

Huber, Jacques. "Observações sobre as árvores de borracha da região Amazonica." *Boletim do Museu Paraense Emilio Goeldi* 3, nos. 1–4 (1902): 345–69.

Huertas Castillo, Beatriz. *Indigenous Peoples in Isolation in the Peruvian Amazon: Their Struggle for Survival and Freedom*. Somerset, NJ: Transaction, 2004.

Huggins, Martha. *From Slavery to Vagrancy in Brazil: Crime and Social Control in the Third World*. New Brunswick, NJ: Rutgers University Press, 1985.

Hunter, James M. *Perspective on Ratzel's Political Geography*. Lanham, MD: University Press of America, 1983.

Huntington, Ellsworth. *Civilization and Climate*. New Haven, CT: Yale University Press, 1915.

Instituto Brasileiro do Café. *O café no Brasil*. Rio de Janeiro: Ministério da Indústria e do Comércío, Instituto Brasileiro do Café, 1978.

Ireland, Gordon. *Boundaries, Possessions, and Conflicts in South America*. Cambridge, MA: Harvard University Press, 1938.

Iribertegui, Ramón. *Amazonas, el hombre y el caucho*. Caracas: Vicariato Apostólico de Puerto Ayacucho, Servicios Educativos Alfa Omega, 1987.

Jackson, Joe. *The Thief at the End of the World: Rubber, Power, and the Seeds of Empire*. New York: Viking, 2008.

James, C. L. R. *The Black Jacobins: Toussaint l'Ouverture and the San Domingo Revolution*. London: Secker and Warburg, 1938.

James, David, Louis Agassiz, and Pedro II. *O imperador do Brasil e os seus amigos da Nova Inglaterra*. Petrópolis, Brazil: Ministério da Educação e Saúde Museu Imperial, 1956.

James, William. *Brazil through the Eyes of William James*. Edited by Maria Helena de Monteiro. Cambridge, MA: Harvard Press, 2007.

Joffily, Geraldo Irenêo. *O quebra-quilo: A revolta dos matutos contra os doutores, 1874*. Brasília: Thesaurus, 1977.

Johan, Maurits, E. van den Boogart, Hendrik Richard Hoetink, and Peter James Palmer Whitehead. *Johan Maurits Van Nassau-Siegen, 1604–1679, a Humanist Prince in Europe and Brazil: Essays on the Occasion of the Tercentenary of His Death*. The Hague: Johan Maurits van Nassau Stichting, Govt. Pub. Office, 1979.

Johannsen, Robert Walter, Sam W. Haynes, and Christopher Morris. *Manifest Destiny and Empire: American Antebellum Expansionism*. College Station: Texas A&M University Press, 1997.

Johnson, Adriana M. C. "Subalternizing Canudos." *Modern Language Notes* 120, no. 2 (2005): 355–82.

Johnson, Allen W. *Sharecroppers of the Sertão; Economics and Dependence on a Brazilian Plantation*. Stanford, CA: Stanford University Press, 1971.

Johnson, Walter. *The Chattel Principle: Internal Slave Trades in the Americas*. New Haven, CT: Yale University Press, 2004.

———. *Soul by Soul: Life inside the Antebellum Slave Market*. Cambridge, MA: Harvard University Press, 1999.

Jullian, V., et al. "Validation of Use of a Traditional Antimalarial Remedy from French Guiana, Zanthoxylum Rhoifolium Lam." *Journal of Ethnopharmacology* 106, no. 3 (2006): 348–52.

Kadir, Aasa. "Natural Rubber: Strategies and Challenges for the 21st Century." *Kautschuk Gummi Kunststoffe* 50, no. 2 (1997): 88–90.

Kalliola, R., et al. "Upper Amazon Channel Migration: Implications for Vegetation Perturbance and Succession Using Bitemporal Landsat Mss Images." *Naturwissenschaften* 79, no. 2 (1992): 75–79.

Karasch, Mary C. "Os quilombos do ouro na Capitania de Goias." In *Liberdade por um fio*, edited by João José Reis and Flávio dos Santos Gomes. São Paulo: Companhia das Letras, 1996.

———. *Slave Life in Rio de Janeiro, 1808–1850*. Princeton, NJ: Princeton University Press, 1987.

Katzer, Friedrich. *Grundziige der Geologie des Unteren Amazonasgebietes*. Leipzig: Max Weg, 1903.

Keller-Leuzinger, Franz. *The Amazon and Madeira River*. New York: D. Appleton, 1874.

Kennedy, Donald, and Marjorie Lucks. "Rubber, Blight, and Mosquitoes: Biogeography Meets the Global Economy." *Environmental History* 4, no. 3 (1999): 369–83.

Kern, Arno Alvarez. *Missões, uma utopía política*. Porto Alegre, Brazil: Mercado Aberto, 1982.

Kidder, Daniel. *Brazil and the Brazilians*. London: Sampson Low, Marston, Searle and Rivington, 1845.

Klieman, Kairn A. *"The Pygmies Were Our Compass": Bantu and Batwa in the History of West Central Africa, Early Times to c. 1900 C.E.* Portsmouth, NH: Heinemann, 2003.

Kohlhepp, Gerd. "Scientific Findings of Alexander von Humboldt's Expedition into the Spanish-American Tropics (1799–1804) from a Geographical Point of View." *Anais da Academia Brasileira de Ciências* 77, no. 2 (2005): 325–42.

Kraay, Hendrik. *Afro-Brazilian Culture and Politics: Bahia, 1790s to 1990s*. Armonk, NY: M. E. Sharpe, 1998.

———. "As Terrifying as Unexpected: The Bahian Sabinada, 1837–1838." *Hispanic American Historical Review* 72, no. 4 (1992): 501–27.

———. "Patriotic Mobilization in Brazil: The Zoavos and Other Black Companies." In *I Die with My Country*, edited by Hendrik Kraay and Thomas Whigham, 61–81. Lincoln: University of Nebraska, 2004.

———. *Race, State, and Armed Forces in Independence-Era Brazil: Bahia, 1790s–1840s*. Stanford, CA: Stanford University Press, 2004.

Kraay, Hendrik, and Thomas Whigham, eds. *I Die with My Country: Perspectives on the Paraguayan War, 1864–1870*. Lincoln: University of Nebraska Press, 2004.

Kristal, Efraín. *Invisible Work: Borges and Translation*. Nashville: Vanderbilt University Press, 2002.

Kroemer, Gunter. *Cuxiúara, o Purús dos indígenas: Ensaio etno-histórico e etnográfico sobre os índios do Médio Purús*. São Paulo: Edições Loyola, 1985.

La Combe, Ernesto, Jorge von Hassel, and Luis Pesche. *El Istmo de Fitscarrald*. Lima: Junta de Vías Fluviales, 1903.

Lagos, Ovidio. *Arana, Rey del Caucho: Terror y atrocidades en el Alto Amazonas*. Buenos Aires: Emecé, 2005.

Landers, Jane, and Barry Robinson. *Slaves, Subjects, and Subversives: Blacks in Colonial Latin America*. Albuquerque: University of New Mexico Press, 2006.

Landes, Ruth. *The City of Women*. Albuquerque: University of New Mexico Press, 1994.

Lane, Kris E. *Pillaging the Empire: Piracy in the Americas, 1500–1750*. Armonk, NY: M. E. Sharpe, 1998.

Lang, Helen S. *The Order of Nature in Aristotle's "Physics": Place and the Elements*. Cambridge: Cambridge University Press, 1998.

Lange, Alcot. *In the Amazon Jungle: Adventures in Remote Parts of the Upper Amazon River, Including a Sojourn among Cannibal Indians*. New York: Knickerbocker, 1912.

Langfur, Hal. *The Forbidden Lands: Colonial Identity, Frontier Violence, and the Persistence of Brazil's Eastern Indians, 1750–1830*. Stanford, CA: Stanford University Press, 2006.

———. "Moved by Terror: Frontier Violence as Cultural Exchange in Late-Colonial Brazil." *Ethnohistory* 52, no. 2 (2005): 255–89.

———. "Myths of Pacification: Brazilian Frontier Settlement and the Subjugation of the Bororo Indians." *Journal of Social History* 32, no. 4 (1999): 879–905.

Lanternari, Vittorio. *As religiões dos oprimidos: Um estudo dos modernos cultos messiânicos*. São Paulo: Perspectivo, 1974.

Lapa, José Roberto do Amaral. *Livro da visitação do Santo Ofício da Inquisição ao Estado do Grão-Pará, 1763–1769*. Petrópolis, Brazil: Vozes, 1978.

LaRosa, Michael, and Germán Mejía P. *The United States Discovers Panama: The Writings of Soldiers, Scholars, Scientists, and Scoundrels, 1850–1905*. Lanham, MD: Rowman and Littlefield, 2004.

Larrabure y Correa, Carlos. *Perú y Colombia en el Putumayo: Réplica de una publicación aparecida, con fecha 27 de mayo último, en el Suplemento Sud-Americano del Times de Londres*. Barcelona: Impr. viuda de L. Tasso, 1913.

Larue, Isaac-Étienne. *Histoire du dix-huit Fructidor: La déportation des députés à la Guyane, leur évasion et leur retour en France*. Paris: E. Plon, Nourrit, 1895.

Lathrap, Donald Ward. *The Upper Amazon*. New York: Praeger, 1970.

Latimer, Jon. *Buccaneers of the Caribbean: How Piracy Forged an Empire*. Cambridge, MA: Harvard University Press, 2009.

Lavalle, B. "Frontiers, Colonization and Indian Manpower in Andean Amazonia (16th–20th Centuries): The Construction of Amazon Socioeconomic Space in Ecuador, Peru and Bolivia (1792–1948)." *Caravelle–Cahiers du Monde Hispanique et Luso-Bresilien*, no. 73 (1999): 315–16.

Lavergne, Barbara. "Quilombismo: The Case of Cafundó," pt. 1, "Quilombo Cafundó: Today's Cultural Resistance in Brazil, Struggle against Its Disappearance." *Journal of Black Studies* 11, no. 2 (1980): 217–22.

Leite, Serafim. *História de Companhia de Jesus no Brasil*. Lisbon: Livraria Portugalia, 1938.

Lemos, Maria Alzira Brum. *O doutor e o jagunço: Ciência, mestiçagem e cultura em "Os Sertões."* São Paulo: Editora Unimar, 2000.

Lemos, Renato. *Benjamin Constant: Vida e história*. Rio de Janeiro: Topbooks, 1999.

———. "Benjamin Constant: The Truth behind the Paraguayan War." In *I Die with My Country*, edited by Hendrik Kraay and Thomas L. Whigham, 81–105. Lincoln: University of Nebraska Press, 2004.

Leonard, Thomas M. *United States–Latin American Relations, 1850–1903: Establishing a Relationship*. Tuscaloosa: University of Alabama Press, 1999.

Lestringant, Frank. *Jean de Léry, ou L'invention du sauvage: Essai sur l'histoire d'un voyage faict en la terre du Brésil*. 2nd rev. and aug. ed. Paris: H. Champion, 2005.

Leuchars, Chris. *To the Bitter End: Paraguay and the War of the Triple Alliance*. Westport, CT: Greenwood, 2002.

Levine, Robert M. "Canudos in the National Context: A Religious Community in the Bahian Sertão." *Americas* 48, no. 2 (1991): 207–22.

———. "Mud-Hut Jerusalem: Canudos Revisited." *Hispanic American Historical Review* 68, no. 3 (1988): 525–72.

———. *Vale of Tears: Revisiting the Canudos Massacre in Northeastern Brazil, 1893–1897*. Berkeley: University of California Press, 1992.

Liais, Emmanuel. *Climats, géologie, faune et géographie botanique du Brésil*. Paris: Garnier Frères, 1872.

Lienhard, Martín. "Kalunga, or Recalling the Transatlantic Slave Trade in Some Afro-Latin-American Songs (African Oral Narrative, Religion, Brazil, Cuba)." *Revista Iberoamericana* 65, nos. 188–89 (1999): 505–17.

Lima, Nísia Trindade. *Um sertão chamado Brasil: Intelectuais e representação geográfica da identidade nacional*. Rio de Janeiro: Editora Revan, 1999.

Linden, Horacio de. *Por las tierras del caucho*. Buenos Aires: Sociedad Geográfica Americana, 1946.

Lindoso, Dirceu. *A utopía armada: Rebeliões de pobres nas Matas do Tombo Réal (1832–1850)*. Rio de Janeiro: Paz e Terra, 1983.

Lindsay-Poland, John. *Emperors in the Jungle: The Hidden History of the U.S. in Panama*. Durham, NC: Duke University Press, 2003.

Linebaugh, Peter, and Marcus Buford Rediker. *The Many-Headed Hydra: Sailors, Slaves, Commoners, and the Hidden History of the Revolutionary Atlantic*. Boston: Beacon, 2000.

Lisboa, Karen Macknow. *A nova Atlântida de Spix e Martius: Natureza e civilização na viagem pelo Brasil (1817–1820)*. São Paulo: Editora Hucitec, 1997.

Littlewood, Ian. *Sultry Climates: Travel and Sex*. Cambridge, MA: Da Capo, 2001.

Look Lai, Walton. *Indentured Labor, Caribbean Sugar: Chinese and Indian Migrants to the British West Indies, 1838–1918*. Baltimore: Johns Hopkins University Press, 1993.

Lopes, N. P., et al. "Antimalarial Use of Volatile Oil from Leaves of *Virola surinamensis* (Rol.) Warb. by Waiapi Amazon Indians." *Journal of Ethnopharmacology* 67, no. 3 (1999): 313–19.

López Austin, Alfredo. *Textos de medicina náhuatl*. 2nd ed. Mexico City: Universidad Nacional Autónoma de México, Instituto de Investigaciones Históricas, 1975.

Lorimer, Joyce. *English and Irish Settlement on the River Amazon, 1550–1646*. London: Hakluyt Society, 1989.

Loureiro, Antonio José Souto. *"Gazeta do Purús," scenas de uma épocha (Sena Madureira, 1918/1924)*. Manaus, Brazil: Imprensa Oficial, 1986.

Lovejoy, Paul E. *Transformations in Slavery: A History of Slavery in Africa*. New York: Cambridge University Press, 2000.

Lovejoy, Paul E., and David Vincent Trotman. *Trans-Atlantic Dimensions of Ethnicity in the African Diaspora*. London: Continuum, 2003.

Lovell, W. George, et al. "1491: In Search of Native America." *Journal of the Southwest* 46, no. 3 (2004): 441–61.

Lowenthal, David. "Colonial Experiments in French Guiana, 1760–1800." *Hispanic American Historical Review* 32, no. 1 (1952): 22–43.

———. "Nature and Nation: Britain and America in the 19th Century." *History Today* 53, no. 12 (2003): 18–25.

Lowie, Robert Harry. "The Cariri." In *Handbook of South American Indians*. Washington, DC: Smithsonian Institution, 1946.

Lubrich, Oliver. "'Egypt Everywhere': Alexander von Humboldt's 'Orientalist Gaze' across America." *Germanisch-Romanische Monatsschrift* 54, no. 1 (2004): 19–39.

Lurie, Edward. *Nature and the American Mind: Louis Agassiz and the Culture of Science*. New York: Science History Publications, 1974.

Luz, Nicia Vilela. *A Amazônia para os negros americanos*. Rio de Janeiro: Saga, 1968.

Lyra, Maria de Lourdes Viana. *A utopía do poderoso império: Portugal e Brasil, bastidores da política, 1798–1822*. Rio de Janeiro: Sette Letras, 1994.

Macedo, José. *Belo Monte: Uma história da Guerra de Canudos*. São Paulo: Expressão e Cultura, 2003.

Machado, Christina Mata. *As táticas de guerra dos cangaceiros*. Rio de Janeiro: Laemmert, 1969.

Machado Filho, Aires da Mata. *Negro e o garimpo em Minas Gerais*. Rio de Janeiro: Jose Olímpio, 1951.

Machado, Humberto F. *Escravos, senhores e café: A crise da cafeicultura escravista do Vale do Paraíba fluminense, 1860–1888*. Niterói, Brazil: Editora Cromos Clube de Literatura, 1993.

Machado, Maria Helena P. T. "From Slave Rebels to Strikebreakers: The Quilombo of Jabaquara and the Problem of Citizenship in Late-Nineteenth-Century Brazil." *Hispanic American Historical Review* 86, no. 2 (2006): 247–74.

MacLachlan, Colin M. "African Slavery and Economic Development in Amazonia, 1700–1800." In *Slavery and Race Relations in Latin America*, edited by Robert Brent Toplin. Westport, CT: Greenwood, 1973.

———. "The Indian Directorate: Forced Acculturation in Portuguese America." *Americas* 28, no. 4 (1972).

———. "The Indian Labor Structure in the Portuguese Amazon." In *Colonial Roots of Modern Brazil*, edited by Dauril Alden. Berkeley: University of California Press, 1973.

MacQueen, Norrie. *The Decolonization of Portuguese Africa: Metropolitan Revolution and the Dissolution of Empire*. London: Longman, 1997.

Mam Lam Fouck, Serge. *Guyane Français*. Paris: Ibis, 1999.

Mam Lam Fouck, Serge, and Jacqueline Zonzon. *L'histoire de la Guyane depuis les civilisations amérindiennes*. Matoury, French Guiana: Ibis Rouge, 2006.

Mann, Charles C. *1491: New Revelations about the Americas before Columbus*. New York: Alfred A. Knopf, 2006.

———. "Ancient Earthmovers of the Amazon." *Science* 321 (2008): 1148–52.

Mann, Thomas. *The Magic Mountain (Der Zauberberg)*. New York: Alfred A. Knopf, 1928.

Mansfield, Charles. *Paraguay, Brazil and the Plate*. Cambridge, UK: Macmillan, 1856.

Markham, Clements R. *The Fifty Years' Work of the Royal Geographical Society*. London: J. Murray, 1881. Microform.

———. *The Incas of Peru*. New York: Dutton, 1910.

———. *Peruvian Bark: A Popular Account of the Introduction of Chinchona Cultivation into British India*. London: J. Murray, 1880.

Markham, Clements R., and Peter Blanchard. *Markham in Peru: The Travels of Clement R. Markham, 1852–1853*. Austin: University of Texas Press, 1991.

Markham, Clements R. *A Memoir on the Indian Surveys*. London: W. H. Allen, 1871.

Marques, Maria Eduarda, Castro Magalhães, and Leslie Bethell. *A Guerra do Paraguai: 130 anos depois*. 2nd ed. Rio de Janeiro: Relume Dumará, 1995.

Martin, Florence, and Isabelle Favre. *De la Guyane à la diaspora africaine: Écrits du silence*. Paris: Karthala, 2002.

Martin, Percy Alvin. "The Influence of the United States in the Opening of the Amazon to the World's Commerce." *Hispanic American Historical Review* 1, no. 2 (1918): 146–62.

Martinelli, Pedro. *Mulheres da Amazônia*. Carapicuiba, Brazil: Jaraqui, 2003.

Mascarenhas, Maria Lucia. "Toda nação em Canudos, 1893–1897, indios em Canudos: Memorias e tradição oral da participação dos Kariris e Kaimbês na Guerra de Canudos." *Revista Canudos* 2, no. 2 (1997).

Mathews, Edward Davis. *Up the Amazon and Madeira Rivers, through Bolivia and Peru*. London: Sampson Low, Marston, Searle and Rivington, 1879.

Mattoso, Kátia M. de Queirós. *To Be a Slave in Brazil, 1550–1888*. New Brunswick, NJ: Rutgers University Press, 1986.

Maury, Mathew Fontaine. *The Amazon and the Atlantic Slopes of South America*. Washington, DC: F. Taylor, 1853.

———. "Commercial Prospects for the South." *Southern Literary Messenger* 17 (1851): 688–98.

———. "Great Commercial Advantages of the Gulf of Mexico." *De Bow's Commercial Review* 7 (1849): 510–23.

———. Letter to Ann Maury. 1850. Archived at Library of Congress, Wasington, DC.

———. Letter to Mrs William Blackford. 1851. Maury Papers 6, Library of Congress, Washington, DC.

———. "Our Gulf States and the Amazon." *DeBow's Review* 18 (1855): 364–66.

Maxwell, Kenneth. "The Spark: Pombal, the Amazon, and the Jesuits." *Portuguese Studies* 17 (2001): 168-83.

May, Robert E. *John A. Quitman: Old South Crusader*. Baton Rouge: Louisiana State University Press, 1985.

———. *Manifest Destiny's Underworld: Filibustering in Antebellum America*. Chapel Hill: University of North Carolina Press, 2002.

———. *The Southern Dream of a Caribbean Empire, 1854–1861*. Baton Rouge: Louisiana State University Press, 1973.

———. *The Union, the Confederacy, and the Atlantic Rim*. West Lafayette, IN: Purdue University Press, 1995.

Mayle, F. E., R. P. Langstroth, R. A. Fisher, and P. Meir. "Long-Term Forest-Savannah Dynamics in the Bolivian Amazon: Implications for Conservation." *Philosophical Transactions of the Royal Society B–Biological Sciences* 362, no. 1478 (2007): 291–307.

McCann, Frank D. *A nação armada: Ensaios sobre a história do exército brasileiro*. Recife: Editora Guararapes, 1982.

McClintock, Anne. *Imperial Leather: Race, Gender and Sexuality in the Colonial Context*. New York: Columbia University Press, 1995.

McDougall, Walter A. *Throes of Democracy: The American Civil War Era, 1829–1877*. New York: Harper, 2008.

McEwan, Colin. *Unknown Amazon: Culture in Nature in Ancient Brazil*. London: British Museum Press, 2001.

McKey, Doyle., S. Rostain, J. Iriarte, B. Glaser, J. J. Birk, I. Holst, and D. Renard. "Pre-Columbian Agricultural Landscapes, Ecosystem Engineers, and Self-Organized Patchiness in Amazonia." *Proceedings of the National Academy of Sciences of the United States of America* 107, no. 17 (2010): 7823–28.

McNeill, John Robert. *Mosquito Empires: Ecology and War in the Greater Caribbean, 1620–1914*. Cambridge: Cambridge University Press, 2009.

Meggers, Betty Jane. *Amazonia: Man and Culture in a Counterfeit Paradise*. Chicago: Aldine Atherton, 1971.

Meggers, Betty Jane, and Clifford Evans. *Archeological Investigations at the Mouth of the Amazon*. Washington, DC: US Government Printing Office, 1957.

Meira, Sílvio Augusto de Bastos. *Fronteiras sangrentas: Heróis do Amapá*. Rio de Janeiro: Meira, 1975.

Méndez, José Armando, and Tribunal Arbitral Boliviano-Brasileño. *Defensa de los derechos de Bolivia ante el Tribunal Arbitral Boliviano-Brasileño*. Buenos Aires: Imprenta de L. Mirau, 1906.

Mendonça, Joseli Maria Nunes. *Cenas da abolição: Escravos e senhores no Parlamento e na justiça*. São Paulo: Editora Fundação Perseu Abramo, 2001.

Mendonça, Marcos Carneiro de. *A Amazônia na era pombalina; Correspondência Inédita do governador e capitão-general do Estado do Grão-Pará e Maranhão*. Rio de Janeiro: n.p., 1963.

Merrell, Floyd. *Capoeira and Candomblé: Conformity and Resistance in Brazil*. Princeton, NJ: Markus Wiener, 2005.

Mesters, Carlos, and Paulo Suess. *Utopía cativa: Catequese indigenista e libertação indígena*. Petrópolis, Brazil: Editora Vozes, 1986.

Metcalf, Alida C. "Millenarian Slaves? The Santidade de Jaguaripe and Slave Resistance in the Americas." *American Historical Review* 104, no. 5 (1999): 1531-59.

Michel, Jacques. *La Guyane sous l'ancien régime: Le désastre de Kourou et ses scandaleuses suites judiciaires*. Paris: L'Harmattan, 1989.

Michelot, Jean Claude. *La guillotine sèche: Histoire des Bagnes de Guyane*. Paris: Fayard, 1981.

Milhou, Alain. "Messianic and Utopian Currents in the Indies of Castille." In *Utopia: The Search for the Ideal Society in the Western World*, edited by Roland Schaer, Gregory Claeys and Lyman Tower Sargent. New York: Oxford University Press, 2000.

Millard, Candice. *River of Doubt: Theodore Roosevelt's Darkest Journey*. New York: Doubleday, 2005.

Miller, David Philip, and Peter Hanns Reill. *Visions of Empire: Voyages, Botany, and Representations of Nature*. New York: Cambridge University Press, 1996.

Miller, Joseph Calder. *Way of Death: Merchant Capitalism and the Angolan Slave Trade, 1730–1830*. Madison: University of Wisconsin Press, 1988.

Millet, Henrique Augusto. *Os quebra-quilos e a crise de Lavoura*. São Paulo: Instituto Nacional do Livro, 1876.

Mills, James H., and Patricia Barton. *Drugs and Empires: Essays in Modern Imperialism and Intoxication, c. 1500–c. 1930*. Basingstoke, UK: Palgrave Macmillan, 2007.

Missal, Alexander. *Seaway to the Future: American Social Visions and the Construction of the Panama Canal*. Madison: University of Wisconsin Press, 2008.

Moniz, Edmundo. *Canudos: A guerra social*. 2nd ed. Rio de Janeiro: Elo Editora, 1987.

———. *Canudos, a luta pela terra*. 3rd ed. São Paulo: Global, 1984.

———. *A guerra social de Canudos*. Rio de Janeiro: Civilização Brasileira, 1978.

Monteiro, Hamilton de Mattos. *Nordeste insurgente, 1850–1890*. São Paulo: Editora Brasiliense, 1981.

Monteiro, John M. *Negros da terra: Índios e bandeirantes nas origens de São Paulo*. São Paulo: Companhia das Letras, 1994.

Montgomery, William. "Global Enigma and Marvel of Life: Ernst Haeckel—Work, Impact and Legacy." *Isis* 91, no. 3 (2000): 601–2.

Mooney, James. *The Ghost-Dance Religion and the Sioux Outbreak of 1890*. Washington, DC: US Government Printing Office, 1896.

Moore, John Bassett. *Brazil and Peru Boundary Question*. New York: Knickerbocker, 1905.

———. *A Digest of International Law*. 8 vols. Washington, DC: US Government Printing Office, 1906.

———. "The Franco-Brazilian Boundary Dispute: A Practical Vindication of the Monroe Doctrine." *New York Times*, December 16, 1900.

———. *The Principles of American Diplomacy*. New York: Harper and Brothers, 1918.

More, Thomas. *Utopia*. Oxford World Classics. Orig.1516. Oxford: 1999.

Mörner, Magnus. *The Political and Economic Activities of the Jesuits in the La Plata Region: The Hapsburg Era*. Stockholm: Victor Pettersons Bokindustri Aktiebolag, 1953.

———. *Race Mixture in the History of Latin America*. Boston: Little, Brown, 1967.

Morse, Richard. *Bandeirantes: The Historical Role of the Brazilian Pathfinders*. New York: Alfred A. Knopf, 1965.

Mosquera, Claudia, Mauricio Pardo, and Odile Hoffmann. *Afro-descendientes en las Américas: Trayectorias sociales e identitarias; 150 años de la abolición de la esclavitud en Colombia*. Bogotá: Universidad Nacional de Colombia, Instituto Colombiano de Antropología e Historia I, 2002.

Mott, Luiz. *Homossexuais da Bahia: Dicionário biográfico, séculos XVI-XIX*. Salvador, Brazil: Editora Grupo Gay da Bahia, 1999.

———. "Os indios e a pecuaria nas fazendas de gado no Piauí." *Revista de Antropologia* 22 (1979).

Moura, Clóvis. *Os quilombos e a rebelião negra*. São Paulo: Brasiliense, 1986.

———. *Os quilombos na dinâmica social do Brasil*. Maceió, Brazil: Editora Universidade Federal de Alagoas, 2001.

———. *Quilombos: Resistência a escravismo*. São Paulo: Atica, 1987.

Moya, Alba. *Auge y crisis de la cascarilla en la Audiencia de Quito, siglo XVIII*. Quito: Facultad Latinoamericana de Ciencias Sociales, Sede Ecuador, 1994.

Muir, John, and Michael P. Branch. *John Muir's Last Journey: South to the Amazon and East to Africa; Unpublished Journals and Selected Correspondence*. Washington, DC: Island Press/Shearwater Books, 2001.

Mukerji, Chandra. "Dominion, Demonstration and Domination: Religious Doctrine, Territorial Politics and French Plant Collection." In *Colonial Botany: Science, Commerce, and Politics in the Early Modern World*, edited by Londa L. Schiebinger and Claudia Swan, 1933. Philadelphia: University of Philadelphia Press, 2007.
Murphy, Robert Francis. *The Rubber Trade and the Mundurucú Village*. Ann Arbor, MI: University Microfilms, 1954. Microform.
Myers, Kathleen Ann, Nina M. Scott, and Gonzalo Fernández de Oviedo y Valdés. *Fernández de Oviedo's Chronicle of America: A New History for a New World*. Austin: University of Texas Press, 2007.
Nabokov, Vladimir. *Notes on Prosody from the Commentary to His Translation of Pushkin's "Eugene Onegin."* London: Routledge and Kegan Paul, 1965.
Naipaul, V. S. *The Loss of El Dorado: A History*. New York. Knopf. 1967.
Nakashima, Mary. *Chico Mendes por ele mesmo*. São Paulo: Martin Claret, 1992.
Naro, Nancy. *Blacks, Coloureds and National Identity in Nineteenth-Century Latin America*. London: University of London Institute of Latin American Studies, 2003.
Necker, Louis. *Indios Guaraníes y chamanes Franciscanos: Las primeras reducciones del Paraguay, 1580–1800*. Asunción: Centro de Estudios Antropológicos Universidad Católica, 1990.
Needell, Jeffrey D. *The Party of Order: The Conservatives, the State, and Slavery in the Brazilian Monarchy, 1831–1871*. Stanford, CA: Stanford University Press, 2006.
———. "The Revolta contra Vacina of 1904: The Revolt against 'Modernization' in Belle-Epoque Rio de Janeiro." *Hispanic American Historical Review* 67, no. 2 (1987): 233–69.
———. *A Tropical Belle Époque: Elite Culture and Society in Turn-of-the-Century Rio de Janeiro*. Cambridge: Cambridge University Press, 1987.
Neller, R. J., J. S. Salo, and M. E. Rasanen. "On the Formation of Blocked Valley Lakes by Channel Avulsion in Upper Amazon Foreland Basins." *Zeitschrift für Geomorphologie* 36, no. 4 (1992): 401–11.
Nery, A. Constantino, and Algenir Ferraz Suano da Silva. *A quarta expedição contra Canudos, 1er phase das operações, cem léguas através do sertão, de Aracajú a Queimadas, via Canudos: Diário de campanha*. Facsimile ed. Manaus: Governo do Estado do Amazonas, Editora da Universidade do Amazonas, 1997.
Neto, Manoel. "Canudos na boca do Povo." *Revista Canudos* 1, no. 1 (1996).
Neves, Luiz Felipe Baêta. *Vieira e a imaginação social jesuítica: Maranhão e Grão-Pará no século XVII*. Rio de Janeiro: Topbooks, 1997.
Newman, James L. *Imperial Footprints: Henry Morton Stanley's African Journeys*. Washington, DC: Brassey's, 2004.
Nicolson, Malcolm. "Humboldtian Science and the Origins of the Study of Vegetation." *History of Science* 25, no. 68 (1987): 167–94.
Nimuendaju, Curt. *In Pursuit of a Past Amazon: Archaeological Researches in the Brazilian Guyana and in the Amazon Region; A Posthumous Work*. Translated by Stig Rydén and Per Stenborg. Ethnological Studies 45. Gothenborg, Sweden, 2004.
Nimuendajú, Curt, and Jurgen Riester. *Los mitos de creación y de destrucción del mundo como fundamentos de la religión de los Apapokuva-Guaraní*. Lima: Centro Amazónico de Antropología y Aplicación Práctica, 1978.
Nina Rodrigues, Raimundo. *Os africanos no Brasil*. 7th ed. Brasília: Editora Universidade de Brasília, 1988.

———. *O alienado no direito civil brasileiro*. 3rd ed. São Paulo: Companhia Editora Nacional, 1939.
———. *O animismo fetichista dos negros bahianos*. Rio de Janeiro: Civilização Brasileira, 1935.
———. "Mestissage, degenerance et crime." *Archives de Antropologie Crimenelle*, 1899.
———. *As raças humanas e a responsabilidade penal no Brasil*. Bahia: Imprensa Popular, 1894.
Nina Rodrigues, Raimundo, Arthur Ramos, and Alfredo Britto. *As collectividades anormaes*. Rio de Janeiro: Civilização Brasileira, s.a., 1939.
Nishida, Mieko. *Slavery and Identity: Ethnicity, Gender, and Race in Salvador, Brazil, 1808–1888*. Bloomington: Indiana University Press, 2003.
Nobre, Freitas. *A epopéia acreana*. 3rd ed. São Paulo: Emprêsa Gráfica da *Revista dos Tribunais*, 1939.
Nogueira, J. C. Ataliba. *Antônio Conselheiro e Canudos: Revisão histórica*. São Paulo: Companhia Editora Nacional, 1974.
Northrup, David. *Indentured Labor in the Age of Imperialism, 1834–1922*. Cambridge: Cambridge University Press, 1995.
Nugent, Stephen. *Amazonian Caboclo Society: An Essay on Invisibility and Peasant Economy*. Providence, RI: Berg, 1993.
———. *Scoping the Amazon: Image, Icon, Ethnography*. Walnut Creek, CA: Left Coast, 2007.
Núñez Cabeza de Vaca, Alvar. *Relations et naufrages d'Alvar Nuñez Cabeça de Vaca. Valladolid: De l'Imprimerie de Francisco Fernandez de Cordoue, 1555*. Paris: A. Bertrand, 1837.
Núñez Cabeza de Vaca, Alvar, Pedro Hernández, William Mills Ivins, and Grenville Kane. *La relacion y comentarios del Governador Alvar Nuñez Cabeça de Vaca, de lo acaescido en las dos jornadas que hizo a las Indias*. Valladolid: Francisco Fernandez de Cordoua, 1555.
Nyhart, Lynn K. *Biology Takes Form: Animal Morphology and the German Universities, 1800–1900*. Chicago: University of Chicago Press 1995.
Olwig, Kenneth. *Landscape, Nature and the Body Politic: From Britain's Renaissance to America's New World*. Madison: University of Wisconsin, 2002.
Orme, Antony R. "American Geomorphology at the Dawn of the 20th Century." *Physical Geography* 25, no. 5 (2004): 361–81.
———. "The Rise and Fall of the Davisian Cycle of Erosion: Prelude, Fugue, Coda, and Sequel." *Physical Geography* 28, no. 6 (2007): 474–506.
Orton, James. *The Andes and the Amazon, or Across the Continent of South America*. New York: Harper and Brothers, 1870.
Ourique, Jacques. *O Amazonas e o Acre: Artigos publicados no "Jornal do Commercio."* Rio de Janeiro: Typ. do *Jornal do Commercio*, Rodrigues & Comp., 1907.
Pacheco B., Pablo. *Integración económica y fragmentación social: El itinerario de las barracas en la Amazonia boliviana*. La Paz: Cedla, 1992.
Padoch, Christine. *Várzea: Diversity, Development, and Conservation of Amazonia's Whitewater Floodplains*. New York: New York Botanical Garden Press, 1999.
Pagden, Anthony. *The Fall of Natural Man: The American Indian and the Origins of Comparative Ethnology*. Cambridge: Cambridge University Press, 1986.
———. *Lords of All the World: Ideologies of Empire in Spain, Britain and France c. 1500–c. 1800*. New Haven, CT: Yale University Press, 1995.
———. *Spanish Imperialism and the Political Imagination: Studies in European and Spanish-*

American Social and Political Theory, 1513–1830. New Haven, CT: Yale University Press, 1990.

Pakenham, Thomas. *The Scramble for Africa: White Man's Conquest of the Dark Continent from 1876 to 1912*. New York: Avon Books, 1992.

Paret, Peter. *Internal War and Pacification: The Vendée*. Princeton, NJ: Woodrow Wilson School of Public and International Affairs, 1961.

Parssinen, M., D. Schaan, and A. Ranzi. "Pre-Columbian Geometric Earthworks in the Upper Purus: A Complex Society in Western Amazonia." *Antiquity* 83, no. 322 (2009): 1084–95.

Payne, Doris L. *Amazonian Linguistics: Studies in Lowland South American Languages*. Austin: University of Texas Press, 1990.

Pearson, Henry C. *The Rubber Country of the Amazon: A Detailed Description of the Great Rubber Industry of the Amazon Valley, Which Comprises the Brazilian States of Pará, Amazonas and Mato Grosso, the Territory of the Acre, the Montaña of Peru and Bolivia, and the Southern Portions of Colombia and Venezuela*. New York: India Rubber World, 1911.

Pedlowski, M. A., V. H. Dale, E. A. T. Matricardi, and E. P. da Silva. "Patterns and Impacts of Deforestation in Rondônia, Brazil." *Landscape and Urban Planning* 38, nos. 3–4 (1997): 149–57.

Pendergrast, Mark. *Uncommon Grounds: The History of Coffee and How It Transformed Our World*. New York: Basic Books, 1999.

Pennano A., Guido. *La economía del caucho*. Iquitos, Peru: Centro de Estudios Teológicos de la Amazonía, 1988.

Péret, Benjamin, and Júlio Henriquez. *O quilombo de Palmares: Crónica da "República dos Escravos," Brasil, 1640–1695*. Lisbon: Fenda Edições, 1988.

Perez, Berta E. "The Journey to Freedom: Maroon Forebears in Southern Venezuela." *Ethnohistory* 47, nos. 3–4 (2000): 611–34.

Perkins, Chris. "Cultures of Map Use." *Cartographic Journal* 45, no. 2 (2008): 150–58.

Perna, S. J. Q., G. L. Cardoso, and J. F. Guerreiro. "Duffy Blood Group Genotypes among African-Brazilian Communities of the Amazon Region." *Genetics and Molecular Research* 6, no. 1 (2007): 166–72.

Peru and Víctor M. Maurtua. *Juicio de límites entre el Perú y Bolivia, contestación al alegato de Bolivia*. Barcelona: Henrich, 1907.

Peru. Ministerio de Relaciones Exteriores. *Perú: Interoceanic Communication across the Peruvian Andes, the Amazon River and Its Great Navigable Affluents*. Lima: Gobierno del Perú, 1903.

Pessar, Patricia R. *From Fanatics to Folk: Brazilian Millenarianism and Popular Culture*. Durham, NC: Duke University Press, 2004.

———. "Millenarian Movements in Rural Brazil: Prophecy and Protest." *Religion* 12, no. 3 (1982): 187–213.

Peters, Charles M., Michael J. Balick, Francis Kahn, and Anthony B. Anderson. "Oligarchic Forests of Economic Plants in Amazonia: Utilization and Conservation of an Important Tropical Resource." *Conservation Biology* 3, no. 4 (1989): 341–49.

Petersen, James B., Eduardo Neves, and Michael J. Heckenberger. "Gift from the Past: Terra Preta and the Prehistoric Occupation of the Amazon." In *Unknown Amazon*, edited by Colin McEwan, Christiana Barreto, and Eduardo Neves, 86–108. London: British Museum, 2001.

Petot, Jean. *L'Or de Guyane: Son histoire, ses hommes*. Paris: Editions Caribéenes, 1986.

Pickering, Barbara. "Domestication of Plants in the Americas: Insights from Mendelian and Molecular Genetics." *Annals of Botany* (2007): 1–16.

Pinaud, João Luiz Duboc. *Insurreição negra e justiça: Pati dos Aleferes, (Brazil) 1838*. Rio de Janeiro: Editora Expressão e Cultura, Ordem dos Advogados do Brasil–RJ, 1987.

Pineda Camacho, Roberto. *Holocausto en el Amazonas: Una historia social de la Casa Arana*. Bogotá: Planeta Colombiana Editorial, 2000.

Pinheiro, Luís Balkar Sá Peixoto. *Visões da Cabanagem: Uma revolta popular e suas representações na historiografia*. Manaus, Brazil: Valer Editora, 2001.

Piperno, Dolores, and Deborah Pearsall. *The Origins of Agriculture in the Lowland Tropics*. New York: Academic Press, 1998.

Pires, Maria Idalina da Cruz. *Guerra dos bárbaros: Resistência indígena e conflitos no nordeste colonial*. Recife: Governo do Estado de Pernambuco, Secretaria de Educação Companhia Editora de Pernambuco, 1990.

Pitman, Nigel C. A., John Terborgh, Miles R. Silman, and Percy Nuñez V. "Tree Species Distributions in an Upper Amazonian Forest." *Ecology* 80, no. 8 (1999): 2651–61.

Pitts, Jennifer. *A Turn to Empire*. Princeton, NJ: Princeton University Press, 2005.

Plane, August. *L'Amazonie: A travers l'Amerique equatorial*. Paris: Plon, 1903.

Plotkin, M. J., B. M. Boom, and M. Allison. "The Ethnobotany of Aublet's *Histoire des plantes de la Guiane Française* (1775)." *Monographs in Systematic Botany* 35 (1991): 1–108.

Pombal, Sebastião José de Carvalho e Melo, and Júlio Ricardo Quevedo dos Santos. *República jesuítica ultramarina*. Porto Alegre, Brazil: Martins Livreiro Editora, 1989.

Pomeroy, Samuel. "Information for Persons Proposing to Join the Free Colored Colony to Central America." Washington, DC: SC Pomeroy, 1862.

Pontes, Eloy. *Vida dramática de Euclides da Cunha*. Rio de Janiero: Jose Olímpio, 1938.

Pontes, Kassius Diniz da Silva. *Euclides da Cunha, o Itamaraty e a Amazônia*. Brasília: Instituto Rio Branco, Fundação Alexandre de Gusmão, 2005.

Pontes, Mário. *Doce como o diabo: Demônio, utopía e liberdade na poesia de Cordel Nordestina*. Rio de Janeiro: Editora Codecri, 1979.

Pöppig, Eduard Friedrich. *Viaje al Perú y al Río Amazonas, 1827–1832*. Iquitos, Perú: Centro de Estudios Teológicos, 2003.

Porro, Antonio. *As crônicas do Rio Amazonas: Tradução, introdução e notas etno-históricas sobre as antigas populações indígenas da Amazônia*. Petrópolis: Vozes, 1993.

———. *O povo das águas: Ensaios de etno-história amazônica*. Petrópolis, Brazil: Vozes, Editora da Universidade de São Paulo (EDUSP), 1996.

Porro, R. "Palms, Pastures, and Swidden Fields: The Grounded Political Ecology of 'Agro-Extractive/Shifting-Cultivator Peasants' in Maranhão, Brazil." *Human Ecology* 33, no. 1 (2005): 17–56.

Posada Carbó, Eduardo. *Wars, Parties and Nationalism: Essays on the Politics and Society of Nineteenth-Century Latin America*. London: Institute of Latin American Studies, University of London, 1995.

Posey, Darrell Addison, and William L. Balée. *Resource Management in Amazonia: Indigenous and Folk Strategies*. Bronx, NY: New York Botanical Garden, 1989.

Posey, Darrell Addison, and Michael J. Balick. *Human Impacts on Amazonia: The Role of Traditional Ecological Knowledge in Conservation and Development*. New York: Columbia University Press, 2006.

Posey, Darrell Addison, and Kristina Plenderleith. *Kayapó Ethnoecology and Culture*. London: Routledge, 2002.

Power, Marcus, and James D. Sidaway. "The Degeneration of Tropical Geography." *Annals of the Association of American Geographers* 94, no. 3 (2004): 585–601.

Prance, G. "The Poisons and Narcotics of the Amazonian Indians." *Journal of the Royal College of Physicians of London* 33, no. 4 (1999): 368–76.

Prance, Ghillean T., Derek Chadwick, and Joan Marsh. *Ethnobotany and the Search for New Drugs*. Chichester, UK: John Wiley, 1994.

Pratt, Mary Louise. *Imperial Eyes: Travel Writing and Transculturation*. New York: Routledge, 1992.

Price, Richard. *Alabi's World*. Baltimore: Johns Hopkins University Press, 1990.

———. *First-Time: The Historical Vision of an Afro-American People*. Baltimore: Johns Hopkins University Press, 1983.

———. "Liberdade, fronteiras, e deuses: Saramacas no Oiyapoque." In *Quase-cidadão: Historias e antropologias da pós-emancipação no Brasil*, edited by Olivia Gomes da Cunha and Flávio dos Santos Gomes, 119–46. São Paulo: Fundação Getulio Vargas, 2007.

———. *Maroon Societies: Rebel Slave Communities in the Americas*. 3rd ed. Baltimore: Johns Hopkins University Press, 1996.

———. *To Slay the Hydra: Dutch Colonial Perspectives on the Saramaka Wars*. Ann Arbor, MI: Karoma, 1983.

Projeto Vida de Negro (Maranhão, Brasil). *Vida de negro no Maranhão: Uma experiência de luta, organização e resistência nos territórios quilombolas*. São Luís, Brazil: Sociedade Maranhense de Direitos Humanos, Centro de Cultura Negra do Maranhão, 2005.

Pyenson, Lewis. *Civilizing Mission: Exact Sciences and French Overseas Expansion, 1830–1940*. Baltimore: Johns Hopkins University Press, 1993.

Quinn, W. H., V. T. Neal, and S. E. A. Demayolo. "El Niño Occurrences over the Past 4½Centuries." *Journal of Geophysical Research—Oceans* 92, no. C13 (1987): 14,449–61.

Rabassa, Gregory. *If This Be Treason: Translation and Its Dyscontents*. New York: New Directions, 2006.

Rabello, Sylvio. *Euclides da Cunha*. 2nd ed. Rio de Janeiro: Civilização Brasileira, 1966.

Raffles, Hugh. *In Amazonia: A Natural History*. Princeton, NJ: Princeton University Press, 2002.

Raimondi, Antonio. *Notas de viajes para su obra "El Perú."* Lima: Imprenta Torres Aguirre, 1942.

Raimondi, Antonio, José Balta, and Sociedad Geográfica de Lima. *El Perú*. Lima: Imprenta del Estado, 1874.

Raleigh, Walter, Vincent T. Harlow, and N. M. Penzer. *The Discoverie of the Large and Bewtiful Empire of Guiana*. London: Argonaut, 1928.

Rangel, Alfredo. *Inferno verde: Scenas e scenarios do Amazonas*. 4th ed. Tours, France: Typ. Arrault, 1927.

Rasanen, M., R. Neller, J. Salo, and H. Jungner. "Recent and Ancient Fluvial Deposition Systems in the Amazonian Foreland Basin, Peru." *Geological Magazine* 129, no. 3 (1992): 293–306.

Ratzel, Friedrich, and Arthur John Butler. *The History of Mankind*. London: Macmillan, 1896.

Raynal, Thomas Payne, and Libraires Associés (Paris, France). *Histoire philosophique et politique des établissements et du commerce des européens dans les deux Indes*. Neuchatel, Switzerland: Chez les Libraires associés, 1783.

Reber, Vera B. "The Demographics of Paraguay: A Reinterpretation of the Great War, 1864–1870." *Hispanic American Historical Review* 68 no. 920 (1988): 289–319.

Rebok, Sandra. "Alexander von Humboldt's Perceptions of Colonial Spanish America." *Dynamis* 29 (2009): 49–72.

Redfield, Peter. *Space in the Tropics: From Convicts to Rockets in French Guiana*. Berkeley: University of California Press, 2000.

Rediker, Marcus Buford. *Villains of All Nations: Atlantic Pirates in the Golden Age*. Boston: Beacon, 2004.

Reis, Arthur Cezar Ferreira. *A Amazônia e a cobiça internacional*. 3rd ed. Rio de Janeiro: Gráfica Record Editora, 1968.

———. *Limites e demarcações na Amazônia Brasileira*. 2nd ed. 2 vols. Belém: Secretaria de Estado da Cultura, 1993.

———. *Portugueses e brasileiros na Guiana Francesa*. Rio de Janeiro: Departamento de Imprensa Nacional, 1953.

———. *Território do Amapá: Perfil histórico*. Rio de Janeiro: Departamento de Imprensa Nacional, 1949.

Reis, Jaime. *Abolition and the Economics of Slaveholding in North East Brazil*. Glasgow: Institute of Latin American Studies, University of Glasgow, 1974.

Reis, João José. *A invenção de liberdade: O negro no Brasil*. São Paulo: Companhia das Letras, 2003.

———. *Rebelião escrava no Brasil: A história do levante dos Malês, 1835*. São Paulo: Brasiliense, 1986.

———. "The Revolution of the Ganhadores: Urban Labour, Ethnicity and the African Strike of 1857 in Bahia, Brazil." *Journal of Latin American Studies* 29 (1997): 355–93.

———. *Slave Rebellion in Brazil: The Muslim Uprising of 1835 in Bahia*. Baltimore: Johns Hopkins University Press, 1993.

Reis, João José, and Flávio dos Santos Gomes. *Liberdade por um fio: História dos quilombos no Brasil*. São Paulo: Companhia das Letras, 1996.

———. "Repercussions of the Haitian Revolution in Brazil." In *The World of the Haitian Revolution*, edited by David Geggus and Norman Fiering, 284–312. Bloomington: University of Indiana Press, 2008.

Renner, Susanne. *A History of Botanical Exploration in Amazonian Ecuador, 1739–1988*. Washington, DC: Smithsonian Institution, 1993.

Restall, Matthew. *Beyond Black and Red: African-Native Relations in Colonial Latin America*. Albuquerque: University of New Mexico Press, 2005.

Rey de Castro, Carlos. *Los escándalos del Putumayo. Carta al director del "Daily News & Leader," de Londres*. Barcelona: Imprenta Vda. de L. Tasso, 1913.

Rey de Castro, Carlos, Andrew Gray, and Alberto Chirif. *La defensa de los caucheros*. Iquitos, Peru: CETA; Copenhagen: IWGIA, 2005.

Rey Fajardo, José del. *Misiones jesuíticas en la Orinoquía*. Caracas: Universidad Católica Andrés Bello, 1977.

Ribeiro, Napoleão. *O Acre e os seus heroes: Contribuição para a história do Brasil*. Maranhão, Brazil: Typ. Rabell, 1930.

Ricardo, Cassiano, Government of Brazil, and Government of Bolivia. *O Tratado de Petrópolis*. Rio de Janeiro: Ministério das Relações Exteriores, 1954.
Richardson, M. K., et al. "Haeckel, Embryos, and Evolution." *Science* 280 (1997): 983–84.
Ricupero, Bernardo. *O romantismo e a idéia de nação no Brasil (1830–1870)*. São Paulo: Martins Fontes, 2004.
Rio Branco, José Maria da Silva Paranhos Júnior. *Mémoire sur la question des limites entre les États-Unis du Brésil et la Guyane Britannique*. Brussels: Imprimerie des Travaux Publics, 1897.
———. *Questões de limites*. Rio de Janeiro: Ministério das Relações Exteriores, 1945.
Risério, Antonio. *A utopia brasileira e os movimentos negros*. São Paulo: Editora 34, 2007.
Rival, Laura M. "Amazonian Historical Ecologies." *Journal of the Royal Anthropological Institute* (2006): S79–S94.
———. *Trekking through History: The Huaorani of Amazonian Ecuador*. New York: Columbia University Press, 2002.
Rival, L., and D. McKey. "Domestication and Diversity in Manioc (*Manihot esculenta* Crantz sp Esculenta, Euphorbiaceae)." *Current Anthropology* 49, no. 6 (2008): 1116–25.
Rivas, Darlene. *Missionary Capitalist: Nelson Rockefeller in Venezuela*. Chapel Hill: University of North Carolina Press, 2002.
Rivera, José Eustasio, and Earle Kenneth James. *The Vortex/La Vorágine*. New York: G. P. Putnam, 1935.
Rivero, Juan, and Ramón Guerra Azuola. *Historia de las misiones de los llanos de Casanare y los Rios Orinoco y Meta*. Bogotá: Imprenta de Silvestre, 1883.
Rivière, Peter. *Absent-Minded Imperialism: Britain and the Expansion of Empire in Nineteenth-Century Brazil*. London: I. B. Tauris, 1995.
———. *The Forgotten Frontier: Ranchers of North Brazil*. New York: Holt, Rinehart and Winston, 1972.
Robuchon, Eugenio. *En el Putumayo y sus afluentes*. Lima: Imprenta La Industria, 1907.
Rocco, Fiammetta. *The Miraculous Fever Tree: Malaria and the Quest for a Cure That Changed the World*. New York: HarperCollins, 2003.
Rocha, Joaquin. *Memorandum de viaje*. Bogotá: El Mercurio, 1905.
Rolle, Andrew F. *The Lost Cause: The Confederate Exodus to Mexico*. Norman: University of Oklahoma Press, 1965.
Romero, Silvio. *O evolucionismo e positivismo no Brasil*. Rio de Janeiro: Livraria Clássica, 1895.
Rondon, Cândido Mariano da Silva, and Esther de Viveiros. *Rondon conta sua vida*. Rio de Janeiro: Livraria São José, 1958.
Roosevelt, Anna C. *Amazonian Indians from Prehistory to the Present: Anthropological Perspectives*. Tucson: University of Arizona Press, 1994.
———. *Moundbuilders of the Amazon: Geophysical Archaeology on Marajó Island, Brazil*. San Diego: Academic Press, 1991.
Roosevelt, Anna C., et al. "8th Millennium Pottery from a Prehistoric Shell Midden in the Brazilian Amazon." *Science* 254, no. 5038 (1991): 1621–24.
Roosevelt, Theodore. *Through the Brazilian Wilderness*. New York: Charles Scribner and Sons, 1914.
Rostain, Stephen. "Agricultural Earthworks on the French Guiana Coast." In *Handbook of*

South American Archaeology, edited by Helaine Silverman and William Isbell. New York: Springer, 2008.

———. "Archaeology of the Guianas: An Overview." In *Handbook of South American Archaeology*, edited by Helaine Silverman and William Isbell. New York: Springer, 2008.

———. "Between the Orinoco and the Amazon: Ceramic Age in the Guyanas." In *Anthropologies of Guyana*, edited by Neil Whitehead and Stephanie Aleman, 36–54. Tucson: University of Arizona Press, 2009.

Rothschild, Emma. "A Horrible Tragedy in the French Atlantic." *Past and Present*, no. 192 (2006): 67–108.

Roumy, V., et al. "Isolation and Antimalarial Activity of Alkaloids from *Pseudoxandra cuspidata*." *Planta Medica* 72, no. 10 (2006): 894–98.

Ruiz-Perez, M., et al. "Conservation and Development in Amazonian Extractive Reserves: The Case of Alto Juruá." *Ambio* 34, no. 3 (2005): 218–23.

Rupke, Nicolaas A. *Alexander von Humboldt: A Metabiography*. Frankfurt am Main: Peter Lang, 2005.

Ruse, Michael. *The Darwinian Revolution: Science Red in Tooth and Claw*. 2nd ed. Chicago: University of Chicago Press, 1999.

Russell-Wood, A. J. R. *The Portuguese Empire, 1415–1808: A World on the Move*. Baltimore: Johns Hopkins University Press, 1998.

Safier, Neil. *Measuring the New World: Enlightenment Science and South America*. Chicago: University of Chicago Press, 2008.

Salamone, Frank A. *The Yanomami and Their Interpreters: Fierce People or Fierce Interpreters?* Lanham, MD: University Press of America, 1997.

Salles, Ricardo. *Guerra do Paraguai: Escravidão e cidadania na formação do exército*. São Paulo: Paz e Terra, 1990.

Salles, Vicente. *Memorial da Cabanagem: Esboço do pensamento político-revolucionário no Grão-Pará*. Belém: Edições Cejup 1992.

———. *O Negro no Pará sob o regime da escravidão*. 3rd rev. and updated ed. Rio de Janeiro: Fundação Getulio Vargas, 2005.

Sampaio, Consuelo Novais. *Canudos: Cartas para o Barão*. São Paulo: Editora da Universidade de São Paulo (EDUSP), Imprensa Oficial, 1999.

Sampaio Neto, José Augusto Vaz. *Canudos: Subsídios para a sua reavaliação histórica*. Rio de Janeiro: Fundação Casa de Rui Barbosa, 1986.

Sampaio, Teodoro. "A memoria de Eudlides da Cunha no décimo anniversario do seu morte." *Revista do Instituto Geográfico da Bahia* 26, no. 45 (1919): 247–55.

———. *O Rio de S. Francisco: Trechos de um diário de viagem; e a Chapada Diamantina*. São Paulo: Escolas Profissionães Salesianas, 1905.

———. *O Rio São Francisco e a Chapada Diamantina*. Salvador, Brazil: Progresso, 1955.

———. *O Tupí na geographia nacional: Memoria lida no Instituto Historico e Geographico de S. Paulo*. São Paulo: Casa Eclectica, 1901.

San Román, Jesús Víctor, Martha Rodríguez, and Joaquín García. *Perfiles históricos de la Amazonía Peruana*. 2nd ed. Iquitos, Peru: Centro de Estudios Teológicos de la Amazonía, Centro Amazónico de Antropología y Aplicación Práctica, Instituto de Investigaciones de la Amazonía Peruana, 1994.

Sanjad, Nelson. *Emílio Goeldi (1859–1917): Aventura de um naturalista entre a Europa e o Brasil*. Rio de Janeiro: EMC, 2009.

Sant'Anna, Sonia. *Barões e escravos do café: Uma história privada do Vale do Paraíba*. Rio de Janeiro: Jorge Zahar Editor, 2001.

Santa-Anna Néry, Frederico José de. *Le pays des Amazones: L'El-Dorado, les terres à Caoutchouc*. Paris: L. Frinzine, 1885.

Santos, Maria Januária Vilela. *A Balaiada e a insurreição de escravos no Maranhão*. São Paulo: Atica, 1983.

Santos, Roberto. *História econômica da Amazônia (1800–1920)*, São Paulo: T. A. Queiroz, 1980.

Santos-Granero, Fernando. "Boundaries Are Made to Be Crossed: The Magic and Politics of the Long-Lasting Amazon/Andes Divide." *Identities: Global Studies in Culture and Power* 9, no. 4 (2002): 545–69.

———. *Opresión colonial y resistencia indígena en la alta Amazonía*. Quito: Facultad Latinoamericana de Ciencias Sociales (FLACSO) and Abya-Yala, 1992.

———. *Vital Enemies: Slavery, Predation and the Amerindian Political Economy of Life*. Austin: University of Texas, 2009.

Santos-Granero, Fernando, and Frederica Barclay. *Órdenes y desordenes en la selva central: Historia y economía de un espacio regional*. Lima: (Instituto Económico del Perú (IEP) and Facultad Latinoamericana de Ciencias Sociales (FLACSO), 1995.

———. *Selva Central: History, Economy and Land Use in Peruvian Amazonia*. Washington, DC: Smithsonian Institution Press, 1998.

———. *Tamed Frontiers: Economy, Society, and Civil Rights in Upper Amazonia*. Boulder, CO: Westview, 2000.

Santos-Granero, Fernando, Frederica Barclay, and Adriana Soldi. *La frontera domesticada: Historia económica y social de Loreto, 1850–2000*. Lima: Pontificia Universidad Católica del Peru, 2002.

Sarmiento, Domingo Faustino, and Kathleen Ross. *Facundo: Civilization and Barbarism*. Berkeley: University of California Press, 2003.

Savage, John. "'Black Magic' and White Terror: Slave Poisoning and Colonial Society in Early 19th Century Martinique." *Journal of Social History* 40, no. 3 (2007): 635–62.

Savage, Victor R. "Tropicality Imagined and Experienced: A Commentary on Felix Driver's 'Imagining the Tropics: Views and Visions of the Tropical World.'" *Singapore Journal of Tropical Geography* 25, no. 1 (2004): 26–31.

Sawyer, Donald R., and Diana Reiko Tutiya Sawyer. *Malaria on the Amazon Frontier: Economic and Social Aspects of Transmission and Control*. Belo Horizonte, Brazil: Centro de Desenvolvimento e Planejamento Regional, Faculdade de Ciências Econômicas, Universidade Federal Minas Gerais, 1987.

Schaan, Denise P. "Recent Investigations on Marajoara Culture, Marajó Island, Brazil." *Antiquity* 74, no. 285 (2000): 460–70.

Schaer, Roland, Gregory Claeys, and Lyman Tower Sargent. *Utopia: The Search for the Ideal Society in the Western World*. New York: Oxford University Press, 2000.

Scheper-Hughes, Nancy. *Death without Weeping: The Violence of Everyday Life in Brazil*. Berkeley: University of California Press, 1992.

Schidrowitz, Philip, and Thomas Dawson. *History of the Rubber Industry*. Cambridge, UK: W. Heffer, 1952.

Schiebinger, Londa L. *Plants and Empire: Colonial Bioprospecting in the Colonial World*. Cambridge, MA: Harvard University Press. 2004.

Schiebinger, Londa L., and Claudia Swan. *Colonial Botany: Science, Commerce, and Politics in the Early Modern World*. Philadelphia: University of Pennsylvania Press, 2005.

Schrepfer, Susan R. *Nature's Altars: Mountains, Gender, and American Environmentalism*. Lawrence: University Press of Kansas, 2005.

Schroth, G., V. Moraes, and M. S. S. da Mota. "Increasing the Profitability of Traditional, Planted Rubber Agroforests at the Tapajós River, Brazilian Amazon." *Agriculture Ecosystems & Environment* 102, no. 3 (2004): 319–39.

Schultes, Richard. "The Amazon Indian and Evolution in *Hevea* and Related Genera." *Journal of the Arnold Arboretum* 37, no. 2 (1956): 123–52.

———. "Odyssey of Cultivated Rubber Tree." *Endeavour* 1, nos. 3–4 (1977): 133–38.

———. "Richard Spruce and the Potential for European Settlement of the Amazon: Unpublished Letter." *Botanical Journal of the Linnean Society* 77, no. 2 (1978): 131–39.

———. "Several Unpublished Ethnobotanical Notes of Richard Spruce." *Rhodora* 87 (1985): 439–41.

Schultes, Richard, and Robert F. Raffauf. *The Healing Forest: Medicinal and Toxic Plants of the Northwest Amazonia*. Portland, OR: Dioscorides, 1990.

Schultz, Kristen. *Tropical Versailles: Empire and the Portuguese Royal Court in Rio de Janeiro, 1808–1821*. New York: Routledge, 2001.

Schwarcz, Lilia Moritz. *The Emperor's Beard: Dom Pedro II and the Tropical Monarchy in Brasil*. New York: Farrar, Straus and Giroux, 1998.

———. *The Spectacle of the Races: Scientists, Institutions, and the Race Question in Brazil, 1870–1930*. New York: Hill and Wang, 1999.

Schwarcz, Lilia Moritz, and Letícia Vidor de Sousa Reis. *Negras imagens: Ensaios sobre cultura e escravidão no Brasil*. São Paulo: Estação Ciência Universidade de São Paulo and Editora da Universidade de São Paulo (EDUSP), 1996.

Schwartz, Stuart B. "Black Latin America: Legacies of Slavery, Race, and African Culture—Introduction." *Hispanic American Historical Review* 82, no. 3 (2002): 429–33.

———. "Black Slaves in Palmares, Brazil: The Mocambos Revolt." *Histoire*, no. 41 (1982): 38–48.

———. "Cantos e quilombos numa conspiração de escravos Haussas: Bahia, 1814." In *Liberdade por um fio: História dos quilombos no Brasil*, edited by João José Reis and Flávio dos Santos Gomes. São Paulo: Companhia das Letras, 1998.

———. "Indian Labor and New World Plantations: European Demands and Indian Responses in Northeastern Brazil." *American Historical Review* 83, no. 1 (1978): 43–79.

———. "New Peoples and New Kinds of People: Adaptation, Readjustment and Ethnogenesis in South American Indigenous Societies." In *Cambridge History of Native Peoples of the Americas*, edited by Frank Salomon and Stuart B. Schwartz, 443–501. Cambridge: Cambridge University Press, 1999.

———. *Slaves, Peasants, and Rebels: Reconsidering Brazilian Slavery*. Urbana: University of Illinois Press, 1992.

Schwartz, Stuart B., and Hal Langfur. "Tapahuns, Negros da Terra and Curibocas: Common Cause and Confrontation between Blacks and Natives in Colonial Brazil." In *Beyond Black and Red*, edited by Matthew Restall, 81–114. Albuquerque: University of New Mexico, 2005.

Scott, James C. *Moral Economy of the Peasant*. New Haven, CT: Yale University Press, 1978.

———. *Weapons of the Weak*. New Haven, CT: Yale University Press, 1985.

Seed, Patricia. *American Pentimento: The Invention of Indians and the Pursuit of Riches*. Minneapolis: University of Minnesota Press, 2001.

———. *Ceremonies of Possession in Europe's Conquest of the New World, 1492–1640*. New York: Cambridge University Press, 1995.

Seibert, R. J. "A Study of *Hevea* (with Its Economic Aspects) in the Republic of Peru." *Annals of the Missouri Botanical Garden* 34, no. 3 (1947): 261–352.

———. "The Uses of *Hevea* for Food in Relation to Its Domestication." *Annals of the Missouri Botanical Garden* 35, no. 2 (1948): 117–21.

Semple, Ellen Churchill, and Friedrich Ratzel. *Influences of Geographic Environment, on the Basis of Ratzel's System of Anthropo-geography*. New York: H. Holt, 1911.

Şenocak, Zafer, and Leslie A. Adelson. *Atlas of a Tropical Germany: Essays on Politics and Culture, 1990–1998*. Lincoln: University of Nebraska Press, 2000.

Serier, Jean-Baptiste. *Les barons du caoutchouc*. Paris: Karthala, 2000.

Serrão, Joel. *Do sebastianismo ao socialismo*. Lisbon: Livros Horizontes, 1983.

Shakespeare, William. *Hamlet*. New York: Washington Square Press, 1992.

Show, Paul V., and Gilberto Freyre. *Gilberto Freyre's "Luso-Tropicalism."* Lisbon: Centro de Estudos Políticos e Sociais, 1957.

Silberling, Louise S. "Displacement and Quilombos in Alcantara, Brazil: Modernity, Identity, and Place." *International Social Science Journal* 55, no. 1 (2003): 145–56.

Silva, Rogério Souza. *Antônio Conselheiro: A fronteira entre a civilização e a barbárie*. São Paulo: Annablume, 2001.

Simón, Pedro, William Bollaert, and Clements R. Markham. *The Expedition of Pedro de Ursúa & Lope de Aguirre in Search of El Dorado and Omagua in 1560–1*. New York: B. Franklin, 1971.

Singelman, Peter. "Political Structure and Social Banditry in Northeast Brazil." *Journal of Latin American Studies* 7, no. 1 (1975): 59–83.

Singleton-Gates, Peter, and Maurice Girodias. *The Black Diaries: An Account of Roger Casement's Life and Times, with a Collection of His Diaries and Public Writings*. New York: Grove, 1959.

Siqueira, Baptista. *Os Cariris do Nordeste*. Rio de Janeiro: Livraria Editora Cátedra, 1978.

Skidmore, Thomas E. *Black into White: Race and Nationality in Brazilian Thought*. Durham, NC: Duke University Press, 1993.

———. "Raizes de Gilberto Freyre." *Journal of Latin American Studies* 34 (2002): 1–20.

Slater, Candace. *Entangled Edens: Visions of the Amazon*. Berkeley: University of California Press, 2002.

———. "Representations of Power in Pilgrim Tales from the Brazilian Northeast." *Latin American Research Review* 19, no. 2 (1984): 71–91.

Slenes, Robert W. "The Brazilian Internal Slave Trade, 1850–1888." In *The Chattel Principle*, edited by Walter Johnson and David Brown, 325–69. New Haven, CT: Yale University Press, 2004.

———. "Comments on Slavery in a Non Export Economy." *Hispanic American Historical Review* 63, no. 3 (1983): 569–81.

Smallman, Shawn C. *Fear and Memory in the Brazilian Army and Society, 1889–1954*. Chapel Hill: University of North Carolina Press, 2002.

Smith, Anthony. *Explorers of the Amazon*. London: Viking, 1990.

Smith, Herbert H. *Brazil, the Amazons and the Coast*. New York: C. Scribner's Sons, 1879.

Smith, Nigel J. H. *The Amazon River Forest: A Natural History of Plants, Animals, and People*. New York: Oxford University Press, 1999.

Smoak, Gregory E. *Ghost Dances and Identity: Prophetic Religion and American Indian Ethnogenesis in the Nineteenth Century*. Berkeley: University of California Press, 2006.

Soares, Jose de Souza. *O militarismo na república*. São Paulo: Monteiro Lobato, 1925.

Soares, Orlando. *A evolução do status jurídico-social da Mulher*. Rio de Janeiro: Editora Rio, 1978.

Sodré, Nelson Werneck. *História militar do Brasil*. Rio de Janeiro: Editora Civilização Brasileira, 1965.

Sola, José Antonio. *Canudos, uma utopía no sertão*. São Paulo: Editoria Contexto, 1989.

Sommer, Barbara A. "Cracking Down on the Cunhamenas: Renegade Amazonian Traders under Pombaline Reform." *Journal of Latin American Studies* 38 (2006): 767–91.

Soublin, Jean. *Cayenne 1809: La conquête de la Guyane par les Portugais du Brésil*. Paris: Karthala, 2003.

Souza, Márcio. *The Emperor of the Amazon*. New York: Avon Books, 1980.

Souza, Paulo Cesar. *A Sabinada: A revolta separatista da Bahia, 1837*. São Paulo: Editora Brasiliense, 1987.

Spary, Emma C. "Of Nutmegs and Botanists: The Colonial Cultivation of Botanical Identity." In *Colonial Botany: Science, Commerce, and Politics in the Early Modern World*, edited by Londa L. Schiebinger and Claudia Swan. Philadelphia: University of Pennsylvania Press, 2005.

———. *Utopia's Garden: French Natural History from Old Regime to Revolution*. Chicago: University of Chicago Press, 2000.

Spruce, Richard, and Alfred Russel Wallace. *Notes of a Botanist on the Amazon and Andes: Being Records of Travel on the Amazon and Its Tributaries, the Trombetas, Rio Negro, Uaupés, Casiquiari, Pacimoni, Huallaga and Pastasa; As Also to the Cataracts of the Orinoco, along the Eastern Side of the Andes of Peru and Ecuador, and the Shores of the Pacific, during the Years 1849–1864*. 2 vols. London: Macmillan, 1908.

St. Aubyn, Giles. *A Victorian Eminence: The Life and Works of Henry Thomas Buckle*. London: Barrie, 1958.

Staden, Hans, and Malcolm Henry Ikin Letts. *Hans Staden: The True History of His Captivity, 1557*. New York: R. M. McBride, 1929.

Stahl, Peter W. *Archaeology in the Lowland American Tropics: Current Analytical Methods and Applications*. Cambridge: Cambridge University Press, 1995.

Stamos, David N. *Darwin and the Nature of Species*. Albany: State University of New York Press, 2007.

Stanfield, Michael Edward. *Red Rubber, Bleeding Trees: Violence, Slavery, and Empire in Northwest Amazonia, 1850–1933*. Albuquerque: University of New Mexico Press, 1998.

Stanley, Henry M. *The Congo and the Founding of Its Free State: A Story of Work and Exploration*. London: Sampson Low Marston Searle and Rivington, 1885. Microform.

———. *Through the Dark Continent, or The Sources of the Nile around the Great Lakes of Equatorial Africa and down the Livingstone River to the Atlantic Ocean*. 2 vols. London: Sampson Low Marston Searle and Rivington, 1878.

Stedman, John Gabriel. *Narrative of a Five Years Expedition against the Revolted Negroes of Surinam: Transcribed for the First Time from the Original 1790 Manuscript*. Baltimore: Johns Hopkins University Press, 1988.

———. *Stedman's Surinam: Life in Eighteenth-Century Slave Society*. Edited by Richard Price and Sally Price. Baltimore: Johns Hopkins University Press, 1992.

Stepan, Alfred C. *Authoritarian Brazil: Origins, Policies, and Future*. New Haven, CT: Yale University Press, 1973.

———. *The Military in Politics: Changing Patterns in Brazil*. Princeton, NJ: Princeton University Press, 1974.

Stepan, Nancy. *Beginnings of Brazilian Science: Oswaldo Cruz, Medical Research and Policy, 1890–1920*. New York: Science History, 1976.

———. *The Hour of Eugenics: Race, Gender, and Nation in Latin America*. Ithaca, NY: Cornell University Press, 1991.

———. *The Idea of Race in Science: Great Britain, 1800–1960*. Hamden, CT: Archon Books, 1982.

———. *Picturing Tropical Nature*. Ithaca, NY: Cornell University Press, 2001.

———. *The Rise of the Brazilian Sciences*. Ithaca, NY: Cornell University Press, 1987.

Stern, Theodore. *Rubber Ball Games of the Americas*. Seattle: University of Washington Press, 1949.

Stocking, George W. *Race, Culture, and Evolution: Essays in the History of Anthropology*. Chicago: University of Chicago Press, 1982.

Stoian, Dietmar. "Making the Best of Two Worlds: Rural and Peri-urban Livelihood Options Sustained by Nontimber Forest Products from the Bolivian Amazon." *World Development* 33, no. 9 (2005): 1473–90.

Stols, Eddy. "Les belges au Mato Grosso et en Amazonie: Ou la recidive de l'aventure congolaise (1895–1910)." In *La Belgique et l'étranger*, edited by Michel Dumoulin and Eddy Stols, 76–111. Brussels: Louvain la Neuve, 1987.

———. "Les investiments belges aux Brasil (1830–1914)." Paper presented at the Colloques Centre National Recherche Scientifique, Paris 1971.

Stone, Andrea. "Spirals, Ropes, and Feathers: The Iconography of Rubber Balls in Mesoamerican Art." *Ancient MesoAmerica* 13 (2002): 21–39.

Stroebel, Michel. *Les gens de l'or*. Paris: Ibis, 1998.

Summerhill, William Roderick. *Order against Progress: Government, Foreign Investment, and Railroads in Brazil, 1854–1913*. Stanford, CA: Stanford University Press, 2003.

Sweet, James H. *Recreating Africa: Culture, Kinship, and Religion in the African-Portuguese World, 1441–1770*. Chapel Hill: University of North Carolina Press, 2003.

Taine, Hippolyte. *Histoire de la littérature anglaise*. 5 vols. Paris: Hachette, 1899.

Taine, Hippolyte, and John Durand. *The French Revolution*. 3 vols. Gloucester, UK: P. Smith, 1962.

Tambs, Lewis A. "Rubber, Rebels and Rio Branco." *Hispanic American Historical Review* 46, no. 3 (1966): 254–73.

Tarkanian, Michael.J., and Dorothy Hosler. "America's First Polymer Scientists: Rubber Processing, Use, and Transport in Ancient Mesoamerica." *American Antiquity*, 2011, forthcoming.

Taussig, Michael T. "Culture of Terror—Space of Death—Casement, Roger: Putumayo Report and the Explanation of Torture." *Comparative Studies in Society and History* 26, no. 3 (1984): 467–97.

———. *Shamanism, Colonialism, and the Wild Man: A Study in Terror and Healing*. Chicago: University of Chicago Press, 1986.

Tavares Bastos, Aureliano Cândido. *O Valle do Amazonas: Estudo sobre a livre navegação do Amazonas, estatistica, producções, commercio, questões fiscães do Valle do Amazonas, com um prefacio contendo o decreto que abre aos navios de todas as nações os rios Amazonas, Tocantins e S. Francisco*. Rio de Janeiro: B. L. Garnier, 1866.

Terán, Oscar. *América Latina, positivismo y nación*. Mexico City: Editorial Katún, 1983.

Terraciano, Kevin. *The Mixtecs of Colonial Oaxaca: Ñudzahui History, Sixteenth through Eighteenth Centuries*. Stanford, CA: Stanford University Press, 2001.

Thomas, Helen. *Romanticism and Slave Narratives: Transatlantic Testimonies*. Cambridge: Cambridge University Press, 2000.

Thompson, Krista A. *An Eye for the Tropics*. Durham, NC: Duke University Press, 2006.

Thorlby, Tiago. *A Cabanagem na fala do povo*. São Paulo: Edições Paulinas, 1988.

Thornton, John K. "The Art of War in Angola." *Comparative Studies in Society and History* 30, no. 3 (1988): 360–78.

———. *The Congolese Saint Anthony*. Cambridge: Cambridge University Press, 1998.

———. "Warfare in Atlantic Africa, 1500–1800." In *Warfare and History*. London: UCL Press, 1999.

Thouar, Arthur. *Explorations dans l'Amérique du Sud: I. A la recherche de la mission Crevaux; II. Dans le Delta du Pilcomayo; III. De Buenos Aires à Sucre; IV. Dans le Chaco boréal*. Paris, Hachette, 1891.

Tierney, Patrick. *Darkness in El Dorado: How Scientists and Journalists Devastated the Amazon*. New York: W. W. Norton, 2000.

Tocantins, Leandro. *Euclides da Cunha e o Paraíso Perdido*. Manaus, Brazil: Governo do Estado do Amazonas, 1966.

———. *Formação histórica do Acre*. 3rd ed. 3 vols. Rio de Janeiro: Civilização Brasileira, 1979.

Toivonen, Tuuli, Sanna Mäki, and Risto Kalliola. "The Riverscape of Western Amazonia: A Quantitative Approach to the Fluvial Biogeography of the Region." *Journal of Biogeography* 34, no. 8 (2007): 1374–87.

Toplin, Robert Brent. *The Abolition of Slavery in Brazil*. New York: Atheneum, 1972.

Toral, André. "Os indios negros ou os Carijó de Goiás: A história das Avá-Canoeiro." *Revista de Antropologia* 27/28 (1984–85): 287–325.

Torres, Haroldo, and George Martine. *Amazonian Extractivism: Prospects and Pitfalls*. Brasília: Instituto Sociedade, População e Natureza, 1991.

Torres, Victor F. *The Canudos War Collection*. Albuquerque: University of New Mexico, 1990.

Tostes, Joel B., and Adelino Brandão. *Aguas de amargura*. Rio de Janeiro: Rio Fundo Editora, 1990.

Touchet, Julien. *Botanique et colonisation en Guyane Française, 1720–1848: Le Jardin des Danaïdes*. Petit-Bourg, France: Ibis Rouge, 2004.

Treccani, Girolamo. "Os diferentes caminhos para a resgate dos territórios quilombolas." *Boletim Informativo do NUER* (Nucleo de Etnicididade e Relações Interétnicas) 2 (2005): 111–21.

———. *Territórios quilombolas*. Belém: Instituto de Terras do Pará (INTERPA), 2007.

Treece, David. *Exiles, Allies and Rebels: Brazil's Indianist Movement, Indigenous Politics and the Imperial Nation State*. Westport, CT: Greenwood, 2000.

Twain, Mark. *Life on the Mississippi*. New York: P. F. Collier and Son, 1923.

———. *Roughing It*. New York: Harper and Brothers, 1959 (orig. 1913).
———. "The Turning Point of My Life." In *The Favorite Works of Mark Twain*, edited by Albert Paine. New York: Harper Brothers, 1935.
Tyacke, Sarah. "English Charting of the River Amazon c. 1595–c. 1630." *Imago Mundi* 32 (1980): 73–89.
United States, William Cullen Dennis, Permanent Court of Arbitration, and Venezuela. *The Counter Case of the United States of America on Behalf of the Orinoco Steamship Company against the United States of Venezuela: With Appendix*. Washington, DC: Government Print Office, 1910.
Urbano, Henrique, and Ana de Zaballa Beascoechea. *Utopía, mesianismo y milenarismo: Experiencias latinoamericanas*. Lima: Universidad San Martín de Porres, 2002.
Useche Losada, Mariano. *El proceso colonial en el Alto Orinoco–Río Negro: Siglos XVI a XVIII*. Bogotá: Fundación de Investigaciones Arqueológicas Nacionales Banco de la República, 1987.
Vainfas, Ronaldo. *Confissões da Bahia: Santo Ofício da Inquisição de Lisboa*. São Paulo: Companhia das Letras, 1997.
———. *A heresia dos índios: Catolicismo e rebeldía no Brasil colonial*. São Paulo: Companhia das Letras, 1995.
———. *História e sexualidade no Brasil*. ed. Rio de Janeiro: Graal, 1986.
———. *Ideologia e escravidão: Os letrados e a sociedade escravista no Brasil colonial*. Petrópolis, Brazil: Vozes, 1986.
———. *Trópico dos pecados: Moral, sexualidade e inquisição no Brasil*. Rio de Janeiro: Editora Campus, 1989.
Valcárcel, Carlos A. *El proceso de Putumayo y sus secretos inauditos*. Lima: Imprenta Comercial de H. La Rosa, 1915.
van Young, Eric. *The Other Rebellion: Popular Violence, Ideology and the Mexican Struggle for Independence*. Palo Alto, CA: Stanford University Press, 2001.
Vanderwood, Paul. *The Power of God against the Guns of Government*. Palo Alto, CA: Stanford University Press, 1998.
Varela Marcos, Jesús, Juan Vernet Ginés, and Sociedad V Centenario del Tratado de Tordesillas (Spain). *El Tratado de Tordesillas en la cartografía histórica*. Valladolid: Junta de Castilla y León, V Centenario Tratado de Tordesillas, 1994.
Vargas Llosa, Mario. *El sueño del celta*. Doral, FL: Alfaguara/Santillana, 2010.
Vattell, Emer de, Joseph Chitty, and Edward D. Ingraham. "The Law of Nations, or Principles of the Law of Nature Applied to the Conduct and Affairs of Nations and Sovereigns." Philadelphia: T. and J. W. Johnson, 1867.
Ventura, Maria da Graça Mateus. *Viagens e viajantes no Atlântico quinhentista*. Lisbon: Edições Colibri, 1996.
Ventura, Roberto. *Estilo tropical: História cultural e polêmicas literárias no Brasil, 1870–1914*. São Paulo: Companhia das Letras, 1991.
———. "'A Nossa Vendéia': Canudos, o Mito da Revolução Francesa e a constituição de identidade nacional (1897–1902)." *Revista de Crítica Literaria Latinoamericana* 12, no. 24 (1986): 109–25.
Ventura, Roberto, José Carlos Barreto de Santana, and Mario Cesar Carvalho. *Retrato interrompido da vida de Euclides da Cunha*. São Paulo: Companhia das Letras, 2003.
Versteeg, A. H. "Peuplements et environnements dans les Guyanes entre 10.000 et 1.000

BP." Paper presented at the Seminaire Atelier Peuplements anciens et actuels des Forêts tropicales, Orléans, France, 1998.

Vidal, Laurent. *Mazagão, la ville qui traversa l'Atlantique: Du Maroc à l'Amazonie, 1769–1783*. Paris: Aubier, 2005.

Vidal, Silvia M. "Amerindian Cartography as a Way of Preserving and Interpreting the Past." In *History and Historicities in Amazonia*, edited by Neil L. Whitehead. Lincoln: University of Nebraska Press, 2003.

———. "Kuwe Duwakalumi: The Arawak Sacred Routes of Migration, Trade, and Resistance." *Ethnohistory* 47, nos. 3–4 (2000): 635–67.

Vigneron, M., X. Deparis, E. Deharo, and G. Bourdy. "Antimalarial Remedies in French Guiana: A Knowledge of Attitudes and Practices Study." *Journal of Ethnopharmacology* 98, no. 3 (2005): 351–60.

Villela Júnior, Marcos Evangelista da Costa. *Canudos: Memórias de um combatente*. São Paulo: Marco Zero, 1988.

Virgili, Carmina. "Charles Lyell and Scientific Thinking in Geology." *Comptes Rendus Geoscience* 339, no. 8 (2007): 572–84.

Voeks, Robert A. *Sacred Leaves of Candomblé: African Magic, Medicine, and Religion in Brazil*. Austin: University of Texas Press, 1997.

Volpato, Luiza Rios Ricci. *Cativos do sertão: Vida cotidiana e escravidão em Cuiabá em 1850–1888*. São Paulo: Editora Marco Zero; Cuiabá: Editora da Universidade Federal de Mato Grosso, 1993.

———. *A conquista da terra no universo da pobreza: Formação da fronteira oeste do Brasil, 1719–1819*. São Paulo: Editora Hucitec, 1987.

Voltaire, and Burton Raffel. *Candide, or, Optimism: Translated from the German of Dr. Ralph with Additions Found in the Doctor's Pocket When He Died, at Minden, in the Year of Our Lord 1759*. New Haven, CT: Yale University Press, 2005.

von Humboldt, Alexander, Aimé Bonpland, and Helen Maria Williams. *Personal Narrative of Travels to the Equinoctial Regions of the New Continent: During the Years 1799–1804*. 1st–3rd ed. London: Longman Hurst Rees Orme Brown and Green, 1821.

von Martius, Karl Friedrich Philipp. "Como se deve escrever a história do Brasil." *Revista do Instituto Histórico e Geográfico Brasileiro* 6, no. 24 (1845): 381–403.

von Spix, Johann Baptist, and Karl Friedrich Philipp von Martius. *Viagem pelo Brasil*. São Paulo: Melhoramentos, 1938.

von Spix, Johann Baptist, Karl Friedrich Philipp von Martius, and Hannibal Evans Lloyd. *Travels in Brazil, in the Years 1817–1820: Undertaken by Command of His Majesty the King of Bavaria*. 2 vols. London: Longman Hurst Rees Orme Brown and Green, 1824.

Wagner de Almeida, Alfredo. "Terra do preto, terra do santo e terras do Indio." *Humanidades* 4, no. 15 (1988): 42–48.

Walcott, Derek. *The Ghost Dance*. Electronic ed. Alexandria, VA: Alexander Street, 2002.

Walker, James. "Canudos Revisited: Cunninghame, Graham, Vargas-Llosa and the Messianic Tradition." *Symposium: A Quarterly Journal in Modern Literatures* 41, no. 4 (1987): 308–16.

Wallace, Alfred Russel. *A Narrative of Travels on the Amazon and Rio Negro, with an Account of the Native Tribes and Observations on the Climate, Geology and Natural History of the Amazon Valley*. London: Ward, Lock, 1889.

Weaver, Blanche. "Confederate Immigrants and Evangelical Churches in Brazil." *Journal of Southern History* 18, no. 4 (1952): 446–68.

Webb, James Watson. *Slavery and Its Tendencies: A Letter from General J. Watson Webb to the New York "Courier and Enquirer."* Washington, DC: Buell and Blanchard, 1856.

Weinstein, Barbara. *The Amazon Rubber Boom, 1850–1920*. Stanford, CA: Stanford University Press, 1983.

———. "Not the Republic of Their Dreams: Historical Obstacles to Political and Social Democracy in Brazil." *Latin American Research Review* 29, no. 2 (1994): 262–73.

———. "Racializing Regional Difference: São Paulo vs. Brazil, 1932." In *Race and Nation in Modern Latin America*, edited by Nancy Appelbaum, Anne Macpherson, and Karin Rosemblatt, 237–62. Chapel Hill: University of North Carolina Press, 2003.

Werlich, David P. *Admiral of the Amazon: John Randolph Tucker, His Confederate Colleagues, and Peru*. Charlottesville: University Press of Virginia, 1990.

Wesseling, H. L., and Arnold J. Pomerans. *Divide and Rule: The Partition of Africa, 1880–1914*. Westport, CT: Praeger, 1996.

Wey Gomez, Nicolas. *The Tropics of Empire*. Cambridge, MA: MIT Press, 2008.

Whigham, Thomas. *The Paraguayan War*. Lincoln: University of Nebraska Press, 2002.

———. *The Politics of River Trade: Tradition and Development in the Upper Plata, 1780–1870*. Albuquerque: University of New Mexico Press, 1991.

Whigham, Thomas, and Barbara Potthast. "The Paraguayan Rosetta Stone: New Insights into the Demographics of the Paraguayan War, 1864–1870." *Latin American Research Review* 34, no. 1 (1999): 174–86.

Whitaker, Robert. *The Mapmaker's Wife: A True Tale of Love, Murder, and Survival in the Amazon*. New York: Basic Books, 2004.

White, Richard. *The Middle Ground: Indians, Empires, and Republics in the Great Lakes Region, 1650–1815*. Cambridge: Cambridge University Press, 1991.

Whitehead, Neil L. "Black Read as Red: Ethnic Transgression and Hybridity in Northeastern South America and the Caribbean." In *Beyond Black and Red*, edited by Matthew Restall, 223–44. Albuquerque: University of New Mexico, 2005.

———. "Carib Ethnic Soldiering in Venezuela, the Guianas, and the Antilles, 1492–1820." *Ethnohistory* 37, no. 4 (1990): 357–85.

———. "The Historical Anthropology of Text: The Interpretation of *Raleghs Discoverie of Guiana*." *Current Anthropology* 36, no. 1 (1995): 53–74.

———. *Histories and Historicities in Amazonia*. Lincoln: University of Nebraska Press, 2003.

———. *Lords of the Tiger Spirit: A History of the Caribs in Colonial Venezuela and Guyana, 1498–1820*. Dordrecht, Netherlands: Foris, 1988.

———. "Native Peoples Confront Colonial Regimes in Northeastern South America." In *Cambridge History of Native Peoples of the Americas*, edited by Frank Salomon and Stuart B. Schwartz, 382–442. Cambridge: Cambridge University Press, 1999.

———. "Recent Research on the Native History of Amazonia and Guayana." *L'Homme* 33, nos. 2–4 (1993): 495–506.

Whitehead, Neil L., and Robin Wright. *In Darkness and Secrecy: The Anthropology of Assault, Sorcery and Witchcraft in Amazonia*. Durham, NC: Duke University Press, 2004.

Whittington, E. Michael, and Mint Museum of Art. *The Sport of Life and Death: The Mesoamerican Ballgame*. New York: Thames and Hudson, 2001.

Widmer, Edward D. *Young America: The Flowering of Democracy in New York City*. New York: Oxford University Press, 1999.

Wilcox, Robert W. "Cattle and Environment in the Pantanal of Mato-Grosso, Brazil, 1870–1970." *Agricultural History* 66, no. 2 (1992): 232–56.

Wilentz, Sean. *The Rise of American Democracy: Jefferson to Lincoln*. New York: W. W. Norton, 2005.

Wilkinson, David M. "Ecology before Ecology: Biogeography and Ecology in Lyell's *Principles*; A Critical Re-assessment of the Significance of Charles Lyell (1832), *Principles of Geology*, Vol. II. John Murray, London." *Journal of Biogeography* 29, no. 9 (2002): 1109–15.

Williams, Donn Alan. "Brazil and French Guiana: The Four-Hundred Year Struggle for Amapá." PhD diss., Texas Christian University, 1975.

Wilson, Raymond Jackson. *Darwinism and the American Intellectual*. Homewood, IL: Dorsey, 1967.

Woell, Edward J. *Small-Town Martyrs and Murderers: Religious Revolution and Counterrevolution in Western France, 1774–1914*. Milwaukee, WI: Marquette University Press, 2006.

Wolff, Cristina Scheibe. *Mulheres da floresta: Uma história do Alto Juruá, Acre, 1890–1945*. São Paulo: Editora Hucitec, 1999.

Wolpert, Lewis. *The Triumph of the Embryo*. Oxford: Oxford University Press, 1991.

Woodroffe, Joseph Froude. *The Upper Reaches of the Amazon*. London: Methuen, 1914.

Woodward, Ralph Lee. *Positivism in Latin America, 1850–1900: Are Order and Progress Reconcilable?* Lexington, VA: Heath, 1971.

Wright, Robin M., and Manuela Carneiro da Cunha. "Destruction, Resistance, and Transformation: Southern Coastal and Northern Brazil." In *Cambridge History of Native Peoples of the Americas*, edited by Frank Salomon and Stuart B. Schwartz, 287–381. Cambridge: Cambridge University Press, 1999.

Wright-Rios, Edward N. "Indian Saints and Nation-States: Ignacio Manuel Altamirano's Landscapes and Legends." *Mexican Studies/Estudios Mexicanos* 20, no. 1 (2004): 47–68.

Xavier, Regina Célia Lima. *A conquista da liberdade: Libertos em Campinas na segunda metade do século XIX*. Campinas, Brazil: Centro de Memória Unicamp, 1996.

Xerez, Francisco Lopez de. *Verdadera relación de la conquista del Perú: Segun la primera edición impresa en Sevilla en 1534*. Lima: Sanmartí, 1917.

Yacou, Alain, and Jacques Adélaïde-Merlande. *La découverte et la conquête de la Guadeloupe*. Pointe-à-Pitre, Guadeloupe, Paris: Karthala, 1993.

Zeballos, Estanislao Severo. *Diplomacia desarmada*. Buenos Aires: Editorial Universitaria de Buenos Aires, 1974.

Zimmerer, Karl S. "Humboldt's Nodes and Modes of Interdisciplinary Environmental Science in the Andean World." *Geographical Review* 96, no. 3 (2006): 335–60.

Index

Italicized page numbers indicate illustrations.